Disaster Resilience and Green Growth

Over the years, the relationship between environment and disasters has received significant attention. This is largely due to the emerging recognition that environmental changes – climate change, land-use and natural resource degradation make communities more vulnerable to disaster impacts. There is a need to break this nexus through environment based and sustainability inclusive interventions. Science – technology and economic measures for disaster risk management, hence, need to adapt more integrated approaches for infrastructure and social resilience. Environmental and anthropogenic factors are key contributors to hazard, risk, and vulnerability and, therefore, should be an important part of determining risk-management solutions.

Green growth approaches have been developed by emphasizing sustainability inclusion and utilizing the benefits of science-technology interventions along policy-practice linkages with circular economy and resource efficiency. Such approaches recognize the perils of traditional material-oriented economy growth models that tend to exploit natural resources, contribute to climate change, and exacerbate disaster vulnerabilities, Green growth integrated approaches are rapidly becoming as preferred investment avenue for mitigating climate change and disaster risks and for enhancing resilience. This includes ecosystem-based and nature-based solutions with potential to contribute to the resilience of infrastructure, urban, rural and peri-urban systems, livelihoods, water, and health. They can lead to food security and can further promote people-centric approaches.

Some of the synergistic outcomes of green growth approaches include disaster risk reduction, climate change mitigation and adaptation, resilient livelihoods, cities, businesses and industry. The disaster risk reduction and resilience outcome of green growth approaches deserve special attention, both for the academic and policy communities. Scholars and professionals across the domains of DRR, CCA, and green growth are in need of publications that fulfill their knowledge needs concerning the disaster resilience outcomes of green growth approaches. Keeping the above background in view, the book series offers comprehensive coverage combining the domains of environment, natural resources, engineering, management and policy studies for addressing disaster risk and resilience in the green growth context in an integrated and holistic manner. The book series covers a range of themes that highlight the synergistic outcomes of green growth approaches.

The book series aims to bring out the latest research, approaches, and perspectives for disaster risk reduction along with highlighting the outcomes of green growth approaches and including Science-technology-research-policy-practice interface, from both developed and developing parts of the world under one umbrella. The series aims to involve renowned experts and academicians as volume-editors and authors from all the regions of the world. It is curated and developed by authoritative institutions and experts to serve global readership on this theme.

More information about this series at http://www.springer.com/series/16471

Manish Kumar Goyal • Anil Kumar Gupta
Editors

Integrated Risk of Pandemic: Covid-19 Impacts, Resilience and Recommendations

Springer

Editors
Manish Kumar Goyal
Civil Engineering
Indian Institute of Technology
Indore, Madhya Pradesh, India

Anil Kumar Gupta
Environment and Disaster Risk Management
and Centre for Excellence on Climate Resilience
National Institute of Disaster Management
New Delhi, Delhi, India

ISSN 2662-4885 ISSN 2662-4893 (electronic)
Disaster Resilience and Green Growth
ISBN 978-981-15-7678-2 ISBN 978-981-15-7679-9 (eBook)
https://doi.org/10.1007/978-981-15-7679-9

This Springer imprint is published by the registered company Springer Nature Singapore Pte Ltd.
The registered company address is: 152 Beach Road, #21-01/04 Gateway East, Singapore 189721, Singapore

Foreword

The novel corona virus (COVID-19) was proclaimed a pandemic on March 11, 2020. It has spread over the world prompting a lockdown on half of the total population. As a consequence of outbreak of this kind of pandemic, the world has faced enormous loss of lives. In addition, economic activities have suffered acutely. Scientists and researchers around the world are trying their best to combat this unprecedented situation, to contain the impact on mankind.

By applying an integrated approach, the impact of outbreak can be assessed more effectively. It also helps in policymaking and decision planning to handle such adverse situations in future. Although there is a lot of literature available on outbreak of viruses, there is hardly any book available which incorporates societal, psychological, environmental and economic impact of this pandemic outbreak on various sectors for the integrated risk assessment. This book is a welcome step in that direction. Additionally, it provides an excellent introduction to non-specialist readers to the field.

This book is a perfect compendium of recent issues, problems and their possible solutions faced across different sectors such as social, economic, technology and environment during the COVID-19 emergency. It covers several case studies to address a range of issues spanning different aspects associated with COVID-19. All major aspects concerning the challenges and responses of the public, at local and global levels and scale, have been presented in organized sections, so as to provide a much better vision of this pandemic outbreak. The authors of the book are well-known experts in their respective fields, thereby providing the readers a studied and encapsulated version of the recent issues, challenges and developments. The content is presented in a well-written and engaging form.

I compliment the editors of the book—Dr. Manish Kumar Goyal and Prof. Anil K. Gupta—for conceptualizing and taking this timely initiative. I also congratulate the contributing authors for the time they have taken to prepare detailed methods as also to provide practical hints and tips that are often essential to get a new working protocol. I am sure the book would be a significant contribution in the area of health and disaster resilience, towards achieving the UN Sustainable Developments Goals (SDGs).

IIT Bombay, Mumbai, Maharashtra, India

IIT Indore, Indore, Madhya Pradesh, India

Deepak Phatak

Acknowledgements

The editors acknowledge the support and hard work put in by all the contributors under the tight timelines. Special thanks are due to Prof. Deepak B. Pathak, a Padma Awardee (A High Honour of Government of India), a noted technologist and knowledge power promoter of India, for writing the Foreword of this book. Thanks are due to the encouragements from Mr. G. V. V. Sarma, Secretary, National Disaster Management Authority of India, Maj. Gen. Manoj K. Bindal, Executive Director, NIDM, Prof. Neelesh Kumar Jain, Officiating Director, IIT Indore and the entire team of professionals and researchers associated with the volume editor's team. A range of literature, reports, data and publications were used by the chapter authors in drawing academic inferences, and sources are cited to the best possible extents. Their original sources and contributors are gratefully acknowledged. Fast track processing of the publication by the Springer Nature team, especially comprising Ms. Aakansha Tyagi (Responsible Editor) and Ms. Vaishnavi Venkatesh (Production Editor), is acknowledged for their support and cooperation.

Introduction

Environmental emergencies were rarely considered at the forefront of disaster management, developmental sustainability or business continuity. In the recent years, anthropogenic, climate change induced or technological mishaps associated with ecological or biological systems have occupied significant portion of risk-space we inhabit. Health emergencies of the scale this COVID-19 pandemic brought before the world to deal with call for systematic understanding of processes and integrated risks of such environmental and health outbreaks. This is important in understanding the emerging issues concerning low-frequency high-intensity disasters that may enlist pandemics, tsunamis, locust swarms, water contamination emergencies, devastating dust storms or volcanic eruptions resulting in loss of sunshine for over months stopping photosynthesis leading to food crisis in geographies.

Looking to the need to comprehend from the lessons of this COVID-19 health emergency that took the form of major global disaster, the edited volume on the theme has been strived for. Experiences of this pandemic disaster are not valuable only for health or social systems but equally for economic and business sectors, environmental safeguards and the journey toward achieving the sustainable development goals. This book aims to provide comprehension on both fundamentals as well as key factors of global disease outbreaks. The objective is to systematically describe the integrated risk of virus outbreaks in terms of social, environmental and economic impressions. The book brings together broad range of topics including basic concepts, role of governance, isolation measure and key technical advancements in the race of containing diseases since Pre-modern era. In addition, it also covers resilience analysis to grasp the influence disease outbreaks on bio-diversity, ecosystem services and agricultural food production. The motive is to define key exit strategies from the lessons learned from failure and success stories of historical disease outbreaks.

As humankind spreads across the globe, so do the infectious diseases, with fatality being related to virus outbreaks. Therefore, the focus of the book is on the assessment of integrated risk pertaining to global virus outbreaks with simultaneous attention on recent COVID-19 outbreak. It provides a detailed assessment on timeline of historical epidemics with a preliminary focus on their impacts on various sectors. It illustrates new opportunities of human and animal interaction relating to enhanced trade, diversified civilizations, eased commute summing to a widespread outbreak in the recent centuries.

The book is presented in four parts, wherein Part I reviews the fundamentals of virus outbreaks including the review of past pandemics and the threat of pandemic to humans and nature. Part II describes the integrated impact and the risk of outbreaks on environment, agriculture, air pollution, economy, food industry and transportation sector. It also describes the governance challenges globally in face of the pandemic. Part III aims at understanding the role of various measures like public health interventions, technological advances and nanotechnology for building the healthcare systems and water body resilience to outbreaks. This section even talks about the multi-hazard and climate risk management during the pandemic. Finally, Part IV of the book suggests key lessons and recommendations for dealing with such outbreaks by enhancing epidemic resilience as well as the lessons drawn from trans-domain assessment of COVID-19 outbreak. Authors are drawn from across the developed and developing world, encompassing varied experience of dealing with pandemics in historical times and the present health crisis.

The concept of resilience evolved recently and is the process of adapting well in the face of adversity, here we attempted to enumerate the virus outbreaks in terms of natural, climatic and healthcare system resilience with a focus on defining pandemic exit strategies. The book sums up with lessons and recommendations from trans-domain assessment of global virus outbreaks and is going to play an important role in building disease resilient communities. The book is expected to serve as knowledge support system and reading reference to the scholars across science, humanities, engineering, management and public policy domains across the world. This pandemic is leading to curriculum amendments across various disciplines, and the book would serve as drivers of change vis-à-vis the content provider for teaching–learning process.

Department of Civil Engineering, Indian Institute of Technology, Indore, Madhya Pradesh, India — Manish Kumar Goyal

Department of Environment and Disaster Risk Management and Centre for Excellence on Climate Resilience, National Institute of Disaster Management, New Delhi, Delhi, India — Anil Kumar Gupta

Contents

About the Editors

Manish Kumar Goyal is an Associate Professor in the Discipline of Civil Engineering, Indian Institute of Technology, Indore. His research interests include water resources engineering, GIS and remote sensing applications in water and environment and climate change. He worked as a postdoctoral fellow at Nanyang Technological University Singapore and McGill University, Canada. He serves as an Associate Editor for the ASCE-Journal of Hazardous, Toxic and Radioactive Waste. His contribution fetched him Recipient of ASCE-VIF (2019), Prof R J Garde Research Award, India Society of Hydraulics (2019), Recipient of American Society of Civil Engineers (ASCE)-Best Theoretical-Oriented Paper Award (2018), NVIDIA GPU Research Grant (2018), IEI Young Engineers Award in Civil Engineering Division (2017-18), Indo-US WARI Fellowship Award (2016), DST – SERB Young Scientist-fast track grant (2015), Inspire Faculty award, 2014, Erasmus Mundus Interweave Award, Europe 2014, JSPS fellowship award, University of Tokyo, Japan (2013) and Canadian Commonwealth Scholarship Award.

Anil Kumar Gupta is a resilience and sustainability strategist, full Professor with India's National Institute of Disaster Management, New Delhi. He Heads the Division of Environment and Disaster Risk Management and Centre for Excellence on Climate Resilience, projects in hand: CAP-RES (DST-GOI), HER-CAP (WHO), National Agriculture DM Plan (MOAF) and Crisis Management Plan (CPCB), Nodal Officer for Central Ministries, UNCCD, UNFCCC. Previously, he founded Bundelkhand University Institute of Environment and Development Studies as Reader, Head & Director I/C. His pioneering contributions were to national and sub-national policy-planning for disaster management, climate adaptation integration, ecosystem based/NBS for DRR, human resource planning and National contributions towards HFA-2/SFDRR. He implemented many international (GIZ, CDKN, UNDP, UNEP, WHO, NORAD, ISET, ICIMOD) and national projects

(MOEF, ICSSR, CSIR, DST), supervised 06 PhDs, has over 140 publications including 20 books/manuals and 50 research/policy papers. He is WMO expert team member on climate statement, IUCN-CEM South Asia CG Member and Vice-chair of Association of Occupational and Environmental Health.

Part I
Review of Virus Outbreaks

Chapter 1
Introduction to Virus Outbreaks

Atisha Sood, Anjali Barwal, Anil Kumar Gupta, and Manish Kumar Goyal

Abstract Human health is significantly affected time and again by emerging and re-emerging incidences of viral infections in spite of exceptional progress in the field of biomedical research. A great example of the same being recent pandemic of COVID 19 across the globe. Even reappearing outbreaks of chikungunya and dengue in tropical and sub-tropical regions, Zika virus outbreak in America and SARS, MERS and influenza A epidemics indicate the same. The natural reservoirs of human viruses are typically farm animals and occasionally the wild animals as well as arthropods. The key to understanding this emergence and re-emergence of these pathogenic viruses lies in the complex 'host-pathogen-environment' interaction. Selective pressure is put on these reservoirs due to changing human habits, increasing population density, stress on mother nature, poor sanitation and changing climate. For protection from these infections, new approaches like consumption of thoroughly cooked meat and animal products only are the most promising control measures. Although substantial progress has been made in human immunodeficiency virus and hepatitis virus control, the arbitrary disposition of evolving viruses and the occasional outbreaks critically restrain the prevention and control process.

Keywords Virus outbreak · Epidemic · Pandemic · Viral transmission

A. Sood (✉) · A. Barwal · A. K. Gupta
National Institute of Disaster Management, Ministry of Home Affairs, Government of India, New Delhi, Delhi, India

M. K. Goyal
Department of Civil Engineering, Indian Institute of Technology, Indore, Madhya Pradesh, India

M. K. Goyal, A. K. Gupta (eds.), *Integrated Risk of Pandemic: Covid-19 Impacts, Resilience and Recommendations*, Disaster Resilience and Green Growth,
https://doi.org/10.1007/978-981-15-7679-9_1

1.1 Introduction

Human health is significantly affected time and again by emerging and re-emerging incidences of viral infections (Krause 1992). The emerging viruses which have just been introduced lately in a population are called novel etiological agents. Since the finding of human immunodeficiency virus (HIV) in 1980s began a global recognition of importance of more research in this field of novel pathogens. New outbreaks of infections in the past several decades has aided in detection of multiple varieties of very infectious viruses (Feng et al. 2008; Padgett et al. 1971).

Viruses are non-cellular pathogens in nature. They are all astrict parasites occurring within a cell because their source of energy and basic composition is the host cell. The consequences of viral infection in humans and animals can vary from unapparent (no disease) to serious diseases and even death. While some cause acute infections in host body for a few days or weeks other viruses may result in a lifelong infection. A distinct characteristic of viruses which increases their perseverance is their ability to adapt quickly to changing environment and subsequently evolve (Gardner et al. 1971).

The effects of viral infection on cells individually can be seen in terms of changed cell functions and structures, even the overall health and fitness of the affected being is influenced noticeably. When a virus comes in contact with a cell one of these four outcome is likely:

- *Permissive (/Productive) infection*: In this case, viral proteins and nucleic acids are synthesised and virions are released.
- *Non-permissive infection.* The cell resists the infection, completely.
- *Abortive or non-productive infection.* The virus is able to enter the cell but before anything is synthesised the replication becomes irreversibly blocked at some step.
- *Latent infection.* It is the condition when although a viral genetic code is present in the cell, but only a few or no viral proteins are synthesised. It means that the virus can actively replicate under suitable conditions.

1.2 Types of Viral Diseases

Disease is caused as a consequence of tissue or organ injury. Viral infections can cause diseases which are acute, chronic or latent (Boldogh et al. 1996).

Acute diseases have a sudden onset which can last between a few days to several months, and the pathogen may be cleared, controlled or even cause the host death. Common cold is one of the many examples of acute viral infections.

Chronic diseases have a steady development and the treatment of can take years or a lifetime altogether. In some cases these infections can even cause death of the host but that is not always the case. Because of longer time taken for resolution, these infections are also called persistent infections. The virus is continuously produced and shed by the host. Hepatitis C virus (HCV), hepatitis B virus (HBV), and human immunodeficiency virus (HIV) have a tendency to cause chronic infections in human beings.

Latent diseases is caused when viral genome is maintained without production of detectable virus. Chickenpox/Shingles virus is an example of virus that causing latent disease. Chickenpox is usually a mild infection, with characteristic pustules (like blisters), that are cured in around a week. Although, even long after the pustules have disappeared physically the virus remains in the body. These virus genomes sustain in the neurons silently, for decades. When this virus exits latency stage by producing blister-like lesions by travelling down to the skin via neuron network, a throbbing and weakening disease called shingles occurs in adults. These lesions in turn contain infectious virus capable of giving chickenpox to another host with low immunity (Mandell et al. 2009).

1.3 Origin of Viral Outbreaks

Viruses cannot replicate outside of living cells like most bacteria. About 80% of all viruses, known to cause human infections, mostly exist naturally in 'reservoirs' that are non-human like farm animals and poultry birds while some continue to be found in wild animals and some arthropods (Cleaveland et al. 2011). The zoonotic pathogens account for almost 60% of the 'known' infectious agents and 75% of 'novel' pathogens in human beings (Taylor et al. 2013; Woolhouse and Gowtage-Sequeria 2005; Kilpatrick and Randolph 2012). We have, however, limited understanding of these zoonosis and the varieties of such virus in known reservoirs. The information we have on domesticated animals maintaining a bunch of these viruses is little as well as data on wild animals hosting a number of viruses is insufficient (Cleaveland et al. 2007). The new influenza virus strains, human coronavirus (h) CoV, Hendra virus, Nipah virus and Zika virus are all associated with zoonotic transmission. Even the recent deadly outbreak of MERS-CoV was thought to be zoonotic as it was found to be genetically closer to bat CoV than any other hCoV (Corman et al. 2012). Data currently shows that bats harbour substantial varieties of CoVs that range from species-to-species and region-to-region (Anderson and Tong 2010).

In nature the existence of a virus is dependent on the perpetuation of a sequent infections, which is a chain of transmission of virus; the incidence of a disease is neither essential nor even beneficial. Although, clinical cases in fact give rise to more infectious viruses than unapparent cases, but the latter produces a larger number and, because there is no restriction on host movement it provides a great opportunity for viral dissemination. The survival of viruses in humans occurs through three different patterns according to epidemiologists, which is differentiated based on the use of reservoirs: acute self-limiting infections with no reservoir, persistent infections with a reservoir in humans, and involvement of an animal reservoir. A basic method is used by all viruses for survival (Table 1.1), and if by chance this system is obstructed for instance by abrupt fall in host species population because of some other infection of a weather event, then substitute pattern may be deployed (Klimpel 1996).

Table 1.1 Different patterns of survival used by viruses in nature

Infection pattern	Mechanism of survival	Examples of viruses
Acute self-limiting infection (lifelong immunity)	Reservoir absent; large population is needed with continuous chain of transmission	Polio, hepatitis A, enteroviruses, dengue, measles, mumps, rubella
Acute self-limiting infection (short-span immunity)	Reservoir absent; reinfection occurs, small-sized population needed	Rotavirus, influenza, coronaviruses, rhinoviruses, respiratory syncytial virus
Persistent infection (intermittent replication +/− shedding)	Reservoir in humans; lifelong source of virus- infected individuals	Varicella-zoster, Herpes simplex, other herpesviruses
Persistent infection (continuous replication)	Human reservoir; lifelong source of virus- infected individuals	Human immunodeficiency virus, Hepatitis-B, C, human papillomavirus
Zoonosis (human–human spread absent)	Endemic infection in animal reservoir and transmission to humans	Most arboviruses except dengue, yellow fever (urban cycle). Avian influenza, rabies, Hendra
Zoonosis (significant human–human spread)	Endemic infection in animal reservoir and transmission to and between humans	Marburg/Ebola, Hantaan, Nipah, dengue, yellow fever (urban cycle)

It is necessary to understand the clinical features as well as the disease pattern of an infection for structuring and actualizing its control programs. For instance, the reason behind successful eradication of smallpox was the understanding that an acute self-limiting infection where bulk of individuals show clinical symptoms was caused by a variola virus which had no animal reservoirs. A large portion of viral infections in humans fall into the acute self-limiting infection category. It is crucial to have optimal transmission and the viruses causing systemic infections with lifelong immunity need large dense populations to survive. While the ones causing surface infections with short- span immunity can even survive in smaller populations, and this ability to persist in limited population can be built up by the antigenic drift (Burrell et al. 2017).

1.3.1 Virus Infection Transmission

For transmission cycle to occur, the entry of virus into the body, its replication inside and shedding with successive spread to another host, is required. Viral transmission (Fig. 1.1) can take place horizontally or vertically, although mostly horizontal transmission occurs which is in between the individual at risk. Unlike most bacteria, different viruses use specified pathways for transmission. This selection is based on a number of factors like physical characteristics of the virus, course of shedding and other features of the pathogenic process. These ways are basically described by separate techniques used to enter the upper epithelial body lining by the viruses.

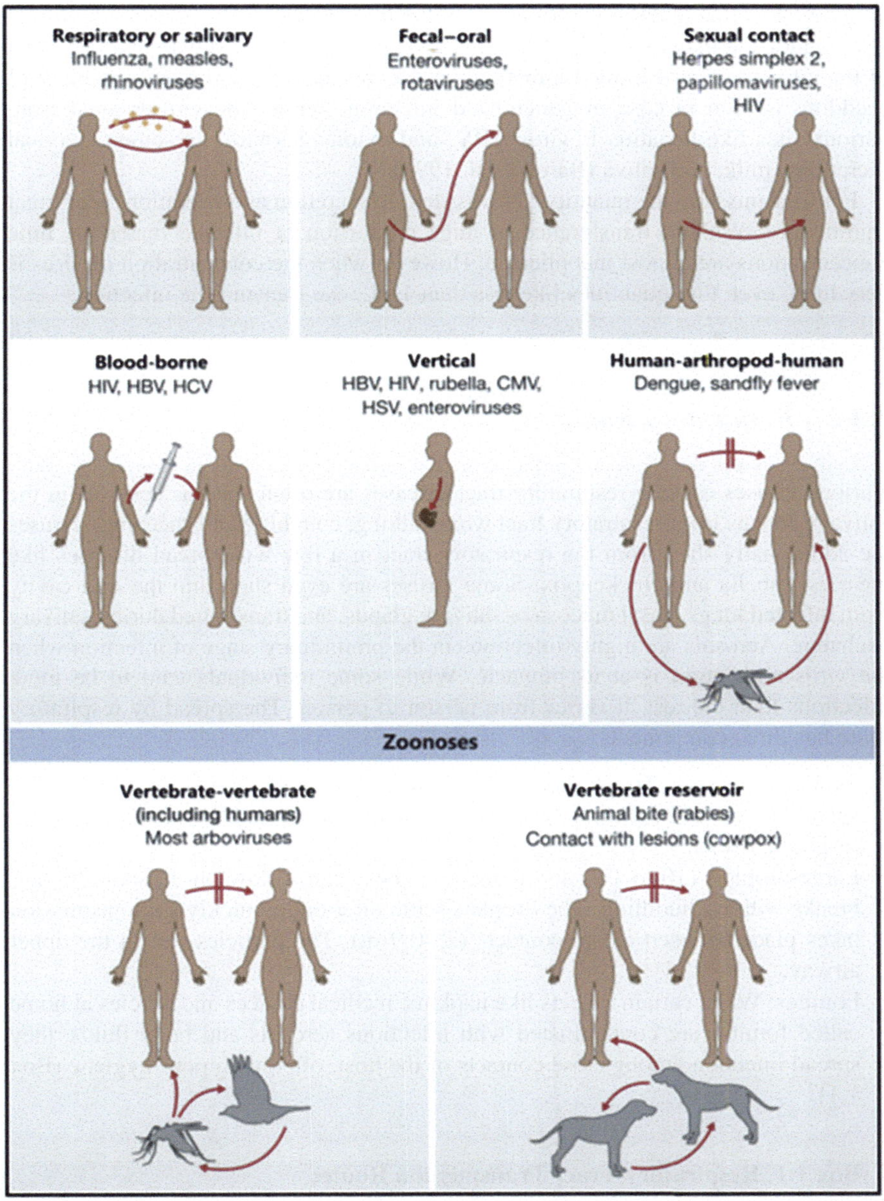

Fig. 1.1 Modes of transmission of human viral diseases. (Source: Burell et al., Epidemiology of Viral Infections. Fenner and White's Medical Virology, 2017)

Virus shedding takes place from one of the body surfaces or opening which was also included in the virus entry. The same body part is concerned with entry and exit of the virus in case of limited infections, but a considerable variation in manner of shedding is seen in case of generalized infections while few viruses shed from various sites like hepatitis B virus, HIV, and cytomegalovirus in semen, cervical secretions, milk, and saliva (Baron et al. 1996).

For transmission, the quantity of virus shed in secretion and excretion, is of great significance. Without transference of huge proportion of infected materials, little concentrations are almost inapplicable. However when the concentration of virus is very high, even tiny quantities like less than 1 μL, can transmit the infection.

1.3.2 Respiratory Route

Various viruses causing respiratory tract diseases are often shed as aerosols in the saliva or mucus from respiratory tract while talking, coughing and sneezing. Viruses are additionally shed from the respiratory tract in a few widespread diseases like measles, rubella and chickenpox. Some viruses are even shed into the oral cavity from infected lungs, nasal mucosa or salivary glands, and transmitted during salivary exchange. Aerosols are highly infectious in the preliminary stage of infection when the virus replication is at its pinnacle. While some individuals tend to be more infectious than the rest, it varies from person to person. The spread by respiratory route has three components:

- Small-droplets (<10 μm diameter): These cause quick sudden outbreaks and spread to far off contacts (>1.8 m). The droplets are likely settle in lower respiratory tract.
- Large-droplets (10 to 100 μm diameter): These cause slow onset, sporadic outbreaks with no huddling. The droplets settle on ground quickly so, transmission takes place between closer contacts (<~0.9 m). The particles rest in the upper airway.
- Fomites: When certain objects like napkins, medical devices and articles at home called fomites are contaminated with infectious aerosols and body fluids, they spread infection among close contacts of the host, often with poor hygiene (Box 1.1).

Box 1.1 Respiratory Tract Transmission Routes

Respiratory transmission is an effective and fast way to cause infection in large number of host contacts and to spread virus across the globe.

There are two methods:

(continued)

Box 1.1 (continued)

1. *Breathing in aerosols*. When breathing at rest, human beings usually filter about 600 L of air per hour.
   Aerosols are formed while coughing and sneezing, lesser by talking.
   Large droplets settle on the floor, while small droplets usually spread along but once dry the virus is inactivated; and therefore the atmosphere would not be infective for long.
   The seasonality of these infections is affected significantly by increase in virus survival in cool temperature, variation in social activities like school vacations etc.
2. *Touching fomites*. Infection spreads though napkins, hands, surfaces and objects that have been contaminated by aerosols. Act of shaking individual's hands, touching fomites etc. gives a clear route for virus to reach another person's nose and mouth and enter their body (Bloom-Feshbach et al. 2013).

1.3.3 Gastrointestinal Route

Viruses of enteric system are shed via vomiting and faeces from the body. Larger the quantity of fluid released more is the environment contaminated. These kind of viruses are harder and more resilient, so they tend to survive outside the body for longer even in harsh conditions, than the respiratory viruses. Two transmission patterns are seen in this route:

1. Point-source outbreak: It usually happens when large number of people actually consume the contaminated water or food in events like parties etc. The sources are usually raw salad, shellfishes, or due to consuming unsafe water contaminated with sewage, and
2. Person-to-person spread: It usually occurs via faeco–oral route, gradually in homes which lack proper toilets, running water and other hygiene facilities that are more often poor and lack education.

1.3.4 Cutaneous Route

Healthy skin provides an inviolable hurdle for virus entry to the body. Virus shedding occurs at an almost insignificant rate from the skin so much so that even for people with blood-borne diseases their unbroken skin poses no threat of infection to their close contacts. Nonetheless, even minute skin scrapping is a great opportunity for viral entry by direct contact. Infected blood from blood borne infections like hepatitis B can shed via scrapped skin and spread horizontally between individuals, specifically those living in overcrowded, low socio-economic settings with a history of skin diseases. Although, these kind of infections are spread usually by various other ways like introduction of infected blood parenterally (through injections, sharing needles or blood transfusion).

Some poxviruses are also spread to humans from animals or vice versa due to contat with the skin lesions. While these lesions or rashes are formed in many diseases, viruses are not usually shed in infections like measles or picornavirus, flavivirus etc., but in herpesvirus infection, vesicular lesions are formed where virus is plenty in the lesion fluid. Even in this case, shedding by saliva and aerosols is more significant for transmission of the disease. A very important virus spread by cutaneous route is that of rabies virus transmission through bite of an infected animal on the skin.

1.3.5 Genitourinary Route

The genital secretions of found in semen and vagina contain various viruses. Transmission of infection by sexual route through mucosal contact is very effective as the virus remains moist and there is no need to survive outside the body for long. The spread by this route however takes more time than respiratory route because association with several contacts is slow. Research from HIV has demonstrated that transmission is increased when multiple consecutive partners are involved, when multiple consecutive mucosal abrasion are present, other infections also occur and the involved male is non circumcised. Some examples are HIV, HBV and herpes simples type 2.

1.3.6 Blood-Borne Route

It is a significant route of disease spread both within an individual as well as between two individuals. Diseases like Hepatitis B, C, and D viruses and HIV all disseminated by blood transfusion, but now due to thorough testing of donated blood the risk has been significantly reduced. Transmission of blood borne infections due to shared needles and injecting paraphernalia is an issue amongst intravenous drug users in most of the countries today. Blood is also the most typical way how arthropods like sandflies, ticks and mosquitoes pick up viruses while having a blood meal from a host. Rarely, few arthropods like horseflies etc. even passively pass on viruses by interspersed blood meals from various hosts with contaminated mouth parts.

1.3.7 Ophthalmic Route

A common route of virus introduction into the body is through the eye by touching dirty fingers, using contaminated swimming pools, from non-sterilized ophthalmic articles or even aerosol droplets.

1.3.8 Milk-Borne Route

A number of viruses that are secreted in milk are transmitted to new born infants. There is also an added risk of introduction of an infectious agent to the infant while breastfeeding from an infected mother. Although, chances of this is a lot less than the risk of vertical transmission of an infection while childbirth. In countries and communities where malnutrition or infectious diseases are a common cause of infant mortality, breastfeeding may still be prescribed despite the added risk of transmission that it poses.

1.3.9 Vertical Transmission

This is the route of transmission of a virus from an infected mother to her embryo, foetus or the infant. It is a significant route of transmission across generations which enables the survival of virus in nature. There are three conditions where such a transmission occurs:

- Through the amalgamation of viral DNA directly into the germline DNA of the gametes and/or the fertilized egg,
- Through trans-placental transmission during pregnancy, and
- Through peri-natal or post-natal transmission by milk, saliva or other body fluids.

This transmission may cause severe infection, some congenital diseases or abnormalities or even death of the foetus and subsequent abortion in some cases. Vertical transmission establishes the infection into a new generation, in cases of hepatitis B and HIV diseases that are then competent to pass on the infection to subsequent birth cohorts for many more generations. For continuation of several arthropod- borne infections. vertical transmission in the vectors is very important (Knipe and Howley 2013).

1.4 Phases of an Outbreak

Several viruses continuously flow amongst animals, mostly birds. Although these viruses are capable of causing viral outbreaks theoretically, but in *Phase 1* of an outbreak there are no reported human infections caused by these viruses circulating amongst animals (Fig. 1.2).

In *Phase 2*, there is a potential epidemic threat from an animal virus circulating among domesticated or wild animals which has caused an infection in humans.

In *Phase 3*, although still no human-to-human transmission enough to sustain community-level outbreaks has been established but even so an animal or human-animal virus has caused sporadic cases or small clusters of disease in people. However in case of close contact between an infected person and an unprotected caregiver there may occur limited human-to-human transmission. The transmission under such constricted circumstances does not show that the virus has gained a level of transmissibility among humans required to cause an epidemic.

The *Phase 4* is said to begin once community- level outbreaks of human-to-human transmission of an animal or human-animal virus have been verified. A substantial increase in the risk of causing a pandemic is marked by the potential of a virus to cause perpetual disease outbreaks in the community. World Health Organisation should immediately be consulted if any country has verified or suspects such an episode such that the circumstances may be assessed together and a decision can be sought by the country that has been affected to implement a quick containment operation. While this phase stipulates a noteworthy hike in the risk of a pandemic it surely does not mean that a pandemic is certain in the future.

Phase 5 is finally where virus has spread via human-to-human route in at least two countries of one WHO region (Fig. 1.3). Although majority of the countries would not be affected at this time, the announcement of Phase 5 is a powerful indicator that a pandemic is absolute and there is just a short time to work out the organization, communication, and implementation strategy of the planned mitigation measures (World Health Organization 2009).

1.5 Factors Affecting Virus Outbreak

1.5.1 Transmissibility

The physical properties of a virus like the nature and extent of shedding from the viral body, social interaction among hosts etc. significantly affect the transmissibility of the virus. The human-to-human transmission is also enhanced by shedding of large volumes of infectious virions. Specifically in case of respiratory viruses, they are shed in high concentrations in a relatively shorter period of some days, as transmission to close contacts is almost certain when explosive aerosols are

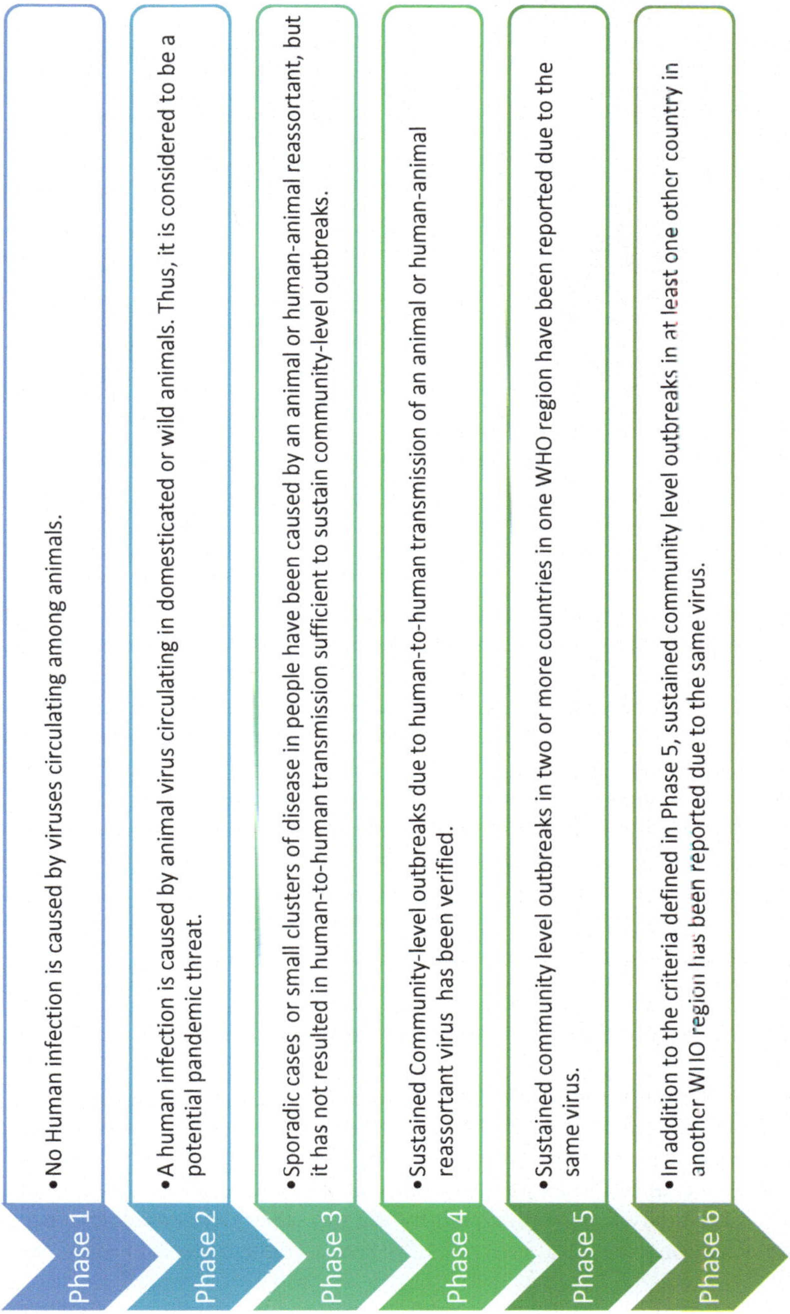

Fig. 1.2 Phases of an outbreak

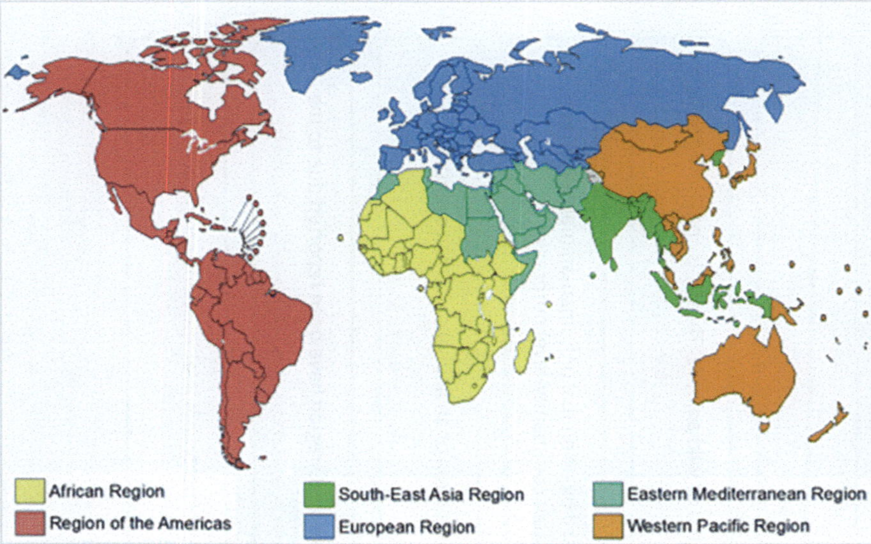

Fig. 1.3 Distribution of the WHO regions

produced while coughing or sneezing. The viral transmissibility of different influenza viruses between humans is determined by their complex use of varied receptors.

Enteric viruses are shed for a prolonged period of time (about a week or more) and released in high numbers in the faecal matter. Viruses also contaminate fomites, water, food and hands. The enveloped respiratory viruses are liable to change during summer or year-round in tropics. While several enteric viruses do not have an envelope and can survive in water, dust or on fomites for several days or even weeks like the hepatitis A and B viruses. However the transmission has slowed down with improved sanitary conditions, socio-economic status and educating the masses, shifting the prime age of acquiring the infections to an older age. Several infections give rise to clinical diseases in older age groups, therefore these improvements might affect the increase in clinical cases absurdly (Box 1.2).

Box 1.2 Impact of Sanitation Measures on Enteric Disease Transmission

In the twentieth century, as the level of hygiene and sanitation advanced in several places around the world, the dissemination of enteric virus infections became more and more inefficient increasing the usual age at which infection previously occurred. Subsequently, many individuals did not possess the acquired immunity for that infection by the time they reached adolescence. For some reason, the typical infection of poliovirus or hepatitis A in adults and

(continued)

Box 1.2 (continued)
older children apparently cause a *clinical* disease, in comparison to infection at a younger age. The result of these improvements in living standards was an absurd *increase* in clinical diseases. It could be seen most notably in "virgin soil" epidemics that occurred long before the era of vaccinations, in outlying communities without previous viral exposure; where majority of fatalities were seen in adults. Ultimately, the countries having high standards of living started seeing a fall in the total numbers of clinical diseases and infections, when the virus circulation fell even more.

Figure 1.4 describes the various steps involved in progress of a typical acute transient infection. In several diseases like chickenpox and measles the individual becomes contagious a day or so before they become ill themselves. While for zoonotic infections either involving wild or domestic animal reservoirs, there always has to be close contact of humans with the animals or a transmission of virus by arthropod vectors.

1.5.2 Seasonality

The incidence of several viral infections shows a noticeable seasonal variation. The arbovirus infection transmission by mosquito or sand fly vectors occurs during summer season in the temperate regions when vector population is abundant and they are most active. While ticks transmit infectious diseases mostly in the spring or early summer season. It is also interesting to notice the seasonal variation in incidence of infections where humans are the only hosts.

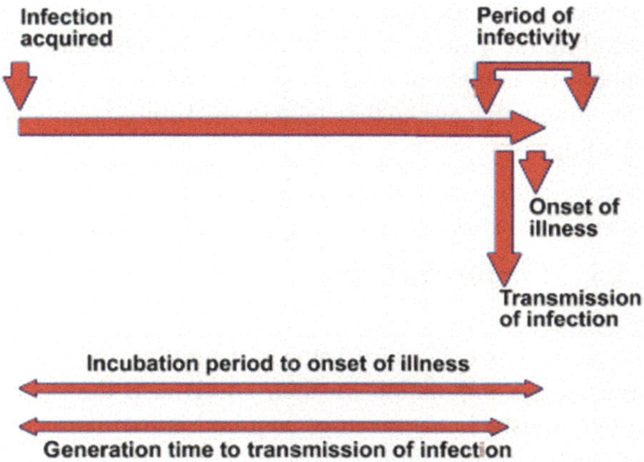

Fig. 1.4 Sequence of infection transmission

The respiratory viral infections are however most prevalent during the winter months and to some extent in autumn or spring in temperate climates. An important characteristic of temperate climates is the annual winter outbreaks of severe respiratory viral infections in infants, even the epidemics of influenza occur mostly during the winter season. During the spring season most of childhood rash diseases that are spread by respiratory route reach the peak.

The seasonality in enteric virus infections actually depend on etiological agents like the incidence of majority of enteric infections is highest during the summers while calciviruses in fact have an irregular seasonal pattern and rotavirus infections occur predominantly during winters. Herpes virus infections are all spread through close contact with infected saliva or other body fluids. The incidence of these kind of infections and even other sexually transmitted diseases all show no seasonal dissimilarity of incidence. In the tropical areas however, summer and winter variation is in fact replaced by the wet and dry seasons that tend to govern the seasonality of diseases. Here the dry season typically witnesses a peak in incidence of measles and chickenpox followed by a swift fall with the beginning of the rainy season, while peak incidences of rhinovirus and influenza infections are noted during that season.

The seasonality mainly depends on biological as well as sociological factors viz. low humidity favours the survival of influenza, vaccinia and measles viruses, rhinoviruses, adenoviruses and polioviruses tend to survive better at high humidity. However, lesser temperatures and aerosols help viruses survive for longer periods. Seasonality also plays a major part in determining the host susceptibility to infections like nasal and oro-pharyngeal mucous lining changes due to smoke, heat or air conditioning. Even the social activities of the host affect the transmission opportunity of respiratory viruses. For instance in Antarctic and Arctic regions, the crowding behaviour into stuffy vehicles, cramped areas and buildings located in temperate regions, greatly enhance the respiratory virus transmission which cold weather alone cannot do. In the locations receiving monsoon rains early in summers, often the movement of people for social gatherings as well as daily life is significantly reduced. While this reduces the chances of getting infection from individuals of other locations, the opportunity of respiratory virus transmission among family members is greatly enhanced. Young children are the most remarkable sources of spreading the viruses in urban settings, because they interact with potential hosts at schools or neighbourhood playtime, and they lack acquired immunity for many infections at that age so they tend to shed considerably larger volumes of viruses than the adults.

1.5.3 Community Size

The viruses which cause acute self-limiting infections need dense and relatively larger at-risk population for their survival. If the prospective supply of vulnerable hosts is finished as a result of acquired immunity to re-infection, there are chances that such viruses might vanish altogether from the population. However, chronic viruses can survive in a handful of population too by going from generation to generation. The community size critical for virus survival depends on its pattern of transmission as well as the time needed to develop immunity for it.

1.5.4 Host Immunity

Acquired immunity through either an earlier infection or a vaccination is very crucial in the viral disease epidemiology. For general infections the acquired immunity is exhibit primarily by IgG antibodies which is often permanent. Even when there are no repeated sub clinical infections the immunity is acquired for life like in case of poliomyelitis and measles infections (Evans 1989).

In case of localised respiratory tract infections, the immunity is somewhat short spanned. Many varieties of rhinoviruses and some types of coronaviruses and enteroviruses can cause upper respiratory tract infections. The apparent continuous series of common colds infections in community actually indicate towards a chain of minor epidemics, which are all due to distinct varieties of virus. The antibody IgA present in nasal secretion provides defence against the chances of reinfection. While individuals do possess short spanned specific immunity, the immunity for other types of viruses does not exist. Majority of people catch the cold two to four times each year. For most of the respiratory viruses the span of shedding is short (few days to a week after symptoms appear) but for rhinoviruses this period can last up to 3 weeks even after common symptoms are unapparent. Research has shown that in order to maintain the herd immunity it is essential to have a continuous supply of vulnerable hosts for harbouring novel viral strains so that the diseases is maintained in nature and even repeat sub-clinical infections continue to occur (Fenner 1996).

1.5.5 Persistent Infections

Persistent viral infections amplify the perpetuation mechanisms of viruses even when they do not directly cause a clinical infection. The hosts carrying a persistent infection can shed the virus sporadically or continually in severe cases which in turn may re-establish the virus in a community where a lot of people were born after the last clinical episode of the infection was seen. Herpes virus survives by this mechanism in smaller populations. The disease causation, viral persistence and transmission pattern of viruses are not inevitably linked with each other. Consequently a persistent infections might have adverse effects on associated reservoir hosts but these are effective in continuing the infection chain. However, in case of sclerosing panencephalitis (SSPE) measles virus, the persistent infection in central nervous system is lethal for host but of barely any importance in virus propagation.

1.5.6 Non-Human Reservoirs

As in the case of zoonotic infections, a constant re-establishment of viruses from a non-human reservoir helps the virus in spreading in human population as well as it controls the distribution and the extent of viral infection in the community like arboviruses, rabies virus and hantaviruses. The level of infection in humans is dependent on the prevalence of the infection in the reservoir and the extent of contact with that animal reservoir. While planning for global eradication of any human viral infection it is important to consider the presence and the possible degree of animal reservoir (Beran 1994).

1.5.7 Arthropod Vectors

Transmission by arthropod vectors is one of the most composite transmission modes for viruses environmentally. Arbovirus is the term used to refer to a virus that has a vertebrate host and an intermittent blood feeding arthropod (typically mosquitoes or ticks) in its life cycle. The virus enters into the body of the vector via blood meal of a viremic individual or animal. The virus then replicates inside the gut initially and then moves to the salivary glands (Incubation period- extrinsic) depending on ambient temperature and the kind of virus. When the vector finally bites a vertebrate host to feast on its blood, the virus travels into the host bloodstream along with the salivary secretions of the vector. Additionally, mechanical contamination of vectors' biting parts ("flying pin") may also transmit the virus.

Vectors give viruses a straight opportunity to break the cross-species barrier because the vector will bite many organisms, like reptiles, mammals and even some birds, which are unlikely to ever come in each other's contact naturally. Some mammals or mostly birds are usually the vertebrate reservoir hosts that help in maintaining the virus in the vertebrate-arthropod-vertebrate life cycle. In very rare circumstances only humans are included in this virus maintenance cycle called the enzootic cycle until they voluntarily come in contact with the infected host (Parvez and Parveen 2017).

1.5.8 Hospital Acquired Infection

The spread of infection when an individual is in a hospital or clinic called the nosocomial transmission and "by the hand of the doctor" called iatrogenic transmission is a very important means for virus perpetuation in community. The best example of an iatrogenic and nosocomial infection was the lethal outbreak of Ebola virus in 1976 in Zaire. Other infections which propagate and maintain themselves in the community through hospitals spread are the common respiratory viruses like influenza, chickenpox and respiratory syncytial virus infections in hospital settings. While infections from Hepatitis B and C viruses, and fewer cases of HIV, can also be spread by medical professionals, acupuncturists, tattoo artists, etc. to the general population and at the same time even the attending medical and laboratory staff are at risk of contracting the virus by contaminated needles stick and other similar injuries. There is an added risk of hospital spread of an infection due to collection of all cases at one place where even intrusive techniques and blood exposures are being done. So, even health professionals take appropriate preventive measures to protect themselves and others.

References

Anderson LJ, Tong S (2010) Update on SARS research and other possibly zoonotic coronaviruses. Int J Antimicrob Agents 36:S21–S25

Baron S, Fons M, Albrecht T (1996) Viral pathogenesis. In: Baron S (ed) Medical microbiology, 4th edn. University of Texas Medical Branch at Galveston, Galveston, TX. Chapter 45

Beran GW (1994) Handbook of zoonosis. In: Section B. Viral, 2nd edn. CRC, Boca Raton, FL

Bloom-Feshbach K, Alonso WJ, Charu V, Tamerius J, Simonsen L, Miller MA, Viboud C (2013) Latitudinal variations in seasonal activity of influenza and respiratory syncytial virus (RSV): a global comparative review. PLoS One 8(2):e54445. https://doi.org/10.1371/journal.pone.0054445

Boldogh I, Albrecht T, Porter DD (1996) Persistent viral infections. In: Baron S (ed) Medical microbiology, 4th edn. University of Texas Medical Branch at Galveston, Galveston, TX. Chapter 46

Burrell CJ, Howard CR, Murphy FA (2017) Epidemiology of viral infections. In: Fenner and White's medical virology. Academic Press, Cambridge, MA, pp 185–203. https://doi.org/10.1016/B978-0-12-375156-0.00013-8

Cleaveland S, Haydon DT, Taylor L (2007) Overviews of pathogen emergence: which pathogens emerge, when and why? Curr Top Microbiol Immunol 315:85–111

Cleaveland S, Laurenson MK, Taylor LH (2011) Diseases of humans and their domestic mammals: pathogen characteristics, host range and the risk of emergence. Phil Trans R Soc Lond B 356:991–999

Corman VM, Eckerle I, Bleicker T et al (2012) Detection of a novel human coronavirus by real-time reverse-transcription polymerase chain reaction. Euro Surveill 17:20285

Evans AS (ed) (1989) Viral infections of humans: epidemiology and control, 3rd edn. Plenum Medical, New York

Feng H, Shuda M, Chang Y, Moore PS (2008) Clonal integration of a polyomavirus in human Merkel cell carcinoma. Science 319:1096–1100

Fenner F (1996) Epidemiology and evolution. In: Baron S (ed) Medical microbiology, 4th edn. University of Texas Medical Branch at Galveston, Galveston, TX. Chapter 43

Gardner SD, Field AM, Coleman DV, Hulme B (1971) New human papovavirus (B.K.) isolated from urine after renal transplantation. Lancet 1:1253–1257

Kilpatrick AM, Randolph SE (2012) Drivers, dynamics, and control of emerging vector-borne zoonotic diseases. Lancet 380:1946–1955

Klimpel GR (1996) Immune defenses. In: Baron S (ed) Medical microbiology, 4th edn. University of Texas Medical Branch at Galveston, Galveston, TX. Chapter 50

Knipe DM, Howley P (2013) Fields virology. Lippincott Williams & Wilkins, Philadelphia

Krause RM (1992) The origin of plagues: old and new. Science 257:1073–1078

Mandell G, Dolin R, Bennett J (2009) Mandell, Douglas, and Bennett's principles and practice of infectious diseases, 7th edn. Elsevier, Amsterdam

Padgett BL, Walker DL, ZuRhein GM et al (1971) Cultivation of papova-like virus from human brain with progressive multifocal leucoencephalopathy. Lancet 1:1257–1260

Parvez MK, Parveen S (2017) Evolution and emergence of pathogenic viruses: past, present, and future. Intervirology 60(1–2):1–7

Taylor LH, Latham SM, Woolhouse MEJ (2013) Risk factors for human disease emergence. Phil Trans R Soc Lond B 356:983–989

Woolhouse MEJ, Gowtage-Sequeria S (2005) Host range and emerging and re-emerging pathogens. Emerg Infect Dis 11:1842–1847

World Health Organization (2009) Pandemic influenza preparedness and response: a WHO guidance document. World Health Organization, Geneva

Chapter 2
A Review of Pandemics

Lorenz Hilfiker and Shashwat Ganguly

Abstract In light of the current SARS-CoV-2 pandemic, we review the history of past pandemics. We draw on existing literature to characterise historical and contemporary infectious agents by their transmission modes and dynamics and their mortality impact. We then give a concise summary of 20 historical pandemics, their origins, spread and impact. Moreover, we reflect on the semantic shifts which the term pandemic has undergone, discuss the role of human mobility and other factors in the worldwide spread of infectious disease, and outline some historical trends in public-health response to pandemics. Finally, we touch on recent pandemic threats such as SARS, MERS, and Ebola, elaborate on their zoonotic origin, and compile some lessons that were drawn from these outbreaks.

2.1 Introduction

Pandemics have shaped human history (McNeill 1976; Diamond 1999; Crosby 1972), but the term pandemic itself is of relatively recent origin (Merriam-Webster n.d.-b). As a consequence of several centuries of inexorably advancing globalisation, modern societies have the dubious privilege of experiencing near simultaneous epidemics worldwide. In the past two decades, an increasing number of emerging diseases has fuelled fear of another looming severe pandemic. Nevertheless, when the World Health Organisation (WHO) declared the emerging coronavirus

---

The original version of this chapter was revised. A correction to this chapter can be found at https://doi.org/10.1007/978-981-15-7679-9_25

L. Hilfiker
Independent Researcher, Belp, Switzerland

S. Ganguly (✉)
Indian School of Business and Finance, New Delhi, India
e-mail: Shashwat.ganguly@isbf.edu.in

M. K. Goyal, A. K. Gupta (eds.), *Integrated Risk of Pandemic: Covid-19 Impacts, Resilience and Recommendations*, Disaster Resilience and Green Growth,
https://doi.org/10.1007/978-981-15-7679-9_2

SARS-CoV-2 a pandemic on March 11, 2020, much of the world was taken by surprise. The aim of this chapter is to sketch out some historical context for the currently unfolding pandemic. Since knowledge around SARS-CoV-2 evolves at a rapid pace as new evidence emerges, we will refrain from making any explicit comparisons involving the current pandemic.

In Sect. 2.2, we provide some preliminaries about infectious diseases, their emergence, and their transmission mechanisms. We discuss commonly employed measures for transmission dynamics and mortality impact such as the basic reproduction number R_0, the serial interval and the case fatality rate. Some well-known historical and contemporary examples are tabulated and compared with respect to these characteristics.

Section 2.3 addresses the definition of a pandemic and offers a lightning review of 20 historical pandemics from Antiquity to the twenty-first Century. For each pandemic, we compile both historical and recent microbiological evidence in little snippet biographies to give a brief overview of what is known about its origin, spread and impact. The section is rounded off by a discussion of factors that have historically facilitated the global spread of disease, as well as a discussion of historical trends in public-health responses to pandemics.

In Sect. 2.4, we look at pandemics and pandemic threats of the past 20 years. The zoonotic origin of emerging diseases is outlined, and finally, we consider some lessons that have been drawn from recent global outbreaks, infectious diseases and their characteristics.

2.2 Preliminaries About Infectious Diseases

2.2.1 Agents, Hosts, and Transmission

Pathogens, or micro-organisms that cause infection, are usually grouped into five types—Bacteria, Viruses, Pathogenic Fungi and Protozoa, and Helminths (worms). Protozoa and Helminths are also referred to as parasites.

The characteristic features of each pathogen involve the mode of transmission, the way of replication, the means by which it causes disease, and the response it triggers. The pathogen, in its attempt to establish an infection, first colonizes the skin or the internal mucosal surfaces of the respiratory, gastrointestinal, or urogenital tracts of the host (Janeway et al. 2001). The infection is established once the pathogen overcomes or bypasses the preliminary immune defences triggered inside the host. Different pathogens accomplish disease transmission inside the host in different ways, requiring an equally diverse set of defensive responses from the host immune system.

The majority of infections are caused by viruses and bacteria (City and Guilds 2014). Fungal infections are rare but occur frequently in immunocompromised patients (Kohler et al. 2016). While bacterial and fungal infections tend to be local and inflict strong localised pain, viral infections typically affect many different parts of the body at same time, causing a characteristic itching pain or burning sensation (Nutrition Health Review 2018). The mutation rate of a pathogen is a critical parameter for understanding the pathogenic evolution and has important practical implications for example in vaccine development. Viruses in general are the pathogens that have the highest mutation rate for their genome size, followed by bacteria (Sanjuán et al. 2010). Given their limited survival capabilities outside a host, the life spans of viruses are generally shorter than those of bacteria. There are notable exceptions such as Variola virus (the pathogen responsible for smallpox) and Poliovirus (Walther and Ewald 2004), which survive quite well outside a host. While a virus must invade cells of a living host to reproduce, a bacterium is self-reliant. Some protozoa are encapsulated in cysts, which help them live outside a host's body and in harsh environments for long periods of time (Merikanto et al. 2012)

When a pathogen invades a host, an immune response takes place naturally within the host's body via a complex array of protective mechanisms that rely on detecting structural features of the pathogen which mark it as distinct from the host cells (Sompayrac 2019). The adaptive part of the immune system has the ability to create immunological memory, allowing for a more targeted response against specific pathogens or toxins when they are encountered a second time (Chaplin 2010). This can lead to temporary or lasting immunity of a host to a specific pathogen. In many cases, immunity extends at least partially to other strains of the same pathogen or to closely related pathogens. This cross-protection (cross-immunity) can shape the epidemiological dynamics of multi-strain pathogens when an epidemic of one strain temporarily suppresses the transmission of another (Bhattacharyya et al. 2015).

An agent of an infectious disease is a necessary but not a sufficient condition for the disease to occur. Asymptomatic carriers, that is hosts which become infected with a pathogen but display no signs or symptoms, are common for many diseases. The most iconic example is perhaps 'Typhoid Mary', a New York woman in the early days of the twentieth century, who unwittingly infected dozens of people around her with Typhoid fever over the course of several decades, until she was diagnosed to be an asymptomatic carrier (Filio et al. 2013).

Infectious disease transmission between two human hosts can occur either directly or indirectly through the environment or intermediate hosts. Some of the most frequent infectious diseases in humans are transmitted by dissemination of respiratory droplets (when an infected person coughs, sneezes or even speaks) during interactions at close proximity or in confined spaces. Another direct mode of transmission is through the exchange of body fluid. Examples include Sexually Transmitted Diseases (STDs) and pathogenic transmission from mother to unborn children via placenta. Indirect transmission occurs in several ways. A limited number of pathogens can be transmitted through aerosols of small droplets and particles that remain suspended in air for an extended period of time (airborne transmission) (Stadnytskyi et al. 2020). Many pathogens can be transmitted through contact with contaminated objects or surfaces. Some pathogens thrive in environmental reservoirs, such as water or soil, which serve as a source of infection, for example through contaminated food or drinking water via the faecal-oral route. Animal populations may function as a reservoir as well. Diseases which transmit from animals to humans are called zoonotic. Finally, a number of pathogens are transmitted through insect bites (vector-borne transmission). This can involve

Table 2.1 Some classical and modern diseases with their infectious agents and mode of transmission

Disease	Infectious agent	Transmission mechanism(s)
Influenza(s)	Virus	Human to human (direct, indirect, and airborne)
Smallpox	Virus	Human to human (direct, indirect, and airborne)
Measles	Virus	Human to human (direct, indirect, and airborne)
SARS/MERS	Virus	Human to human (direct, indirect, and airborne)
AIDS	Virus	Human to human (sexual contact)
Ebola	Virus	Human to human (bodily fluids, direct or indirect)
Encephalitis lethargica	Virus	Unknown
Poliomyelitis	Virus	Environmental reservoir (faecal-oral), human to human (oral-oral)
Yellow fever	Virus	Vector-borne (mosquitoes)
Malaria	Parasite	Vector-borne (anopheles' mosquito)
African trypasonomiasis	Parasite	Vector-borne (tsetse fly)
Plague	Bacteria	Vector-borne (rat flea, maybe others in the past) with animal reservoir (rodents), human to human (more rarely, only in pneumonic form)
Typhus	Bacteria	Vector-borne (lice [epidemic version], flea [endemic version])
Cholera	Bacteria	Environmental reservoir (faecal-oral)
Dysentery	Bacteria	Environmental reservoir (faecal-oral)
Syphilis	Bacteria	Human to human (sexual contact)
Tuberculosis	Bacteria	Human to human (mostly airborne)
Diphtheria	Bacteria	Human to human (mostly airborne)
Leprosy	Bacteria	Human to human (mode unclear, prolonged close contact needed)

transmission from one human to another via insects (vectors), or a more complex interaction between vectors, humans, and an animal reservoir. Table 2.1 lists some diseases with their infectious agents and modes of transmission.

2.2.2 *Emergence of New Infectious Diseases in Humans*

Humans acquired their pathogens both by descent (i.e. from a common hominin ancestor) and from wild and domesticated animal sources (Morand et al. 2014). There are three (not mutually exclusive) hypothesised epidemiological transitions by which this acquisition may have occurred in the course of human history: the *out of Africa* hypothesis (Hoberg et al. 2001) holds that many pathogens followed the dispersal of modern humans out of Africa, while new pathogens were gradually acquired through contact with new ecological spaces. The *domestication* hypothesis (McNeill 1976) suggests that pathogens were acquired in the early domestication centres from where they dispersed more widely. Finally, the *globalisation* hypothesis emphasises the role of trade and increased connectivity between settled societies as the major contributor to the modern human disease pool and its geographical variation.

Emergence of new diseases continues today, which is the source for potential virgin soil pandemics, global epidemics caused by a novel pathogen introduced into populations with no prior exposure and consequently no immunity against it. Since 1940 around 400 emerging infectious diseases have been identified, more than 60% of which are zoonotic (Morse et al. 2012). The non-zoonotic cases primarily concern vector-borne diseases and drug-resistant microbes (Jones et al. 2008). In total, around 80% of viruses, 50% of bacteria, 40% of fungi, 70% of protozoa and 95% of helminths that infect humans are zoonotic. Most identified reservoirs are mammalian (around 80%) or avian (Morse et al. 2012). The vector-borne diseases are confined to invertebrates serving as intermediate hosts before passing the pathogens to others.

Several socio-economic and environmental factors influence human-animal interaction, and thus drive the likelihood of zoonosis. These factors include decline in wildlife leading to a population increase in disease spreading species such as rodents, urbanisation leading to shrinking natural habitat of wild animals and associated risks of infection transfer to domesticated animals and to humans, subsistence farming, bushmeat hunting, and overuse of antibiotics in livestock raising leading to antibiotic resistance. These factors are elaborated in Sect. 2.4.

2.2.3 Transmission Dynamics and Fatality Rates

The epidemic threat posed by an infectious disease is a function of how easily it is transmitted and how fatal it is. There are several empirical quantities which are used to describe the transmission dynamics and the mortality impact of an infectious disease.

A common measure for the deadliness of a disease is the *case fatality ratio* (CFR), which corresponds to the proportion of diagnosed cases which die due to the disease (Last 2001).[1] The CFRs of different epidemic diseases vary widely. Fatality ratios need to be seen in context with incidence levels. On the one hand, a CFR of 1% would be considered high for influenza, since the disease is easily transmissible to a large share of the population in a short time period and can thus cause substantial excess mortality. On the other hand, for cholera a CFR of 1% is even today considered relatively low (but achievable). Of course, in the nineteenth century

[1]This is to be distinguished from the *infection fatality ratio* (IFR), the proportion of infected people which die due to the disease. For diseases with a large number of asymptomatic carriers these two quantities can differ widely, and especially in a historical setting the CFR is often the only one which can be determined with some precision. The CFR is, of course, prone to substantial systematic biases.

Table 2.2 Serial intervals for a number of infectious diseases as reported in the literature

Disease	SI	Source	Disease	SI	Source
Influenza	1–5 days	Vink et al. (2014)	Cholera	3–5 days	Phelps et al. (2018)
Measles	7–17 days	Vink et al. (2014)	Bubonic plague	1–20 days	Dean et al. (2019)
SARS	3–16 days	Lipsitch et al. (2003)	Pneumonic plague	1–13 days	Nishiura (2008)
MERS	~11 days	Cauchemez et al. (2016), Park et al. (2018)	Smallpox	10–23 days	Nishiura et al. (2008)
Ebola	~15 days	van Kerkhove et al. (2015)	Tuberculosis	>0.5 years	Ma et al. (2018)

The given time intervals correspond roughly to the 2.5% and 97.5% percentiles of reported or inferred serial interval distributions. If only one number is given, it represents a compromise based on widely varying estimates

CFRs for cholera were closer to 50%, which highlights the advances in treatment of bacterial diseases (Sack et al. 2004).

The transmission dynamics of a human-to-human or vector-borne infectious disease are governed by the *serial interval* (SI) and the *basic reproduction number* R_0 or the *effective reproduction number* R_e.

The SI measures the time which elapses between the symptom onset of a primary case and the symptom onset of a secondary case infected by the primary case (Fine 2003). This time interval is typically longer than the incubation period but can also be substantially shorter (even negative) if asymptomatic transmission is widespread. The mean SI may change during the course of an outbreak, for example when infected individuals are more readily isolated, and most infections happen early.

The basic reproductive number R_0 is defined as the number of secondary infections produced by a primary case in a completely susceptible population (Diekmann and Heesterbeek 2000; Dietz 1993). The effective reproduction number R_e denotes the number of secondary cases in situations where a substantial level of immunity already exists among the population, or when control interventions to limit the spread have been put in place. If $R_e > 1$, the disease spreads exponentially (epidemic phase), and if $R_e < 1$, the disease is in decline (self-limiting phase). In a totally homogeneous population, R_e drops below 1 when a proportion equal to $1 - 1/R_0$ of the total population has acquired immunity.

Tables 2.2 and 2.3 compile some values of CFR, R_0 and SI for historical and contemporary disease outbreaks as reported in the literature. Such values can give a good first impression of the virulence of a disease, but always need to be interpreted with caution.

It should be kept in mind that fatality rates and reproductive numbers can rarely be measured directly, but usually depend on some model assumptions. For instance, fatality rates of influenza are often estimated by considering excess mortality due to pneumonia or other related causes of death (Olson et al. 2005; Wong et al. 2013), and the computation of R_0 normally relies on epidemiological models with a number

Table 2.3 Case fatality rate and basic reproduction number for a number of epidemic events

Disease	Outbreak event	CFR (%)	R_0	Sources
Smallpox	18/19. century outbreaks in cities (UK/US)	15–25	3.5–6	Gani and Leach (2001)
Cholera	Third cholera pandemic (Denmark 1853)	54–68[a]	1.7–2.6	Phelps et al. (2018)
Bubonic plague	Third plague pandemic (Glasgow 1900)	43	0.9–2.9	Dean et al. (2019)
Measles	2003 measles epidemic in Niamey (Niger)	8–12	5–16	Grais et al. (2006), Nandy et al. (2006)
Influenza H3N8?	1889/90 Russian flu (first wave, European and US cities)	0.1–0.3	1.9–2.4	Valleron et al. (2010)
Influenza H1N1(18)	1918 Spanish flu pandemic (first wave, UK/US)	1–3	1.7–2.0	Ferguson et al. (2006), Frost (1920)
Influenza H2N2(57)	1957 Asian flu pandemic (UK)	0.15	1.5–1.7	Ferguson et al. (2006), Jackson (2009)
Influenza H1N1(09)	2009 swine flu pandemic (UK/US)	0.2–0.5	1.3–1.7	Biggerstaff et al. (2014), WHO (2010a)
Influenza (seasonal)	Annual seasonal influenza (Europe/US)	0.1	1.2–1.4	Biggerstaff et al. (2014), Li et al. (2008)
SARS	2003 SARS outbreaks (Hong Kong, Beijing, Taiwan, Singapore, Toronto)	17	3.1–4.2	Wallinga and Teunis (2004), Yip et al. (2005)
MERS	2012/2013 outbreaks (Arabian Peninsula)	35	0.3–0.6	Chowell et al. (2014)
Ebola	2013–2016 Ebola epidemic (West Africa)	40	0.9–2.1	Camacho et al. (2014), Shultz et al. (2016)

[a]This number is based on a definition of cases which does not include patients with mild symptoms

of parameters, for example SI, which are themselves hard to estimate (Delamater et al. 2019; Heffernan et al. 2005).

It is also important to recognise that many factors play a role in determining the dynamics of the spread and the mortality impact of an infectious disease. Some of them are intrinsic to the pathogen, others depend on the biology of the affected population, and yet others relate to social behaviour and organisation, or to environmental circumstances. The empirical values for CFR, R_0 (and to a lesser extent SI) are thus always subject to local variation. Neither of them is purely a biological constant.

For instance, R_0 may depend on the population density (Delamater et al. 2019; Gani and Leach 2001; Hu et al. 2013; Valleron et al. 2010), on local customs which influence contact rates between people, or on the extent to which infections are reported and cases are isolated (Agusto et al. 2015). Humidity and temperature are known to have a strong impact on the transmission of airborne viral infections such as seasonal influenza (Tamerius et al. 2013) or smallpox (Nishiura and

Kashiwagi 2009), and temperature is correlated with the reproduction number of several vector-borne diseases (Peña-García and Christofferson 2019; Vogels et al. 2017).

As Table 2.3 shows, pandemic influenza is generally characterised by higher fatality rates and higher R_0 than seasonal influenza. However, significant regional variation is observed both in fatality rates and transmission dynamics (Chowell et al. 2008), and one can speculate whether the trend of decreasing R_0 from one pandemic to the next is due to an intrinsic trend in the evolution of influenza, or whether it simply reflects an observed decline in household size over the last century (Ferguson et al. 2006). Interestingly, Valleron et al. (Valleron et al. 2010) found R_0-values for a sample of US cities during the pandemics of 1889 and 1918 to be positively correlated, underscoring the reality of long-lasting systematic local biases.

To end this section with another caveat, it should be noted that on the level of individual humans, R_0 is a random variable that follows a non-trivial distribution around its population mean. This additional level of detail can significantly influence the transmission dynamics if the distribution of R_0 is wide. For example, outbreaks of SARS-CoV-1 and MERS-CoV are known to be fuelled by superspreading events in which one person transmits the virus to dozens while the typical patient transmits the disease to one other person at most (Kucharski and Althaus 2015; Lloyd-Smith et al. 2005). This scenario corresponds to distributions of the reproduction number with a long tail to the right.

2.3 Past Pandemics: A Historical Overview

2.3.1 Semantics

An epidemic is commonly understood to be "an outbreak of disease that spreads quickly and affects many individuals at the same time" (Merriam-Webster n.d.-a). The term outbreak refers to a (localised) "sudden rise in the incidence of a disease". A pandemic is an epidemic which has spread over a "wide geographic area and affects an exceptionally high proportion of the population".

While these definitions provide a smallest common denominator, it is worth pointing out both the considerable historical shift in meaning that the terms epidemic and pandemic have undergone, as well as competing definitions in use today.

The term epidemic derives from old Greek *epidemios* (meaning: on the people, on the country). Hippocrates (460–370 BC) was the first to use it in the context of disease, but to him it signified a collection of syndromes or diseases commonly occurring in a given place or during a given season (such as summer diarrhoeas on some specific islands). It is only in the course of the medieval plague epidemics that the Latin (and Italian) word *epidemia* and its French derivatives gradually came to be associated with one clearly distinguishable disease at a time (Martin and Martin-Granel 2006).

Epidemic in its modern usage has to be distinguished from endemic disease. An endemic disease is characterised by its continuous prevalence among a specified population, while an epidemic implies a significantly increased frequency of disease occurrence "relative to usual frequency of the disease in the same area, among the specified population, at the same season of the year" (Last 2001).

There is no sharp distinction between the terms outbreak and epidemic as well as between the terms—epidemic and pandemic. They indicate a progressively larger demographic and geographic spread of the disease. A pandemic is usually understood to involve at least several countries or a continent (CDC 2012).

The advent of modern genomic analysis has resulted in another semantic shift over the past decades. Severe outbreaks tend to be caused by particularly virulent mutations of existing pathogens to which populations have little or no pre-existing immunity, so-called epidemic strains (Martin and Martin-Granel 2006). Following this line of thought, the WHO characterises a pandemic as "the worldwide spread of a new disease" (WHO 2010b). However, the WHO has also been vague on what constitutes a new disease (Doshi 2011). For instance, the pandemic influenza guidelines from 1999 (WHO 1999) seemed to require a novel virus subtype as the source of an influenza pandemic. In 2005, the guidelines were altered to include virus subtypes that have "not circulated in humans for at least several decades and to which the great majority of the human population therefore lacks immunity" (WHO 2005).

With the increasing number of novel virus outbreaks over the past two decades (Nipah, SARS, MERS, bird flu, swine flu, Ebola, Zika), the term pandemic has also acquired a loaded meaning in public discourse. In popular usage, the terms pandemic and epidemic convey a sense of heightened danger and lethality, which raises the question if severity of a disease should be a criterium when assessing if those terms are appropriate. The 1999 WHO pandemic influenza guidelines indicate that high likelihood of "serious morbidity and mortality in at least one segment of the population" is a necessary criterium to declare a pandemic (WHO 1999). In 2009 the WHO was widely criticised as having prematurely declared the H1N1 influenza ("swine flu") a pandemic, before sufficient evidence on the severity of the disease was available (Collignon 2011; Doshi 2011).

Another ambiguity lies in the distinction between contagious and non-contagious diseases. While we are exclusively concerned with outbreaks of infectious diseases in this chapter, the term epidemic is sometimes applied to a wider range of diseases and conditions (Green et al. 2002). Examples include the notion of "lung cancer epidemic", "opioid epidemic" or "obesity epidemic". These phenomena may in fact be quite similar to infectious disease outbreaks. It has been argued that conditions such as obesity or drug abuse spread in many ways like contagious diseases (Christakis and Fowler 2009).

2.3.2 List of Pandemics

In this section, we give a brief historical overview of past pandemics. We roughly distinguish three phases:

(a) until 1850: Reliable data and records for this pre-modern period often tend to be scarce, while at the same time severe epidemics were a regular part of human history. Although theories abounded, physicians were largely in the dark as to the root causes of disease.
(b) 1850–1950: Ground-breaking advances in microbiology during the second half of the nineteenth century made the identification of many infectious agents, particularly bacteria, possible. At the same time, tighter links between different parts of the world in the wake of European colonial expansion amplified the potential for truly global pandemics. Moreover, a growing trend towards public data collection in this period allows for a more granular reconstruction of epidemic events.
(c) 1950–present: This is the age of antibiotics, whose invention coincided with the advent of modern air traffic. Correspondingly, this phase is dominated by pandemics and pandemic scares due to novel viruses which have the potential to be disseminated ever faster in a globalised world.

The following selection of historical pandemics is non-exhaustive. For the period before 1850 we consider three of the most iconic and well-researched pandemics. For the period 1850–1950 we select pandemics with estimated death tolls >100,000 and a documented epidemic spread across at least two continents. A case could be made to include others, notably the typhus epidemic in Russia, Central Asia and Eastern Europe (1918–1922) (Patterson 1993), as well as the creeping emergence of Poliomyelitis as an epidemic force since the end of the nineteenth century (Nathanson and Kew 2010). Conversely, the inclusion of the Franco-Prussian smallpox pandemic (1870–1875) is clearly somewhat Eurocentric, given that the incidence levels in Europe and North America during the pandemic would have been just normal in most other parts of the world at the time (Henderson 2011). For the period since 1950, we consider all commonly accepted worldwide pandemics.

2.3.2.1 Major Pandemics Before 1850

The Justinian Plague (541–544): During the reign of Emperor Justinian (ca. 527–565), a severe pandemic swept the Roman Empire as well as adjacent lands in the Middle East, North Africa, and Northern Europe. Mortality was high, with contemporary observers estimating up to 10,000 deaths a day in the capital Constantinople. This first wave was followed by at least 14 waves over the next two centuries, which are collectively referenced as the first plague pandemic (Hays 2005; Little 2007). Recent genomic analysis confirmed the long-standing hypothesis that a strain of yersinia pestis (causing the plague) was indeed the causative agent (Wagner et al. 2014). While historical sources tend to favour Africa as the immediate origin of the pandemic (Stathakopoulos 2007), further genomic analysis suggests the disease may have had its origin in the Tian Shan mountains of Kyrgyzstan (Eroshenko et al. 2017). The social, economic, cultural and political disruptions of the first plague pandemic are thought to have had a lasting impact on the Byzantine empire (Sarris 2007) and contributed to the transition from the ancient to the medieval period of European history (Hays 2005).

The Black Death (1347–1353): Introduced by traders via the Silk Road from Central Asia where it had emerged by zoonosis from local rodent reservoirs (Morelli et al. 2010; Wild et al. 2013; Yue et al. 2017), a dreadful plague struck Europe and North Africa in 1347. From the Crimean Peninsula it spread in a matter of weeks along the Mediterranean shipping routes from port to port. During the following 5 years, there were very few pockets of land in Europe and around the Mediterranean which escaped the scourge unscathed. Incidence during this initial wave often approached 100%, and case fatality rates were around 50% (Benedictow 2004; Welford 2018). After three centuries of steady population growth, Europe lost around half of its populace to what became much later known as the Black Death. The demographic cataclysm had far-reaching socio-economic consequences, among them an increase in labour cost accompanied by falling land costs, which strengthened the position of peasants with respect to their landlords (Hays 2005). The subsequent three centuries saw a series of increasingly localised plague epidemics, together with the Black Death referred to as the second plague pandemic, before the disease eventually disappeared from the continent. In spite of repeated controversies over the nature of the disease, genomic evidence has confirmed yersinia pestis as the culprit (Haensch et al. 2010). The relevant bacteria belong to a different biovar than those identified as the root cause of the first plague pandemic as well as those which cause plague today (Achtman et al. 1999). There are ongoing disputes about the mode of transmission as well as the role that the pneumonic form of plague or other diseases may have played (Dean et al. 2018; Zimbler et al. 2015). Moreover, it remains unclear whether the disease was introduced once or multiple times from Asia during the second pandemic (Bos et al. 2016; Schmid et al. 2015), and why it eventually receded.

The "Columbian exchange" pandemics (1492–eighteenth century): When the Spanish conquerors set foot on the American continent, they brought along with them a plethora of infectious diseases that were endemic to the Eurasian-African land mass but to which indigenous populations had little to no immunity (Cook 1998; Crosby 1972; Diamond 1999; McNeill 1976). These included smallpox, measles, mumps, and influenza, but many other involved diseases have still not been identified with certainty (Vågene et al. 2018). Drastic changes imposed by the Spanish rulers on the way people lived (e.g. forced labour, relocations, new diets due to altered farming practices) may have exacerbated the impact of these pathogens (Hays 2005). Moreover, ships engaged in the African slave trade established a steady influx of exogenous diseases, especially smallpox. All of this caused a sheer interminable wildfire of fragmentarily documented epidemics and pandemics (Dobyns 1993) which decimated the indigenous population of the Americas. Current estimates of population loss over the first 150 years of European contact range from 80 to 95% of the original population (Koch et al. 2019; Nunn and Qian 2010). The term virgin soil epidemics has been coined to describe these events (Crosby 1976). They are believed to have played a crucial role in the sudden and rapid conquest of the Americas as well as in the vast transfer of African slaves to the new continent as a labour force replacement (Cook 1998; Hays 2005). It has also been argued that the ensuing abandoning of agricultural land and reforestation significantly impacted global climate (Koch et al. 2019).

2.3.2.2 Pandemics Between 1850 and 1950

Cholera pandemics (1817–1923): In the century between 1817 and 1923, the world saw six cholera pandemics which lasted roughly a decade each. Cholera had been endemic to the Bengal region of India but was new to most of the world. Each of the pandemics started in India from where the disease spread in a roughly concentric wave on the increasingly tighter global transportation network, following trade routes as well as movements of troops, migrants and pilgrims (Barua 1992; Hays 2005; Pollitzer 1954). It fell on fertile ground in the densely populated, poverty-stricken, and poorly sanitised cities of the early industrial age. Case fatality rates were high at around 50%, but incidence in most countries fell short of those experienced in earlier pandemics and often even of those from contemporary endemic diseases such as tuberculosis (Hays 2009). After decades of little reliable evidence on what caused cholera, the underlying bacterium and its mode of transmission were identified by Robert Koch in 1884 during the fifth pandemic, signifying a breakthrough towards eventual control of the disease (Lippi and Gotuzzo 2013). Cholera largely vanished from Western Europe and North America before the turn of the century but continued to thrive elsewhere.

The Franco-Prussian smallpox pandemic (1870–1875): Since the beginning of the nineteenth century, widespread and often state-mandated vaccination had brought down smallpox mortality in Europe by roughly an order of magnitude (Edwardes 1902). However, the disease continued to be endemic. The Franco-Prussian war 1870–1871 saw a resurgence of cases in France, which had been among the countries more resistant to vaccination. Prisoners of war were instrumental in spreading the disease to Germany (Hays 2005). Eventually, the epidemic engulfed much of Europe and spread to the Americas. As populations with high vaccination and revaccination rates were found to be much less severely affected, this pandemic proved to be a milestone towards mandatory and efficient vaccination programmes (Edwardes 1902; Rolleston 1933).

The Third Plague pandemic (1894–1912): Plague was endemic in China since the late eighteenth century, slowly spreading through the land for several decades. Only when the disease reached the ports of Guangzhou and Hong Kong in 1894 where it claimed on the order of 100,000 lives, a pandemic was triggered (Hays 2005; Welford 2018). Within the next couple of years, steamships dispatched infected humans, rodents, and fleas to other port cities all around the world. India was hit hardest and saw around 10–12 million succumb to the disease. Outside Asia and a few regions of Africa, incidence was relatively low, and the disease remained localised mostly around ports (Echenberg 2002). As early as 1894, Alexandre Yersin identified the causative agent of the disease that was subsequently named after him (Hawgood 2008). In the following years evidence for rats and fleas as reservoirs and vectors accumulated, offering new points of attack for public health interventions (Crawford 1996). Moreover, Waldemar Haffkine developed a vaccine in 1897 which offered partial protection and was administered to many millions, mostly in India, over the next decades (Hawgood 2007). While the relative control of the pandemic in many parts of the world has been hailed as a triumph of modern science (Hirst 1953; McNeill 1976), recent accounts put more emphasis on the already increased hygiene standards in the Western World as well as, in the case of Europe, the absence of local rodent reservoirs (Bramanti et al. 2019; Echenberg 2002).

The Russian Flu (1889–1892): In late 1889, an influenza virus originating in the central Asian regions of the Russian Empire swept over Eurasia and then across most of the world. Two more waves followed in 1890 and 1892. Fatality rates were low and focussed on the elderly. However, the high incidence made this one of the deadliest pandemics of the century for Europe (Hays 2005; Patterson 1987; Taubenberger and Morens 2006). This was at least the fourth influenza pandemic since 1700 (Potter 2001) and it stands out as the first which experienced rapid global dispersion due to modern steamships and railroads (Valleron et al. 2010). The type of influenza which caused the pandemic has not been identified to date, although an avian H3N8 type virus has been put forward as the most likely based on serologic and epidemiological data (Taubenberger et al. 2007).

The Spanish Flu (1918–1919): Three waves of a novel influenza hit the war-torn world first in spring 1918, then in fall 1918 and yet again in early 1919. Incidence was high, and fatality rates were staggering compared to any other influenza pandemic in recorded history. Severe pulmonary complications were frequent and young adults experienced excessive fatality rates (Taubenberger and Morens 2006). Older adults seem to have had some level of immunity based on previous exposure to historical influenza types, but the details of this are debated (Worobey et al. 2014). The Spanish flu claimed an estimated 25–100 million lives worldwide, significantly more than the 17 million which were lost on the battlefields during the war. In contrast to the war casualties, most of those deaths occurred in Asia and Africa, while Europe mourned around 2.3 million influenza deaths and the Americas around 1.5 million (Johnson and Mueller 2002; Patterson and Pyle 1991). The underlying virus has only been reconstructed in 2005 from frozen autopsy material of 1918 second wave influenza victims and is of avian H1N1 type (Taubenberger et al. 2005). The geographical and virologic origins of the pandemic are still contested. According to a long held theory it started in Kansas in the United States in spring 1918 and was brought to Europe by troops fighting in World War I (Crosby 2003). However, evidence has been mounting that Spanish flu or some closely related virus were circulating earlier, possibly as early as 1915 (Olson et al. 2005; Worobey et al. 2019). As the second wave saw significantly higher fatality rates than the first, it has been speculated that different variants of the virus may have been responsible for the different waves, but autopsy of additional victims showed very little evolution of the virus between May and October 1918 (Sheng et al. 2011).

Encephalitis Lethargica (1917–1930): First discovered in 1917 in Austria and France, encephalitis lethargica (EL) was a mysterious neurological condition which over the course of a decade afflicted between hundred thousand and several million people worldwide. Besides most of Europe and North America, countries with a significant number of reported cases include Japan, South Africa, Australia, and New Zealand, but global records are patchy (Foley 2018; Lancet Editorial 1981; Ravenholt and Foege 1982). Characteristic for the acute phase of the disease were mild flu symptoms followed by stupor, intense sleepiness and a mask-like facial appearance. The acute phase killed around one third of patients, but only few could hope for a full recovery. The remaining patients were left with often severe neurological damage including personality changes, spasms, and parkinsonism, which could appear years after the infection (Dourmashkin 1997). EL all but disappeared by 1930, precluding its identification by modern virologic methods developed in the years thereafter. There was little doubt that EL was an infectious disease, although very few direct transmission events were observed, and several other routes of transmission were excluded. A possible connection with the Spanish flu has been considered and rejected many times (McCall et al. 2008). A plausible candidate for the infectious agent is an enterovirus which was recently detected in brain tissue of some victims (Dourmashkin et al. 2012). Attempts have been made to explain the sporadic impact of EL by an extremely high ratio of asymptomatic carriers, which would simultaneously explain the end of the pandemic by means of acquired herd immunity (Foley 2018).

2.3.2.3 Pandemics Since 1950

Influenza pandemics (1957, 1968–1970, 1977): Spanish flu lost its excessive virulence after 1919, but all influenza A viruses circulating ever since are genetically descended from it. Influenza outbreaks have been well-documented and characterised since the advent of virology in the 1930s. H1N1 strains continued to dominate seasonal flu until 1957, when they were abruptly and entirely replaced by a modified virus of type H2N2 which caused the Asian flu pandemic. In 1968 a new H3N2 type virus started the Hong Kong flu pandemic and in turn replaced the H2N2 strains (Kilbourne 2006; Taubenberger and Morens 2010). Both these pandemic viruses entered the human population through zoonosis from unidentified animal sources in South East Asia (Scholtissek et al. 1978). They spread worldwide and caused on the order of a million deaths each. As is characteristic for pandemic influenza strains, death rates were elevated among the young and middle-aged (Simonsen et al. 1998). Global and national mitigation attempts were minor (Honigsbaum 2020). One marked difference between the two pandemics was the comparatively slow dissemination and more sporadic impact of the Hong Kong flu, the reason for which has been debated (Cox and Subbarao 2000; Reichert and Christensen 2005; Viboud et al. 2005). Another mild pandemic, the (second)

Russian flu, occurred in 1977, when an old H1N1 strain from the 1950s re-entered circulation in obscure circumstances. It affected almost exclusively children and adults in their twenties, since older persons had substantial immunity (Kilbourne 2006; Rozo and Gronvall 2015).

The seventh cholera pandemic (1961–): In 1961, the El Tor biotype of cholera emerged from Indonesia where it had been sporadically present for a few decades. El Tor is characterised by a high number of asymptomatic or mild cases, which makes monitoring outbreaks more challenging. In the first 5 years, it spread across South East Asia, the Indian subcontinent, and the Middle East. In 1970/1971 it made a rapid advance into Eastern Europe, Africa and onto the Iberian Peninsula (Barua 1992; Cvjetanovic and Barua 1972). After becoming endemic in many countries, the disease had a pandemic revival in 1991 when it appeared in South America and spread across most of the continent within 3 years (Colwell 1996; Tauxe et al. 1994). Case fatality rates during the pandemic were generally an order of magnitude below nineteenth century levels (Glass and Black 1992). This is primarily due to the successful application of rehydration therapy. Vaccines showed limited success, and although widespread prophylactic use was made of antibiotics in the first two decades of the pandemic, this led to high levels of antibiotic resistance (Barua 1992; Sack et al. 2004). El Tor has all but replaced the classical type of cholera, but a new type has appeared alongside (Colwell 1996). El Tor is today endemic in many countries and continues to trigger local epidemics when introduced into vulnerable environments such as Haiti following the earthquake in 2010 (Lantagne et al. 2013) and war-torn Yemen in 2016 (Weill et al. 2019).

HIV/AIDS (1981–): Acquired Immune Deficiency Syndrome (AIDS) was recognised as a new disease in the United States in 1981 after a growing number of young homosexual men died from infections that normally pose no threat to a functioning immune system (Greene 2007). Only after 3 years did news of a related epidemic in a primarily heterosexual demographic in Africa emerge (Piot et al. 1984). The retrovirus now termed human immunodeficiency virus type 1 (HIV-1) was identified in 1984 as the causative agent. It is closely related to the simian immunodeficiency viruses (SIVs) endemic in wild ape populations of West Central Africa, and it is characterised by its extremely rapid evolution (Sharp and Hahn 2011). The earliest direct evidence of an HIV infection dates back to 1959, but statistical analysis of the genetic spectrum of currently circulating HIV-1 virus strains points to a common ancestor around 1920 near Kinshasa (Faria et al. 2014). Most HIV infections in Europe and the Americas arose from a single strain that appears to have passed from Africa to Haiti in the 1960s and then onward to the US and other Western countries (Gilbert et al. 2007). The disease has since spread worldwide, claimed an estimated 40 million lives (UNAIDS 2006, 2019) and is the leading cause of death in sub-Saharan Africa (Danforth et al. 2017), where rates of co-infections with tuberculosis, cryptococcosis, malaria and hepatitis B/C are particularly high (Chang et al. 2013). Early association of AIDS with gay men, intravenous drug users and other marginalised groups lead to a persistent stigmatisation of the disease and its victims (ECDC 2017; Herek and Capitanio 1999). Despite decreasing incidence and fatality rates due to a raised public awareness and availability of antiretroviral therapy since the 1990s, HIV still accounted for around 770,000 deaths in 2018 (WHO 2019a).

2.3.3 From Epidemic to Pandemic: The Global Spread of Disease

What drives the long-range geographical spread of an infectious disease? How does a local outbreak turn into an epidemic? In medieval and early modern societies with little individual mobility, three major facilitators of global disease spread stand out: traders, soldiers, and pilgrims. Examples are abundant. The extensive silk road trade network did not only connect the economies of medieval Europe and Asia but was the major artery through which the plague and other infectious diseases were passed from east to west and vice versa (Morelli et al. 2010; Welford 2018). Towns near overland or riverine trade routes in Europe had a significantly increased likelihood of plague outbreaks (Yue et al. 2017). Mercenaries in early modern Europe ravaged the lands as much by means of the many diseases they carried with them as by the violence they exercised (Hays 2009). Cholera invaded Europe five times in the nineteenth century, and every time it followed remarkably similar pathways along major colonial trade and Muslim pilgrimage routes (Barua 1992; Boyle 2015; Pollitzer 1954).

Until the twentieth century, shipping traffic was a major mode of transport by which diseases spread across long distances. Both medieval and modern plague were first and foremost disseminated through port cities (Echenberg 2002). However, oceans remained at least partly a barrier to disease transmission until the mid-nineteenth century. Ocean crossings by sail ships in the early modern age took many weeks or months, and the number of passengers on a ship was limited. Substantial herd immunity to endemic diseases such as smallpox or measles meant that outbreaks on ships typically ended before the final destination was reached. For instance, it seems to have taken several decades from Columbus' first voyage until smallpox and measles reached America (Hays 2009). However, the advent of large ocean steamers from the 1850s onwards reduced crossing times, increased the number of passengers on a single vessel as well as the total volume of shipping trade (Pascali 2017). The direct and substantial effect this had on disease dissemination has been demonstrated in studies of the disease history of islands (Cliff and Haggett 2004).

At the same time as the steamship, railways conquered much of the world. As already contemporary observers emphasised (Pyle 1986), the Russian flu pandemic of 1889 clearly shows the impact of railways and steamships: first, it arrived from Central Asia to St. Petersburg by means of the recently completed trans-Caspian railway line (Honigsbaum 2009). Then, both within Europe as well as across oceans, it spread as fast as a modern flu pandemic (Valleron et al. 2010), which stands quite in contrast to flu pandemics in the first half of the nineteenth century (David

Patterson 1985). The expanding nineteenth century railway network connecting North America also visibly impacted the way in which cholera spread through the United States and Canada: while the early cholera pandemics followed the coastlines and canals, the fourth pandemic spread through the new intercity railway links (Pyle 1969).

The advent of modern commercial air traffic has changed the global patterns of disease spread again, albeit not in a completely straight-forward way. Worldwide hub-to-hub dissemination is faster, but some peripheral locations have experienced a decrease in connectivity (Cliff and Haggett 2004). The new long-range links generally lead to more erratic patterns of geographical spread (Colizza et al. 2006).

Air travel was the major driver behind the quick global spread of the 2009 swine flu (Khan et al. 2009), the 2003 SARS pandemic and behind the initial spread of the 1981 HIV epidemic in the United States (Tatem et al. 2006). However, other means of transport continue to play an important role. For example, HIV was first introduced to Europe by a Norwegian sailor who had contracted the virus in a Western African seaport (Tatem et al. 2006).

Overall, individual human mobility in developed countries has increased by a factor of 1000 in the past two centuries (Cliff and Haggett 2004). But for some diseases, human mobility is not the only way to spread far. The epidemic potential of vector-borne or environmentally mediated diseases depends on complex ecological networks which are both influenced by human geography as well as factors outside of human control.

Cholera as a water-borne disease provides a good example for such interactions with the environment. Only recently has it become clear that cholera bacteria are a natural part of the aquatic flora independent of human faeces (Hood and Ness 1982). The seventh cholera pandemic provided ample evidence that contaminated seafood also functions as a transmission route (Estrada-García 1996). In fact, cholera bacteria have the ability to survive in water for years by using for example plankton or shellfish as hosts, or to get themselves transported thousands of kilometres by ocean currents (Colwell 1996). Sensitivity of the hosts to changes in temperature, salinity and other environmental variables can explain seasonal variations in cholera incidence as well as its potential for epidemic resurgence after many years (Islam et al. 2020; Lutz et al. 2013). It has been argued that the near instantaneous spread of the seventh cholera pandemic along the Peruvian coast in 1991 was related to the El Niño (Colwell 1996).

2.3.4 Historical Trends in Public Health Response

Public responses to pandemics are intimately linked with how societies conceptualise disease. Historical interventions have to be understood in the context of prevailing paradigms about the nature of disease spread. In Western history, there were two competing schools of thought on this matter. One, going back to Hippocrates, held that diseases are caused by miasma, some form of poisonous air that contaminated the environment. The other, contagionism, argued that diseases are spread through close contact between humans. In the absence of much scientific evidence, both theories were prevalent until the end of the nineteenth century.

The repeated medieval plague visitations gradually strengthened the contagionists' views (Hays 2009). The Italian city states were the first to impose mandatory quarantines on incoming ships and travellers by land from the late fourteenth century onwards. In fact, the word quarantine is derived from Italian quaranta (meaning: 40), referring to the 40 days the maximal incubation period of plague was believed to be. Moreover, cities instated public health boards with sweeping surveillance and enforcement powers, including the quarantining of infected individuals and their families in their homes or in public pest houses, the confiscation and destruction of private property and trading goods, and even torture to obtain information about cases of plague in their city.

From the mid-sixteenth century onwards, travel restrictions of varying rigour became a common measure taken to combat the plague not only by cities in Italy, but throughout Western Europe (Welford 2018). For instance, travellers wishing to enter a city would have to present evidence that they did not come from affected regions, while beggars and vagabonds were often categorically denied entry. On a larger scale, the seventeenth and eighteenth century saw surveillance systems put in place between cities to stay informed about the whereabouts of the plague, and ports started to regulate shipping trade. The Hapsburg empire manned its Balkan border with up to 11,000 civilian and military patrol who denied anyone from the Ottoman empire entry in times of plague.

In the nineteenth century, after the disappearance of plague from Western Europe, miasma theory had a comeback. On the one hand, Ackerknecht (1948) argues, liberals and reformers tended to view contagionism as a vehicle to exert bureaucratic state power and unreasonably limit the freedom of citizens. Moreover, traditional contagionist approaches such as quarantine measures were a nuisance to growing international trade. On the other hand, the idea that disease was primarily caused by the corrupting influence of specific environments was in line with the prevailing enlightenment view of the natural environment as something to be conquered and controlled (Hays 2009).

According to Baldwin (2004), the polarisation between contagionists and miasma advocates was not quite as stark in practice. However, the appearance of cholera in Europe certainly delivered a blow to purely contagionist approaches. Neither the massively reinforced Austrian cordon sanitaire at the Ottoman border nor the Russian lockdown of cities under martial law in 1830 succeeded in keeping the little understood disease at bay. Nor did any of the quarantine and isolation measures that were employed in most European states. After 1851, international *sanitary conferences* were held to coordinate responses against cholera and later the Plague, with most of them failing to reach consensus on what the right approach was (Howard-Jones 1974).

Theories of social causation of disease abounded. European colonialists were quick to associate diseases such as cholera with the 'filthy' and 'backward' way of life of their subjects in Asia, Africa, and the Middle East (Boyle 2015). Similarly, attention moved to the dense and unhygienic living conditions of the new urban working class in industrial cities as a source of disease. The nineteenth century thus saw a major shift towards sanitary and hygiene measures against disease spread (Hays 2009). The sanitationist movement successfully pushed governments to completely rethink urban planning. By the end of the century, modern sewage, and water supply systems had been built all over the old continent, in North America and in some colonies. These improvements in infrastructure went hand in hand with an increased awareness of the role of personal hygiene, promoted by public health campaigns. In medical practice, the importance of disinfection was realised even before germs were discovered. While this attitude, in conjunction with class or racial prejudice, sometimes lead to discrimination of the poor or non-Western peoples (and served as a pretext for segregation in some of these cases), it did contribute to an enormous reduction of classical diseases associated with a lack of hygiene, such as cholera and typhus.

With the advent of microbiology and modern germ theory in the 1870s and 1880s, new options to fight outbreaks arose. For the first time, it was possible to identify the causative agents of many diseases. Moreover, insects responsible for the transmission of vector-borne diseases such as yellow fever and malaria were identified and targeted by chemical extermination campaigns. With unrelenting optimism and varying degree of success, vaccines were being developed against many bacterial infections, among them cholera and plague. And soon came the first pharmaceutical drugs designed to kill germs inside the human body (often with considerable side effects) (Hays 2009).

By the time the Spanish flu struck in 1918, scientific and public consensus were firmly back in the contagionist camp. Health authorities across the world recommended or imposed a plethora of non-pharmaceutical measures, most of them focused on social distancing and imposing physical barriers to stop the spread of germs. Physical isolation of patients as well as quarantine of hospitals were common practice. While the usefulness of gauze or muslin masks was the subject of medical discourse, their use in public was widespread and in some instances legally mandated, for example by the city of San Francisco and in the Japanese police and military (Rice and Palmer 1993; Crosby 2003). Public mass gatherings were banned in many places, and schools, theatres, churches, and dancehalls were temporarily shut. Some of the most effective measures were put in place in the United States, but there was a considerable geographic variation between different states and cities. Studies comparing interventions in different American cities found a strong correlation between excess mortality and the timing of measures (Bootsma and Ferguson 2007; Hatchett et al. 2007). For example, Philadelphia and St. Louis implemented similar interventions, but St. Louis acted proactively while Philadelphia reacted only 2 weeks into the outbreak as hospitals were already overwhelmed. Correspondingly, excess mortality in Philadelphia was more than double as large as in St. Louis.

Pharmaceutical interventions to prevent infection with influenza were more erratic and ranged from the gargling of antiseptic solutions to the installation of zinc sulphate inhalation chambers (Rice and Palmer 1993). As contemporary knowledge about viruses was still in its infancy, Spanish flu was assumed to be caused by yet another bacterium. With bacterial coinfections being common among influenza patients, several bacterial candidates were readily isolated. Attempts to develop vaccines based on these certainly did advance science but failed to stop the pandemic (Eyler 2010).

The disappointing experience with vaccines against the Spanish flu is characteristic of two trends for the twentieth century pandemic responses. First, while the discovery of antibiotics in 1928 has fundamentally turned the tables in humanity's fight with bacteria, viral infections remain still to this day a formidable challenge for medical science. Secondly, the rapid development and deployment of vaccines has become a go-to strategy of epidemic management, but their impact on pandemic control has remained small. To be sure, vaccines deserve enormous credit for such feats as stopping the international Polio epidemic in the 1950s (Nathanson and Kew 2010), in eradicating smallpox from the planet in the 1970s (Henderson 2011), and in reducing the annual global death toll of measles by over 90% (WHO 2019b). But those diseases were well-known and had been extensively studied, and smallpox vaccines had been in use for centuries. In contrast, influenza vaccines were first developed in the 1930s, but in the absence of a universal vaccine still need to be adapted every year to include protection against new strains (Barberis et al. 2016). As the past influenza pandemics have shown, such vaccine updates usually come too late to avert the main wave of a new pandemic strain. Similarly, after the identification of the HIV virus in 1984 in the middle of the AIDS pandemic, the US government displayed optimism about the possibility of a quick vaccine development (Greene 2007). Yet 35 years later it is clear that any vaccine, should development still succeed, will come too late to avert most of the millions of deaths the pandemic has caused and will continue to cause. Instead, non-pharmaceutical interventions such as safe sex practices were the only reliable means to fight the spread of the disease in the first decades of its global presence.

2.4 Recent Pandemics and Pandemic-Threats

In the years after 2000 and before the emergence of SARS-Cov-2, the swine flu pandemic had the greatest impact on the global human population as a whole in terms of morbidity and mortality. In addition to that, a series of emerging infectious diseases caused outbreaks in various parts of the world and raised concerns about global preparedness for a severe pandemic. Among them SARS, MERS, and Ebola stood out as the major pandemic threats.

Severity and ease of transmission of a disease influence the rigor of national and international containment efforts. While SARS, MERS and Ebola were highly lethal, swine flu was not much more fatal than normal seasonal influenza. Correspondingly, swine flu was the only one among them which was able to run its course, while the other outbreaks were contained before reaching undisputed pandemic status. Among the above mentioned very lethal infections, SARS was the only one with a basic

Table 2.4 Some of the recent pandemics and pandemic-threats

Event	Year	Location(s)	Cases	Deaths	Source(s)
SARS outbreak	2002–2004	China +25 other countries worldwide	>8000	>800	WHO (2004b)
Swine flu pandemic	2009–2010	Worldwide	700 mil–1.4 bil	150,000–600,000	Dawood et al. (2012)
MERS outbreaks	2012–2019	Middle-East, Africa, Southern Asia	2519	866	WHO (2020b)
West African Ebola outbreak	2013–2016	Mainly Guinea, Liberia, Sierra Leone	>28,000	>11,000	Shultz et al. (2016)
Zika virus outbreaks	2015–2016	North and South Americas, South-East Asia	Unknown	50–60	Kindhauser et al. (2016)
Nipah virus outbreaks	2001–2018	India, Bangladesh	344	258	Chattu et al. (2018)
H5N1 outbreaks	2003–2019	Mostly Indonesia, Vietnam, Egypt	861	455	WHO (2020a)

reproduction number R_0 significantly above 1 and which was transmissible by droplets and through the air. This made it particularly challenging to contain, and the lessons learned from that outbreak still resonate strongly in light of the current pandemic of the novel and closely related coronavirus SARS-CoV-2. The 2003/2004 global SARS outbreak could be called the first successfully aborted pandemic in human history (Table 2.4).

The following sections shed light on the zoonotic origin of the major recent outbreaks, and the possible explanations for their common geographic birthplace. Finally, we discuss a few key lessons that can be drawn from these recent pandemics and pandemic threats.

2.4.1 *Pandemics and Zoonosis*

Diseases that transmit from animals to humans are called zoonoses or zoonotic diseases. According to the WHO, an emerging (or re-emerging) zoonosis is a "zoonosis that is newly recognized or newly evolved, or that has occurred previously but shows an increase in incidence or expansion in geographical, host or vector range" (WHO 2004a).

Most of the major outbreaks that occurred after the year 2000 are strongly suspected to have emerged through zoonosis. For instance, all known human coronaviruses have zoonotic origins in bats, mice, or domesticated animals where they are typically non-pathogenic (Ye et al. 2020). The source of SARS-CoV-1 in humans is believed to have emerged from infected bats (GLEWS—WHO 2020) before passing on to secondary hosts, civet cats (Walker 2016). The coronavirus MERS-CoV which emerged in 2012 in the Middle East is also zoonotic, with camels being implicated either as the source or as an intermediate animal reservoir facilitating the transmission of the disease to humans (Han et al. 2016). The viruses giving the 2009 swine flu its name have emerged from pigs (Thacker and Janke 2008; Isaacs 2010). Pigs are deemed ideal mixing vessels since they can be infected both with human as well as avian influenza viruses (influenza viruses endemic in birds), facilitating exchange of genes amongst the viruses to produce new and dangerous strains (Thacker and Janke 2008). The repeated Ebola outbreaks on the African continent are suspected to emerge from bats or non-human primates (GLEWS—WHO 2020).

Three stages can be distinguished from the emergence to the eventual spread among humans of a zoonotic disease (Morse et al. 2012; Wolfe et al. 2007). Stage 1 is the preemergence stage, during which, typically, environmental disruptions caused by human activity (e.g. change of land use) lead to altered transmission dynamics of the pathogen, such that it is able to expand within its animal reservoir, spread to a new region or transmit to another animal population (for example domesticated animals). Stage 2 is characterised by localised spillovers where a series of minor outbreaks among humans in a particular region paves the way for more widespread human to human transmission in geographical clusters. Stage 3 is the pandemic stage in which the disease transmission transcends national and continental boundaries and moves to a global level.

For example, the 2009 swine flu reached Stage 3, while H5N1 Bird Flu was mostly contained at Stage 2. Most pathogens evolve during emergence. For instance, HIV took around 40 years from the initial spillover to evolve into a more virulent form with pandemic potential (Faria et al. 2014). Other pathogens such as Ebola virus do not require any further evolution in order to spread effectively between humans, and thus enter the human population directly at stage 2 (Morse et al. 2012).

In order to discover and contain emerging diseases during stage 1, various approaches have been suggested. On the one hand, mathematical modelling considering factors such as population density, wildlife reservoirs, existing human-animal interfaces as well as previous evidence of emergence events can provide some hints as to where zoonotic diseases are most likely to emerge. As a general rule, zoonotic spillovers are most likely to occur from animal species that are either closely related to humans (primates) or that are in close contact with humans (e.g. domesticated animals). On the other hand, molecular techniques have become more refined and are increasingly able to assess the host range and evolutionary potential of certain infectious agents. However, the prediction of an infectious agent's pathogenicity and transmissibility in humans remains a challenge. Obvious candidates to evolve into pandemic pathogens are Stage 2 infectious agents which have proven to be able to infect humans and cause illness, but which have a basic reproduction number below the epidemic threshold ($R_0 < 1$) and are therefore not yet able to sustain an epidemic (Morse et al. 2012).

The WHO has listed all the prevalent diseases that are of zoonotic origin, the animals that are usually affected and the way of transmission to humans (Table 2.5).

Table 2.5 List of prevalent diseases of zoonotic origin as classified by the WHO

Zoonotic disease	Animals affected	Human contraction
Anthrax	Herbivorous mammals—fatal disease in cattle, sheep, goats, camels, horses, and pig	Occupational exposure to infected or contaminated animal products
Bovine spongiform encephalopathy	Cattle	Close proximity, Meat as food consumption
Brucellosis	Cattle, swine, goats, sheep and dogs	**Direct**—infected materials like afterbirth. **Indirect**—ingestion of animal products and by inhalation of airborne agents.
Crimean-Congo haemorrhagic fever (CCHF)	Transmitted by ticks	Contact with viraemic animal tissues (during and post-slaughter)
Ebola haemorrhagic fever	Unknown. Likely to be bats or nonhuman primates with bats or nonhuman primates (chimpanzees, apes, monkeys, etc.)	Contact with infected animals (usually following butchering, cooking or eating), contact with the bodily fluids of infected humans
Foodborne Diseases	Food-producing animals	Preparation, consumption
Avian influenza	Birds, primarily poultry	Close proximity
Japanese Encephalitis	Transmission cycle between mosquitoes, pigs and/or water birds	Bites from infected mosquitoes
Marburg haemorrhagic fever	Unknown. Likely to be bats or nonhuman primates with bats or nonhuman primates (chimpanzees, apes, monkeys, etc.)	Prolonged exposure infectious zones, direct contact with blood, secretions, bodily fluids
Nipah Virus (NiV) Infection	Primarily pigs. Also dogs, goats, cats, horses, and sheep	Contaminated food or directly between people
Q Fever	Goats, sheep, and cattle	Breathing in dust that has been contaminated by infected animal faeces, urine, milk, and birth products
Rabies	Mammals	Animal bites

2.4.2 Current Risk Factors for Zoonosis

One clear pattern emerges regarding the geographic origin of the post-2000 pandemic threats—they tend to emerge from either Africa or South East Asia. While the source of all the recent pandemic threats are strongly suspected to be zoonotic, these parts of the world remain particularly prone to zoonotic spillovers owing to high population density and a substantial number of people living in close proximity to animals. Here, we outline some of the contributing factors.

2.4.2.1 Population Growth and Urbanisation

Asia and Africa are the two most populous continents with 61% of the global population living in Asia (4.7 billion) and 17% in Africa (1.3 billion) (UN-WPP 2019; Goh 2009). The prospects of employment and improved living standards on these continents have led to rapid urbanisation. For instance, in the first decade of the twenty-first century, almost 200 million people moved to urban areas in East Asia. Africa is projected to have the fastest urban growth rate in the world (OECD/SWAC 2020) with the share of African urban residents among the global urban population projected to grow from 11.3% in 2010 to over 20% by 2050 (Saghir and Santoro 2018).

To sustain migration on such a large scale, urban development leads to an expansion of cities and the shrinking of forest land. Wild animals deprived of their natural habitat are now closer to cities and towns, inevitably encountering domesticated farm animals and the human population. Wildlife is an important source of novel pathogens, and domestic animals act as amplifiers of pathogens emerging from the wild (Morand et al. 2014).

2.4.2.2 Subsistence Farming

The practice of subsistence farming involves raising crops and livestock primarily to maintain the farmer and the farmer's family with minimal leftover for trade or selling. It is still prevalent today in parts of Africa, Asia, and Latin America. For instance, around 62% of Bhutan's population lives in rural areas and depends on subsistence agriculture and livestock farming (Rinchen et al. 2019). In China, 98% of all farmers cultivate farms smaller than two hectares (Rapsomanikis 2015). The livestock raised, such as poultry, sheep, and goats, are often kept under poor hygienic conditions with little attention to disease control, housing, or feed supplementation (Kimman et al. 2013). Livestock animals also suffer from a high burden of endemic diseases that they may contract from one another or through contact with a variety of non-domestic animals (Tomley and Shirley 2009).

2.4.2.3 Wet/Live Markets

Prevalent in South-East Asia, especially China, wet markets are live animal markets where vendors slaughter animals upon customer purchase (Woo et al. 2006; Wan 2012). Despite the widespread availability of affordable refrigeration, many Asian consumers prefer live animals for fresh produce. Daily introduction of new animals provides optimum conditions for amplification and perpetration of infectious agents. Similar to how pigs act as vessels for reassortment of multiple viruses to create new and deadly strains, multiple different subtypes of influenza viruses (H1N1, H2N9, H3N2, H3N3, H3N6, and H4N6) have been identified and possibly circulate in live-poultry markets (Webster 2004). A 2004 study conducted in a live animal market in Guangdong after the SARS outbreak found non-human coronaviruses in several animal species as well as humans working in the market (Worobey et al. 2014). Various direct and indirect infection transmission modes in wet markets have been identified, such as close proximity to infected animals, their urine or excreta (Rahman et al. 2018).

2.4.2.4 Bushmeat Hunting

Bushmeat hunting is the practice of killing wild animals for food and is practised in rural parts of Africa. Owing to the prevalence of zoonotic diseases in wild animals, the odds of contracting diseases are high in this hunting practice. Bushmeat hunting has been associated with outbreaks such as Ebola virus and HIV. The practice has been discouraged by various African governments through various efforts such as strengthening local communities, developing alternate food sources, deploying community officers and stewards for supervision and selective hunting (Marcus et al. 2004; Hackel 1998). There are a number of potential reasons why bushmeat hunting is nevertheless still widely practised: tradition and acquired taste developed for wild animal meats, the inherent attractiveness among poor communities of a readily accessible high-protein food source, changing landscape favouring human proximity to wild animals, the opportunity of generating extra income, and civil wars making people reliant on these alternative sources of food (Subramanian 2012). According to studies, awareness of the pathogenic risks involved with bushmeat hunting is low amongst hunters and traders (Subramanian 2012). In addition, 55% of people involved in a study in Nigeria knew the risks associated but are still willing to follow the traditions of this practice (Friant et al. 2015).

2.4.3 Lessons Learned

A number of lessons have been drawn from the various pandemic threats of the twenty-first century. One of them is, as discussed above, an increased awareness of the risk that emanates from emerging zoonotic diseases. In this final section, we selectively highlight a few more lessons from these recent outbreaks that are emphasised in the literature.

2.4.3.1 Strain on Healthcare Systems

One of the most consistent findings is that Healthcare Workers (HCWs) are a particularly exposed and vulnerable group when it comes to infectious diseases (Haagsma et al. 2012), be it Influenza H1N1 (Lietz et al. 2016), SARS (Knobler et al. 2004; Heymann and Rodier 2004), MERS (Ha 2015), Avian Influenza (H5N1)

(Bernard et al. 2009), Ebola (Weber et al. 2015; Piot et al. 2019), HIV (Wyżgowski et al. 2016), Hepatitis (Lewis et al. 2015) or Tuberculosis (Nienhaus et al. 2014; Ho et al. 2007).

This became particularly apparent during the SARS and MERS outbreaks, in which hospital outbreaks played a disproportionately large role. HCWs accounted for 37–63% of all SARS infections in highly affected countries (Park et al. 2004), while one fifth of all laboratory-confirmed MERS infections were attributed to HCWs. By far the majority of affected HCWs were women (Elkholy et al. 2020; WHO 2003). In addition, HCWs dealing with SARS patients experienced significant levels of burnout and post-traumatic stress (Maunder et al. 2006). More generally, the risk for HCWs extends to the healthcare system as a whole. During the SARS outbreaks, several hospitals around the globe were forced to temporarily shut down as a result of internal outbreaks (CDC 2003; Gopalakrishna et al. 2004; Low 2004).

A WHO study (Heymann and Rodier 2004) identified two main contraction risks amongst HCWs in the case of SARS—close face-to-face contact with patients and procedures that brought HCWs into contact with respiratory secretions. These procedures include intubation, (Knobler et al. 2004), the use of nebulizers, nasopharyngeal aspiration, bronchoscopy, airway suction, and non-invasive ventilation procedures (Lee et al. 2003). Non-respiratory exposure risk factors involve proximity to faeces and urine of infected patients (Abdullah et al. 2004). Even after the implementation of adequate hygiene and personal protection measures, such as wearing gloves, gowns, and face masks, new infections among HCWs kept occurring (Lee and Sung 2003). These were attributed to a combination of complacency to follow routine actions like hand-wash after examining each patient and the initial substantial uncertainty about the transmission mechanisms of the novel virus (Abdullah et al. 2004). As became increasingly clear, a key infection control strategy to reduce transmission in healthcare settings is to identify patients with severe acute respiratory infections at the first point of contact in order to minimise exposure to others (Park et al. 2020).

More generally, the global outbreaks of the past 20 years have underlined the importance of a resilient healthcare system. Even a comparably mild pandemic such as the 2009 swine flu brought healthcare systems, and in particular ICU units, in many wealthy countries to the limits of their capacity (Collignon 2011). On the other hand, the 2014 West African Ebola outbreaks put a spotlight on the potential global implications of weak and underfunded healthcare systems in developing countries (Piot et al. 2019).

2.4.3.2 Tried and Tested Measures: Infection Control, Contact Tracing, and Isolation

While there is ample evidence that infection control protocols in healthcare settings were instrumental in eventually containing the SARS and MERS outbreaks, hygienic measures played a role beyond the confines of hospital wards. During the SARS outbreak in Hong Kong, the country most affected outside mainland China, the government in collaboration with various organisations was engaged in an intense public hygiene campaign (Mackenzie et al. 2004). A survey assessing public engagement in those hygiene measures concluded that the "majority of the population wore a face mask (76%), washed their hands after contact with potentially contaminated objects (65%), used soap when washing hands (75%), covered their mouths when sneezing or coughing (78%), and used diluted bleach for household cleaning (>50%)" (Lo et al. 2005). The outbreak in Hong Kong was successfully contained in less than 4 months. One study concluded that the public hygiene measures alongside similar measures in hospitals contributed substantially to the control of the outbreak (Lau et al. 2004).

Contact tracing and prompt isolation of suspected cases were instrumental in stopping the unfolding SARS pandemic, especially in the absence of rapidly deployable diagnostic tools at the time. In the words of Brian Doberstyn who was in charge of the WHO response in the Western Pacific Region: "Most important in controlling SARS were the 19th-century public-health strategies" (Doberstyn 2006). Several countries implemented even more stringent and resource intensive quarantine regimes, sometimes with considerable adverse mental health effects on the affected (Brooks et al. 2020). For instance, in Taiwan a total of 150,000 persons (0.7% of the country's population) were quarantined during the outbreak. Only 46 of them turned out to have probable SARS (Chen et al. 2005). Studies later concluded that contact tracing along with effective isolation of symptomatic individuals were sufficient to contain any SARS outbreak (Kwok et al. 2019). It was acknowledged, however, that this finding might be different for diseases with higher transmissibility or significant asymptomatic transmission. Therefore, quarantine was still the means of choice when MERS hit South Korea in 2015. The South Korean MERS outbreak also initiated the development of the first major electronic contact tracing app (Park 2019).

Meticulous contact tracing has been credited as a major factor in the eventual control of the 2014 West African Ebola outbreak as well (Swanson et al. 2018; Bell et al. 2016; WHO 2015). In this context, sources emphasise the importance of social mobilisation and community involvement in order to build awareness and acceptance for intrusive epidemiological investigations and measures. Difficulty to engage with local communities, together with a lack of resources, had initially been an obstacle to containing the spread of Ebola.

The importance of adequate surveillance was echoed during the swine flu pandemic, but from the perspective of a missed opportunity. The spread of the novel influenza around the world exposed how reliance on passive surveillance leads to serious under-reporting, which can make an outbreak quickly get out of control (Scoones 2009). Instead, it was argued, there should be a shift to active surveillance through general practices and/or vaccine clinics to detect cases adequately (Collignon 2011).

The swine flu pandemic also underscored Doberstyn's point about SARS, namely that traditional non-pharmaceutical interventions often outperform modern scientific interventions in their immediate influence on disease control during a pandemic. On the one hand, governments around the world invested large funds in stockpiling antiviral medications and in the development of a swine flu vaccine that arrived too late (Collignon 2011). On the other hand, presumably more lives were saved by temporary school closures in Hong Kong and India which resulted in a reduction of the effective reproduction number R_e by around 25% and 20%, respectively (Wu et al. 2010; Ali et al. 2013).

2.4.3.3 Global Cooperation and Transparency

In the aftermath of SARS, it was emphasised how efficiently the global network of scientists collaborated in order to identify the causative agent of the disease and develop diagnostic tools (WHO 2003). Global coordination and social cohesion were celebrated as having crucially contributed to the control of SARS, which was achieved in the absence of curative drugs or vaccines (Knobler et al. 2004). Similarly, Ebola was finally brought under control by engaging every level from local communities up to international actors such as the US Center for Disease Control and by international pharmaceutical companies developing vaccines (Piot et al. 2019).

But that is only one side of the medal. In the case of Ebola, for instance, a lack of surveillance and healthcare resources on the ground contributed to the situation getting out of control in the first place. Furthermore, even though Ebola was not a new disease, a relative disinterest of pharmaceutical research in pathogens that affect predominantly poorer countries effectively meant that vaccines and treatments were only developed once the regional epidemic threatened to affect other parts of the globe (Piot et al. 2019).

In the case of SARS, the initial alert of an outbreak in China reached the international community with a delay of nearly 4 months, at a point when much of the damage was already done and the spread across international borders was in full swing. The Chinese government at the time tried to deny and then downplay the spread of SARS within its borders. Only upon considerable internal pressure was the policy changed abruptly. After that, drastic measures eliminated the virus in China within a few months (Ahmad et al. 2009; Huang 2004). The thinly veiled admonition of the WHO still rings as true today as it did in 2003: "Attempts to conceal cases of an infectious disease, for fear of social and economic consequences, must be recognized as a short-term stop-gap measure that carries a very high price: the potential for high levels of human suffering and death, loss of credibility in the eyes of the international community, escalating negative domestic economic impact, damage to the health and economies of neighbouring countries, and a very real risk that outbreaks within the country's own territory will spiral out of control" (WHO 2003).

References

Abdullah ASM, Tomlinson B, Thomas GN, Cockram CS (2004) Impacts of SARS on health care systems and strategies for combating future outbreaks of emerging infectious diseases. In: Learning from SARS: preparing for the next disease outbreak: workshop summary. National Academies Press (US), Washington, DC

Achtman M, Zurth K, Morelli G, Torrea G et al (1999) Yersinia pestis, the cause of plague, is a recently emerged clone of Yersinia pseudotuberculosis. Proc Natl Acad Sci U S A 96 (24):14043–14048. https://doi.org/10.1073/pnas.96.24.14043

Ackerknecht EH (1948) Anticontagionism between 1821 and 1867: the Fielding H. Garrison Lecture. 1948. Bull Hist Med 22:562–593. https://doi.org/10.1093/ije/dyn254

Agusto FB, Teboh-Ewungkem MI, Gumel AB (2015) Mathematical assessment of the effect of traditional beliefs and customs on the transmission dynamics of the 2014 Ebola outbreaks. BMC Med 13(1):1–17. https://doi.org/10.1186/s12916-015-0318-3

Ahmad A, Krumkamp R, Reintjes R (2009) Controlling SARS: a review on China's response compared with other SARS-affected countries. Trop Med Int Health. https://doi.org/10.1111/j.1365-3156.2008.02146.x

Ali ST, Kadi AS, Ferguson NM (2013) Transmission dynamics of the 2009 influenza A (H1N1) pandemic in India: the impact of holiday-related school closure. Epidemics 5(4):157–163. https://doi.org/10.1016/j.epidem.2013.08.001

Baldwin P (2004) Contagion and the State in Europe 1830-1930. Cambridge University Press, Cambridge. Available from http://repositorio.unan.edu.ni/2986/1/5624.pdf

Barberis I, Myles P, Ault SK, Bragazzi NL et al (2016) History and evolution of influenza control through vaccination: from the first monovalent vaccine to universal vaccines. J Prev Med Hyg 57(3):E115–E120

Barua D (1992) History of cholera. In: Barua D, Greenough WB III (eds) Cholera. Plenum, New York, pp 1–36

Bell BP, Damon IK, Jernigan DB, Kenyon TA et al (2016) Overview, control strategies, and lessons learned in the CDC response to the 2014–2016 Ebola epidemic. Available from https://www.cdc.gov/mmwr/volumes/65/su/su6503a2.htm

Benedictow OJ (2004) The Black death: a complete history. Woodbridge, Boydell & Brewer

Bernard H, Fischer R, Mikolajczyk RT, Kretzschmar M et al (2009) Nurses' contacts and potential for infectious disease transmission. Emerg Infect Dis 15(9):1438–1444. https://doi.org/10.3201/eid1509.081475

Bhattacharyya S, Gesteland PH, Korgenski K, Bjørnstad ON et al (2015) Cross-immunity between strains explains the dynamical pattern of paramyxoviruses. PNAS 112(43):13396–13400. https://doi.org/10.1073/pnas.1516698112

Biggerstaff M, Cauchemez S, Reed C, Gambhir M et al (2014) Estimates of the reproduction number for seasonal, pandemic, and zoonotic influenza: a systematic review of the literature. BMC Infect Dis 14. https://doi.org/10.1186/1471-2334-14-480

Bootsma MCJ, Ferguson NM (2007) The effect of public health measures on the 1918 influenza pandemic in U. S. cities. Proc Natl Acad Sci U S A 104:7588–7593

Bos KI, Herbig A, Sahl J, Waglechner N et al (2016) Eighteenth century Yersinia pestis genomes reveal the long-term persistence of an historical plague focus. eLife 5:1–11. https://doi.org/10.7554/eLife.12994

Boyle SA (2015) Cholera, colonialism, and pilgrimage: exploring global/local exchange in the Central Egyptian Delta, 1848–1907. J World Hist. 26(3):581–604

Bramanti B, Dean KR, Walløe L, Stenseth NC (2019) The third plague pandemic in Europe. Proc R Soc B Biol Sci 286(1901):1–8. https://doi.org/10.1098/rspb.2018.2429

Brooks SK, Webster RK, Smith LE, Woodland L et al (2020) The psychological impact of quarantine and how to reduce it: rapid review of the evidence. Lancet 395(10227):P912–P920. https://doi.org/10.1016/S0140-6736(20)30460-8

Camacho A, Kucharski AJ, Funk S, Breman J et al (2014) Potential for large outbreaks of Ebola virus disease. Epidemics 9(2014):70–78. https://doi.org/10.1016/j.epidem.2014.09.003

Cauchemez S, Nouvellet P, Cori A, Jombart T et al (2016) Unraveling the drivers of MERS-CoV transmission. Proc Natl Acad Sci U S A 113(32):9081–9086. https://doi.org/10.1073/pnas.1519235113

CDC (2003) CDC | morbidity and mortality weekly report. Available from https://www.cdc.gov/mmwr/preview/mmwrhtml/mm5220a1.htm

CDC (2012) Introduction to epidemiology. In: Principles of epidemiology in public health practice, 3rd edn, pp 1–72. https://doi.org/10.1201/9781315381848.

Chang CC, Crane M, Zhou J, Mina M et al (2013) HIV and co-infections. Immunol Rev 254 (1):114–142. https://doi.org/10.1111/imr.12063.HIV

Chaplin DD (2010) Overview of the immune response. J Allergy Clin Immunol 125(2):S3–S23. https://doi.org/10.1016/j.jaci.2009.12.980

Chattu VK, Kumar R, Kumary S, Kajal F et al (2018) Nipah virus epidemic in southern India and emphasizing "One Health" approach to ensure global health security. J Family Med Prim Care 7 (2):275–283. https://doi.org/10.4103/jfmpc.jfmpc_137_18

Chen KT, Twu SJ, Chang HL, Wu YC et al (2005) SARS in Taiwan: an overview and lessons learned. Int J Infect Dis 9(2):77–85. https://doi.org/10.1016/j.ijid.2004.04.015

Chowell G, Bettencourt LMA, Johnson N, Alonso WJ et al (2008) The 1918-1919 influenza pandemic in England and Wales: spatial patterns in transmissibility and mortality impact. Proc R Soc B Biol Sci 275(1634):501–509. https://doi.org/10.1098/rspb.2007.1477

Chowell G, Blumberg S, Simonsen L (2014) Synthesizing data and models for the spread of MERS-CoV, 2013: key role of index cases and hospital transmission. Epidemics. 9:40–51

Christakis NA, Fowler JH (2009) Connected: the surprising power of our social networks and how they shape our lives. Little, Brown, New York

City and Guilds (2014) Causes and spread of infection. In Health & social care (level 2). City and Guilds. Available from https://www.cityandguilds.com/-/media/productdocuments/health_and_social_care/care/4222/additional_documents/handsc_level_2_textbook_sample_chapters-pdf.ashx

Cliff A, Haggett P (2004) Time, travel and infection. Br Med Bull 69:87–99. https://doi.org/10.1093/bmb/ldh011

Colizza V, Barrat A, Barthelemy M, Vespignani A (2006) The role of the airline transportation network in the. PNAS. 103(7):2015–2020

Collignon P (2011) Swine flu: lessons we need to learn from our global experience. Emerg Health Threats J 4:7169. https://doi.org/10.3402/ehtj.v4i0.7169

Colwell RR (1996) Global climate and infectious disease: the cholera paradigm. Science 274:2025–2031. 10.1126/science.274.5295.2025

Cook ND (1998) Born to die: disease and New World conquest, 1492–1650. Cambridge University Press, Cambridge

Cox NJ, Subbarao K (2000) Global epidemiology of influenza: past and present. Annu Rev Med 51 (1):407–421. https://doi.org/10.1146/annurev.med.51.1.407

Crawford EA (1996) Paul-Louis Simond and his work on plague. Persp Biol Med 39(3):446–458. https://doi.org/10.1353/pbm.1996.0031

Crosby A (1972) The Columbian exchange: biological and cultural consequences of 1492. Greenwood, Westport

Crosby AW (1976) Virgin soil epidemics as a factor in the aboriginal depopulation in America. William Mary Q 33:289–299. https://doi.org/10.2307/1922166

Crosby AW (2003) America's forgotten pandemic: the influenza of 1918, 2nd edn. Cambridge University Press, Cambridge, pp 1–337. https://doi.org/10.1017/CBO9780511586576

Cvjetanovic B, Barua D (1972) The seventh pandemic of cholera. Nature 239(5368):137–138. https://doi.org/10.1038/239137a0

Danforth K, Granich R, Wiedeman D, Baxi S et al (2017) Global mortality and morbidity of HIV/AIDS. In: Holmes KK, Bertozzi S, Bloom BR, Jha P (eds) Disease control priorities, vol 6:

major infectious diseases, 3rd edn. The World Bank, Washington, DC, pp 29–44. https://doi.org/10.1596/978-1-4648-0524-0_ch2

David Patterson K (1985) Pandemic and epidemic influenza, 1830-1848. Soc Sci Med 21 (5):571–580. https://doi.org/10.1016/0277-9536(85)90042-5

Dawood FS, Iuliano AD, Reed C, Meltzer MI et al (2012) Estimated global mortality associated with the first 12 months of 2009 pandemic influenza A H1N1 virus circulation: a modelling study. Lancet Infect Dis 12(9):P687–P695. https://doi.org/10.1016/S1473-3099(12)70121-4

Dean KR, Krauer F, Walløe L, Lingjærde OC et al (2018) Human ectoparasites and the spread of plague in Europe during the Second Pandemic. Proc Natl Acad Sci U S A 115(6):1304–1309. https://doi.org/10.1073/pnas.1715640115

Dean KR, Krauer F, Schmid BV (2019) Epidemiology of a bubonic plague outbreak in Glasgow, Scotland in 1900. Roy Soc Open Sci 6(1):181695. https://doi.org/10.1098/rsos.181695

Delamater PL, Street EJ, Leslie TF, Yang YT et al (2019) Complexity of the basic reproduction number (R0). Emerg Infect Dis 25(1):1–4. https://doi.org/10.3201/eid2501.171901

Diamond JM (1999) Guns, germs, and steel: the fates of human societies. W.W. Norton & Cole, New York

Diekmann O, Heesterbeek JAP (2000) Mathematical epidemiology of infectious diseases: model building, analysis and interpretation. Wiley, New York

Dietz K (1993) The estimation of the basic reproduction number for infectious diseases. Stat Methods Med Res 2(1):23–41. https://doi.org/10.1177/096228029300200103

Doberstyn B (2006) What did we learn from SARS? In: SARS: how a global pandemic was stopped. World Health Organization, Geneva, pp 241–254

Dobyns HF (1993) Disease transfer at contact. Ann Rev Anthropol 22:273–291

Doshi P (2011) The elusive definition of pandemic influenza. Bull World Health Organ 89 (7):532–538. https://doi.org/10.2471/blt.11.086173

Dourmashkin RR (1997) What caused the 1918–30 epidemic of encephalitis lethargica? J R Soc Med 90(9):515–520. https://doi.org/10.1177/014107689709000916

Dourmashkin RR, Dunn G, Castano V, Mccall SA (2012) Evidence for an enterovirus as the cause of encephalitis lethargica evidence for an enterovirus as the cause of encephalitis lethargica. BMC Infect Dis 12(1):1. https://doi.org/10.1186/1471-2334-12-136

ECDC (2017) Evidence brief: impact of stigma and discrimination on access to HIV services in Europe. Available from https://ecdc.europa.eu/sites/portal/files/documents/Dublin-EB-Stigma

Echenberg M (2002) Pestis redux: the initial years of the third bubonic plague pandemic, 1894-1901. J World Hist 13(2):429–449

Edwardes EJ (1902) A concise history of small-pox and vaccination in Europe. H.K Lewis, London

Elkholy AA, Grant R, Assiri A, Elhakim M et al (2020) MERS-CoV infection among healthcare workers and risk factors for death: retrospective analysis of all laboratory-confirmed cases reported to WHO from 2012 to 2 June 2018. J Infect Public Health 13(3):418–422. https://doi.org/10.1016/j.jiph.2019.04.011

Eroshenko GA, Nosov NY, Krasnov YM, Oglodin YG et al (2017) Yersinia pestis strains of ancient phylogenetic branch 0.ANT are widely spread in the high-mountain plague foci of Kyrgyzstan. PLoS One 12(10):1–10. https://doi.org/10.1371/journal.pone.0187230

Estrada-García T (1996) Cholera: foodborne transmission and its prevention. Eur J Epidemiol 12 (5):461–469. https://doi.org/10.1007/bf00143997

Eyler JM (2010) The state of science, microbiology, and vaccines circa 1918. Public Health Rep 125:27–36. https://doi.org/10.1177/00333549101250S306

Faria NR, Rambaut A, Suchard MA, Baele G et al (2014) The early spread and epidemic ignition of HIV-1 in human populations. Science 346(6205):56–61. https://doi.org/10.1126/science.1256739.The

Ferguson NM, Cummings DAT, Fraser C, Cajka JC et al (2006) Strategies for mitigating an influenza pandemic. Nature 442(7101):448–452. https://doi.org/10.1038/nature04795

Filio M, Gregory T, Marianna K, George A (2013) Mary Mallon (1869-1938) and the history of typhoid fever. Ann Gastroenterol 26(2):132–134

Fine PEM (2003) The interval between successive cases of an infectious disease. Am J Epidemiol 158(11):1039–1047. https://doi.org/10.1093/aje/kwg251

Foley PB (2018) Encephalitis lethargica: the mind and brain virus. Springer, New York. https://doi.org/10.1136/bmj.2.3334.937

Friant S, Paige SB, Goldberg TL (2015) Drivers of bushmeat hunting and perceptions of zoonoses in Nigerian hunting communities. PLoS Negl Trop Dis 9(5):e0003792. https://doi.org/10.1371/journal.pntd.0003792

Frost WH (1920) Statistics of influenza morbidity: with special reference to certain factors in case incidence and case fatality. Public Health Rep (1896-1970) 35(11):584. https://doi.org/10.2307/4575511

Gani R, Leach S (2001) Transmission potential of smallpox in contemporary populations. Nature 414(6865):748–751. https://doi.org/10.1038/414748a

Gilbert MTP, Rambaut A, Wlasiuk G, Spira TJ et al (2007) The emergence of HIV/AIDS in the Americas and beyond. Proc Natl Acad Sci U S A 2007:18566–18570. https://doi.org/10.1073/pnas.0705329104

Glass RI, Black RE (1992) The epidemiology of cholera. In: Barua D, Greenough WB III (eds) Cholera. Plenum, New York, pp 129–154

GLEWS - WHO (2020) Zoonoses and veterinary public health. World Health Organisation. Available from https://www.who.int/zoonoses/outbreaks/glews/en/index2.html. Accessed 31 March 2020

Goh VH (2009) Aging in Asia: a cultural, socio-economical and historical perspective. Aging Male 8(2):90–96. https://doi.org/10.1080/13685530500088472

Gopalakrishna G, Choo P, Leo YS, Tay BK et al (2004) SARS transmission and hospital containment. Emerg Infect Dis 10(3):395–400. https://doi.org/10.3201/eid1003.030650

Grais RF, Ferrari MJ, Dubray C, Bjørnstad ON et al (2006) Estimating transmission intensity for a measles epidemic in Niamey, Niger: lessons for intervention. Trans R Soc Trop Med Hyg 100 (9):867–873. https://doi.org/10.1016/j.trstmh.2005.10.014

Green MS, Swartz T, Mayshar E, Lev B et al (2002) Perspective when is an epidemic an epidemic? IMAJ 4(January):3–6

Greene WC (2007) A history of AIDS: looking back to see ahead. Eur J Immunol 37:94–102. https://doi.org/10.1002/eji.200737441

Ha K-M (2015) A lesson learned from the MERS outbreak in South Korea in 2015. J Hosp Infect 92:232–234. https://doi.org/10.1016/j.jhin.2015.10.004

Haagsma JA, Tariq L, Heederik DJ, Havelaar AH (2012) Infectious disease risks associated with occupational exposure: a systematic review of the literature. Occup Environ Mede 69:140–146. https://doi.org/10.1136/oemed-2011-100068

Hackel J (1998) Community conservation and the future of Africa's wildlife. Conserv Biol 13:726–734

Haensch S, Bianucci R, Signoli M, Rajerison M et al (2010) Distinct clones of Yersinia pestis caused the black death. PLoS Pathog 6(10):e1001134. https://doi.org/10.1371/journal.ppat.1001134

Han H-J, Hao Y, Yu X-J (2016) Evidence for zoonotic origins of Middle East respiratory syndrome coronavirus. J Gen Virol 97:274–280. https://doi.org/10.1099/jgv.0.000342

Hatchett RJ, Mecher CE, Lipsitch M (2007) Public health interventions and epidemic intensity during the 1918 influenza pandemic. Proc Natl Acad Sci U S A 2007:7582–7587. https://doi.org/10.1073/pnas.0610941104

Hawgood BJ (2007) Waldemar Mordecai Haffkine, CIE (1860–1930): prophylactic vaccination against cholera and bubonic plague in British India. J Med Biogr 15(1):9–19. https://doi.org/10.1258/j.jmb.2007.05-59

Hawgood BJ (2008) Alexandre Yersin (1863-1943): discoverer of the plague bacillus, explorer and agronomist. J Med Biogr 16(3):167–172. https://doi.org/10.1258/jmb.2007.007017

Hays JN (2005) Epidemics and pandemics: their impacts on human history. ABC-CLIO, Santa Barbara

Hays JN (2009) The burdens of disease: epidemics and human response in western history, Revised ed. Rutgers University Press. Available from file:///C:/Users/User/Downloads/fvm939e.pdf

Heffernan JM, Smith RJ, Wahl LM (2005) Perspectives on the basic reproductive ratio. J R Soc Interface 2(4):281–293. https://doi.org/10.1098/rsif.2005.0042

Henderson DA (2011) The eradication of smallpox - AN overview of the past, present, and future. Vaccine 29(suppl. 4):D7. https://doi.org/10.1016/j.vaccine.2011.06.080

Herek GM, Capitanio JP (1999) AIDS stigma and sexual prejudice. Am Behav Sci 42:1126–1143. https://doi.org/10.1177/00027649921954804

Heymann DL, Rodier G (2004) SARS: lessons from a new disease. In: Learning from SARS: preparing for the next disease outbreak: workshop summary. National Academies Press (US), Washington, DC

Hirst F (1953) The conquest of plague: a study of the evolution of epidemiology. Oxford University Press, Oxford

Ho PL, Becker M, Chan-Yeung MM (2007) Emerging occupational lung infections. Int J Tuberc Lung Dis 7:710–721

Hoberg EP, Alkire NL, de Queiroz A, Jones A (2001) Out of Africa: origins of the Taenia tapeworms in humans. Proc R Soc B Biol Sci 268(1469):781–787. https://doi.org/10.1098/rspb.2000.1579

Honigsbaum M (2009) Living with Enza. Palgrave Macmillan, Basingstoke

Honigsbaum M (2020) Revisiting the 1957 and 1968 influenza pandemics. Lancet 395 (10240):1824–1826. https://doi.org/10.1016/S0140-6736(20)31201-0

Hood MA, Ness GE (1982) Survival of Vibrio cholerae and Escherichia coli in estuarine waters and sediments. Appl Environ Microbiol 43(3):578–584. https://doi.org/10.1128/aem.43.3.578-584.1982

Howard-Jones N (1974) The scientific background of the International Sanitary Conferences 1851-1938. WHO Chronicle 28:159–171, 229–247, 348–369, 414–426, 455–470, 495

Hu H, Nigmatulina K, Eckhoff P (2013) The scaling of contact rates with population density for the infectious disease models. Math Biosci 244(2):125–134. https://doi.org/10.1016/j.mbs.2013.04.013

Huang Y (2004) The SARS epidemic and its aftermath in China: a political perspective. In: Learning from SARS: preparing for the next disease outbreak: workshop summary. National Academies Press (US), Washington, DC

Isaacs D (2010) Lessons from the swine flu: pandemic, panic and/or pandemonium? J Paediatr Child Health 46(11):623–626. https://doi.org/10.1111/j.1440-1754.2010.01912.x

Islam MS, Zaman MH, Islam MS, Ahmed N et al (2020) Environmental reservoirs of Vibrio cholerae. Vaccine 38:A52–A62. https://doi.org/10.1016/j.vaccine.2019.06.033

Jackson C (2009) History lessons: the Asian flu pandemic. Br J Gen Pract 59(565):622–623. https://doi.org/10.3399/bjgp09X453882

Janeway CJ, Travers P, Walport M (2001) Infectious agents and how they cause disease. In: Immunobiology: the immune system in health and disease, 5th edn. Garland Science, New York. Available from https://www.ncbi.nlm.nih.gov/books/NBK27114/

Johnson NPAS, Mueller J (2002) Updating the accounts: global mortality of the 1918-1920 "Spanish" influenza pandemic. Bull Hist Med 76(1):105–115. https://doi.org/10.1353/bhm.2002.0022

Jones KE, Patel NG, Levy MA, Storeygard A et al (2008) Global trends in emerging infectious diseases. Nature 451:990–993. https://doi.org/10.1038/nature06536

van Kerkhove MD, Bento AI, Mills HL, Ferguson NM et al (2015) A review of epidemiological parameters from Ebola outbreaks to inform early public health decision-making. Sci Data 2:1–10. https://doi.org/10.1038/sdata.2015.19

Khan K, Arino J, Hu W, Raposo P, Sears J, Calderon F et al (2009) Spread of a novel influenza A (H1N1) virus via global airline transportation. N Engl J Med 361(2):212–214

Kilbourne ED (2006) Influenza pandemics of the 20th century. Emerg Infect Dis 12(1):9–14. https://doi.org/10.3201/eid1201.051254

Kimman TG, Hoek M, de Jong MCM (2013) Assessing and controlling health risks from animal husbandry. NJAS Wageningen J Life Sci 66:7–14. https://doi.org/10.1016/j.njas.2013.05.003
Kindhauser MK, Allen T, Frank V, Santhana RS et al (2016) Zika: the origin and spread of a mosquito-borne virus. Bull World Health Organ 94:675–686. https://doi.org/10.2471/BLT.16.171082
Knobler S, Mahmoud A, Lemon S, Mack A et al (2004) Learning from SARS: preparing for the next disease outbreak. National Academies Press (US), Washington, DC
Koch A, Brierley C, Maslin MM, Lewis SL (2019) Earth system impacts of the European arrival and Great Dying in the Americas after 1492. Q Sci Rev 207:13–36. https://doi.org/10.1016/j.quascirev.2018.12.004
Kohler M, Blond N, Clappier A (2016) A city scale degree-day method to assess building space heating energy demands in Strasbourg Eurometropolis (France). Appl Energy 184:40–54. https://doi.org/10.1016/j.apenergy.2016.09.075
Kucharski AJ, Althaus CL (2015) The role of superspreading in middle east respiratory syndrome coronavirus (Mers-CoV) transmission. Eurosurveillance. 20(25):1–5
Kwok KO, Tang A, Wei VWI, Park WH et al (2019) Epidemic models of contact tracing: systematic review of transmission studies of severe acute respiratory syndrome and middle east respiratory syndrome. Comput Struct Biotechnol J 17:186–194. https://doi.org/10.1016/j.csbj.2019.01.003
Lancet Editorial Article (1981) *Encephalitis Lethargica*. The Lancet, pp 1396–1397
Lantagne D, Balakrish Nair G, Lanata CF, Cravioto A (2013) The cholera outbreak in Haiti: where and how did it begin? Curr Topics Microbiol Immunol 379:145–164. https://doi.org/10.1007/82_2013_331
Last JM (2001) A dictionary of epidemiology, 4th edn. Oxford University Press, New York. https://doi.org/10.1136/jech.47.5.430
Lau JTF, Hiyi T, Lau M, Yang X (2004) SARS transmission, risk factors, and prevention in Hong Kong. Emerg Infect Dis 10(4):587–592. https://doi.org/10.3201/eid1004.030628
Lee N, Sung JJY (2003) Nosocomial transmission of SARS. Nat Public Health Emerg Collection 5 (6):473–476. https://doi.org/10.1007/s11908-003-0089-4
Lee N, Hui D, Wu A, Chan P et al (2003) A major outbreak of severe acute respiratory syndrome in Hong Kong. N Engl J Med 348(20):1986–1994. https://doi.org/10.1056/NEJMoa030685
Lewis JD, Enfield KB, Sifri CD (2015) Hepatitis B in healthcare workers: transmission events and guidance for management. World J Hepatol 7(3):488–497. https://doi.org/10.4254/wjh.v7.i3.488
Li FCK, Choi BCK, Sly T, Pak AWP (2008) Finding the real case-fatality rate of H5N1 avian influenza. J Epidemiol Commun Health 62(6):555–559. https://doi.org/10.1136/jech.2007.064030
Lietz J, Westermann C, Nienhaus A, Schablon A (2016) The occupational risk of influenza A (H1N1) infection among healthcare personnel during the 2009 pandemic: A systematic review and meta-analysis of observational studies. PLoS One 11(8):1–19. https://doi.org/10.1371/journal.pone.0162061
Lippi D, Gotuzzo E (2013) The greatest steps towards the discovery of Vibrio cholerae. Eur Soc Clin Infect Dis 20(3):191–195. https://doi.org/10.1111/1469-0691.12390
Lipsitch M, Cohen T, Cooper B, Robins JM et al (2003) Transmission dynamics and control of severe acute respiratory syndrome. Science 300(5627):1966–1970. https://doi.org/10.1126/science.1086616
Little LK (ed) (2007) Plague and the end of antiquity: the pandemic of 541-750. Cambridge University Press, Cambridge
Lloyd-Smith JO, Schreiber SJ, Kopp PE, Getz WM (2005) Superspreading and the effect of individual variation on disease emergence. Nature 438(7066):355–359. https://doi.org/10.1038/nature04153
Lo JYC, Tsang THF, Leung Y-H, Yeung EYH et al (2005) Respiratory infections during SARS outbreak, Hong Kong, 2003. Emerg Infect Dis 11(10):1738–1741

Low DE (2004) SARS: lessons from Toronto. In: Learning from SARS: preparing for the next disease outbreak: workshop summary. National Academies Press (US), Washington, DC

Lutz C, Erken M, Noorian F, Sun S et al (2013) Environmental reservoirs and mechanisms of persistence of Vibrio cholerae. Front Microbiol 4:375. https://doi.org/10.3389/fmicb.2013.00375

Ma Y, Horsburgh CR, White LF, Jenkins HE (2018) Quantifying TB transmission: a systematic review of reproduction number and serial interval estimates for tuberculosis. Epidemiol Infect 146(12):1478–1494. https://doi.org/10.1017/S0950268818001760

Mackenzie JS, Drury P, Ellis A, Grein T et al (2004) The WHO response to SARS and preparations for the future. In: Learning from SARS: preparing for the next disease outbreak: workshop summary. National Academies Press (US), Washington, DC

Marcus R, de Merode E, Cowlishaw G (2004) Do wildlife laws work? Species protection and the application of a prey choice model to poaching decisions. Proc R Soc Lond 2004:2631–2636

Martin PM, Martin-Granel E (2006) 2500-Year evolution of the term epidemic. Emerg Infect Dis 12 (6):976–980. https://doi.org/10.3201/eid1206.051263

Maunder RG, Lancee WJ, Balderson KE, Bennett JP et al (2006) Long-term psychological and occupational effects of providing hospital healthcare during SARS outbreak. Emerg Infect Dis 12(12):1924–1932. https://doi.org/10.3201/eid1212.060584

McCall S, Vilensky JA, Gilman S, Taubenberger JK (2008) The relationship between encephalitis lethargica and influenza: a critical analysis. J Neurovirol 14(3):177–185. https://doi.org/10.5040/9781501351105.part-002

McNeill WH (1976) Plagues and peoples. Garden City, New York

Merikanto I, Laakso J, Kaitala V (2012) Outside-host growth of pathogens attenuates epidemiological outbreaks. PLoS One 7(11):1–8. https://doi.org/10.1371/journal.pone.0050158

Merriam-Webster (n.d.-a) Epidemic | definition of epidemic by Merriam-Webster. Merriam-Webster. Available from https://www.merriam-webster.com/dictionary/epidemic. Accessed 29 May 2020

Merriam-Webster (n.d.-b) Pandemic | definition of pandemic by Merriam-Webster. Merriam-Webster. Available from https://www.merriam-webster.com/dictionary/pandemic#faqs. Accessed 29 May 2020

Morand S, McIntyre KM, Baylis M (2014) Domesticated animals and human infectious diseases of zoonotic origins: domestication time matters. Infect Genet Evol 24:76–81. https://doi.org/10.1016/j.meegid.2014.02.013

Morelli G, Song Y, Mazzoni CJ, Eppinger M et al (2010) Phylogenetic diversity and historical patterns of pandemic spread of Yersinia pestis. Nat Genet 42(12):1140–1143. https://doi.org/10.1038/ng.705.Phylogenetic

Morse SS, Mazet JA, Woolhouse M, Parrish CR et al (2012) Prediction and prevention of the next pandemic zoonosis. Lancet 380(9857):1956–1965. https://doi.org/10.1016/S0140-6736(12)61684-5

Nandy R, Handzel T, Zaneidou M, Biey J et al (2006) Case-fatality rate during a measles outbreak in Eastern Niger in 2003. Clin Infect Dis 42(3):322–328. https://doi.org/10.1086/499240

Nathanson N, Kew OM (2010) From emergence to eradication: the epidemiology of poliomyelitis deconstructed. Am J Epidemiol 172(11):1213–1229. https://doi.org/10.1093/aje/kwq320

Nienhaus A, Schablon A, Preisser AM, Ringshausen FC et al (2014) Tuberculosis in healthcare workers – a narrative review from a German perspective. J Occup Med Toxicol. https://doi.org/10.1186/1745-6673-9-9

Nishiura H (2008) Back calculation of the disease-age specific frequency of secondary transmission of primary pneumonic plague. Asian Pac J Trop Med 1(3). Available from http://arxiv.org/abs/0810.1606

Nishiura H, Kashiwagi T (2009) Smallpox and season: reanalysis of historical data. Interdiscip Perspect Infect Dis 2009:591935. https://doi.org/10.1155/2009/591935

Nishiura H, Brockmann SO, Eichner M (2008) Extracting key information from historical data to quantify the transmission dynamics of smallpox. Theor Biol Med Modell 5:1–12. https://doi.org/10.1186/1742-4682-5-20

Nunn N, Qian N (2010) The Columbian exchange: a history of disease, food, and ideas. J Econ Perspect 24(2):163–188. https://doi.org/10.1257/jep.24.2.163

Nutrition Health Review (2018) Viral infection or bacterial infection? Know the difference. Vegetus Publications. Available from https://link.gale.com/apps/doc/A534317826/ITOF?u=ull_ttda&sid=ITOF&xid=1a972a6b

OECD/SWAC (2020) Africa's urbanisation dynamics 2020: Africapolis, mapping a new urban geography. In: West African studies. OECD Publishing, Paris. https://doi.org/10.1787/b6bccb81-en

Olson DR, Simonsen L, Edelson PJ, Morse SS (2005) Epidemiological evidence of an early wave of the 1918 influenza pandemic in New York City. Proc Natl Acad Sci U S A 102:11059–11063. https://doi.org/10.1073/pnas.0408290102

Park C (2019) MERS-CoV infection in South Korea and strategies for possible future outbreak: narrative review. J Glob Health Rep 3:e2019088. https://doi.org/10.29392/joghr.3.e2019088

Park BJ, Peck AJ, Kuehnert MJ, Newbern C et al (2004) Lack of SARS Transmission among healthcare workers, United States. Emerg Infect Dis 10(2):217–224. https://doi.org/10.3201/eid1002.030793

Park JE, Jung S, Kim A (2018) MERS transmission and risk factors: a systematic review. BMC Public Health 18(1):574. https://doi.org/10.1186/s12889-018-5484-8

Park M, Thwaites RS, Openshaw PJM (2020) COVID-19: lessons from SARS and MERS. Eur J Immunol 50:308–316. https://doi.org/10.1002/eji.202070035

Pascali L (2017) The wind of change: maritime technology, trade, and economic development. Am Econ Rev 107(9):2821–2854. https://doi.org/10.1257/aer.20140832

Patterson KD (1987) Pandemic influenza, 1700–1900: a study in historical epidemiology. Rowman & Littlefield, Totowa

Patterson KD (1993) Typhus and its control in Russia, 1870-1940. Med Hist 37(4):361–381. https://doi.org/10.1017/S0025727300058725

Patterson KD, Pyle GF (1991) The geography and mortality of the 1918 influenza pandemic. Bull Hist Med 65(1):4–21

Peña-García VH, Christofferson RC (2019) Correlation of the basic reproduction number (R0) and eco-environmental variables in Colombian municipalities with chikungunya outbreaks during 2014-2016. In: Nathan D Grubaugh (ed). PLoS Negl Trop Dis 13(11):e0007878. https://doi.org/10.1371/journal.pntd.0007878

Phelps M, Perner ML, Pitzer VE, Andreasen V et al (2018) Cholera epidemics of the past offer new insights into an old enemy. J Infect Dis 217(4):641–649. https://doi.org/10.1093/infdis/jix602

Piot P, Taelman H, Bila Minlangu K, Mbendi N et al (1984) Acquired immunodeficiency syndrome in a heterosexual population in Zaire. Lancet 324(8394):65–69. https://doi.org/10.1016/S0140-6736(84)90241-1

Piot P, Soka MS, Spencer J (2019) Emergent threats: lessons learnt from Ebola. Int Health 11 (5):334–337. https://doi.org/10.1093/inthealth/ihz062

Pollitzer R (1954) Cholera studies. 1. History of the disease. Bull World Health Organ 10 (3):421–461

Potter CW (2001) A history of influenza. J Appl Microbiol 91:572–579

Pyle GF (1969) Diffusion of cholera in the United States. Geogr Anal 1:59–75. https://doi.org/10.1111/j.1538-4632.1969.tb00605.x

Pyle GF (1986) The diffusion of influenza: patterns and paradigms. Rowman & Littlefield, Totowa

Rahman MHAA, Hairon SM, Hamat RA, Jamaluddin TZMT et al (2018) Seroprevalence and distribution of leptospirosis serovars among wet market workers in Northeastern, Malaysia: a cross sectional study. BMC Infect Dis 18(1):569. https://doi.org/10.1186/s12879-018-3470-5

Rapsomanikis G (2015) The economic lives of smallholder farmers. Available from http://www.fao.org/3/a-i5251e.pdf

Ravenholt RT, Foege WH (1982) 1918 influenza, encephalitis lethargica, parkinsonism. Lancet 2:860–864

Reichert TA, Christensen RA (2005) It's not about smoldering or neuraminidase: there were 2 variants of the A(H3N2) pandemic virus differing in internal genes. J Infect Dis 192 (10):1858–1860. https://doi.org/10.1086/497153

Rice GW, Palmer E (1993) Pandemic influenza in Japan, 1918-19: mortality patterns and official responses. J Jpn Stud 2:389–420. https://doi.org/10.2307/132645

Rinchen S, Tenzin T, Hall D, van der Meer F et al (2019) A community-based knowledge, attitude, and practice survey on rabies among cattle owners in selected areas of Bhutan. PLoS Negl Trop Dis 13(4):e0007305. https://doi.org/10.1371/journal.pntd.0007305

Rolleston JD (1933) The smallpox pandemic of 1870-1874: (Section of Epidemiology and State Medicine). Proc R Soc Med 27(2):177–192. http://www.ncbi.nlm.nih.gov/pubmed/19989604

Rozo M, Gronvall GK (2015) The reemergent 1977 H1N1 strain and the gain-of-function debate. mBio 6(4):e01013. https://doi.org/10.1128/mBio.01013-15

Sack DA, Sack RB, Nair GB, Siddique AK (2004) Cholera (The Lancet). Lancet 363 (9404):223–233. https://doi.org/10.1016/S0140-6736(03)15328-7

Saghir J, Santoro J (2018) CSIR report on urbanization in Sub-Saharan Africa. Centre for Strategic and International Studies. Available at https://www.csis.org/analysis/urbanization-sub-saharan-africa. Accessed 12 May 2020

Sanjuán R, Nebot MR, Chirico N, Mansky LM et al (2010) Viral mutation rates. J Virol 84 (19):9733–9748. https://doi.org/10.1128/jvi.00694-10

Sarris P (2007) Bubonic plague in Byzantium: the evidence of non-literary sources. In: Little LK (ed) Plague and the end of antiquity: the pandemic of 541-750. Cambridge University Press, Cambridge, pp 119–132

Schmid BV, Büntgen U, Easterday WR, Ginzler C et al (2015) Climate-driven introduction of the Black Death and successive plague reintroductions into Europe. Proc Natl Acad Sci 112 (10):3020–3025. https://doi.org/10.1073/pnas.1412887112

Scholtissek C, Rohde W, von Hoyningen V, Rott R (1978) On the origin of the human influenza virus subtypes H2N2 and H3N2. Virology 87(1):13–20. https://doi.org/10.1016/0042-6822(78)90153-8

Scoones I (2009) The lessons of swine flu. Guardian. Available from https://www.theguardian.com/commentisfree/2009/may/10/swine-flu

Sharp PM, Hahn BH (2011) Origins of HIV and the AIDS pandemic. Cold Spring Harbor Perspect Med 1(1):a006841. https://doi.org/10.1101/cshperspect.a006841

Sheng ZM, Chertow DS, Ambroggio X, McCall S et al (2011) Autopsy series of 68 cases dying before and during the 1918 influenza pandemic peak. Proc Natl Acad Sci U S A 108 (39):16416–16421. https://doi.org/10.1073/pnas.1111179108

Shultz JM, Espinel Z, Espinola M, Rechkemmer A (2016) Distinguishing epidemiological features of the 2013–2016 West Africa Ebola virus disease outbreak. Disaster Health 3(3):78–88. https://doi.org/10.1080/21665044.2016.1228326

Simonsen L, Clarke MJ, Schonberger LB, Arden NH et al (1998) Pandemic versus epidemic influenza mortality: a pattern of changing age distribution. J Infect Dis 178(1):53–60. https://doi.org/10.1086/515616

Sompayrac L (2019) How the immune system works, 6th edn. Wiley Blackwell, Hoboken

Stadnytskyi V, Bax CE, Bax A, Anfinrud P (2020) The airborne lifetime of small speech droplets and their potential importance in SARS-CoV-2 transmission. PNAS 117(22):11875–11877. https://doi.org/10.1073/pnas.2006874117

Stathakopoulos D (2007) Chapter: crime and punishment: the plague in the Byzantine Empire, 541–749. In: Plague and the end of antiquity: the pandemic of 541–750. Cambridge University Press, Cambridge, pp 99–118

Subramanian M (2012) Zoonotic disease risk and the bushmeat trade: assessing awareness among hunters and traders in Sierra Leone. EcoHealth 9:471–482. https://doi.org/10.1007/s10393-012-0807-1

Swanson KC, Altare C, Wesseh CS, Nyenswah T et al (2018) Contact tracing performance during the Ebola epidemic in Liberia, 2014-2015. PLoS Negl Trop Regions 12(9):e0006762. https://doi.org/10.1371/journal.pntd.0006762

Tamerius JD, Shaman J, Alonso WJ, Bloom-Feshbach K et al (2013) Environmental Predictors of Seasonal Influenza Epidemics across Temperate and Tropical Climates Steven Riley (ed.). PLoS Pathogens 9(3):e1003194. https://doi.org/10.1371/journal.ppat.1003194

Tatem AJ, Rogers DJ, Hay SI (2006) Global transport networks and infectious disease spread. Adv Parasitol 62(5):293–343. https://doi.org/10.1016/S0065-308X(05)62009-X

Taubenberger JK, Morens DM (2006) 1918 Influenza: the mother of all pandemics. Emerg Infect Dis 12(1):15–22. https://doi.org/10.3201/eid1209.05-0979

Taubenberger JK, Morens DM (2010) Influenza: the once and future pandemic. Public Health Rep 125(suppl 3):15–26. https://doi.org/10.1177/00333549101250s305

Taubenberger JK, Reid AH, Lourens RM, Wang R et al (2005) Characterization of the 1918 influenza virus polymerase genes. Nature 437(7060):889–893. https://doi.org/10.1038/nature04230

Taubenberger JK, Morens DM, Fauci AS (2007) The next influenza pandemic. JAMA 297 (18):2025–2027. https://doi.org/10.1001/jama.297.18.2025

Tauxe R, Seminario L, Tapia R, Libel M (1994) The Latin American epidemic. In: Vibrio cholerae and cholera. American Society of Microbiology, Washington, DC, pp 321–344. https://doi.org/10.1128/9781555818364.ch21

Thacker E, Janke B (2008) Swine influenza virus: zoonotic potential and vaccination strategies for the control of avian and swine influenzas. J Infect Dis Supplement 1:S19–S24. https://doi.org/10.1086/524988

Tomley FM, Shirley MW (2009) Livestock infectious diseases and zoonoses. Philos Trans R Soc B Biol Sci 364(1530):2637–2642. https://doi.org/10.1098/rstb.2009.0133

UNAIDS (2006) 2006 Report on the global AIDS epidemic. Available from https://data.unaids.org/pub/report/2006/2006_gr_en.pdf

UNAIDS (2019) UNAIDS DATA 2019. Available from https://www.unaids.org/sites/default/files/media_asset/2019-UNAIDS-data_en.pdf

UN-WPP (2019) 2019 Revision of world population prospects. United Nations - Department of Economics and Social Affairs. Available from https://population.un.org/wpp/. Accessed 30 May 2020

Vågene ÅJ, Herbig A, Campana MG, Robles García NM et al (2018) Salmonella enterica genomes from victims of a major sixteenth-century epidemic in Mexico. Nat Ecol Evol 2(3):520–528. https://doi.org/10.1038/s41559-017-0446-6

Valleron AJ, Cori A, Valtat S, Meurisse S et al (2010) Transmissibility and geographic spread of the 1889 influenza pandemic. Proc Natl Acad Sci U S A 107(19):8778–8781. https://doi.org/10.1073/pnas.1000886107

Viboud C, Grais RF, Lafont BAP, Miller MA et al (2005) Multinational impact of the 1968 Hong Kong influenza pandemic: evidence for a smoldering pandemic. J Infect Dis 192(2):233–248. https://doi.org/10.1086/431150

Vink MA, Christoffel M, Bootsma J, Wallinga J (2014) Systematic reviews and meta- and pooled analyses serial intervals of respiratory infectious diseases: a systematic review and analysis. Am J Epidemiol 180(9):865–875. https://doi.org/10.1093/aje/kwu209

Vogels CBF, Hartemink N, Koenraadt CJM (2017) Modelling West Nile virus transmission risk in Europe: effect of temperature and mosquito biotypes on the basic reproduction number. Sci Rep 7(1). https://doi.org/10.1038/s41598-017-05185-4

Wagner DM, Klunk J, Harbeck M, Devault A et al (2014) Yersinia pestis and the Plague of Justinian 541-543AD: a genomic analysis. Lancet Infect Dis 14(4):319–326. https://doi.org/10.1016/S1473-3099(13)70323-2

Walker DG (2016) Viral infections. In: Infections and tropical medicine. Elsevier B.V., Amsterdam

Wallinga J, Teunis P (2004) Different epidemic curves for severe acute respiratory syndrome reveal similar impacts of control measures. Am J Epidemiol 160(6):509–516. https://doi.org/10.1093/aje/kwh255
Walther BA, Ewald PW (2004) Pathogen survival in the external environment and the evolution of virulence. Biol Rev Camb Philos Soc 79(4):849–869. https://doi.org/10.1017/S1464793104006475
Wan XF (2012) Lessons from emergence of A/Goose/Guangdong/1996-Like H5N1 highly pathogenic avian influenza viruses and recent influenza surveillance efforts in Southern China. Zoonoses Public Health 59:32–42. https://doi.org/10.1111/j.1863-2378.2012.01497.x
Weber DJ, Fischer WA, Wohl DA, Rutala WA (2015) Protecting healthcare personnel from acquiring ebola virus disease. Infect Control Hosp Epidemiol 36(10):1229. https://doi.org/10.1017/ice.2015.205
Webster RG (2004) Wet markets—a continuing source of severe acute respiratory syndrome and influenza? Lancet 363(9404):234–236
Weill FX, Domman D, Njamkepo E, Almesbahi AA et al (2019) Genomic insights into the 2016–2017 cholera epidemic in Yemen. Nature 565(7738):230–233. https://doi.org/10.1038/s41586-018-0818-3
Welford M (2018) Geographies of plague pandemics: the spatial-temporal behavior of plague to the modern day. Routledge, New York. https://doi.org/10.4324/9781315307435
WHO (1999) Influenza pandemic preparedness plan: the role of WHO and guidelines for national and regional planning. WHO, Geneva
WHO (2003) The world health report 2003: shaping the future. ISBN 92 4 156243 9. ISSN 1020-3311 [Online]. Available from https://www.who.int/whr/2003/en/whr03_en.pdf?ua=1
WHO (2004a) Report of the WHO/FAO/OIE joint consultation on emerging zoonotic diseases. Available from https://apps.who.int/iris/bitstream/handle/10665/68899/WHO_CDS_CPE_ZFK_2004.9.pdf
WHO (2004b) WHO | Summary of probable SARS cases with onset of illness from 1 November 2002 to 31 July 2003. World Health Organisation
WHO (2005) WHO global influenza preparedness plan: the role of WHO and recommendations for national measures before and during pandemics [Online]. Available from https://apps.who.int/iris/bitstream/handle/10665/68998/WHO_CDS_CSR_GIP_2005.5.pdf
WHO (2010a) Clinical aspects of pandemic 2009 influenza A (H1N1) virus infection. N Engl J Med 362:1708–1719. https://doi.org/10.1056/NEJMra1000449
WHO (2010b) WHO | What is a pandemic? World Health Organisation. Available from https://www.who.int/csr/disease/swineflu/frequently_asked_questions/pandemic/en/. Accessed 18 May 2020
WHO (2015) Emergency guideline: implementation and management of contact tracing for Ebola virus disease. Available from https://apps.who.int/iris/bitstream/handle/10665/185258/WHO_EVD_Guidance_Contact_15.1_eng.pdf
WHO (2019a) Progress report on HIV, viral hepatitis and sexually transmitted infections 2019. Available from https://apps.who.int/iris/bitstream/handle/10665/324797/WHO-CDS-HIV-19.7-eng.pdf?ua=1
WHO (2019b) WHO | measles - key facts. Available from https://www.who.int/news-room/fact-sheets/detail/measles. Accessed 26 May 2020
WHO (2020a) Cumulative number of confirmed human cases for avian influenza A(H5N1) reported to WHO. World Health Organisation. Available from https://www.who.int/influenza/human_animal_interface/2020_01_20_tableH5N1.pdf?ua=1. Accessed 29 May 2020
WHO (2020b) MERS situation update, January 2020. World Health Organisation. Available from https://www.emro.who.int/health-topics/mers-cov/mers-outbreaks.html. Accessed: 21 May 2020
Wild S, Benedictow OJ, Benedictow OJ (2013) Yersinia pestis, the bacterium of plague, arose in East Asia. did it spread westwards via the silk roads, the Chinese maritime expeditions of Zheng He or over the vast Eurasian populations of sylvatic (wild) rodents? J Asian Hist 47(1):1–31

Wolfe ND, Dunavan CP, Diamond J (2007) Origins of major human infectious diseases. Nature 447 (7142):279–283. https://doi.org/10.1038/nature05775

Wong JY, Wu P, Nishiura H, Goldstein E et al (2013) Infection fatality risk of the pandemic a (H1N1)2009 virus in Hong Kong. Am J Epidemiol 177(8):834–840. https://doi.org/10.1093/aje/kws314

Woo PC, Lau SK, Yuen K (2006) Infectious diseases emerging from Chinese wet-markets: zoonotic origins of severe respiratory viral infections. Curr Opin Infect Dis 19(5):401–407. https://doi.org/10.1097/01.qco.0000244043.08264.fc

Worobey M, Han GZ, Rambaut A (2014) Genesis and pathogenesis of the 1918 pandemic H1N1 influenza a virus. Proc Natl Acad Sci U S A 111(22):8107–8112. https://doi.org/10.1073/pnas.1324197111

Worobey M, Cox J, Gill D (2019) The origins of the great pandemic. Evol Med Public Health 2019 (1):18–25. https://doi.org/10.1093/emph/eoz001

Wu JT, Cowling BJ, Lau EHY, Ip DKM et al (2010) School closure and mitigation of pandemic (H1N1) 2009, Hong Kong. Emerg Infect Dis 16(3):538–541. https://doi.org/10.3201/eid1603.091216

Wyżgowski P, Rosiek A, Grzela T, Leksowski K (2016) Occupational HIV risk for health care workers: risk factor and the risk of infection in the course of professional activities. Ther Clin Risk Manage 12:989–994. https://doi.org/10.2147/TCRM.S104942

Ye Z-W, Yuan S, Yuen K-S, Fung S-Y et al (2020) Zoonotic origins of human coronaviruses. Int J Biol Sci 16(10):1686–1697. https://doi.org/10.7150/ijbs.45472

Yip PSF, Lam KF, Lau EHY, Chau PH et al (2005) A comparison study of realtime fatality rates: Severe acute respiratory syndrome in Hong Kong, Singapore, Taiwan, Toronto and Beijing, China. J R Stat Soc A Stat Soc 168(1):233–243. https://doi.org/10.1111/j.1467-985X.2004.00345.x

Yue RPH, Lee HF, Wu CYH (2017) Trade routes and plague transmission in pre-industrial Europe. Sci Rep 7:12973. https://doi.org/10.1038/s41598-017-13,481-2

Zimbler DL, Schroeder JA, Eddy JL, Lathem WW (2015) Early emergence of Yersinia pestis as a severe respiratory pathogen. Nat Commun 6:7487. https://doi.org/10.1038/ncomms8487

Chapter 3
Bio-Diversity, Ecosystem-Health and Their Relation with Pandemic

Rama Umesh Pandey, Shruti Nair S. Muralee, and Juhi Sah

Abstract Along with climate change, another foremost concern that can debilitate our lives and impact humankind as a whole happens to be the spread of Infectious zoonotic diseases. Unfortunately, epidemics arising out of such diseases have been on the rise over time. One of the major causes of the outbreak of these diseases is the degradation of forests due to the loss of biodiversity and the pristine ecosystem. The land use and land cover (LULC) changes within and around the forest due to anthropogenic pressures are disturbing the sustenance and resilient capacity of the ecosystem, resulting in loss of habitat for the animals. This chapter highlights few such concerns like deteriorating man and environment relationship leading to forest degradation; the rising zoonotic disease outbreaks and its relationship with land use and land cover changes; and the role of forest plantation in the degradation of forest ecosystem health. These concerns are further analyzed through a case study of Wayanad district known for its rich evergreen and deciduous forest in the State of Kerala, India. The chapter concludes with the need for recognition of establishing regional priorities with the identification of hot spot areas, where several of the drivers of emerging zoonotic diseases are present, with strengthened afforestation and suitable LULC change policies and surveillance systems.

3.1 Introduction

It often takes a crisis to bring about fundamental changes. The current crisis triggered by COVID-19 too may bring about such a change for the larger benefit of society. At a certain level, this pandemic and its aftermath appear to be an outcome of many

R. U. Pandey (✉) · S. N. S. Muralee
School of Planning and Architecture, Bhopal, India
e-mail: ramapandey@spabhopal.ac.in

J. Sah
Ministry of Housing and Urban Affairs, New Delhi, India

M. K. Goyal, A. K. Gupta (eds.), *Integrated Risk of Pandemic: Covid-19 Impacts, Resilience and Recommendations*, Disaster Resilience and Green Growth,
https://doi.org/10.1007/978-981-15-7679-9_3

existing practices, notions, and beliefs covering a whole gamut of social and economic factors prevalent so far. Scientists now are discussing not just the animal to human transmission but about drug-resistant diseases that may get transmitted from plants to humans (Kannan 2020). As such, it is apprehended that the frequency of epidemics in the future may increase. One crucial aspect in this context would be as to how we are managing our resources, especially endowed to us by nature. Amongst others, the policies for forest land management and approach towards maintaining the ecosystem health both in a forest as well as in the transitional areas therefore would have to undergo major transformations to mitigate the possibilities of a contagion.

The anthropogenically driven changes such as biodiversity loss, land use, and land cover changes and changing climatic conditions are affecting the inherent capacity of pristine ecosystems for both resilience and adaptation. This is aggravating the risks of contracting diseases through enhanced exposure of human beings to wildlife hosting diseases that can be transferred to human beings (UNEP 2016). Maintaining biodiversity can protect us from the risks of Pandemics as the high host species diversity dilutes the contact of disease risks (Bett 2020).

This chapter explores the relationship between biodiversity and ecosystem health with disease outbreaks. An overview of this association over time has been documented to highlight the relationship. A case study of Wayanad district in the State of Kerala, India, where the alteration of land use and land cover in the forest area is considered to be the probable cause of outbreaks of infectious disease. The land use and afforestation policies have been discussed for arriving at strategies to build resilience for restraining disease outbreaks.

3.2 Human-Environment Interaction and Health

Ever wondered what all comes under the definition of NATURAL? Does Environment sound synonymous to Natural? Ever thought WE—"the civilized living being" and Environment both belong to the same category of NATURAL? These questions may form the basis of discussing the evolution of the Human–Environment/ Environment–Human Interaction. Yes, the chronology of the words in the phrase "Human–Environment" is important in the present-day functioning of the ecosystem. An overview of the changes in the interaction of humans with the environment and its effect on the trade-off decisions of rational living beings is presented in the following section.

Man-Environment interaction can be dated back to the era of the Stone Age, where Early Man learned to use elements of the environment for his survival. To establish the fact of evolution in literary society, it started with the determinism of geographical factors in the 1700s followed by the theory of Evolution by Natural Choices by Charles Darwin in 1859. The geographic factors explained the human-environment relationship, establishing the fact that nature forces overpower humans. It depicted that humans adjusted to living in harmony with nature and its seasonal events (Turner 2015). Five years after Darwin's theory an inspirational text was published by the name of—Man and Nature or, Physical Geography as Modified by Human Action (Marsh 1864), which challenged the prevailing discourse of overpowering environmental forces over society. George Perkin Marsh in his book warned American ways of invading the ecological balance with the example of Ancient Mediterranean environmental degradation due to human activities such as industries expansion and clearing of forest area, increasing drainage, and soil fertility problems. There is also a reference to climate change due to deforestation activities and their impacts on the characters of vegetation in Greece. Philip Wagner's, The Human Use of Earth: An Examination of the Interactions between Man and His Physical Environment (1960), is one of the first references to study the trajectory of the man's interaction to nature. Man–Environment relationship has been an exploratory topic, given the examples of researchers—Carl O. Sauer, D.N. Livingstone, J. Leighly, R. Hartshorne, S. Hanson, and many more are continuing the studies to warn the society on various facets of the well-being of ecological balance.

Declaration by the United Nation Conference held at Stockholm in June 1972, on Human Environment, stated the concern around the man's activities that poses an incurable effect on air water and land quality and imbalance in ecology. It also stated that the human environment, ecological balance, economic development, and well-being are all weaved together and have reactive effects on each other. To bind the intents of the declaration to the objectives of this chapter, one of the related principles is stated here—"Man has a special responsibility to safeguard and wisely manage the heritage of wildlife and its habitat, which are now gravely imperiled by a combination of adverse factors. Nature conservation, including wildlife, must, therefore, receive importance in planning for economic development" (UNEP, Declaration of the United Nations Conference on the Human Environment 1972). It seems like the present situations (COVID 19 and other environmental calamities) are the real-life nightmare emerging from the problems discussed 48 years back.

Many of the ill effects of the human activities on the environment have been well researched and enlisted but what failed to get limelight until now, is the health impacts on animals and the human population on earth. The Rio Declaration reflects: "Human beings are the center of concern for sustainable development. They are entitled to a healthy and productive life in harmony with nature" (United Nations Conference on Environment and Development 1992). It is imperative to state that health plays a vital role in human ecology. Human Ecology deals with the holistic approach of the man-made environment, interaction with the natural counterparts, and individual responsibility towards the inter-relationships of bio-logical, eco-logical, and economic development. Health Ecology is the interdisciplinary term, concerned with a comprehensive picture of human health at the individual, societal, and global levels (Honari and Boleyn 2005). The key issues of the health ecology are—multi-dimensional characteristics of environmental forces, human interaction with the environment for development activities, non-recognition of sustainable limits to natural resource usage, non-recognition of health as a core element in public welfare and community participation domain and no mechanism for risk identification and its management (Hunarī 1999).

It is evident that due to the increasing population, unplanned urbanization and expansion of the settlements into the sacred habitats of the animals are exposing human civilization to the unforeseen consequences, as Global warming, Pandemics, and recurring Natural Disasters. Increased rate of deforestation and motorized traffic congestion in the city centers has given rise to the worst air quality, which is the cause of respiratory diseases in humans as well as in animals. Environmental damage is responsible for 2–6% of the total burden of disease in OECD countries (OECD 2001). One-fourth of the global burden of disease as per World Health Organization (WHO) is due to the modifiable factors in air, water, soil, and food wherein one-third of the burden is childhood diseases (Prüss-Üstün and Corvalán 2006). The case of Arsenic contamination of Groundwater which is used as drinking water, in Bangladesh has gained the attention of WHO since 1990. The US National Research Council has noted that as many as one in a hundred additional cancer deaths could be expected from a lifetime exposure to drinking-water containing 50 μg/L of arsenic. The exposure to the inorganic chemicals is increased because of the insensible industrial waste disposal and the use of pesticides in the agriculture sector. Lead and Mercury poisoning have presented its devastating effects on fauna which then propagated to human beings through bio-magnification in the food chain process. All the health impacts on the humans and animals are being monitored and studied but the argument here is, human civilization is still not able to address the cause of these life-endangering issues. The irony is that humans are responsible for their problems and still not able to curb the situation because Mankind has become a slave to the aspirations of becoming a new economic power. All economic activity ultimately depends on services provided by nature, making it an immensely valuable component of a nation's wealth. It's estimated that, globally, nature provides services worth around US$125 trillion a year (WWF 2018). All gets pinned down to the fight for Environment Protection and Habitat Conservation, where the decision making and policy initiative needs to be more eco-centric. Human decision-making is considered a key factor in the relationship assessment model, as humans can regulate and control the type of interaction within Human-Environment Systems (Scholz and Binder 2003).

3.3 Biodiversity and Ecosystem Health

Biodiversity and Ecosystem both direct towards the importance of maintaining the balance in Habitat and Biogeochemical cycles of the earth. The biogeochemical cycle is the path by which the substances move between biotic and abiotic compartments of the earth such as the Water cycle, Carbon cycle, Nitrogen cycle, Oxygen cycle, and Phosphorous cycle, etc. Food chain and Food Web are the nutrient cycle where the energy, nutrients / harmful chemicals climb up in the successive living being systems. When certain nutrients or heavy metal gets accumulated in the food chain in the upper tiers (known as bio-magnification) causing dis-balance in the nutrient cycle and affecting the health of all. MeHg (Methyl Mercury) concentrations get magnify through the food web, often increasing two to five folds across trophic levels (such as piscivorous fish) and are potentially dangerous to consumers (Watras et al. 1998). In the United States, fishes from lots of inland streams and lakes have been restricted for consumption due to elevated levels of mercury found in the water through accumulation from industrial effluents (Ward et al. 2010).

In the 13th Conference of Parties to the Convention on Migratory Species, Gandhinagar, it was found that there is a decrease in the number of raptors, migratory shorebirds, gulls, and terns, forest and grassland specialists, long and short distance migrants, Western Ghats endemics, and carnivores. Peafowls have increased in abundance over the past few decades. Research attributes these losses in biodiversity are because of the degradation of the habitats. As a member of the ecosystem, birds play the supporting, provisioning, regulating, and cultural roles in the service delivery system. Birds act as pollinators, seed dispersers, predators, and scavengers. Habitat loss, invading species, and climate change (Human interference) are resulting in the loss of certain species of birds which in-turn is affecting the food web thereby giving rise to uncertain disasters. To quote, the recent disaster of Locust attacks on the farmland of the Indian subcontinent is one direct effect of degraded ecosystem health. These locust fleets that attacked the states of Rajasthan, Madhya Pradesh, Maharashtra, and Uttar Pradesh in May 2020, traveled through Pakistan into the Indian states. The huge number of locusts is due to the loss of birds that feed on these locusts. Thus, the locust reproduction rate increased manifolds giving rise to the migratory tendency in the fleet to find food. Due to the invasion by the locust fleet, Indian farmers have lost the present year's cereal produces, affecting the livelihood of farmers and soil health directly. Early reports from the United States Department of Agriculture (USDA) Division of Ornithology and Mam-Malogy on food habits of the common hawks and owls of the United States, helped to change the negative perception: most hawks and owls were far more helpful than injurious to the farmer or "poulterer" (Whelan et al. 2008).

May 2020 reported the death of 300 Bats in the Gorakhpur district of Uttar Pradesh. Indian Veterinary Research Institute on autopsy assessed the cause of death to be the heatstroke and consumption of insecticides. Bats are a very important link in the food web, as prey and predator, arthropod suppression, seed dispersal, pollination, material, and nutrient distribution, and recycle (Kunz et al. 2011). Such loss of Biodiversity is also the result of the climate change phenomenon. Lots of inconsistency is observed in the migration patterns of the migratory birds and other animal species, due to seasonal shift experienced over 50 years. Such happenings are affecting the ecological balance in nature and can be attributed to the increased anthropogenic activities around the sacred habitats of plants and animals' species.

3.4 Zoonotic Diseases and Their Relationship with Biodiversity

The first five global risks as per the World Economic Forum Global Risks Report of 2020 are from the environmental domain wherein the fourth risk mentioned is 'biodiversity losses'. Humans are responsible for the loss of half of the plants and 83% of all wild animals even though they just account for 0.01% of total living biomass (Quinney 2020). The forests are being opened up for mining and other developmental activities resulting in fragmentation and degradation of forest ecosystems. Due to the fragmentation of the forest, wild animals live close to each other providing ideal conditions for the spread of diseases (Quinney 2020). Increasing urbanization leading to land use and land cover changes brings settlements close to the forest and exposes human beings to wild animals. Deforestation thus decreases the human's proximity to wild animals making them vulnerable to infectious diseases (Scott 2020). Seventy percent of the recognized Emerging Infectious Diseases (EIDs) are, zoonotic, i.e. transmitted between animals and humans (Scott 2020). The ancestry of the pathogens causing these diseases can usually be traced to wildlife (Wilcox and Ellis 2006). Climate change and biodiversity loss are among the most striking cases of disease outbreaks (Scott 2020).

The existing ecosystem boundaries get reshaped due to forest clearance and this modifies the ecosystem dynamics leading to new breeding habitats for disease vectors, such as mosquitoes, fleas, and ticks (da Costa et al. 2015). Such boundaries are often sites of contact between humans and forest pathogens (Gottwalt 2013). There is a well-documented, positive association between the increased deforestation of an area and the emergence of zoonotic, vector-borne diseases (Sheela et al. 2017). Populations living within or near these fragmented forests are at a much higher risk of infection due to increased contact with vectors at forest edges and the reduced biodiversity of the area (Wangdi et al. 2015). As per Scott (2020) within 23 years (1980–2013) 12,012 disease outbreaks involving 44 million people were recorded worldwide. Thirty-one percent of the outbreaks (*Ebola*, *Zika*, and *Nipah* viruses) are linked with deforestations (Scott 2020).

Zoonotic diseases are often categorized according to their route of transmission (i.e. vector-borne or foodborne), pathogen type (i.e. *microparasites, macroparasites, viruses, bacteria, protozoa, worms, ticks, or fleas*), or degree of person-to-person transmissibility (Karesh et al. 2012). Pathogens shared with wild or domestic animals cause more than 60% of infectious diseases in man. Such pathogens and diseases include *leptospirosis, cysticercosis, echinococcosis, toxoplasmosis, anthrax, brucellosis, rabies, Q fever, Chagas disease, type A influenzas, Rift Valley fever, severe acute respiratory syndrome (SARS), Ebola hemorrhagic fever, and the original emergence of HIV*. The historical overview of zoonotic disease occurrences and the causes of the emergence of their hosts are recorded in Table 3.1. This suggests that the causes of disease emergence are linked with deforestation; climate change; hunting and butchering; wildlife trade for human consumption; habitat destruction; land use and land cover changes for anthropogenic purposes; forest

Table 3.1 Linkages of zoonotic diseases events with biodiversity and ecosystem change (Wilcox and Ellis 2006; Patz and Confalonieri 2005; Myers and Patz 2011; McFarlane et al. 2013; Ghanbari and Gómez-Aguilar 2019)

Disease	Year	Host	Emergence mechanism
Yellow fever	1900	Vector non-human primates	Deforestation and expansion of settlements along forest and hunting water and wood collection, domestication of vectors and pathogens
Ebola	1977	Direct, non-human primates, bats	Hunting and butchering, Logging, outbreaks along forest fringes. Agriculture, alteration of natural fauna
Nipah virus/ Hendra	2001	Direct, bats, pigs	Pig and fruit production on forest border, habitat destruction/fragmentation of bats, Industrial food production, climate abnormalities
SARS-CoV	2003	Direct, bats, civets	Harvesting, marketing, and mixing of bats and civet cats. Wildlife trade for human consumption, intensive livestock-related work i.e. mixing wild and domestic animals
SIV		Direct, non-human primates	Deforestation and human expansion into the forest, hunting, and butchering of forest wildlife
Rabies, Australian bat lyssavirus	1995	Direct canines, bats other wildlife	Human expansion of forest, deforestation, mining
Rocky mountain spotted fever	1901	Invertebrate ticks	Human expansion of forest and forest recreation
HIV	1930	Direct, wildlife, NON-human primates	Scientists believe that HIV originally came from a virus particular to chimpanzees in West Africa, bushmeat hunting, forest encroachment, human behaviour
Hantavirus	1977	Vector	Climate variability and variations in population density of natural food sources
Trypanosomiasis	1998	Vector	Habitat alteration, deforestation
Lyme diseases	1975	Vector, humans, deer. Mice	Biodiversity loss, depletion of predators, Possible Association with deforestation and habitat fragmentation. Forest workers at increased risk of disease
Chagas diseases	1909	Indirect, protozoan parasite	Habitat alteration, deforestation, Urban sprawl, and encroachments
Leishmaniasis	1862	Vector, Numerous mammals	Human expansion into the forest, Domestication of zoophilic vectors, Habitat alteration, habitation building near forest, deforestation, agriculture development
Babesiosis	1966	Vector, humans, wildlife	The disease often found among ticks in forested areas
Avian influenza (h5N1)	2004	Indirect/direct, avian origin	Wild waterfowl and can spread into domestic poultry, such as chickens, turkeys, ducks, and geese. The disease is transmitted via contact with an infected bird's faeces, or secretions
Leptospirosis	1999	Indirect, rodents	Watershed alteration and flooding, agriculture development, urban sprawl

(continued)

Table 3.1 (continued)

Disease	Year	Host	Emergence mechanism
Monkey fever	1957	Indirect, direct, ticks, monkeys	Forest degradation, human expansion to the forest, forest-dependent livelihood practices
Spanish flu	1918	Indirect/direct, avian origin	Poultry farming, climate variations, habitat fragmentation
Echinococcus multilocular	1948	Direct, rodents foxes, small mammals	Deforestation, increase in rodent and fox hosts, pathogen spillover to dogs. Human expansion into the forest, exposure to the susceptible population

recreation; etc. The policies and regulations for the management of forest and the transitional areas along with effective implementation mechanisms become important to curb the emergence of zoonotic diseases.

3.5 Ecosystem Health and Its Linkage with Disease Outbreaks: A Case Study of Wayanad District in Kerala, India

Land-Use and Land Cover Change (LULC) is considered as a significant driver of Emerging Infectious Diseases (EIDs) (McFarlane et al. 2013). LULC triggers modification in natural settings causing changes in the ecology of hosts, vectors, and pathogens, changes in the lifecycle, and exposure to various infections (Sheela et al. 2017). These changes in the host and pathogens are influenced by the changes in land cover mostly due to anthropogenic activities occurring at varied spatial-temporal scales; and vary as per the changes in the interactions of biotic and abiotic elements of the particular ecosystem (McFarlane et al. 2013). The largest environmental cause of altered risk for infectious diseases is likely to be the disturbance of habitats due to land cover change (Patz et al. 2008).

To understand the relation between spatial-temporal changes in LULC and disease outbreaks, a study has been conducted in the Wayanad district of Kerala state, India. The role of plantations in causing forest degradation and its link with infectious diseases is further investigated through regression analysis. This case study thus able to present the effect of LULC changes, on the emergence of infectious diseases in a particular geographic region. The analysis does not include the scientific explanations or reasoning between the relationship of LULC change and disease outbreaks but tries to exhibit a possibility of occurrence of such EIDs due to the changes in forest ecosystem through data analysis and logical reasoning.

3.5.1 A Scenario of LULC Change in Wayanad

The district is located in the highland region (11.6854° N, 76.1320° E) as such, its land-use pattern and climate show a significant difference from all other districts of Kerala in Southern India. The land use pattern clearly shows that Wayanad is an agrarian district with a large forest area and resources such as agricultural land and forest (Government of Kerala 2014).

The entire Wayanad district can be partitioned into four activity zones Non-Agricultural (Urban), Agricultural, Plantation, and forest land use. The portion under agricultural land use is 10.94%, the plantation is 39.02% and, forest land use is 39.26% (Government of Kerala 2014). The land use concentration pattern shows that most of the plantation/forest activities are concentrated on the outer regions of urban areas and forests/woods areas. The urbanised areas are located along the major transportation corridors whereas agricultural activities are predominantly concentrated in the central region of the district (Government of Kerala 2014).

The LULC of Wayanad district was mapped using the Survey of India topographical sheets of 1950 whereas the LULC for the year 1982 was adapted from the vegetation map prepared by the French Institute, Pondicherry in 1982 (Easa and Sankar 2001). The change in LULC from 1950 to 1982 is reported in Table 3.2. The forest cover has been reduced by 1086 km^2 in 30 years with the corresponding increase in area under plantations and cultivation (Fig. 3.1).

Temporal Land Use Land Cover Change from 2011 to 2018 is analyzed for the recent changes in 16 land use classifications and its spatial location concerning forest. Land use Land cover (LULC) map of Wayanad district for the years 2011 and 2018 are generated from Bhuvan using Google Earth as a base map by applying the image interpretation technique (Figs. 3.2 and 3.3). The percentage change in land use from 2011 to 2018 as reported in Table 3.3, indicates agriculture plantation has increased from 2011 to 2018 whereas cropland has seen a slight decrease. All Mining area present in the year 2011 is converted into Barren rocky land Cover by the year 2018 as mining activities are now banned.

There is a rapid increase in the built-up area (16% rural and 58% urban) in the region. The plantation, scrubland, scrub forest, evergreen forest, deciduous forest land cover in 2011 got converted into rural land use in 2018. This indicates the land cover change is progressive towards the forest region due to increasing settlements. The area under evergreen and the deciduous forest has decreased (as highlighted in

Table 3.2 Land cover of Wayanad district (area in sq.km) (1950-SOI topographical maps 49M 13 & 14 1982-Vegetation map prepared by the French Institute of Pondicherry) (Easa and Sankar 2001)

Land cover type	1950	1982	Difference
Forest	1811.35	724.54	−1086.81
Agricultural Plantations	63.93	532.75	462.82
Cultivation	255.72	873.71	617.99
Total	2131.00	2131.00	

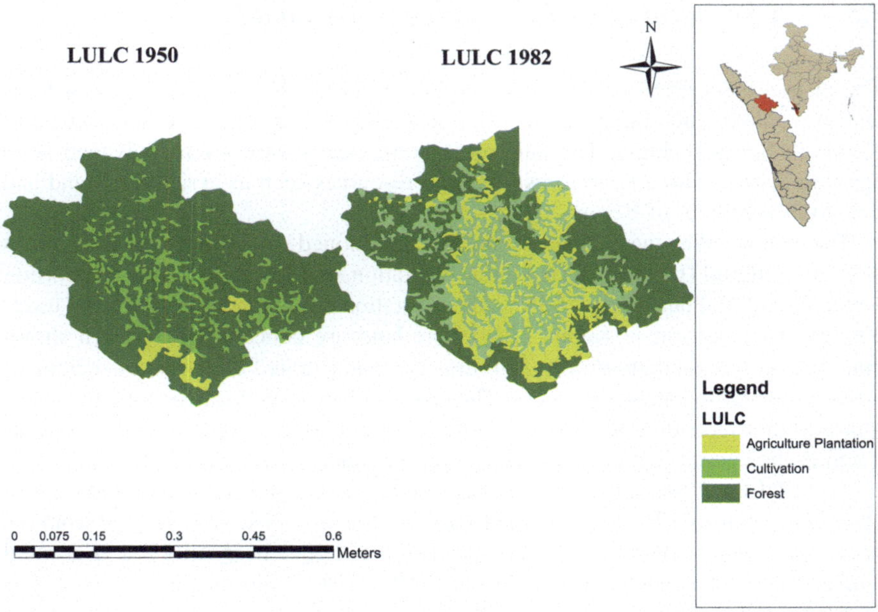

Fig. 3.1 LULC of Wayanad in the year 1950 and 1982 (Easa and Sankar 2001)

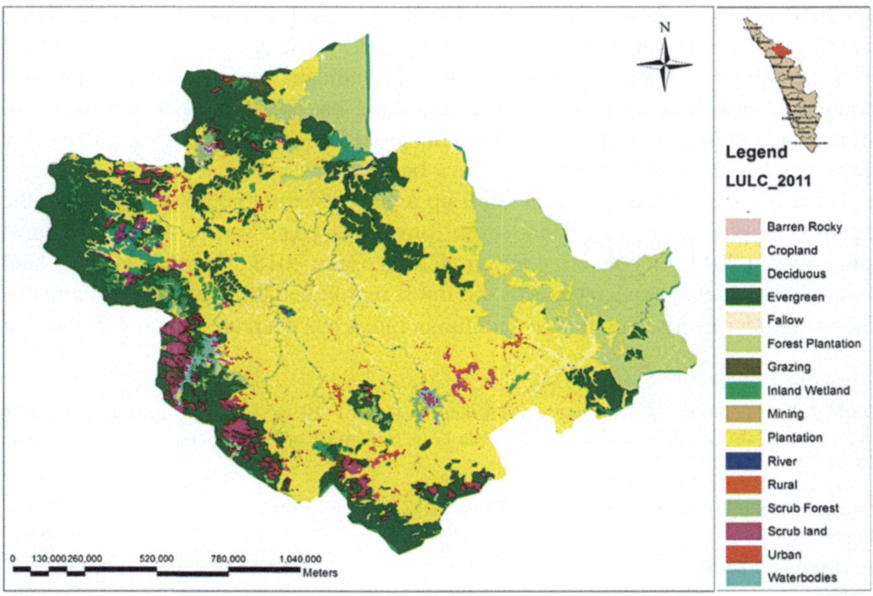

Fig. 3.2 LULC Wayanad year 2011 (Authors)

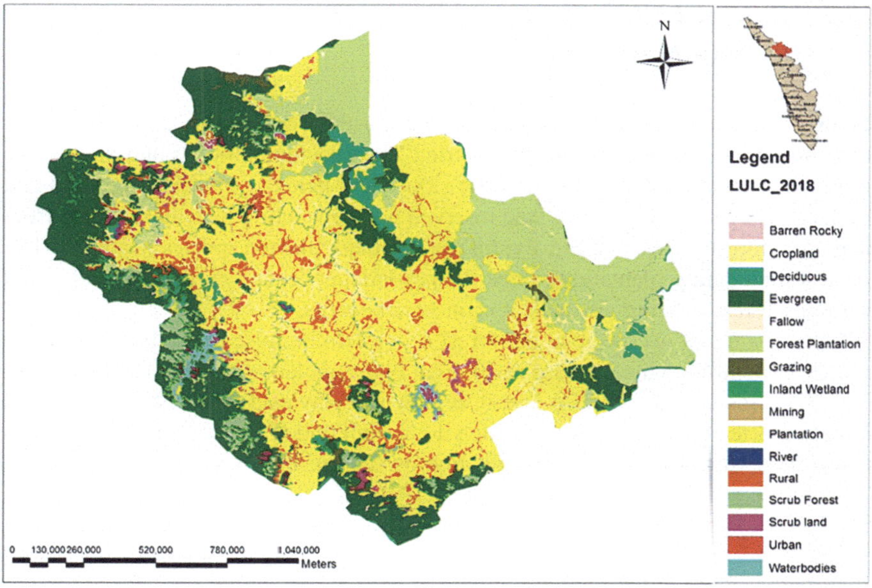

Fig. 3.3 LULC Wayanad year 2018 (Authors)

Table 3.3 Percentage of temporal change from 2011 to 2018 (Authors)

LULC categories	LULC classes	2011 area	2018 area	% Change from 2011 to 2018
Agriculture	Cropland	201.76	200.22	−0.76
	Plantation	868.52	899.67	3.59
Barren	Barren Rocky	2.96	3.29	11.15
	Fallow	3.11	0.37	−88.10
	Scrub Land	70.17	28.55	−59.31
Builtup	Mining	0.06	0	−100.00
	Rural	103.12	120.08	16.45
	Urban	2.95	4.65	57.63
Forest	Deciduous	80	61	−23.75
	Evergreen	580.36	323.32	−44.29
	Forest Plantation	230.64	497.23	115.59
	Scrub Forest	48.66	60.41	24.15
	Grazing	8.32	11.1	33.41
Waterbodies	Inland Wetland	0.24	0.02	−91.67
	River	11.6	10.9	−6.03
	Waterbodies	14.02	5.95	−57.56
Total		2226.49	2226.76	

Table 3.3) due to the increase of forest plantations areas. The land under deciduous forest, scrub forest and evergreen forest cover in 2011 is converted into forest plantation in 2018.

3.5.2 *Role of LULC Change in Forest Degradation of Wayanad*

The evergreen and semi-evergreen forests are mostly located in the mountain zones in the north and western edge of the district. The deciduous forests and plantations on the other hand are located in delicate slopes and fields of the eastern region and also have a huge chunk of forest land reserved for the Wildlife Sanctuary. Apart from the natural forests, Teak and Eucalyptus are the major planted species (Gadgil 2011). For accommodating plantations since the pre-independence era, the deciduous and evergreen forest is being cleared for economic benefits.

There is a massive change in the forest ecosystem because of the transformation in forest type through the increase in different types of plantations from 2011 to 2018 (Fig. 3.4). Inside the wildlife forest region of Wayanad, the maximum forest plantation is teak and sandalwood which do not fall under the category of evergreen or deciduous forest typology of vegetation as seen in Fig. 3.5. Thus, harming the

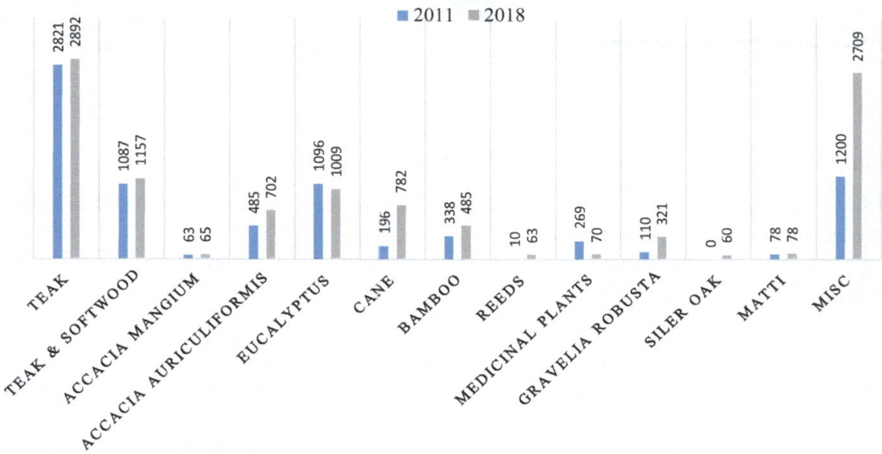

Fig. 3.4 Types of forest plantation in Wayanad in the year 2011–2018 (Statistics Wing-Forest Headquarters 2011, 2018)

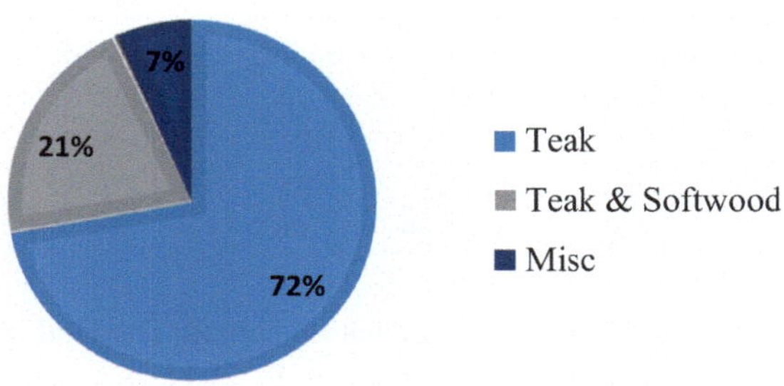

Fig. 3.5 Types of forest plantation in the wildlife area of Wayanad (Statistics Wing-Forest Headquarters 2018)

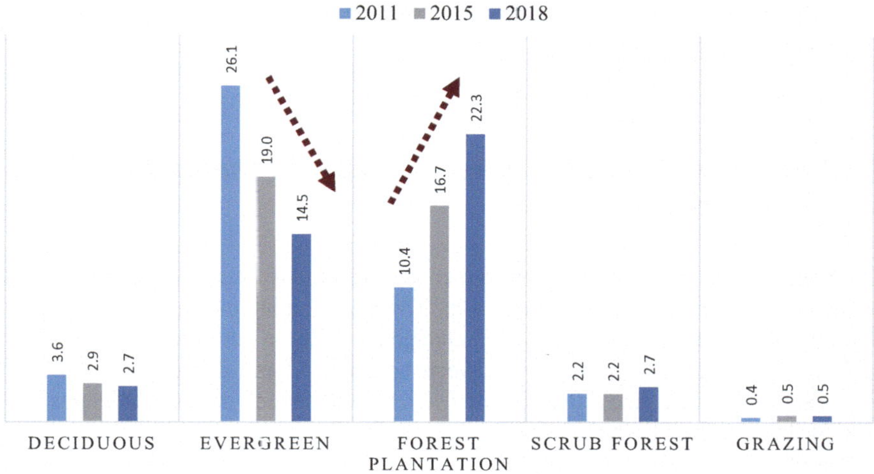

Fig. 3.6 Forest land use land cover change of 2011, 2015 and 2018 (Statistics Wing-Forest Headquarters 2011, 2015, 2018)

ecology of the forest and decreasing resources for the habitats inside the forest. Over the last few decades, extensive natural forests have been replaced with forest plantations. Commercial plantations in the transitional areas further fragment the contiguity of forests. The forest LULC change from 2011, 2015, and 2018 as in Fig. 3.6 shows clearly that there has been a decrease in evergreen vegetation typology while there is a drastic increase in forest plantations. This transformation is changing the characteristics of the forest ecosystem.

The excess nitrogen, acid deposition, and fertilizer application are changing the pace of nutrients input. This additionally prompts adjustments in soil properties, supplement cycling, and microbial network synthesis. The supplement irregularity and nutrient imbalances adversely affect forest ecosystem health and make vegetation vulnerable to other outside stressors, for example, diseases and proliferation of obtrusive species (Dreiss and Volin 2014). The monoculture tree ranches don't help maintain biological diversity and landscape. Controlled, uniform columns of monoculture estates are something contrary to evergreen forests.

The continuous unsettling influence brought by the short revolution, clear-felling, and herbicide splashing is among the most damaging and restricting variables on biodiversity. Plantations to some extent can give natural surroundings to local species, particularly orchids. Exotic monocultures additionally increment fire hazards and can go about as a wellspring of bugs and pathogens that spread into neighbouring indigenous woodland. Around the world, monocultures are susceptible to pests and diseases. Plantations cause environmental change and increments in UV-B (Ultraviolet-B) light that also demonstrates pieces of evidence of climate change. An increase in future environmental stress will likely be from increased

UV-B concentration though very little is known about the effects of UV-B on forest tree species (Rosoman and Grant 1994).

3.5.3 Relation of LULC Change with Disease Outbreaks in Wayanad

The LULC changes mainly from evergreen and deciduous forest to plantations lead to degradation of forest ecosystem health. Therefore, an attempt has been made to find out the correlation between LULC change and disease outbreaks in the context of Wayanad. A correlation matrix between both LULC and disease outbreaks for 3 years 2011, 2015, and 2018 is analysed to understand the probable positive relationship between both the aspects.

It is clear from Table 3.4 that there is a positive correlation with 'forest plantation area' and 'disease outbreaks' as compared to other land-use changes. Therefore, there is a probability of having more chances of disease outbreaks with increasing forest plantations. To investigate how the loss of biodiversity and degradation of ecosystem health due to land cover and vegetation changes triggers diseases outbreaks, a relationship analysis was carried out between the 'area of forest plantations' and 'disease outbreaks' for a timeline of 8 years (from the year 2011 to 2018).

The relationship between two variables 'plantation area' as an independent variable (X-axis) and 'total no of infectious diseases' as the dependent variable are taken for correlation using scatter with straight-line plotting (as depicted in Table 3.5 and Fig. 3.7) to compare the two points of data keeping the time interval as constant.

The scatter plotting indicates that there is a positive correlation between disease outbreak and plantation areas within the set time interval from 2011 to 2018. The scattered plots connecting both the result at the same points indicate that the occurrence of diseases is directly linked to the total plantation areas. Every year with increasing plantations there is an increase in no of cases and with the decrease in plantation areas, there is a decrease in the number of disease cases. To further test the relationship, regression analysis was carried out in Microsoft Excel to find the strength of the relation between both the variables (Table 3.6).

The regression analysis shows a high correlation between 'plantation areas' and 'Total Infectious Diseases Cases'. The strength of the correlation gets ascertained through (1) Multiple R-value which is 0.81 indicating high co-relationship between the two variables; (2) 'P' value is less than 0.05 indicating the significance of the relationship at 95% confidence level. The results thus indicate a strong positive correlation. It can, therefore, be interpreted that there is a high probability of the occurrence of disease outbreaks depending upon the increasing plantation areas in Wayanad's forest. This further corroborates the findings of literature cited in an earlier section that plantations are affecting the ecosystem of forest regions thus leading to possibilities of such vector-borne disease outbreaks.

Table 3.4 Correlation between diseases and factors of LULC (authors)

LULC	Cropland	Plantation	Barren Rocky	Fallow	Scrub Land	Mining	Rural	Urban
Diseases	0.72	0.87	−0.49	−0.78	−0.71	0.84	0.18	−0.15
LULC	Deciduous	Evergreen	Forest Plantation	Scrub Forest	Grazing	Inland Wetland	River	Water bodies
Diseases	−0.55	−0.39	0.9	−0.27	0.72	−0.62	−0.72	0.32

Table 3.5 Year-wise cases of diseases and plantation area (Statistics Wing-Forest Headquarters 2011, 2012, 2013, 2014, 2015, 2016, 2017, 2018; Directorate of Health Services 2011, 2012, 2013, 2014, 2015, 2016, 2017, 2018)

Year	Total no of infectious diseases (y)	Plantation area (m^2) (x)
2011	111	4955.7
2012	169	4620.36
2013	130	4620.36
2014	159	4620.36
2015	523	5228.73
2016	469	5146.73
2017	644	5511.1
2018	226	5253.31

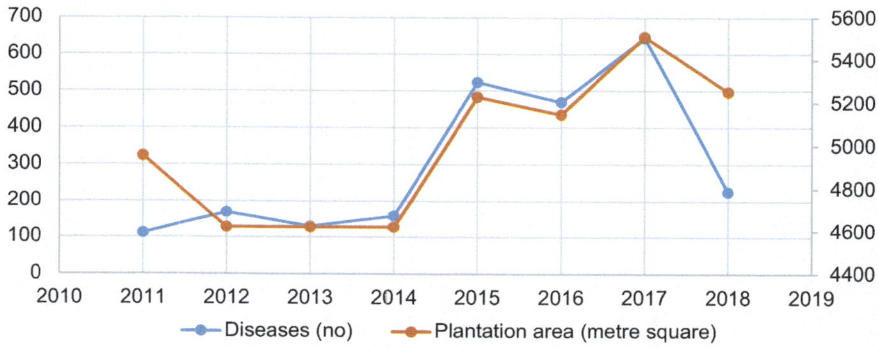

Fig. 3.7 Graph showing plotted relation between total diseases and plantation areas from the year 2011 to 2018 in Wayanad (Authors)

Table 3.6 Regression analysis between total no of diseases and plantation area in Wayanad (Authors)

Variable	Multiple R	R^2	*Coefficients*	P-value
Forest plantation area	0.81	0.66	(+) ve	0.01

3.5.4 Relation of Human-Animal Conflicts and Zoonotic Disease Outbreaks in Wayanad

Wayanad being a victim to both excessive cases of human-animal conflict as well as disease outbreaks, an attempt has been made to find linkages between forest degradation and human-animal conflicts and how it further increases the threat to zoonotic disease outbreaks in the context of Wayanad. The pressure on the natural resources from the industries competing for the resources, demand of the local people along with the fragmentation and degradation of the existing habitat has worsened the issue of human-animal conflict in most of the Protected Areas. The once contiguous and diverse forest ecosystem in the Western Ghats is now deteriorated due to

Table 3.7 Incidences of monkey conflicts reported in the year 2013–2019 (*Data from the Forest department, Wayanad*)

Year	2013	2014	2015	2016	2017	2018	2019
Incidences	300	390	484	430	480	440	300

deforestation. This has further led to the fragmentation of habitat, forming 'islands' affecting the wildlife in general and the larger mammals in particular. These islands are further fragmented/degraded by encroachments, developmental programmes, and the unscientific land use patterns.

Wild animals get strayed into inborn settlements and other local areas when their environments and assets are exhausted. As per authorities, the Protected Areas (PAs) of forest are designated habitat for different wild creatures. In a review, led by the Natural Life Establishment of India in 2016, 29% of tigers and 67% of elephants live outside PAs in India. This leads to tragic events due to the enhanced interaction of humans with wildlife resulting in a conflict. As the populace expanded, the settlements started to interfere with the neighbouring woodlands disturbing the natural living environment. The developmental projects that followed, contributed further to the disintegration of the remaining woods territories (Easa and Sankar 2001).

Kyasanur Forest Disease (KFD), referred to as monkey fever is an irresistible draining infection in monkey and human brought about by a profoundly pathogenic infection called KFD infection. It is an intense *prostrating febrile sickness*, transmitted by infective ticks, particularly, *Haemaphysalis spinigera*. Later KFD infections additionally detailed from sixteen different types of ticks. Rodents, Shrews, Monkeys, and feathered creatures fed upon tick chomp become a store for this infection. The possibility of spreading KFD, therefore, increases in the regions close to the outskirt's zones of monkey's habitat. Wayanad has the enormous populace of monkeys which is the significant host of KFD, co-home of man-tick monkey, and accessibility of various vertebrate hosts to keep up tick populace. The climatic conditions of Wayanad forest are very conducive for the survival of tick plenitude (Easa and Sankar 2001). The high predominance of larval ticks was found in the March–April months under dry leaves of timberland vegetation.

In the context of KFD having monkeys as a source, the study investigates the rising intrusion (conflicts) of monkeys in settlements due to increasing forest plantation in Wayanad. The study is restricted to *Macaca radiata* (the Bonnet Monkey) the species highly susceptible to KFD. The increasing population of monkeys is a major concern in the urban as well as rural areas of Wayanad district. One of the reasons for the rising issue of monkeys intruding into public properties can be the easy availability of resources (food) outside the forest area. The data are shown in Table 3.7, a major case recorded in 2014 and 2015 which were also the years with maximum KFD outbreaks.

The major conflicts as shown in Fig. 3.8, witnessed near the forest regions with more forest plantation areas as well as urban areas. The main concern is that monkeys have accessibility to the entire region, they could become the host carrying

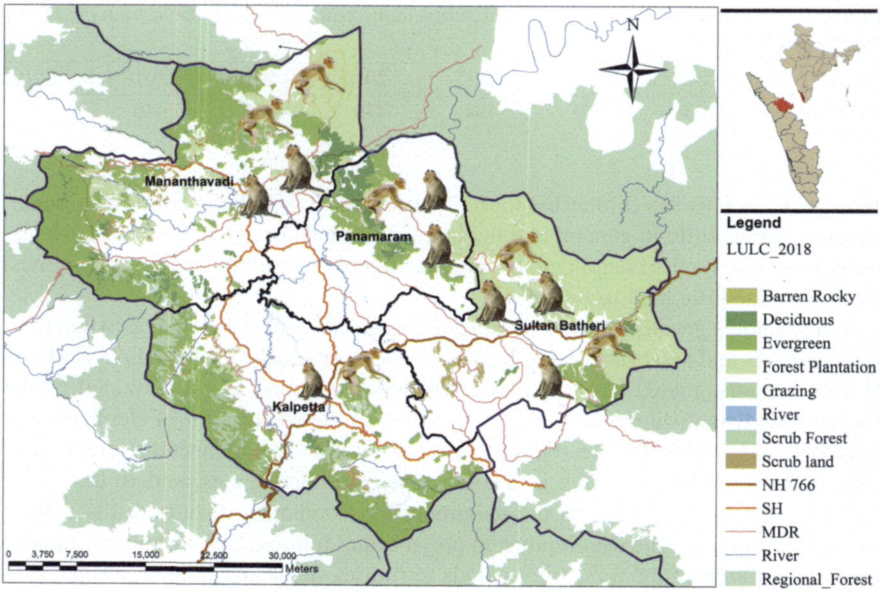

Fig. 3.8 Places with more monkey conflicts as per forest department authorities (*Data from Forest department, Wayanad*)

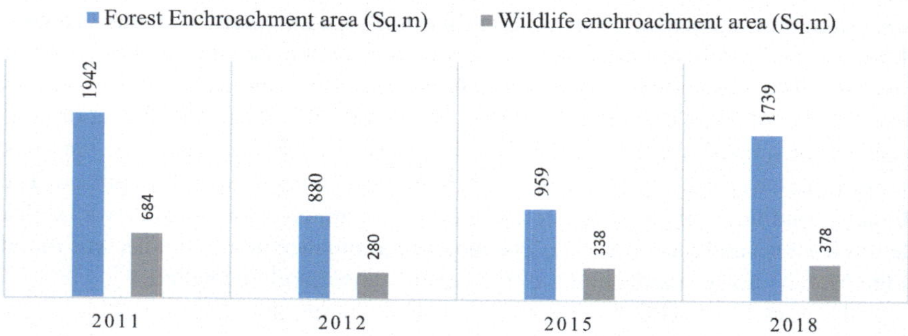

Fig. 3.9 Encroachment area inside the forest of Wayanad from the year 2011 to 2018 (Statistics Wing-Forest Headquarters 2011)

diseases inside the forest region and transmit it to other regions, as contact with human increases.

Wayanad being the land of maximum tribal population and indigenous communities, the encroachments as shown in Fig. 3.9, inside the protected forest land is a constant concern to the authorities. Settlements inside the forest are gradually being relocated and rehabilitated by providing compensations.

Besides settlements, human interventions inside the forest for other economic-based activities are always carried out not only by the local people but also by the

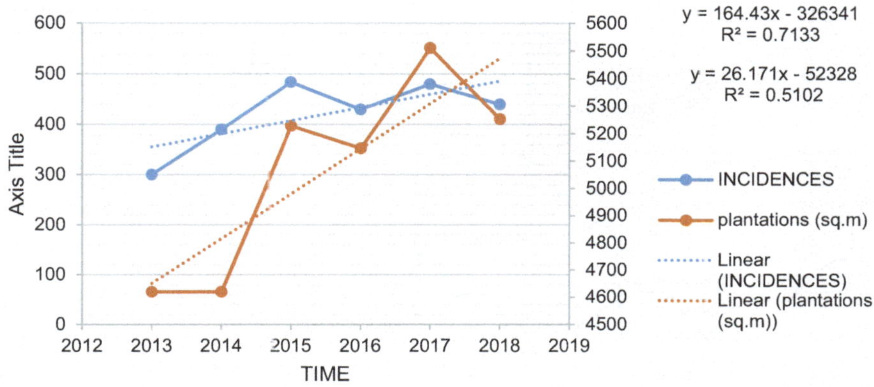

Fig. 3.10 Graph showing the relation between conflict and plantation area in the time interval of 2012–2019 (Authors)

authorities too. Officials of the forest department have a constant concern regarding the relocation of tribal people who start building up their houses inside the forest for their livelihood. Gradually, there is an increase in the number of houses resulting in the formation of an informal settlement. In Wayanad district, it is observed that the organic pattern of settlements has created an unplanned land use planning of the region. This leads to the problem of encroachment in forest areas by the tribal and shaping their spatial organisations causing more illegal intrusion to protected and restricted areas. Thus, such invasion causes habitat disturbances inside the forest ecosystem and increases the occurrence of man-animal conflict.

The frequent disturbances caused by short-rotation, clear-felling, and reestablishment are some of the most destructive and limiting elements of diversity. It prevents the evolution of a range of habitats or any continuity between felling cycles, or any organisms that rely on dead wood. Thus, Fig. 3.10 shows the relation of monkey conflicts statistically correlates with the plantation area positively, explaining that there is a direct relation of increasing plantations with the increasing conflicts in Wayanad.

In an interview, Dr. P. Pugazhendi, Chief Conservator of Forests, highlighted that human-animal conflicts occur due to a lack of resources inside the deep forest. Unavailability of basic needs like water and food forces them to come outside their zone in search of these resources. Agriculture crops and plantation become a major attraction to them.

Plantation forests do not provide enough resources to the wildlife because they are Monoculture plantations (Teak plantations) which are simple systems of one type of tree of the same age, grown at the same distance apart, and clear-felled at the same time. The imbalance caused in the flora and fauna inside the forest ecosystem, directly and indirectly, affect the habitats of this wildlife. Young plantations are particularly poor habitat for many species. The Western Ghats provide ideal topographical and climatological conditions for the vector ticks, thus making these Ghats

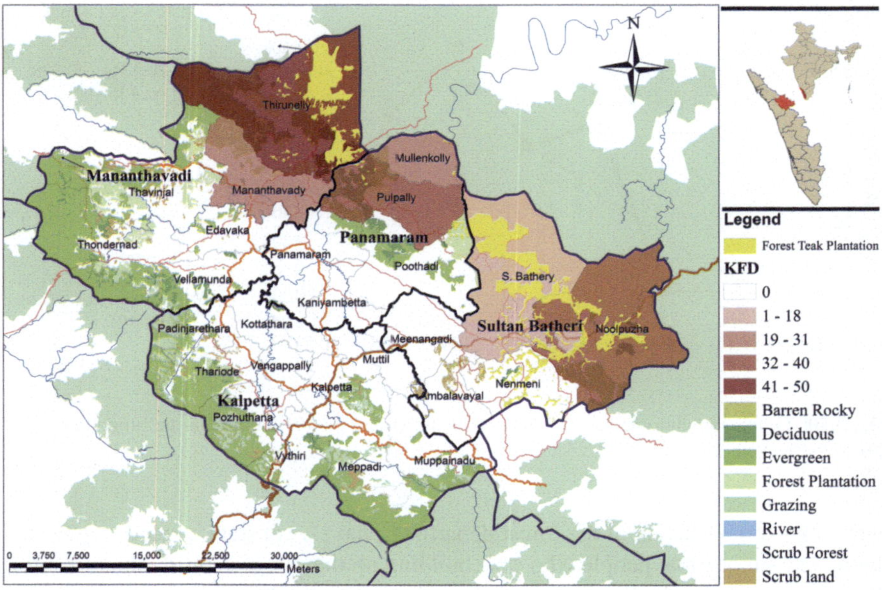

Fig. 3.11 Overlay map of KFD and plantation within the forest boundary

Table 3.8 Correlation analysis

	Teak. Plantation area (m^2)	KFD	Conflicts
Teak. Plantation area (m^2)	1		
KFD	0.705754	1	
Conflicts	0.86937828	0.75474	1

as an epitome for this tick-borne disease. This study is an attempt to further spatially understand the relation of KFD with Plantations and Monkey conflicts to explore the dependency of forest diseases on wildlife and forest ecosystem change. The disease outbreak regions are spatially located and related to factors of forest degradation and its connection with forest area and forest plantations. A correlation between animal conflict, Teak plantations, and disease outbreaks in the context of Wayanad is carried out to find the probability of their dependency on each other.

The analysis hereby concludes that there is a strong relationship between the occurrence KFD with Monkey conflicts as well as the region with more teak plantations near the forest region as highlighted in Fig. 3.11. The animals are also seen in urban areas of Wayanad. But the transmission of diseases started from the forest region to the tribal population with forest dependency. The result as per Table 3.8 shows that in Wayanad, due to more monoculture plantation practices inside the natural forest causes the ecosystem to change resulting in the formation of ticks found in Monkeys. Degraded habitats are likely to carry more viruses that can infect humans. Hence it is a loop as shown in Fig. 3.12 where forest plantation

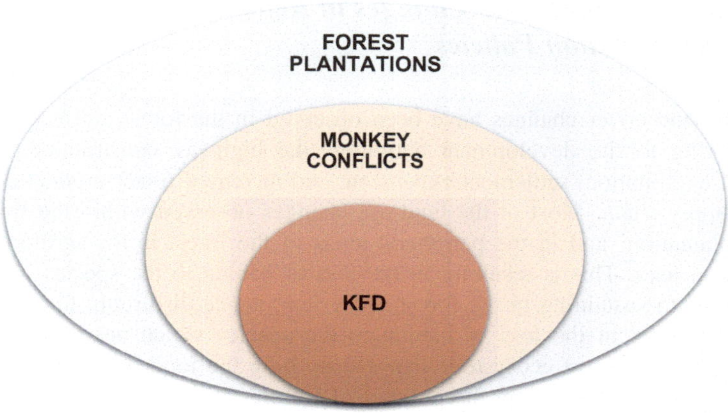

Fig. 3.12 Relationship of forest plantation, conflicts, and diseases (Authors)

causes an increase in the probability of virus-carrying wildlife host which in return gets transmitted to humans because of increasing contact episodes.

No matter, the host be a monkey or any wildlife species, the cause of pathogens and ticks originate from the forest itself due to the ecological imbalances in the ecosystem. Hence, pandemics happen when viruses seek new hosts. The increase in practices like destroying the habitat and organisms leads to more contact with wildlife carrying such viruses in them leading to epidemics and outbreaks at a larger scale. Therefore, contact with wildlife animals with diseases carrying vectors should be controlled.

3.5.4.1 Managing Biodiversity and Ecosystem Health to Improve Resilience

This chapter has discussed the problems being faced today on the environment front. It has been established that the Man–Environment relationship has been about the economy and resultant economic benefits, where the interdependency is being ignored to develop protocols for resiliency, adaptivity, and emergency responses. We are now so much engrossed in managing the after-effects, that planning for a harmonious living (future) is now confined to—*planning for response mechanisms.* Sustainable Goal No. 15—Life on Land, aims at protecting, restoring, and promoting sustainable use of terrestrial ecosystems, sustainably managing forests, combating desertification and halt as well as reverse land degradation and biodiversity loss. Hence, the ecocentric policy framework designed with proper analysis of the tradeoff on the development and economic front is the need of the hour. The resilience to disease outbreaks can be enhanced through various strategies for managing biodiversity and ecosystem health as highlighted hereunder.

3.5.5 Integrating LULC Changes in Reforestation/ Afforestation Policies

Land use land cover changes have been observed in the forest and eco-sensitive areas, owing to the development activities like highway construction, tourism potential exploitation, settlement expansion, and increased forest product demand, in the supply chain. Most of the land-use changes observed within the forest are 'forest plantation' and in the peripheral areas of the forest is for agroforestry or agricultural uses. This is resulting in the loss of habitat to the species which are accustomed to sustaining in the forest ecosystem, hence disturbing the food web. This is resulting in the loss of habitat to the species which are accustomed to sustaining in the forest ecosystem, hence disturbing the food web. So, to address the problem of deforestation, the policies of Reforestation/Afforestation are to be framed with special consideration to Land use change between forest and agriculture, in the existing policies. Lateral coordination between the departments and polices on the forest, agriculture, rural development, urban development, water, and climate change is required to set a strategic target for land-use changes. It has been also observed over the years that despite having the framework for forest protection, legal enforcement in the land-use change or violation cases is not adequately addressed. Integrated land-use planning is important to balance competing land uses among stakeholders.

3.5.6 Forest Plantation for Biodiversity Conservation

Forest plantation is the focal point for ensuring forest sustainability and biodiversity conservation. The effect of these plantations on biodiversity depends upon the type of land uses they replace. These Plantations should replicate the character of the natural forest being replaced. These plantations bring economic benefits to the region but cannot always ecologically support the forest ecosystem. Such changes in fauna can result in an enhanced risk for diseases, which are getting further aggravated with the impacts of climate change.

Therefore, the real cause of such diseases should be understood to avoid the risk of potential outbreaks. The practice of waiting for a disaster to happen to find the necessary solutions should be reconsidered by having a proactive research-based action plan. To control the possibilities of disease outbreaks in the future, different stakeholders at the decision-making level need to go through such research which helps in identifying the root cause of disasters to work on areas of immediate concern.

3.6 Awareness Regarding the Linkages of Environment and Zoonotic Diseases

Awareness regarding the linkages of environment and zoonotic diseases should be made as an important concern for all the stakeholder viz., from the vulnerable sections to the development authorities and decision-makers. The collaboration at the regional level between spatial planning, afforestation policy, forest regulations, environmental policies, and the health sector is important to address these issues. The gap in institutional frameworks and policies should be notified where the facts relating to forest degradation are considered as equally a matter of importance as compared to the economic benefits of forest products. The awareness is required both at the decision-makers level for making policy instruments as well as stakeholders at ground level for managing risks associated with biodiversity loss and ecological health of a forest.

3.7 Inadequacy of Policies to Address Underlying Causes

India has a Forest Policy, Draft National Land Utilization policy, Draft National Urban Policy. All these talks about environment protection, land-use conversion check, forest protection. The draft of National Forest Policy 2018 by the Ministry of Environment, Forest and Climate Change addresses the issue of Human-wildlife conflict by proposing short term and long-term measures of reducing such incidents. Short term measure consists of emergency response initiative and medical facility availability for humans and wildlife simultaneously. Long term measure includes monitoring of wildlife population inside and outside the forest area. The underlying causes of human-animal conflicts and forest degradation are though not addressed explicitly. Monitoring of wildlife population is important but somehow the land-use change issue in the forest peripheral areas due to expanding development infrastructure is still overlooked.

For effective forest conservation, preserving the ecology of habitat including vegetation is as important as restricting deforestation (Koulgi et al. 2019). Forest Conservation (FC) Act, 1980 allows the conversion of forest land to non-forest land for mining, construction of road, and other development activities but with the provision of mandatory compensatory afforestation imposed on each forest land parcel conversion. The afforestation activities become part of CSR (Corporate Social Responsibility) and in many cases end up with plantation of invasive species for landscaping/scenic value enhancement. Such activities disturb the local balance of the ecology.

Ecosystem approaches to enhance the resilience of settlements to pandemics and climate change and other invisible disasters are important. Regional priorities need to be established with the identification of hot spot areas, where several of the drivers of emerging zoonotic diseases are present, with strengthened surveillance and control capabilities. Multi-sectoral involvement to recognize the relationship of diseases with different causative factors is necessary to respond to pre-disaster precautions as well as future preparedness and recovery.

References

Bett B (2020) How nature can protect us from pandemics. https://www.unenvironment.org/news-and-stories/video/how-nature-can-protect-us-pandemics. Accessed 05 06 2020

da Costa AP, Costa FB, Soares HS, Ramirez DG, Araújo AD, da Silva Ferreira JI et al (2015) Environmental factors and ecosystems associated with canine visceral Leishmaniasis in Northeastern Brazil. Vector Borne Zoonotic Dis 15(12):765–774. https://doi.org/10.1089/vbz.2015.1866

Directorate of Health Services (2011) Data on communicable disease Kerala-2011. Directorate of Health Services, Thiruvananthapuram

Directorate of Health Services (2012) Data on communicable disease Kerala-2012. Directorate of Health Services, Thiruvananthapuram

Directorate of Health Services (2013) Data on communicable disease Kerala-2013. Directorate of Health Services, Thiruvananthapuram

Directorate of Health Services (2014) Data on communicable disease Kerala-2014. Directorate of Health Services, Thiruvananthapuram

Directorate of Health Services (2015) Data on communicable disease Kerala-2015. Directorate of Health Services, Thiruvananthapuram

Directorate of Health Services (2016) Data on communicable disease Kerala-20156. Directorate of Health Services, Thiruvananthapuram

Directorate of Health Services (2017) Data on communicable disease Kerala-2017. Directorate of Health Services, Thiruvananthapuram

Directorate of Health Services (2018) Data on communicable disease Kerala-2018. Directorate of Health Services, Thiruvananthapuram

Dreiss LM, Volin JC (2014) Forests: temperate evergreen and deciduous. In: Wang Y (ed) Encyclopedia of natural resources – land – volume I. CRC Press/Taylor & Francis Group, New York, pp 214–223

Easa PS, Sankar S (2001) Study on man – wildlife interaction in wayanad wildlife sanctuary. Kerala Forest Research Institute, Kerala

Gadgil M (2011) Report of the Western Ghats ecology expert panel. Minsirty of Enviornment and Forest, GOI, New Delhi

Ghanbari B, Gómez-Aguilar JF (2019) Analysis of two avian influenza epidemic models involving fractal-fractional derivatives with power and Mittag-Leffler memories. Chaos 29:123113. https://doi.org/10.1063/1.5117285

Gottwalt A (2013) Impacts of deforestation on vector-borne disease incidence. Global J Health Sci 3:16–19

Government of Kerala (2014) IDDP. Department of Town and Country Planning - Government of Kerala, Kerala

Honari M, Boleyn T (2005) Health ecology: health, culture and human-environment interaction. Routledge, London

Hunarī M (1999) Health ecology: health, culture, and human-environment interaction. Routledge, London

Kannan R (2020) Why pathogens travel in search of a host. The Hindu, Newspaper, Chennai. https://www.thehindu.com/sci-tech/health/coronavirus-why-pathogens-travel-in-search-of-a-host/article31434762.ece. Accessed 26 April 2020

Karesh WB, Dobson A, Lloyd-Smith JO, Lubroth J, Dixon MA, Bennett M et al (2012) Ecology of zoonoses: natural and unnatural histories. Zoonoses 380(9857):1936–1945. https://doi.org/10.1016/S0140-6736(12)61678-X
Koulgi PS, Clinton N, Karanth KK (2019) Extensive vegetation browning and drying in forests of India's Tiger Reserves. Scientific Reports
Kunz T, de Torrez EB, Bauer D, Lobova T, Fleming TH (2011) Ecosystem services provided by bats. Ann N Y Acad Sci 1223(1):1–38
Marsh GP (1864) Man and nature; or, physical geography as modified by human action. Charles Scribner, District Court of United States of Souther District, New York
McFarlane RA, Sleigh AC, McMichael AJ (2013) Land-use change and emerging infectious disease on an island continent. Int J Environ Res Public Health 10(7):2699–2719. https://doi.org/10.3390/ijerph10072699
Myers SS, Patz J (eds) (2011) Land use change and human health. Encyclopedia of environmental health. Elsevier, Amsterdam, pp 396–404. https://doi.org/10.1016/B978-0-444-52272-6.00166-5
OECD (2001) Human health and environment. OECD Environmental Outlook, pp 249–254
Patz JA, Confalonieri UE (2005) Human health: ecosystem regulation of infectious. In: Hassan R, Scholes RJ, Ash N (eds) Condition and trends working group. Ecosystems and human well-being: current state and trends: findings of the condition and trends working group of the millennium ecosystem assessment. Island Press, Washington DC, pp 391–415
Patz J, Campbell-Lendrum D, Gibbs H, Rosalie (2008) Health impact assessment of global climate change: expanding on comparative risk assessment approaches for policy making. Annu Rev Public Health 29:27–39
Prüss-Üstün A, Corvalán C (2006) Preventing disease through healthy environments. Towards an estimate of the environmental burden of disease. WHO, Geneva
Quinney M (2020) 5 reasons why biodiversity matters – to human health, the economy and your wellbeing. https://www.weforum.org/agenda/2020/05/5-reasons-why-biodiversity-matters-human-health-economies-business-wellbeing-coronavirus-covid19-animals-nature-ecosystems/. Accessed 30 May 2020
Rosoman, Grant (1994) The plantation effect an ecoforestry review of the environmental effects of exotic monoculture tree plantations in Aotearoa/New Zealand. Greenpeace New Zealand with support from Canterbury Branch Maruia Society
Scholz RW, Binder CR (2003) The paradigm of human-environment systems. https://doi.org/10.3929/ethz-a-004520890
Scott J (2020) How biodiversity loss is hurting our ability to combat pandemics. https://www.weforum.org/agenda/2020/03/biodiversity-loss-is-hurting-our-ability-to-prepare-for-pandemics. Accessed 25 May 2020
Sheela AM, Ghermandi A, Vineetha P, Sheeja RV, Justus (2017) Assessment of relation of land use characteristics with vector-borne diseases in tropical areas. Land Use Policy 63:369–380
Statistics Wing-Forest Headquarters (2011) Forest Statistics-2011. Kerala Forest and Wildlife Deparment, Thiruvananthapuram
Statistics Wing-Forest Headquarters (2012) Forest Statistics-2012. Kerala Forest and Wildlife Department, Thiruvananthapuram
Statistics Wing-Forest Headquarters (2013) Forest Statistics-2013. Kerala Forest and Wildlife Department, Thiruvananthapuram
Statistics Wing-Forest Headquarters (2014) Forest Statistics-2014. Kerala Forest and Wildlife Department, Thiruvananthapuram
Statistics Wing-Forest Headquarters (2015) Forest Statistics-2015. Kerala Forest and Wildlife Department, Thiruvananthapuram
Statistics Wing-Forest Headquarters (2016) Forest Statistics-2016. Kerala Forest and Wildlife Department, Thiruvananthapuram
Statistics Wing-Forest Headquarters (2017) Forest Statistics-2017. Kerala Forest and Wildlife Department, Thiruvananthapuram

Statistics Wing-Forest Headquarters (2018) Forest Statistics-2018. Kerala Forest and Wildlife Department, Thiruvananthapuram

Turner BL (2015) Nature and society in geography. In: International encyclopedia of the social & behavioral sciences, 2nd edn. Elsevier, Amsterdam, pp 334–339

UNEP (1972) Declaration of the United Nations conference on the human environment. United Nations Environment Programme, Stockholm

UNEP (2016) UNEP frontiers 2016 report: emerging issues of environmental concern. United Nations Environment Programme, Nairobi

United Nations Conference on Environment and Development (1992) The Earth Summit: the United Nations Conference on Environment and Development (UNCED). Graham & Trotman/Martinus Nijhoff, London

Wangdi K, Gatton ML, Kelly GC, Clements AC (2015) Cross-border malaria: a major obstacle for malaria elimination. Adv Parasitol 89:79–107. https://doi.org/10.1016/bs.apar.2015.04.002

Ward DM, Nislow KH, Folt CL (2010) Bioaccumulation syndrome: identifying factors that make some stream food webs prone to elevated mercury bioaccumulation. Ann N Y Acad Sci 1195:62–83

Watras CJ, Back RC, Halvorsen S, Hudson RJ, Morrison KA, Wente SP (1998) Bioaccumulation of mercury in pelagic freshwater food webs. Sci Total Environ 219:183–208

Whelan C, Wenny D, Marquis R (2008) Ecosystem services provided by birds. Ann N Y Acad Sci 1134:25–60

Wilcox BA, Ellis B (2006) Forests and emerging infectious diseases of humans. Unasylva 224 57:8

WWF (2018) Living planet report. WWF, Gland

Chapter 4
COVID 19 Threat and the Role of Human and Natural Factors

Srinidhi Jha and Manish Kumar Goyal

Abstract The coronavirus disease 2019 (COVID 19) turned out to be one the most substantial global crisis in the recent times. Researchers all around the world are trying to understand the factors which influence and govern the occurrence and evolution of the pandemic. Earlier understanding of diseases generated by similar family of viruses suggest that climate factors do influence the growth of disease. Similarly, the risk of natural or manmade disaster depends on the vulnerability, exposure and capacity of the population. These factors in turn depend on the socio-economic status of the exposed population. During the past few years, it has been realized that India is highly vulnerable to climate change with the existing socio-economic condition. Given the severity of COVID 19 pandemic, it becomes necessary to investigate the role of climatic factors and socio-economic conditions in augmenting the risk of the disease. This chapter discusses the role of climatic and socio-economic conditions in increasing the risk of COVID 19 pandemic. We first discuss the dependence of climatic variables in augmenting the risk of the similar diseases. Then, the role of socio-economic status of the exposed population is investigated by previous studies. Further, the chapter incorporates a case study which explores the role of four climatic variables (pressure, relative humidity, temperature and wind speed) in governing the risk of COVID 19 in India. The hazard measure in terms of the occurrence of different percentiles of confirmed COVID 19 cases was calculated and then combined with the vulnerability and exposure indicators to estimate the risks. The case study is carried out using extreme value theory in a nonstationary setting to check and incorporate the dynamic nature of climate and COVID 19 dependence.

Keywords Climate · COVID 19 · Exposure · Risk · Socio-economic · Vulnerability

---

S. Jha (✉) · M. K. Goyal
Discipline of Civil Engineering, Indian Institute of Technology, Indore, Madhya Pradesh, India
e-mail: mkgoyal@iiti.ac.in

M. K. Goyal, A. K. Gupta (eds.), *Integrated Risk of Pandemic: Covid-19 Impacts, Resilience and Recommendations*, Disaster Resilience and Green Growth,
https://doi.org/10.1007/978-981-15-7679-9_4

4.1 Introduction

Since the inception of first coronavirus disease 2019 (COVID 19) in the Chinese city of Wuhan in December 2019, the world is now staring at a crisis at several fronts such as public health, social welfare, economic development etc. This global crisis does not seem to be getting slow and the impacts of the disease are putting forth new challenges to the governments all around the globe. General understanding of similar diseases in the past explains that evolution, spread and fatality of such epidemics depend on the natural as well as human factors. Many studies have been conducted to resolve the intricate relationship of the disease with natural and manmade factors. For example, Zhou et al. (2020), in a very significant work suggested that the patients with pre-existing disease conditions are more prone to the COVID 19 disease. Fang et al. (2020) revealed that people with severe disease conditions are more susceptible to the extreme health impacts. These factors, particularly in poor countries where health infrastructure is not up to the mark and the socio-economic condition is not adequate to tackle the impacts of the pandemic, are playing a crucial role in deciding the fatality rates. A study investigating the healthcare system of African countries reveal that COVID 19 pandemic pose great threat to the continent at several levels. (Nkengasong and Mankoula 2020). The continent, not only is poorly prepared to tackle such disease but also, have been seen to be poorly resilient to the impact of such wide scale epidemic. Similarly, the economic implications of the pandemic have been studied by a number of researchers. For instance, Zhang et al. (2020) found that the COVID 19 has turned out to be one of the most significant factors in augmenting the volatility of global financial markets. The investigation of different micro and macro variables in different subjects have revealed that prominent efforts and capital will be needed to reduce the scale of damage particularly in the developing countries with higher population density (McKibbin and Fernando 2020). Further, Qu et al. (2020) suggested for an urgent need of investigating the influence of environmental variables in expansion of the pandemic. A number studies have been performed to explain the complex relationship of environmental and COVID 19, however, there still lacks a general agreement between the researchers and scientists about the same. For instance, Wu et al. (2020) by investigating the cases of more than 150 countries suggested that an increase in the temperature and humidity may aid to limiting the COVID 19 pandemic partially. However, Zhu and Xie (2020) by studying the rise of cases of more than 100 in China, suggested that an increase in 1-degree Celsius in the mean temperature was positively correlated with about 5% of the infected cases. They concluded that there was no strong evidence of COVID 19 confirmed cases with possible warming scenarios. But importantly, previous studies show that the expanse of diseases evolved from the 'severe acute respiratory syndrome' (SARS) family of viruses are related to environmental conditions such as pollution, temperature, relative humidity and several other climatic factors (Bao et al. 2016; Cui et al. 2003; Lin et al. 2006). In light of the above discussion, it is understandable that the dependence of other factors with the COVID 19 should be explored. This will aid the policymakers in framing better risk measurement, preparedness, response and mitigation policies.

It is well known that the threat perception of similar infectious diseases in general, primarily depend on the three factors which are carrier of the transmission, the host of disease and the environment in which it is evolving (Lin et al. 2006). The COVID 19 pandemic now poses a threat to the population all around the globe irrespective of the socio-economic condition or the climatic conditions. India is one of the most populous countries on the earth and is highly vulnerable to changing climatic conditions. The first confirmed case of the pandemic was reported in the Kerala state of the India in the first week of January 2020 (Rawat 2020). It took only 20 days for the number of cases to jump from 1000 to 20,000. Lately, the country has witnessed huge spike in both total number of confirmed cases and the total number of deaths failing all the efforts taken by the government to contain the disease. Some of the quick measures which were adopted by government of India include enforcing strict lockdown, large-scale testing, upgrading the healthcare systems, preparing makeshift hospitals, home quarantine to suspected population and home delivery of essential goods and services. However, regardless of the number of efforts taken by the government of India in containing the disease the net population of COVID 19 confirmed cases spiked to 45,674 as per the data of fourth May 2020. It should be noted that the world-wide fate of the pandemic is heavily dependent on the advancement of the COVID 19 in developing countries like India. Further, several studies have indicated that India is vulnerable to changing climatic conditions and the poor socio-economic capacity of the population makes it more prone to the ill impacts of the pandemic. (Das et al. 2020a; Jha et al. 2019b; Sathaye et al. 2012; Sharma and Goyal 2018; Sinha et al. 2019). Therefore, investigating the complex relationship between the COVID 19 and different natural and anthropogenic factors is important for developing a more suitable risk reduction framework.

Based on the recent literature survey, it can be understood that most of the recent studies have employed simple statistical modeling techniques to investigate the dependence of COVID 19 risk with climatic and human factors. It has also been observed that none of the studies have investigated the nonstationary time varying probabilistic characteristic of COVID 19 cases. It is well known that socio-economic and climatic characteristics of India vary widely in time and space. Therefore, it can be expected that the relationship between the climatic variables, socio-economic conditions and the climatic variables are itself time and space varying (Goyal and Ojha, 2012; Goyal and Ojha 2014). Several studies in different fields have reported that the time-varying probabilistic models prove to be more productive as compared to the stationary probabilistic models (Das et al. 2020b; Dong et al. 2019; Hunter et al. 2017; Sarhadi et al. 2016). In this chapter, we discuss a case study, employing nonstationary extreme value concept to explore the relationship of climatic conditions and socio-economic conditions in the governing the risk of COVID 19 in the country. The study is conducted for more than 450 districts of the India by exploring the climatic and socio-economic conditions of the individual districts to identify the possible drivers of risk and prepare district-wise high resolution maps of the risk distribution.

4.2 Material and Methods

4.2.1 Data Used

4.2.1.1 District and State Boundaries of India

As discussed, this case study was performed at district-wise level in India. The districts are the smallest administrative units of India and now a day, the response measures to COVID19 at ground level is being decided by the district authorities. The study was conducted at district level to provide the risk measure at the smallest possible level for efficient plans and policy making. Further, the results were also summarized at the state level for region-wise characterization of the results. The selection of states and districts was done in accordance with the last available census of India (2011) records according to which there are 35 states and union territories in the country. The details of the state and district boundaries have been given in the Fig. 4.1. The source of administrative boundaries was the global administrative boundaries (GADM) database which has been widely utilized in a number of spatial analysis works (Hijmans et al. 2011; Kugler et al. 2015).

4.2.1.2 Cumulative COVID 19 Confirmed Cases Data

Since the inception of the disease, many governmental and nongovernmental organizations have started to track the expanse and evolution of the COVID 19 infection cases and deaths. For this study, the cumulative COVID 19 confirmed cases data for the duration of March 2–May 2, 2020 were acquired from the web portal https://howindialives.com/gram/metrics.php. The time series of confirmed cases have been prepared by collecting the information from various health, administration and policy organizations.

4.2.1.3 Pressure, Relative Humidity, Temperature and Wind Speed Data

THE NCEP/NCAR reanalysis project is one of the most trustworthy sources of climate data which has widely been used in a variety of studies. We utilized the NCEP/NCAR reanalysis project data set which contains several climatic data starting from 1948 to real time (Kalnay et al. 1996). These data sets have been developed to support the research community in their climate research endeavors. The data product is updated on daily basis and the global gridded data set can be obtained from the web portal https://www.psl.noaa.gov/data/gridded/. The climate variables from the website has been studied and compared with several other data sets and it can be validated that these data sets pose efficient characteristics of the global climatic patterns (Kanamitsu et al. 2002; Sachindra et al. 2014; Sillmann et al.

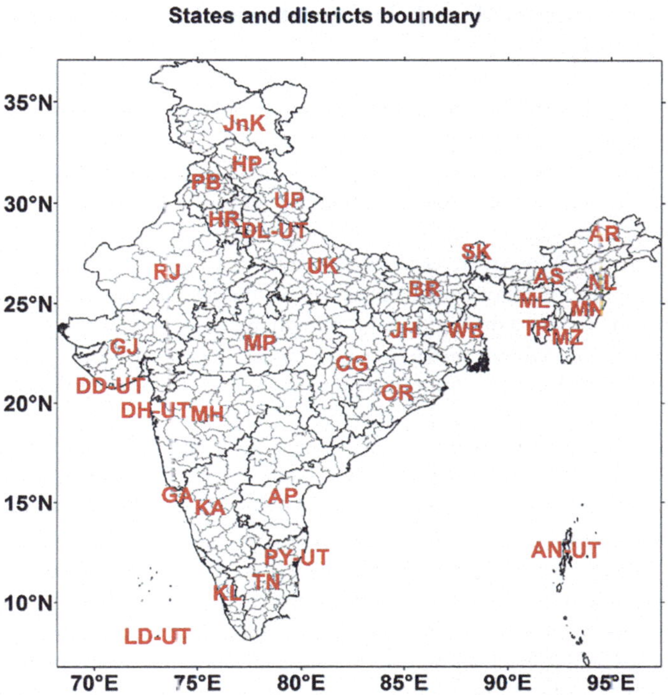

Fig. 4.1 Districts and state boundaries. The abbreviation of the regions are as: *AN-UT* Andaman and Nicobar, *AP* Andhra Pradesh, *AR* Arunachal Pradesh, *AS* Assam, *BR* Bihar, *CH-UT* Chandigarh, *CG* Chhattisgarh, *DH-UT* Dadra and Nagar Haveli, *DD-UT* Daman and Diu, *DL-UT* Delhi, *GA* Goa, *GJ* Gujarat, *HR* Haryana, *HP* Himachal Pradesh, *JnK* Jammu and Kashmir, *JH* Jharkhand, *KA* Karnataka, *KL* Kerala, *LD-UT* Lakshadweep, *MP* Madhya Pradesh, *MH* Maharashtra, *MN* Manipur, *ML* Meghalaya, *MZ* Mizoram, *NL* Nagaland, *OR* Orissa, *PY-UT* Puducherry, *PB* Punjab, *RJ* Rajasthan, *SK* Sikkim, *TN* Tamil Nadu, *TR* Tripura, *UK* Uttar Pradesh, *UP* Uttaranchal, *WB* West Bengal

2013). Four daily surface/near-surface climatic variables—pressure, relative humidity, temperature and wind speed were collected for the period of March 2–May 2, 2020 to support the modeling of COVID 19 confirmed cases. It should be noted that these variables are primarily available in 2.5 × 2.5-degree resolution. For proper analysis, the data was re-gridded to a finer 0.5 × 0.5-degree resolution using the inverse distance weighted (IDW) average method (Snell et al. 2000). After re-gridding the data, the climate data was obtained for each district by averaging the grid points obtained during the extraction.

4.2.1.4 Census Data for Household Population and Socio-Economic Indicator

For relating the COVID 19 risk to the socio-economic parameters, we considered four parameters of vulnerability and exposure. These data sets were obtained from the last available Census of India 2011 data and then utilized to produce the exposure and vulnerability parameters at district level. For exposure elements, we considered the population aged greater than 65 years and population having age less than or equal to 14 years. Moreover, besides the age, we also considered that 'other working' population category as an element of exposure. This was considered to accommodate the risk to working population which could possibly be exposed to the infection once the lockdown restrictions are eased. The term 'other working' population is defined as the class of population working in sectors other than cultivation, agriculture or the household industry sector.

The vulnerability to COVID 19 risk was considered to be function of three significant factors which are crucial in managing the risk to pandemic. We considered that the COVID 19 expanse depends a lot on the fact that how smoothly the knowledge about the disease is received by the population. Therefore, awareness campaigns, capacity building and proper understanding of the disease is easy when the population is literate. Therefore, we considered, illiteracy of population as one of the measures of vulnerability. Similarly, the importance of maintaining proper hygiene and intermittent cleaning and washing are among most crucial measures to save oneself rom the disease. However, this depends on the availability of clean water and sanitation facility in the household. We calculated the district-wise number of households in each district which lack clean water. Moreover, the density of houses with absent electrical and sanitation facilities were added for better risk estimation.

4.2.2 Methodology

In this study, we employ the time varying extreme value theory to model the cumulative COVID 19 confirmed cases in different districts of India. This approach was chosen considering the dynamic nature of climatic and pandemic cases in the country. The evolution of the cases is also known to be extreme in nature. In this investigation, the nonstationary extreme value theory is used to model the parameters of confirmed COVID 19 cases as the linear combinations of climatic covariates which are pressure, relative humidity, pressure and wind speed. We first check the suitability of the nonstationary and stationary extreme value models and then derive the exceedance probabilities of different quantiles of COVID 19 cases. Mean of the obtained probabilities were considered as the measure of COVID 19 hazard which was eventually combined with the socio-economic measures of exposure and vulnerability.

4.2.2.1 Nonstationary Extreme Value Modelling of COVID 19 Cases

The Generalized Extreme Value (GEV) models have been widely used in the nonstationary extreme value analysis of a variety of cases. In this study, we considered the GEV distributions as probable distribution of the cumulative COVID 19 cases. For the sake of computational simplicity, we considered the probability distributions of the confirmed cases as continuous random variables. The well-known mathematical description of the GEV distribution is given below:

$$F(x;\mu,\sigma,\xi) = \begin{cases} \exp\left\{-\left[1+\frac{\xi(x-\mu)}{\sigma}\right]^{-1/\xi}\right\}, & \sigma>0, \quad 1+\frac{\xi(x-\mu)}{\sigma}>0, \quad \xi\neq 0 \\ \exp\left\{-\exp\left[-\frac{(x-\mu)}{\sigma}\right]\right\}, & \sigma>0, \quad \xi=0 \end{cases} \tag{4.1}$$

Here, x is the confirmed COVID 19 cases in individual districts cases and μ, σ, and ξ denote the location, location, scale, and shape parameters of the GEV distribution, respectively. Therefore, $F(x;\mu,\sigma,\xi)$ represents the cumulative probability distribution (CDF) of confirmed COVID 19 cases with the distribution parameters given by μ, σ, and ξ. As discussed, we utilized two different scenarios of the extreme value analysis, which are the stationary and nonstationary cases. As per the nature of stationary modeling, the parameters of the distributions are considered constant, whereas, in the nonstationary distribution modeling, the parameters have been considered as the linear function of climatic covariates. More precisely, the GEV distribution derived from confirmed COVID 19 cases were obtained by modeling their parameters as linear combinations of the climatic variables which are pressure, relative humidity, temperature and wind speed in this case. The modeling of the parameters was done keeping in mind some basic assumptions based on the literature. For instance, the nonstationary modeling was done only for the location and scale parameters of the probability distribution. Coles (2001) and Yilmaz and Perera (2014) recommended that the nonstationarity in the scale parameter cannot be modeled easily and it raises the problem of uncertainty and complexity in the analysis.

As discussed, the location and scale parameters were modeled using the linear combinations of the climatic covariates. In this study we used 28 linear combinations of the climatic covariates for the nonstationary cases. One case, i.e., the stationary case was modeled using the constant parameters. Summarizing, the 29 linear combinations of the location and the scale parameters were formulated using the climatic covariates of pressure, relative humidity, temperature and wind speed (termed as C_1, C_2, C_3 and C_4 respectively in the models). The details about these covariates can be understood from Table 4.1. For illustration purposes, a few examples of the nonstationary GEV models are given below

Table 4.1 Description of the models used in the present study

Model ID	Description
M0	$X \sim GEV[\mu, \sigma, \xi]$
M1	$X \sim GEV[(\mu_0 + \mu_1 C_1, \sigma, \xi]$
M2	$X \sim GEV[(\mu_0 + \mu_2 C_2, \sigma, \xi]$
M3	$X \sim GEV[(\mu_0 + \mu_3 C_3, \sigma, \xi]$
M4	$X \sim GEV[(\mu_0 + \mu_4 C_4, \sigma, \xi]$
M5	$X \sim GEV[(\mu_0 + \mu_1 C_1 + \mu_2 C_2, \sigma, \xi]$
M6	$X \sim GEV[(\mu_0 + \mu_2 C_2 + \mu_3 C_3, \sigma, \xi]$
M7	$X \sim GEV[(\mu_0 + \mu_3 C_3 + \mu_4 C_4, \sigma, \xi]$
M8	$X \sim GEV[(\mu_0 + \mu_4 C_4 + \mu_2 C_2, \sigma, \xi]$
M9	$X \sim GEV[(\mu_0 + \mu_4 C_4 + \mu_1 C_1, \sigma, \xi]$
M10	$X \sim GEV[(\mu_0 + \mu_1 C_1 + \mu_2 C_2 + \mu_3 C_3\ \sigma, \xi]$
M11	$X \sim GEV[(\mu_0 + \mu_1 C_1 + \mu_2 C_2 - \mu_4 C_4\ \sigma, \xi]$
M12	$X \sim GEV[(\mu_0 + \mu_1 C_1 + \mu_2 C_3 + \mu_4 C_4\ \sigma, \xi]$
M13	$X \sim GEV[(\mu_0 + \mu_1 C_2 + \mu_2 C_3 + \mu_3 C_4\ \sigma, \xi]$
M14	$X \sim GEV[(\mu_0 + \mu_1 C_1 + \mu_2 C_2 + \mu_3 C_3 + \mu_4 C_4, \sigma, \xi]$
M15	$X \sim GEV[(\mu_0 + \mu_1 C_1), (\sigma_0 + \sigma_1 C_1), \xi]$
M16	$X \sim GEV[(\mu_0 + \mu_2 C_2), (\sigma_0 + \sigma_2 C_2), \xi]$
M17	$X \sim GEV[(\mu_0 + \mu_3 C_3), (\sigma_0 + \sigma_3 C_3), \xi]$
M18	$X \sim GEV[(\mu_0 + \mu_4 C_4), (\sigma_0 + \sigma_4 C_4), \xi]$
M19	$X \sim GEV[(\mu_0 + \mu_1 C_1 + \mu_2 C_2), (\sigma_0 + \sigma_1 C_1 + \sigma_2 C_2), \xi]$
M20	$X \sim GEV[(\mu_0 + \mu_2 C_2 + \mu_3 C_3), (\sigma_0 + \sigma_2 C_2 + \sigma_3 C_3), \xi]$
M21	$X \sim GEV[(\mu_0 + \mu_3 C_3 + \mu_4 C_4), (\sigma_0 + \sigma_3 C_3 + \sigma_4 C_4), \xi]$
M22	$X \sim GEV[(\mu_0 + \mu_4 C_4 + \mu_2 C_2), (\sigma_0 + \sigma_4 C_4 + \sigma_2 C_2), \xi]$
M23	$X \sim GEV[(\mu_0 + \mu_4 C_4 + \mu_1 C_1), (\sigma_0 + \sigma_4 C_4 + \sigma_1 C_1), \xi]$
M24	$X \sim GEV[(\mu_0 + \mu_1 C_1 + \mu_2 C_2 + \mu_3 C_3), (\sigma_0 + \sigma_1 C_1 + \sigma_2 C_2 + \sigma_3 C_3), \xi]$
M25	$X \sim GEV[(\mu_0 + \mu_1 C_1 + \mu_2 C_2 + \mu_4 C_4), (\sigma_0 + \sigma_1 C_1 + \sigma_2 C_2 + \sigma_4 C_4), \xi]$
M26	$X \sim GEV[(\mu_0 + \mu_1 C_1 + \mu_3 C_3 + \mu_4 C_4), (\sigma_0 + \sigma_1 C_1 + \sigma_3 C_3 + \sigma_4 C_4), \xi]$
M27	$X \sim GEV[(\mu_0 + \mu_2 C_2 + \mu_3 C_3 + \mu_4 C_4), (\sigma_0 + \sigma_2 C_2 + \sigma_3 C_3 + \sigma_4 C_4), \xi]$
M28	$X \sim GEV[(\mu_0 + \mu_1 C_1 + \mu_2 C_2 + \mu_3 C_3 + \mu_4 C_4), (\sigma_0 + \sigma_1 C_1 + \sigma_2 C_2 + \sigma_3 C_3 + \sigma_4 C_4), \xi]$

Here, C_1, C_2, C_3 and C_4 denotes the climatic covariates pressure, relative humidity, temperature and wind speed respectively

$$M2 : XG\tilde{E}V[(\mu_0 + \mu_1 C_1, \sigma, \xi)] \tag{4.2}$$

$$M19 : XG\tilde{E}V[(\mu_0 + \mu_1 C_1 + \mu_2 C_2), (\sigma_0 + \sigma_1 C_1 + \sigma_2 C_2), \xi] \tag{4.3}$$

Here, in the Model M2 as described by Eq. (4.2), μ_1 describes the inclination in the location parameter over climatic covariate C_1 which is the pressure data. Likewise, in Model M19, C_1 and C_2 signifying pressure and relative humidity have been utilized to define the location and scale parameters. The technicality of every nonstationary model in Table 4.1 can be described following identical notion.

4.2.2.2 Best-Fit Model and Parameters

As discussed in the last section, the nonstationary and stationary models were fit using the different climatic covariate combinations. The significance and best fit model was obtained by employing the likelihood ratio test (Coles 2001). According to this test, the difference between the log-likelihood of stationary and nonstationary model is considered which is assumed to following an approximate Chi-squared distribution. Mathematically, the condition at any significance level α (5% here) of L.R. test denotes,

$$2\left[nllh_{(S)} - nllh_{(NS)}\right] > c_\alpha \tag{4.4}$$

Here, c_α is the $(1-\alpha)$ quantile of obtained Chi-squared distribution. Moreover, if the nonstationarity is found true, the best nonstationary model is selected by investigating the p-value obtained from Chi-squared distribution. Conventionally, the null hypothesis, which is stationarity in this case, is rejected when the p-value is more than 0.05 at 95% confidence level. It should be noted that the L.R. test was done to compare 28 nonstationary models against the stationary model and parameter estimation was performed using the well-known maximum likelihood approach.

4.2.2.3 Evaluation of District-Wise COVID 19 Risk

As discussed, the best fit nonstationary and stationary model was decided by the L.R. test and investigating the p-value. Once the best fit model was decided, the exceedance probabilities of cumulative COVID 19 cases at different quantiles were decided. These equally spaced quantiles, i.e. 25th, 50th, 75th and 95th were obtained and averaged to estimate the COVID 19 hazard measure. This was a pre-requisite of the risk calculation formula which basically operates in terms of probabilities. We computed the risk based on very popular formulation of risk given by the Intergovernmental Panel on Climate Change (IPCC) *Risk* = *Hazard* × *Exposure* × *Vulnerability* (Oppenheimer et al. 2015). We derived the risk estimation formula based on this concept and then obtained a risk index value for each district. Mathematically, the formulation can be described as

$$I_C = P_H \times (P_C + P_O + P_W) \times \frac{1}{(P_{ILL} + P_{SE} + P_{WA})} \tag{4.5}$$

Here,
I_C: The measure of COVID 19 risk.
P_H: Average probability of hazard.
P_C: Exposure defined by children population density.
P_O: Exposure defined by old population density.
P_W: Exposure defined by working population density.

P_{VLL}: Vulnerability defined by illiterate population density.

P_{SE}: Vulnerability defined by density households with no electricity and sanitation facility.

P_{WA}: Vulnerability defined by density of household with no clean water availability in the premises.

This measure of risk (I_c) was calculated for all 461 districts of the country and then the values were divided into five equally spaced different categories of risk which are low (0–0.25), moderate (0.25–0.50), high (0.50–0.75), very high (0.75–1) and extreme (>1) risk. After characterization of results on district level, we summarized the results on state level too.

4.3 Results and Discussion

Figure 4.2 shows the district-wise distribution of confirmed COVID 19 cases across India. It can be inferred from the map that almost all districts of the country except a few regions in central east and northeastern India have confirmed cases. It was observed from the analysis that there were no cases in 179 out of 640 selected districts till 2nd May 2020. 68 districts of different states had cases greater or equal to 100. The southern and western India had the most number of cases. The districts with zero cases were exempted from further analysis. Once the valid districts were finalized, the time series for covariate data, including climatic variables, were prepared. The time average distribution of climatic variables pressure, relative

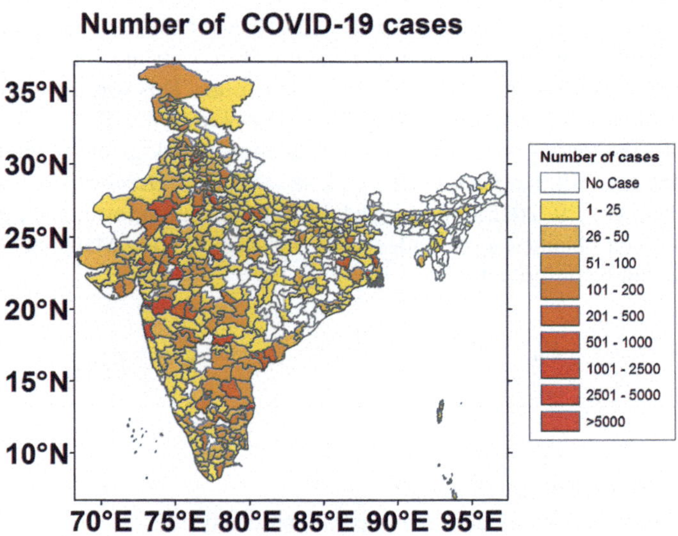

Fig. 4.2 The district-wise distribution of confirmed COVID 19 cases in India. 'No-Case' denotes the districts where no confirmed cases were recorded till 2nd May 2020

humidity, temperature and wind speed were calculated. It should be noted that the climatic covariates used for fitting nonstationary GEV models were in the form of time series. We observed that the pressure distribution shows that the lowest values were obtained in the high altitude districts of the country. Maximum surface pressure was observed in the districts in the northern and coastal area districts of the country. Least relative humidity was observed in the northwestern and central part of the country where the weather is usually drier. Significantly high values of relative humidity were observed in the regions of the northeast, upper north and the southern districts of the country. These regions maintain a good amount of green cover throughout the year; therefore, high relative humidity in these regions is expected (Jha et al. 2019a). The temperature distribution was ranging from −13.58 to 31.20 degree Celsius with the high altitude regions having the lowest temperature and southwestern the lowest. The spatial distribution of mean temperature of the selected months is coherent with the average temperature distribution as suggested by other sources such as India Meteorological Department (IMD) (Srivastava et al. 2009). For wind speed distribution, the magnitude of the u and v of the wind velocity components were estimated and averaged for the grid points falling under the each individual districts. It should be noted that instead of using time average values of the climatic data, we related their time series (2nd March 2020 to 2nd May 2020) with the confirmed COVID 19 cases of each corresponding districts. This modeling was done using the 29 covariate combination, as discussed in the methodology section. The covariate combination was prepared such that location and scale parameters which define the magnitude and variability of the number of confirmed COVID 19 cases could be explained by climatic variables. Further, the combinations were made such that the location and scale parameters are explained by both single and multiple climatic variables (Table 4.1). Strong evidence of nonstationarity due to climatic conditions was observed in the 248 (~54%) districts out of 461 districts. It was found that nonstationarity due to climatic variables in the regions seems to explain both magnitude (through location parameter) and variability (scale parameter). It can be inferred from Table 4.1, that combinations involving the modeling the location parameter as a linear function of climatic variables include model M1 to M14. Figure 4.3 shows the representation of the count of the best fit nonstationary models obtained for 461 districts. It can be understood from the figure that models which include single climatic variable as an explanatory variable for location parameter were best fit for most of the districts. It should also be noted that the combination of more than one climatic variables for explaining the location parameter was generally not suited for location parameter estimation. However, the scale parameter, along with the location parameter, could be explained by such combinations. It should be noted that it is not advisable to model the scale parameter separately. Therefore, we estimated the scale parameter only with the location parameter in respective combinations of M15 to M28. It was found that wind speed was the most dominating climatic factor followed by relative humidity, pressure, and temperature in the evolution of the cases (Fig. 4.3). The results revealed, using temperature as a possible covariate in location as well as scale parameter was best suited only for 12 districts. However, pressure, relative humidity

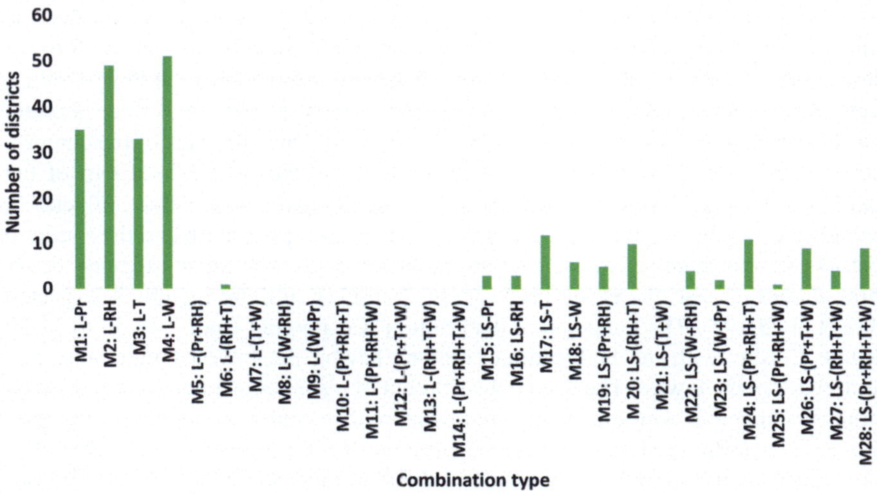

Fig. 4.3 The number of districts and different covariate combination models (M1-M28). Here, L and S denote location and scale, whereas, Pr, RH, T, W denote pressure, relative humidity, temperature and wind speed respectively

and wind speed had significant influence in scale parameter estimation in only 3,3, and 6 districts respectively. However, relative humidity in combination with temperature was best suited for 10 districts.

Once the best covariate combination or the model was obtained, the probabilities at different quantiles (25th, 50th, 75th and 95th) were also estimated using the parameters estimated in the previous step. The average of these probabilities was used as COVID 19 hazard measure. It is essential to understand that estimating different quantiles and then averaging them was performed to transform the occurrence of COVID 19 cases into the probabilistic setting. This probabilistic setting was required to estimate the risk index values, as explained in Eq. (4.5). Further, according to the given formula, risk estimation required the estimation of exposure and vulnerability measures. For exposure, as discussed, the elderly and child population density along with the other working population was calculated for each district. The density here implies the number of people per household. In other words, the elderly, child and working population were calculated and then individually divided by the total number of households in each district. Similarly, the vulnerability measures were also estimated for each district and divided by the number of households. It has been observed that COVID 19 propagation has been distinctive in the rural and urban areas. Considering this, we estimated the risk measures separately for rural, urban and total (combining rural and urban). Therefore, exposure and vulnerability measures were also calculated separately for rural, urban and total population. Normalization of the exposure and vulnerability measures were done to bring all the values at a common scale. Eventually, the risk index was calculated using Eq. (4.5) at the scale of the rural, urban and total population. As

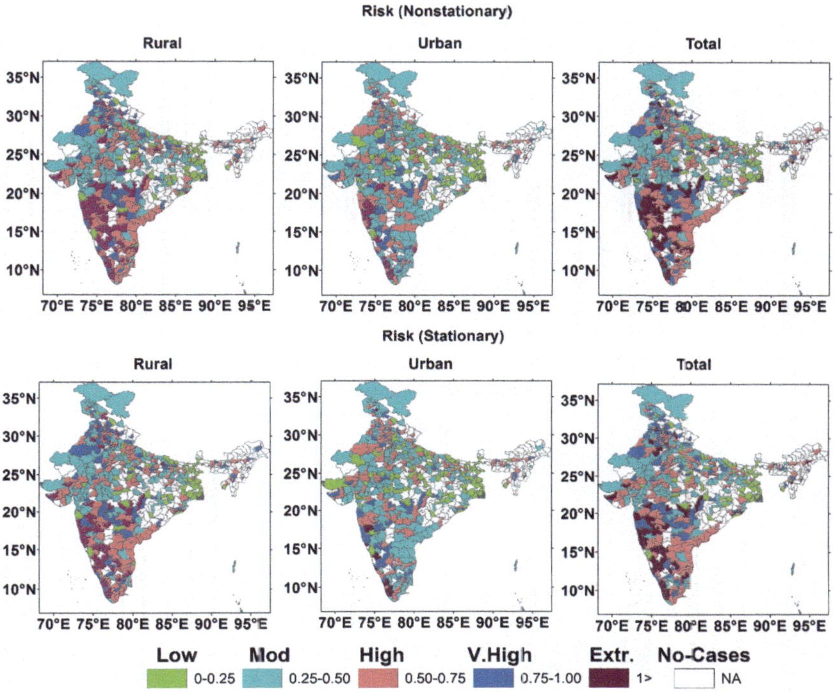

Fig. 4.4 District-wise risk measures at three scales (rural, urban and total) for nonstationary and stationary cases. The risk index values were divided into five different classes, as shown in the figure

discussed, the risk index values were classified into five different levels so that districts could be characterized into different classes of risk. The risk index values and their classification for each district are shown in Fig. 4.4. The risk index was calculated separately for nonstationary and stationary cases to understand the role of climatic factors in terms of risk. As discussed, the stationary models (M0) do not incorporate the climatic influence and while the nonstationary models do so through model M1-M28. The risk index was divided into five classes, as explained earlier. It can be easily inferred from Fig. 4.4 that a good number of districts come under the high, very high or extreme risk category. It was also observed that there is a clear distinction between the risk distribution of stationary and nonstationary cases.

This confirms the role of climatic factors in influencing the occurrence of confirmed COVID 19 cases in the country. For a simpler demonstration of results, we estimated the percentage of districts showing risk index value greater than or equal to 0.5, i.e., the percentage of districts lying in the high, very high and extreme risk category. Figure 4.5 shows the state-wise distribution of the percentage of districts under high risk (or above). It was found that the rural population in 10 out of 35 states and union territories have at least 50% of their districts under high, very

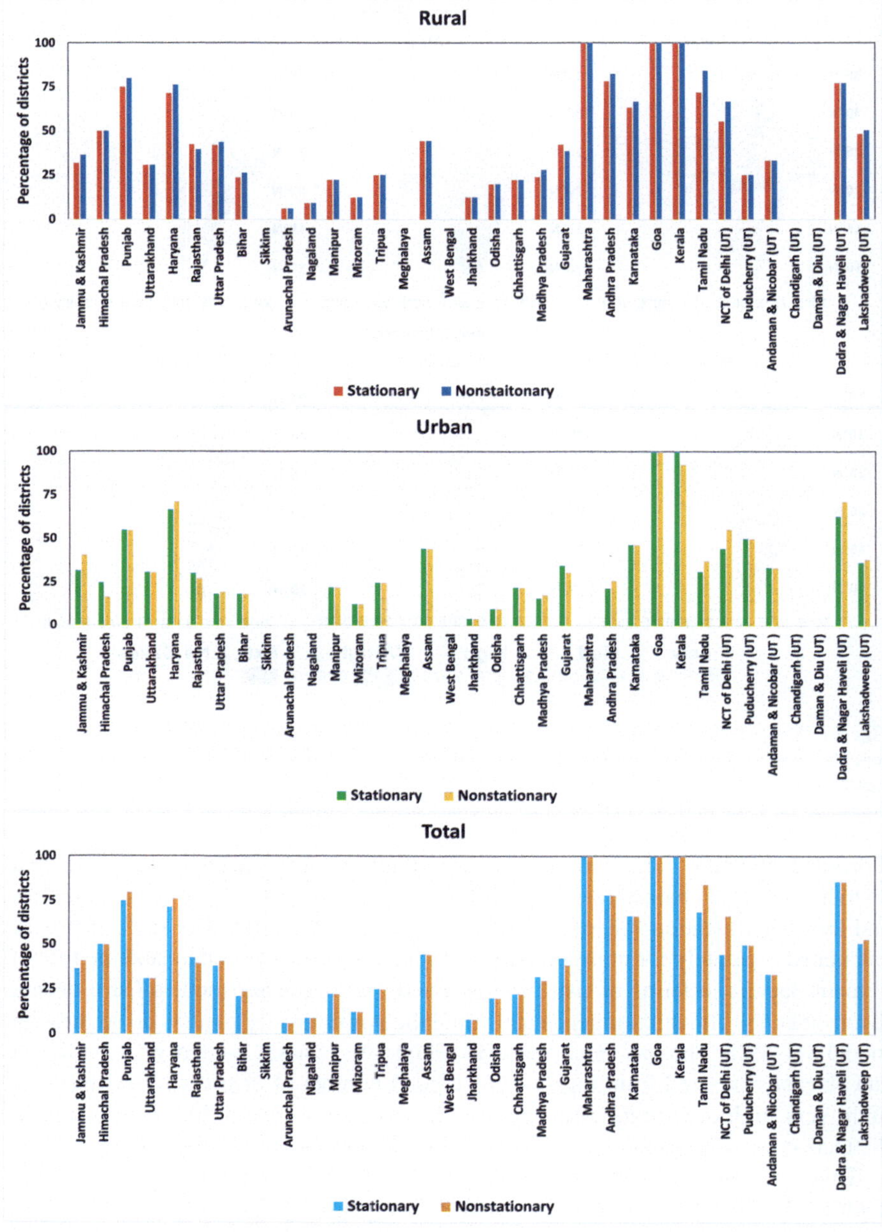

Fig. 4.5 Percentage of districts (state-wise) under high risk at rural, urban and total scale in stationary and nonstationary condition. High risk denotes (Risk index value i.e., $I_c >= 0.5$)

high or extreme risks under the stationary condition. In nonstationary case, rural population in 11 out of 35 states and union territories have at least 50% of the districts under such risk. However, the total urban population in only 5 out of 35 states were at high risk (or above) in stationary and nonstationary condition. The difference in the number of states at high risk (or above) while considering the rural and urban population was significant. This may be due to the fact that about two-thirds of India's population resides in rural areas. Considering the both rural and urban and population, 12 out of 35 states in stationary condition whereas 13 out of 35 states in nonstationary condition were lying in the high, very high or extreme risk category. Most risked states were Kerala, Maharashtra, Andhra Pradesh, Punjab and Haryana. Summarily, when climate influence was considered, i.e. in nonstationary case 278 (~60%), 181 (~39%) and 288 (~62%) districts out of 461 districts were at high risk (or above) at rural, urban and total population scales respectively. Whereas, a less number of districts, 265 (~57%), 174 (~38%) and 277 (~60%) districts (in the same order) were at high risk (or above) when the cases were modeled considering stationarity.

The study provides useful insights for the decision-makers to identify the high-risk hotspots of the pandemic in India and the exposure and vulnerability factors associated with it. Although a few points must also be considered while replicating the risk analysis technique. For instance, using the latest possible data set for real-time risk measures. It should also be noted that the number of cases in different parts of the country also depends on a number of factors such as availability of testing facilities, efficient contact tracing, the success rate of the testing method, therefore, proper validation checks should be performed before utilizing infection data.

4.4 Summary and Conclusions

This study enables the understanding of the influence of climate variables in inducing risk due to COVID 19 in rural, urban and combined population scale at the district level in India. The results reveal there is a significant relationship between climatic factors and COVID 19 risk. It was also found that the risk of the pandemic is greater in rural population. This risk is primarily due to high vulnerability in terms of lack of proper sanitation and clean water availability. The findings of our study are in line with some other recent works which suggest that there could be a possible link between the COVID 19 risk and climate variables. The investigation draws the attention of the decision-makers to strengthen the capacity of the population, especially in rural areas. The vulnerability to the pandemic is also a factor of the density of the educated population. Therefore, the policymakers in India must focus on increasing awareness in climatically vulnerable rural areas in the country which are backward in terms of education, sanitation and clean water availability.

References

Bao J, Wang Z, Yu C, Li X (2016) The influence of temperature on mortality and its lag effect: a study in four Chinese cities with different latitudes. BMC Public Health 16:375

Coles S (2001) An introduction to statistical modeling of extreme values. Springer, London

Cui Y, Zhang Z-F, Froines J, Zhao J, Wang H, Yu S-Z, Detels R (2003) Air pollution and case fatality of SARS in the People's republic of China: an ecologic study. Environ Health 2:15

Das J, Jha S, Goyal MK (2020a) On the relationship of climatic and monsoon teleconnections with monthly precipitation over meteorologically homogenous regions in India: wavelet & global coherence approaches. Atmos Res 238:104889

Das J, Jha S, Goyal MK (2020b) Non-stationary and copula-based approach to assess the drought characteristics encompassing climate indices over the Himalayan states in India. J Hydrol 580:124356

Dong ND, Agilan V, Jayakumar KV (2019) Bivariate flood frequency analysis of nonstationary flood characteristics. J Hydrol Eng 24:4019007. https://doi.org/10.1061/(ASCE)HE.1943-5584.0001770

Fang L, Karakiulakis G, Roth M (2020) Are patients with hypertension and diabetes mellitus at increased risk for COVID-19 infection? Lancet Respir Med 8(4):e21

Goyal MK, Ojha CSP (2012) Downscaling of precipitation on a lake basin: evaluation of rule and decision tree induction algorithms. Hydrol Res 43(3):215–230. https://doi.org/10.2166/nh.2012.040

Goyal MK, Ojha CSP (2014) Evaluation of rule and decision tree induction algorithms for generating climate change scenarios for temperature and pan evaporation on a lake basin. ASCE J Hydrol Engg 19(4):828–835. https://doi.org/10.1061/(ASCE)HE.1943-5584.0000795

Hijmans R, Nell G, Arnel R, Maunahan A, Wieczorek J, Kapoor J (2011) Global administrative areas. GADM v2 global shapefile. http://biogeo.ucdavis.edu/data/gamdm2/gadm_v2_shp.zip

Hunter J, Burke SP, Canepa A (2017) Multivariate modelling of non-stationary economic time series. Springer, Berlin

Jha S, Das J, Goyal MK (2019a) Assessment of risk and resilience of terrestrial ecosystem productivity under the influence of extreme climatic conditions over India. Sci Rep 9:1–12

Jha S, Das J, Sharma A, Hazra B, Goyal MK (2019b) Probabilistic evaluation of vegetation drought likelihood and its implications to resilience across India. Global Planet Change 176:23–35. https://doi.org/10.1016/j.gloplacha.2019.01.014

Kalnay E, Kanamitsu M, Kistler R, Collins W, Deaven D, Gandin L, Iredell M, Saha S, White G, Woollen J (1996) The NCEP/NCAR 40-year reanalysis project. Bull Am Meteorol Soc 77:437–472

Kanamitsu M, Ebisuzaki W, Woollen J, Yang SK, Hnilo JJ, Fiorino M, Potter GL (2002) NCEP-DOE AMIP-II reanalysis (R-2). Bull Am Meteorol Soc 83:1631–1643+1559. https://doi.org/10.1175/BAMS-83-11-1631(2002)083<1631:NAR>2.3.CO;2

Kugler TA, Van Riper DC, Manson SM, Haynes DA II, Donato J, Stinebaugh K (2015) Terra Populus: workflows for integrating and harmonizing geospatial population and environmental data. J Map Geogr Libr 11:180–206

Lin K, Fong DY-T, Zhu B, Karlberg J (2006) Environmental factors on the SARS epidemic: air temperature, passage of time and multiplicative effect of hospital infection. Epidemiol Infect 134:223–230

McKibbin WJ, Fernando R (2020) The global macroeconomic impacts of COVID-19: seven scenarios (March 2, 2020). CAMA Working Paper No. 19/2020. SSRN https://ssrn.com/abstract=3547729. https://doi.org/10.2139/ssrn.3547729

Nkengasong JN, Mankoula W (2020) Looming threat of COVID-19 infection in Africa: act collectively, and fast. Lancet 395:841–842

Oppenheimer M, Campos M, Warren R, Birkmann J, Luber G, O'Neill B, Takahashi K, Brklacich M, Semenov S, Licker R (2015) Emergent risks and key vulnerabilities. In: Climate change 2014 impacts, adaptation and vulnerability: part a: global and Sectoral aspects. Cambridge University Press, Cambridge, pp 1039–1100

Qu G, Li X, Hu L, Jiang G (2020) An imperative need for research on the role of environmental factors in transmission of novel coronavirus (COVID-19). Environ Sci Technol 54(7):3730–3732. https://doi.org/10.1021/acs.est.0c01102

Rawat M (2020) Coronavirus in India: tracking country's first 50 COVID-19 cases; what numbers tell [WWW document]. India Today. https://www.indiatoday.in/india/story/coronavirus-in-india-tracking-country-s-first-50-covid-19-cases-what-numbers-tell-1654468-2020-03-12. Accessed 15 May 2020

Sachindra DA, Huang F, Barton A, Perera BJC (2014) Statistical downscaling of general circulation model outputs to precipitation-part 1: calibration and validation. Int J Climatol 34 (11):3282–3303. https://doi.org/10.1002/joc.3914

Sarhadi A, Burn DH, Concepción Ausín M, Wiper MP (2016) Time-varying nonstationary multivariate risk analysis using a dynamic Bayesian copula. Water Resour Res 52:2327–2349

Sathaye J, Shukla PR, Ravindranath NH (2012) Climate change, sustainable development and India : global and national concerns. Curr Sci 90:314–325

Sharma A, Kumar GM (2018) District-level assessment of the ecohydrological resilience to hydroclimatic disturbances and its controlling factors in India. J Hydrol 564:1048–1057

Sillmann J, Kharin VV, Zhang X, Zwiers FW, Bronaugh D (2013) Climate extremes indices in the CMIP5 multimodel ensemble: part 1. Model evaluation in the present climate. J Geophys Res Atmos 118:1716–1733. https://doi.org/10.1002/jgrd.50203

Sinha J, Jha S, Goyal MK (2019) Influences of watershed characteristics on long-term annual and intra-annual water balances over India. J Hydrol 577:123970

Snell SE, Gopal S, Kaufmann RK (2000) Spatial interpolation of surface air temperatures using artificial neural networks: evaluating their use for downscaling GCMs. J Climate 13:886–895

Srivastava AK, Rajeevan M, Kshirsagar SR (2009) Development of a high resolution daily gridded temperature data set (1969–2005) for the Indian region. Atmos Sci Lett 10:249–254. https://doi.org/10.1002/asl

Wu Y, Jing W, Liu J, Ma Q, Yuan J, Wang Y, Du M, Liu M (2020) Effects of temperature and humidity on the daily new cases and new deaths of COVID-19 in 166 countries. Sci Total Environ 729:139051

Yilmaz AG, Perera BJC (2014) Extreme rainfall Nonstationarity investigation and intensity – frequency–duration relationship. J Hydrol Eng 19:1160–1172. https://doi.org/10.1061/(ASCE)HE.1943-5584.0000878

Zhang D, Hu M, Ji Q (2020) Financial markets under the global pandemic of COVID-19. Financ Res Lett:101528

Zhou F, Yu T, Du R, Fan G, Liu Y, Liu Z, Xiang J, Wang Y, Song B, Gu X (2020) Clinical course and risk factors for mortality of adult inpatients with COVID-19 in Wuhan, China: a retrospective cohort study. Lancet 395(10229):1054–1062

Zhu Y, Xie J (2020) Association between ambient temperature and COVID-19 infection in 122 cities from China. Sci Total Environ 724:138201

Part II
Integrated Impact and Risk of Virus Outbreaks

Chapter 5
Impact of Pandemics

Shashwat Ganguly, Atika Gupta, Sahil Arora, and Neha Arora

Abstract Pandemics, even though being rare events, leave profound footprints in the shaping of the world. The impacts elicit carefully-devised contingency measures to mitigate the spread of the disease and loss of lives. While containment plans are executed sequentially in phases and medical interventions take time, the pandemic leaves significant marks on various aspects, such as economic, financial, social, and environmental, in terms of impacts. Some impacts may even induce irreversible changes to the aspect(s). This chapter reviews the impact of pandemics ranging from the bubonic plague during the Roman Empire to Covid-19. Trade and travel activities has consistently been the common factor in the spread of majority of the pandemics, with subsequent hit succumbed by the economy. Fiscal shocks are comparatively more acute in nature and stock markets usually recover within 3–6 months after the pandemic trigger. While social effects are more varied and localised in different parts of the world, environmental impacts reveal the positive side of restricted human activities along with some negative side-effects towards the environment.

5.1 Introduction

The risk reduction measures followed after various pandemics involving social distancing, lockdowns in clusters, quarantines, travel bans, and halts in the transportation systems causes both direct and indirect impacts on social, economic, and financial aspects in the world on a global scale as well as local scale (Long and Feng 2020). Furthermore, a significant association between contingency measures and improvements in air quality, clean beaches, and environmental noise pollution has been found (Lal et al. 2020). With motivation of the existing literature, this chapter

S. Ganguly (✉) · A. Gupta · S. Arora · N. Arora
Indian School of Business and Finance, New Delhi, India
e-mail: shashwat.ganguly@isbf.edu.in

M. K. Goyal, A. K. Gupta (eds.), *Integrated Risk of Pandemic: Covid-19 Impacts, Resilience and Recommendations*, Disaster Resilience and Green Growth,
https://doi.org/10.1007/978-981-15-7679-9_5

presents a review of the impacts of pandemics in terms of four aspects—economic, financial, social, and environmental. While presenting the impacts in each of these aspects, coverage priority has been assigned to cases where a pandemic has had a pronounced effect on human lives in terms of death and pronounced effect on each of the corresponding aspects.

5.2 Economic Impacts

The world has experienced several pandemics in past. Each pandemic has impacted the economy in several ways, but the scale or magnitude of its impact depends upon factors such as level of economic development in the country and the extent of its integration with the rest of the world.

Below a review of the economic effects of pandemics that had occurred in past and their inter-linkages with trade is presented. This will help the understanding of the effects of modern day pandemics and the likely policy outcomes from that.

5.2.1 *Black Death in Fourteenth Century*

In fourteenth century (1347–1351), the world witnessed the most devastating pandemic—The Black Death. It originated in China in mid-1300s. It spread to the Mediterranean and Northern Africa and Sothern England through various trade routes. By 1350 it reached Northern Britain and Scandinavia. Due to the high fatality rate, the plague received the name of Black Death.

The countries during fourteenth century was much more integrated through trade than they were in sixth century (when another plague outbreak had happened). Thus, it impacted the economies much more. Boerner and Severgnini (2011) highlighted that the speed of trade was much more through seas and rivers than through land trade routes. This validates the fact that shipping routes aid in transmitting the plague as they were the carriers of food and supplies contaminated by infected fleas and rats (Howard 2019). Therefore, in fourteenth century, while on one hand trade was growing and benefitting economies, on the other hand it had also made countries more vulnerable to such epidemics.

Black death killed 33–60% of Europe's total population (Slavicek 2008). Consequently, it had significant impact on the economy. Economies during this time had agriculture as their dominant sector characterized by feudalism and serfdom. Therefore, economic impacts were largely driven by changes in the agriculture sector.

Trade suffered but only for short duration. Loss of population lead to reduction in demand for goods and services and drastic reduction in the amount of land under cultivation. There was shortage of labour which lead to increase in wages and lower rent payments (Britannica Encyclopedia 2020). According to Louise Chipley Slavicek (2008), labourer's wages increased 4–5 times over a period of few years in some parts of England. Several factors lead to rising standard of living. Increase in wages, less population relative to available resources (helped in reducing poverty and malnutrition) were some factors that lead to higher standard of living. Landowners providing workers with free housing, agricultural tools and inputs, a fixed share in profits, workers refusal to work if not provided with such facilities are indicative of a downfall of an exploitative form of agricultural serfdom or feudalism. A rise in standard of living was also evident from a switch of rural families from earthenware pots to costly metal cookware.

Slavicek in his work (2008) also highlighted that there were disruptions in work—"all over Europe shops were boarded up and farm fields and livestock were left unattended". Even healthy workers were not able to carry out their usual duties as they were demoralized by daily deaths. Similar thing happened in the city of Florence where peasants and workers neglected their daily work. Florence experienced high costs of certain goods and services. Goods consumed by sick like sweetmeats and sugar, capons, eggs became very costly. Servants who took care of sick became very expensive. During the pandemic there was high demand and shortage of services of physicians and apothecaries to treat the ill, lawyers and notaries to certify the will, grave diggers (as dozens or hundreds of people dying each day), clergymen to perform riots etc. Due to high demands of wages by workers, government intervened by regulating wages. However, all these attempts were met with a failure due to ongoing shortage of labour. The great mortality brought about by Black death also lead to the decay of serfdom. While workers representing lower or middle class saw a rise in their standard of living, landlords (representing upper class) saw a decline in their income (due to higher wages, reduced rent payments, forgiving of debt to peasants). Many families vanished completely as a result of the pandemic.

5.2.2 *Pandemics During Twentieth Century*

Twentieth century had seen three influenza pandemics—Spanish Influenza, Asian flu and Hong Kong flu. Below a brief review of their economic impacts based on the review of few papers on each are presented.

5.2.2.1 Spanish Flu

Spanish flu of 1918, had impacted world's half of the population and lead to 40–50 million deaths. This pandemic emerged during the last year of World War I. Hence, the disease spread as a result of mass movement of troops/militarists during World War I and not through trade routes. Garrett (2007) in his paper/report mentions "The global magnitude and spread of the pandemic was exacerbated by World War I, which itself is estimated to have killed roughly 10 million civilians and 9 million troops" It is considered to be the most devastating pandemic in the history.

As per the review by Saunders-Hastings and Krewski (2016), there were shutdowns of schools and businesses. Many businesses went bankrupt. There was increased absenteeism of workers from workplace. In Telephone companies' people were absent from their posts. "The Cumberland Telephone Co. reported more than a hundred operators absent from their posts. The telephone company asked that unnecessary calls be eliminated" (Garrett 2007).

The report by Garrett (2007) also mentions that businesses in health care lines experienced an increase in revenue while other businesses especially the service and entertainment industries suffered double-digit losses in revenue. Saunders-Hastings and Krewski (2016) found that Merchants' business declined by 40–70% whereas, retail grocery business declined by one-third. There was 50% decrease in production of coal mine operators and closing down of mines. Increase in demand for beds, mattresses and springs as sick people were recommended bed rest.

Garrett (2007) in his review of the Spanish flu discusses findings from different papers. One of the papers found that the shortage of labour supply lead to higher wages in manufacturing sector. Another study found that states which had experienced higher death rates due to influenza showed higher growth in per capita income. This is because after the pandemic they had higher capital per worker which increased their productivity (that is output per worker) leading to higher incomes. While these papers looked at economic effects of influenza in short run, a paper studied long run effects of this. The paper found that pregnant women who were exposed to this influenza during 1918 gave birth to children who had greater medical problems, lower educational attainment, lower income and higher physical disability. The author found "The children of infected mothers were up to 15% less likely to graduate from high school. Wages of men were 5–9% lower because of infection."

Spanish flu came at a time when countries had poor public health systems and planning. Thus, a positive Impact of Spanish flu was it led to advances in public health system (Garrett 2007).

As per Saunders-Hastings and Krewski (2016), the period from 1920s to 1950s saw advancement in public health sector—development of vaccines and establishment of a health organisation by League of Nations which was replaced by World Health Organisation in 1948. This period also saw growth in trade and travel. People travelled not only for business purposes but also for leisure. "With the arrival of commercial jet aircraft in the 1950s, the number of international travellers began to climb rapidly". The study also mentions that international trade had grown "140-fold from the industrial revolution of the nineteenth century to the twenty-first century". "Though the globalization of trade stalled from 1914 to 1945—limited by WWI, the Great Depression, and WWII—it, too re-emerge in the 1950s, initiating what has been called the *second age of globalization*". This also implies that the pandemics that had emerged during this period had limited impact on trade if any or for a short run period. Hence, on one hand advancement in public health may have increased country's ability to fight any pandemic in future but expansion in trade and travel also increased the risk of being affected by such pandemics in future. One can also say that the massive devastation caused by Spanish flu highlighted the weaknesses in existing health infrastructure and hence opened the way towards advancement of medical science and health practices.

5.2.2.2 Asian Flu

We didn't find much research work on this pandemic as it was a mild pandemic leading between one to two million deaths worldwide. As per Saunders-Hastings and Krewski (2016), this flu emerged in China in 1957. It spread to the United States by June 1957 where it caused about 70,000 deaths. By June the virus had spread to 20 countries. The pandemic spread through land, sea routes and military routes. Air travel played a small part in the spread of the virus. There was absenteeism at workplace and schools. Overall the economic impact was very mild. For example, the paper mentions that Gross Domestic Product (GDP) in US reduced by about 1%. "The economy recovered quickly after the pandemic. Though a mild influenza pandemic, the Asian flu provided a reminder of the persisting threat of the global spread of emergent diseases."

5.2.2.3 Hong Kong Flu

As per Saunders-Hastings and Krewski (2016), the period of 1959–1968 saw further developments in medical science and globalization. This flu was different from previous Asian flu in the sense that it spread rapidly through air routes. But it was milder than Asian flu. It had resulted into 500,000 and two million deaths worldwide. Similar to Asian flu there was increase in the absenteeism in schools and workplaces. However, as it was mild it didn't have any significant economic or social impacts.

5.2.3 Pandemics During Twenty-First Century

During 1970s countries became confident regarding the controlling of any pandemic as they were able to deal with Asian and Hong Kong flu (Saunders-Hastings and Krewski 2016). There were occurrences of other flus—Russian flu (which was reoccurrence of Asian (H1N1) flu) in 1977 and Avian Influenza. They were not widely spread and did not lead to significant illness and deaths (Saunders-Hastings and Krewski 2016). The authors also report that there was significant increase in trade and travel during this period. Air travel became very important due to the invention of the passenger jet—Boeing 747. In 1995, Establishment of World Trade organisation had further boosted the growth of multilateral International Trade.

Hence, one can say that the pandemics of twenty-first century emerged in a world which was much more globally integrated than any other previous pandemics. Hence, their scope of economic impact widens.

5.2.3.1 SARS (Severe Acute Respiratory Syndrome)

As per World Health Organisation, in 2002, there emerged a virus in the Guangdong province of Southern China which was called as SARS-CoV but it was identified in 2003. “An epidemic of SARS affected 26 countries and resulted in more than 8000 cases in 2003” (WHO 2020c). As per Keogh-Brown and Smith et al. (2009), there was decline in travel and tourism income for many effected countries. They studied the macroeconomic impact of SARS on selected countries and found that SARS had largest economic impact on GDP and investment. The sectors investment, retail sales, air transport, hotels, tourism and restaurants were the most affected sectors. The paper also found that China and Hong Kong were affected the most. It was also noted that SARS affected the economy only for a very short term period (often a month or less than quarter). Overall, the negative economic impact was balanced out by positive recovery thereafter. So, on the whole they did not find any significant economic loss due to SARS. The paper also found significant adverse effects on tourism and exports in some Asian countries.

Fan (2003) had also discussed about similar short term macroeconomic impact of SARS. It says that SARS can be considered as a temporary negative demand shock to the economy. There has been loss of consumer confidence, reduction in demand via decline in consumption, fall in investment and reduction in exports and imports in particular of tourism sector.

5.2.3.2 Swine Flu

It originated in Mexico and US In 2009. As per the review by Girard et al. (2010), “The extent of global trade and travel allowed swine flu to spread as widely in 6 weeks as past pandemics had in 6 months. By July, infection was reported in 122 countries, with 134,000 laboratory-confirmed cases and 800 deaths”. The paper by Saunders-Hastings and Krewski (2016), mentions that there have been societal disruptions and economic burden. There was increased expenditure on health care like drugs and hospitalization. For example: Total costs estimated in Canada was around CAD$2 billion. In terms of GDP the paper mentions a loss from 0.5 to 1.5%.

The paper by Smith et al. (2009) uses a general equilibrium model to assess the economic impact of the flu in UK economy under different scenarios. They found that illness causes loss of 0.5–1.0% of gross domestic product for low fatality scenarios and 3.3–4.3% for high fatality scenarios. School closures also impact GDP as it leads to absenteeism from work of family member. The study by Borse et al. (2011) on New York city, found that one adult has to miss work because of closures in 17% households. Global tourism and airlines were also negatively affected by the pandemic (Saunders-Hastings and Krewski 2016). Overall the countries were able to control its spread, but it did put strain in terms of healthcare expenditure.

5.2.3.3 MERS (Middle East Respiratory Syndrome)

In 2012 there emerged another virus called as Middle East Respiratory Syndrome (MERS) which was first identified in Saudi Arabia in 2012 (WHO 2019b). It reached Republic of Korea in 2015 due to an infected traveller coming from the Middle East. We found that most of the research papers have studied economic impact of MERS on Korean economy.

The paper by Joo et al. (2019) estimated loss of US$2.6 billion in tourism revenue which amounted to 0.2% of GDP. However, MERS did not have any significant impact on Korea's annual GDP growth in 2015. This is because travel and tourism accounted for only 5.1% of total GDP. Hence, a temporary decrease in travellers had a small effect on GDP. This implies that the impact on GDP will be greater for those countries which are highly dependent on tourism. The paper estimated losses to other sectors associated with noncitizen travellers like accommodation, food and beverage services and transportation sector. The pandemic has also negatively affected consumption expenditure. A paper by Jung et al. (2016), has examined the effects of pandemic on consumption and consumer behaviour. The paper mentions that "retail shops in South Korea faced a 3.4% decrease in sales, and department stores experienced a 16.5% decrease in sales". They found a significant drop in consumers' expenditures in traditional shopping channels a switch in their expenditure towards e-commerce.

5.2.3.4 Ebola

Ebola was first discovered in 1976. But its largest outbreak took place in West Africa in 2014–2016. It started in Guinea and then moved to several West African countries—Sierra Leone, Liberia, Senegal, Nigeria and Mali (UNDP 2015; WHO 2020b). It also spread to other parts of the world through air travels—Spain, Italy, Germany and US (WHO 2020b). Guinea, Sierra Leone and Liberia had the largest outbreak. As per the report by United Nations Development group (UNDP 2015), it was the longest, largest and deadliest epidemic. Unlike other epidemics it had lasted for more than a year. "In fact, in only around 6 months, there were 3774 cases of people infected by Ebola Virus Disease (EVD) and 1888 deaths in Guinea, Liberia and Sierra Leone, which surpassed the cumulative sum in 32 years (1976–2008) of 2232 infected and 1503 deaths."

The UNDP report also mentions that the costs associated with epidemic are medical expenses both at macro level and at household level (via their use of savings). Indirect costs are loss in productivity or reduction in labour participation. Due to high fatality rates there was reduction in labour supply. The trade impacts are not significant as the three most affected countries Guinea, Liberia and Sierra Leone had a very low share in trade. "Indeed, over the 2010–2013 period, the three countries account only for 1.73% of imports and 1.39% of intra-EU exports annually."

The report further predicted a reduction of 3.4% in average GDP growth in Guinea between 2014 and 2017 in high Ebola scenario. In Liberia, the loss of GDP per capita growth was 5% in high scenario. For Sierra Leone they predicted 8% points reduction in GDP growth. In other less affected countries as well they had predicted reduction in GDP. The report also said that the pandemic has affected the labour market negatively. "Seven out of ten people believed that the labour market shrank. Many workers reduced their weekly hours of work and some decided to temporarily stay at home to prevent themselves from Ebola contamination" There was decrease in household incomes as businesses were not working as usual. There was scarcity of goods which lead to higher prices. The pandemic had adversely affected people's food habits and reduced the access to food. In education sector there has been closure of schools, lower attendance and loss of teachers and students. "on average, four teachers or school staff who have died from the EVD".

5.2.3.5 COVID-19

COVID-19 originated in Wuhan, China. It was first reported to WHO on 31st December 2019. "The outbreak was declared a Public Health Emergency of International Concern on 30 January 2020" (WHO 2020a). The economic impact of COVID-19 can be categorized into five aspects—Supply Side Shocks, Demand Side Shocks, Labour Market, Prices, Trade. The following sections highlight these aspects in more detail.

Supply side shocks—COVID-19 has resulted into a negative supply shock due to supply side disruptions both in the world and in the domestic economy. In India, quarantines, lockdown, closing down of factories, workplaces and businesses have led to supply side disruptions internally. Many producers and traders import their products from China. Closing down of factories in China has reduced export supply affecting the production of goods. "China is the workshop of the world, being central to the entire global network.

So manufacturing disruption there will create secondary supply shocks in manufacturing sectors in almost all nations" (Baldwin and di Mauro 2020). Further, as seen in previous pandemics poor health of people and their family members will impact their productivity. It will also lead to increased absenteeism at workplaces as the lockdown opens. The review of past pandemic also suggest that school closures will further aggravate the supply shock as people chose to stay home to take care of kids rather than going to work. On the whole, the basic macroeconomic theory tells us that a negative supply shock results into lower GDP and higher prices in the economy.

Demand side shocks—The above supply side disruptions have also led to demand side disruptions. Loss of jobs, closing down of businesses and factories worldwide has impacted the income and therefore consumption. In this scenario there are delayed purchases by consumers. This has reduced the demand for both domestic products and foreign products (import demand). A negative demand shock also leads to lower GDP but lower prices.

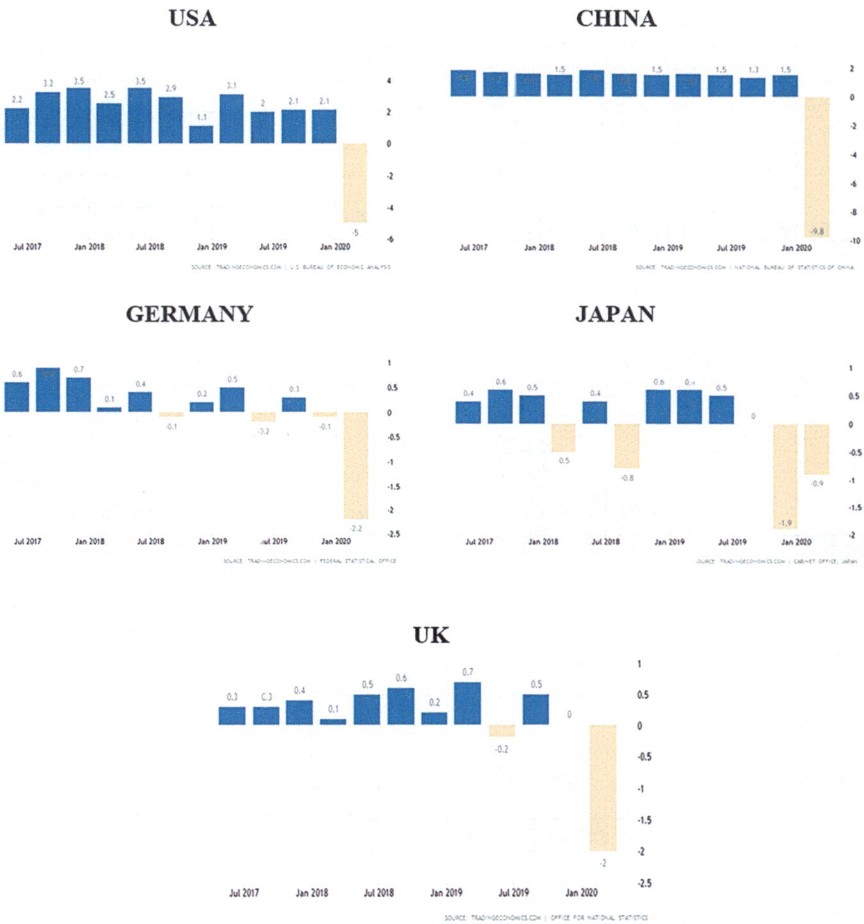

Fig. 5.1 GDP growths in various countries across the period of Covid Pandemic

Baldwin and di Mauro (2020) discuss that these shocks have hit the manufacturing and service sector (movies and restaurants) the hardest. They say that manufactured goods are more susceptible to demand shocks as these goods are 'postpone-able'. The consequence of these shocks is low GDP and growth in GDP. While some businesses are experiencing revenue losses other businesses dealing with Fast Moving Consumer Goods (FMCG) or health care products are experiencing higher revenue.

The above discussion implies that both demand and supply shocks will adversely affect GDP (as was seen in other pandemics too). We can see this through the graphs showing GDP growth rate in USA, China, UK, Germany and Japan (Fig. 5.1). China has shown the highest reduction in GDP by 9.8% followed by USA where GDP has reduced by 5% in the first quarter of 2020. Germany and UK have shown a

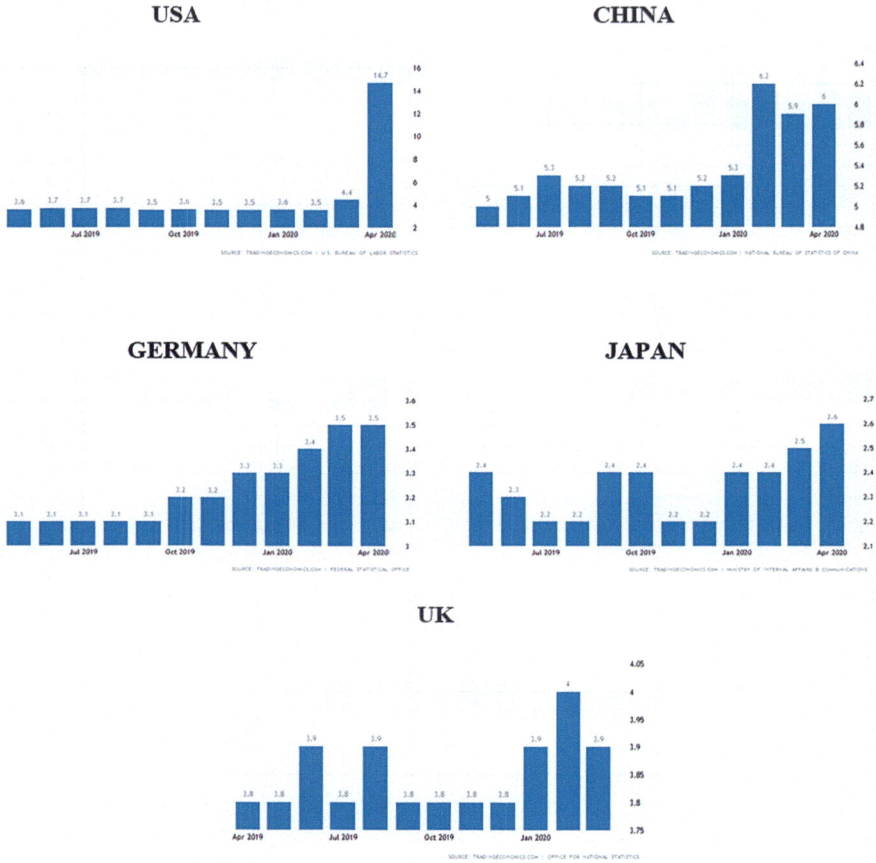

Fig. 5.2 Unemployment rates in various countries across the period of Covid

contraction by 2.2% and 2%, respectively. Japan has shown the least reduction by 0.9% at the beginning of the first quarter of 2020.

Labour market—As businesses, factories etc. are closing down there is less demand for labour resulting into job losses and higher unemployment. As can be seen in graphs (Fig. 5.2), the unemployment rate has increased in all the countries—USA, China, UK, Germany and Japan during the first quarter of 2020 with USA showing the highest increase in the unemployment rate by 14.7%. At the same time, people not being able to go to work due to sickness or school closure and deaths of people may lead to shortage of labour. Basic macroeconomic theory tells us that reduction in labour demand result into lower wages. On the other hand, reduction in labour supply will lead to higher wages which happened in case of Black death and Spanish flu. "If the pandemic causes a shortage of employees, there could be a temporary increase in wages for remaining employees in some industries" (Garrett

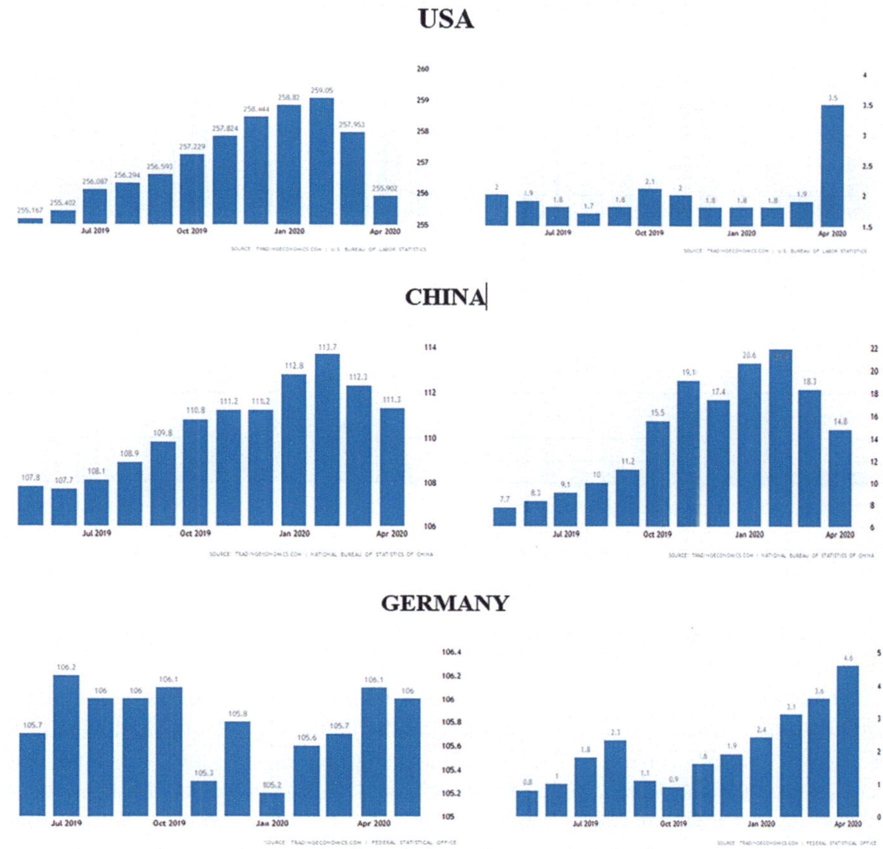

Fig. 5.3 Consumer Price Index (Left) vs. Food Inflation (right) for various countries across Covid

2007). Hence, how wages change will depend upon which effect dominates. For instance, wages in the United States increased to 25.12 USD/h in April from 24.08 USD/h in March of 2020 (Trading Economics 2020). It may be noted that the effect on wages can vary from sector to sector depending upon which factor (labour supply or labour demand is more dominating).

Prices—A negative supply shock results into higher prices and a negative demand shock results into lower prices. Therefore, the effect on overall prices and inflation is ambiguous. The data on Consumer price index (Figs. 5.3 and 5.4) shows decline since February 2020 in USA, China, Japan (since January) and UK. But, in case of Germany CPI has been increasing since February 2020, showing a slight decline after April 2020. But, cost of food has increased in all countries except for China (Figs. 5.3 and 5.4). Another impact that has been observed is decline in petroleum prices which had adversely affected MENA (Middle East and North Africa) countries. However, these countries have very limited participation in global

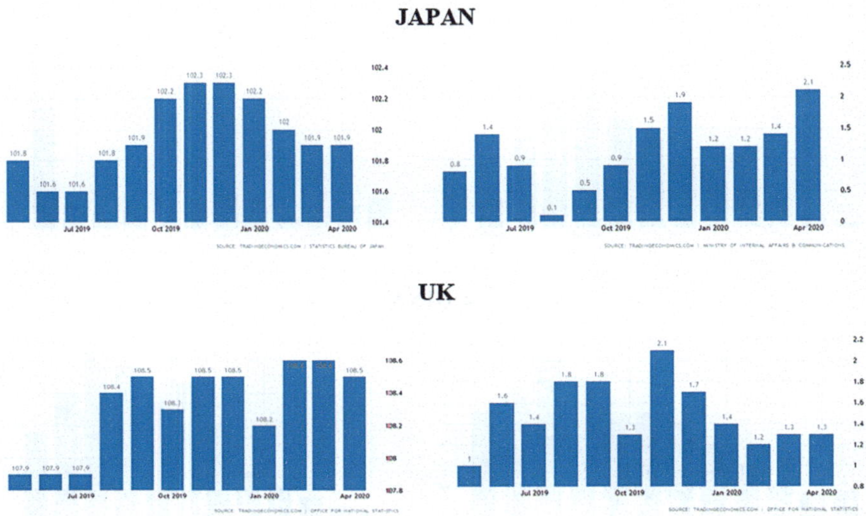

Fig. 5.4 Consumer Price Index (Left) vs. Food Inflation (right) for various countries across Covid (continued)

value chains and therefore were not impacted by disruption in Chinese manufacturing sector (Baldwin and di Mauro 2020).

Trade—Baldwin and di Mauro (2020) discuss that these shocks have significant implications for trade as the pandemic had hit the nations which are significant part of global supply chain. "The manufacturing sector of the six hard hit economies—China, Korea, Italy, Japan, US and Germany are at the heart of a myriad of international supply chains; each is an important supplier of industrial inputs to each other and to third nations". Hence, both supply and demand shock in these nations will slow down global trade.

The world trade organisation has predicted a decline in world trade by 13–32% in 2020. Their main predictions were:

- A fall in world merchandise trade by 13–32% in 2020.
- All regions will suffer double digits decline in trade volumes in 2020.
- Trade is more likely to fall in sectors having complex value chains like electronics and automotive parts
- Transport and travel restrictions will affect trade in services

Trade in services—hotels, tourism and travel has already been impacted. But one cannot say same for trade in health care sector as there is increased demand for healthcare products during the pandemic.

5.3 Financial Impacts

Stocks markets are interlinked and interdependent with the macro-economic variables of the economy. Researchers have discovered the close cross-market correlations during the crisis. Though, it has been observed that the changes in the stock prices indicators are short-lived and temporary in nature, induced primarily by anxiety, fear and rumours. Markets correct themselves and revert back to their previous levels within a period of 6 months.

Stock prices movements in one country spread to another because of being correlated. Negative sentiments induced by the pandemics among the investors strongly affect their investment decisions which changes the market forces for their price determination.

Table 5.1 depicts the percentage Market Returns 1 month after a major pandemic appeared over the period of 1980–2020. This depiction is based on the MSCI World index data which captures the large and mid–cap representation across 21 developed markets countries. With 1646 constituents, the index covers approximately 85% of the free float-adjusted market capitalization in each country. Also captured in the table are the changes in subsequent returns in the market between 1-and-3-months, and 3-and-6-months following the pandemics.

The table clearly indicates that even though the short-term impact of the respective pandemics turned negative in most of the cases, the market returns were positive within a period of 3 months and the returns furthermore improved within a period of 6 months.

The effects of various major pandemics individually on the stock markets are described as follows.

Spanish Flu (1918–1919)—The Spanish Flu and the World War 1, both occurred at the same time and it is reported that the Dow Jones index fell by around 30%.

Table 5.1 Market Returns between 1980 and 2020 with important pandemics and outbreaks

Pandemic	Period	1-Month market return (%)	Change in return (1–3 months)	Change in return (3–6 months)
HIV/AIDS	Jun-1981	−0.46	−4.18	1.39
SARS	Apr-2003	8.64	7.72	5.15
Avian Flu (H5N1)	Jun-2006	−0.18	2.95	7.28
Swine Flu (H1N1)	Apr-2009	10.9	8.83	10.23
MERS	May-2013	−0.29	2.44	6.43
Ebola	Mar-2014	−0.09	2.46	2
	Oct-2018	−7.42	−6.32	10.25

While the pandemic itself has a definite role to play towards this plummet, gauging the actual share of the same towards this change remains a challenge owing to the associated uncertainties in isolating the impact of the war and the pandemic.

Hong Kong Flu (1968–1969)—The effect of Hong Kong Flu has been depicted in the Dow Jones Industrial Average as it is reported that the Dow Jones index fell by 13.24% by the end of 1969 and the pandemic created such a wide-scale negative sentiment that the index further fell by a worst case of up to 21%.

SARS (2002–2004)—As per Liu et al. (2020), it is reported that cost of the 2003 SARS outbreak resulted in losses as high as in the financial crisis of Asia, estimated at $3 trillion value in GDP and $2 trillion value in financial markets equity.

SARS pandemic was selective in its impact on the stock market. Nippani and Washer (2006) examined the effect of SARS on the stock markets of Canada, China, the particular administrative region of Hong Kong, Indonesia, China, Singapore, the Philippines, Vietnam and Thailand, and concluded that SARS only affected the stock markets of China and Vietnam.

Swine Flu (2009–2010)—The virus began in a period when the economy had just came out of the financial crisis of 2008–2009. At the start of the flu, the stocks were already undervalued because of the financial crisis, so when the Swine flu pandemic actually ended around August 2010, its reported by Forbes that the Dow Jones index had risen almost by 40%.

Here, contradictory to other pandemics, it has been observed that the stock market improved because the start of the pandemic already had largely undervalued stocks, which lost their valuations with the bursting of the bubble at that time caused by the financial crisis.

Ebola (2014–2016)—It has been observed that investors may act irrationally with the news of the outbreak and eventually return to their previous levels once they have absorbed the outbreak news. Their changes in preferences actually lead to the changes in the prices of the stock and it is reported that they become risk averse at the time of the announcement of the outbreak and then stabilize themselves by returning back to their previous levels. When the investors in the market first observe the outbreak by the coverage of media, newspapers among other awareness channels, the abnormal market returns sharply decline and it's been reported that the decline represents an average loss of $300 million per event while concluded that the upper bound of the outbreak cost is estimated at $30 billion.

Covid-19 (2019-Present)—Market analysis post-Covid intervention shows that the travel related stocks experience gradual decrease in prices whereas some others such as the education and pharmaceutical sectors improved. It is thus observed that the impact of the pandemic is not uniform among all the sectors. It is also observed that different pandemics induce varying extent of impact on the share prices of a company. For instance, it is reported that the share prices of Lakeland Industries (LAKE), a small-capitalization company that makes personal protective equipment used by health-care professions, after Covid-19 is up more than 25%. The same organisation, jumped by almost six times after the 2014 Ebola problems came to light.

Shah in his article (Shah 2020) described how he used a python program to calculate the daily percentage changes in the prices. While the article describes methods to analyse the market visually using data, the reason behind the observations required revisiting. Once the daily percentage change of S&P500 had been calculated using the market close prices, an increased volatility was observed starting at March 2020. This can be correlated with the rise in Covid cases in the USA since March 2020. The underlying factor behind this correlation being the investors' sentiment changes leading to an increased volatility in the stock markets.

As per the study, Shah found similar volatility in other assets classes as well, including commodities including Crude Oil, Gold, S&P500 and the TLT (20 year US Treasury Bond) along with stock indexed. Once again, the corresponding increased volatility can be attributed to the changes in the investor's sentiments led by the news of the increased cases, due to which there is increase in the levels of anxiety among the behaviour of the investors.

5.4 Social Impacts

Who would have thought that in the modern world, where everyone dreamt of better technology, health infrastructure, quality of schools and colleges, education and availability of better resources, we will come across a pandemic crisis? Since the word "better" is relative, these accomplishments are still not enough. As the world is suffering because of Covid-19, it may be argued that the suffering may be different for different people. However, under the grand scheme of things, a significant majority of us are in the vicious circle of this virus. At these times of distress, one needs to remember the history as history would facilitate against repeating the same mistakes again. In terms of social impacts, we are at the same or perhaps even a step backwards as compared to our previous situations.

The social impacts of pandemics are interrelated with financial, emotional and physical implications in terms of widespread job loss and the related income loss, closing of educational institutes, increase in child labour, sexual exploitation and increase in the incidence of domestic violence and deaths. Social impacts are interlinked with economic impacts. Since we have discussed economic impacts in detail earlier, this section focuses on the social impacts of the various major pandemics.

Closure of educational institutes like school and colleges are often considered the first non-pharmaceutical intervention to reduce the effect of pandemic, as students lack carefulness in maintaining the social distancing along with other actions that may result in increase in the rate at which the virus can spread. Chen et al. (2011) conducted a research on responses coming from the educational institutions following an epidemic and a key observation from the study can be quoted: "Timely school closure and cancellation of public gatherings was significantly associated with reduced mortality related to influenza epidemics during the 1918 influenza epidemic in the United States". On similar lines, another study conducted by Navarro et al.

(2016) found that "More than 1300 public, charter, and private schools in 240 communities across the United States closed during the spring wave of the 2009 H1N1 pandemic".

Covid-19 also brings in similar effects which has led to loss of complete education for so many children around the world. As per the data shared by UNESCO for Covid-19, in mid-February 2020 there is only one country which closed the educational institute nationwide, affecting 0.1% of the total enrolled learners globally (UNESCO 2020). In mid-April 2020, educational institute closure was exercised in 194 countries nationwide, affecting 90.1% of the total enrolled learners. Similarly, because of having different degrees of the facilities at home as a result of varying income levels, it has been reported that the "crisis has exposed vast disparities in countries' emergency preparedness, internet access for children, and availability of learning materials" (Human Rights Watch 2020). Although many of the educational institutes have shifted to online learning platforms, many schools and the associated faculty are not friendly in using them which has brought in greater inefficiencies in delivering quality education to the students enrolled in such schools. It is reported that nearly half of the world has no internet access, which hints at the kind of education quality that we can achieve given the facilities or the time that will be required by us to meet the education standards for such children. Research shows that "households in low income countries, are more likely to reduce children's full-time school attendance and send them back to work when hit by economic shocks, using child labour as a form of risk coping mechanism" (UCW 2020).

A study (Cauchemez et al. 2009) have highlighted that "school closure also raises a range of ethical and social issues, particularly since families from underprivileged backgrounds are likely to be disproportionately affected by the intervention".

Under-privileged families who were already living in poverty before the pandemic may now be struggling even more, because of the additional costs of having the whole family at home, as it means extra expenses for sustaining the additional people. With additional members at home resulting from the lockdown, the family would need more food, gas and electricity. Additionally, with children being in and at home 24/7, the requisite entertainment cost of them is an additional burden. Also, the kids who were getting meals from the schools under the mid-day meal scheme, aren't getting fed in the situation created out of the pandemic (Barnard 2020). It can be noticed that the affected economies have already faced a hike in the unemployment rate. The services on which people on low incomes rely are also at risk of disruption, such as food banks (Stevens 2020). The millions of workers trapped in poverty are more likely to have insecure jobs, with fewer rights and employee benefits, and they are less likely to have savings to help cover additional unplanned costs or gaps in income. People stuck in poverty are more likely to experience anxiety, depression and other mental health difficulties (Barnard 2020).

The increased incidence of domestic violence as reported is indicative of the increased stress levels on the families, particularly those living under quarantines and lockdowns "The United Nations secretary-general has reported a "horrifying" global surge in domestic-based violence linked to COVID-19, and calls to helplines in some countries have reportedly doubled" (Human Rights Watch 2020). Along with the increased domestic violence, the increased number of women and girl abuse hints at the economic and social stresses as well as restrictions on movement, created as a result of the pandemics, in almost all countries. It is reported that the women who experience physical or sexual abuse are twice as likely to have an abortion, and the experience nearly doubles their likelihood of falling into depression. "Violence against women is as serious a cause of death and incapacity among women of reproductive age as cancer, and a greater cause of ill health than traffic accidents and malaria combined" (UN 2020).

To prevent the spread of COVID-19, social-distancing plays a key role, but has resulted in proximity for days-on-end with a partner or spouse one may no longer get along with. "Just as the Coronavirus pandemic increased divorce rates in China and the rest of the world, it seems like India, too, could face this psychosocial crisis after months of lockdown" (Outlook 2020).

5.5 Environmental Impacts

This is a lack of literature considering pandemic response-induced environmental impact studies. The term—Air Quality Index (AQI) only came into being only after 1999 prior to which different countries used their own metrics. Also, digitisation of air quality data only prevalent since the 1980s. Furthermore, post-1980 pandemics did not attract much research work on the environmental impacts owing to the insignificance in visible change to pollution levels. Covid-19, owing to the wide-spread global infections and the large-scale contingency measures to contain the infection, has led to significant visible changes to the environment and motivated towards a highly active research subject. Furthermore, the advancements in digitised information system eases the process of having data to infer from and base conclusions from.

The advancement in ice-core glaciology and laser technologies enabled scientists and researchers to derive vital information from the Arctic, leading to exceptional insights of pandemic-induced environmental impacts from our past. Hence this section is divided in to two groups—pre-Covid and post-Covid environmental impacts.

5.5.1 Historical Insights

This section is entirely based on the results obtained from the advancements in ice-core glaciology and laser technologies.

5.5.1.1 Pandemics and Pollution in the Iron Age

WHO's Institute of Health Metrics and Evaluation (IHME) estimated that in 2017, lead exposure accounted for 1.06 million deaths and 24.4 million years of healthy lives lost (disability-adjusted life years) worldwide due to long-term effects on health (WHO 2019a). Lead damages the central nervous system, linked to violent behaviour developmental problems in children. Lead pollution in air was always assumed to naturally exist at a certain level. However, investigations using ice core glaciology combined with history reveals the direct influence of plagues on air pollution levels during the period of 1100 BC to 800 AD—Late Iron Age to Antiquity and Late Antiquity period of the Early Middle Age (McConnell et al. 2018).

The most prevalent dynasty of this period, the Roman Empire, used silver coins for monetary purpose. These coins were produced in the Iberian Peninsula where the mines unearthed ores of silver, lead, and copper. From this pure silver was extracted using smelting, which filled the air with lead (McConnell et al. 2018). This highlights the significant impact of the bubonic plague during the Roman Empire on the environment in terms of the sharp decline in lead pollution in air.

5.5.1.2 The Black Death and Lead Plummet

Studies highlight a well-established consensus being pre-Industrial Revolution lead production was negligible and it took recent advancements in ice-core glaciology projects to prove how the consensus led to over-estimation of natural level of lead concentration in air (More et al. 2017). The study performed ice core investigation at the Swiss-Italian Alps and observed that a single 4-year period can be discovered when the atmospheric lead plunged to undetectable levels. These years (1349–1353) corresponds to the period of the Black Death. The plummet can be explained by the fact that a significant amount of smelting, mining, and manufacturing industries had to be stopped due to the great plague.

5.5.2 Post-Covid Impact Insights

The novel coronavirus has led to some severe containment measures across the globe. In addition to pan-national travels, the lockdown measures implemented confinement of population, reduction of public transport, and halted economic activities. In Italy, the most massive travel restrictions are being placed since Second World War (Saadat et al. 2020). In London, people are asked to stay inside their homes while the usually busy bars, theatres, and other public places have been closed. The flights are being cancelled in all over the world. This practice of staying at home, working remotely and practicing social distancing has quickly become a norm amongst the majorities (Harapan et al. 2020). While this is happening to control the spread of coronavirus, and to decrease the death rate, some unexpected side-effects of all the various containment measures are observed on the environment. While there has been an increasingly growing account of global records highlighting improvements in air pollution, some of the other segments of the environment have experienced negative impacts. The following sections will further describe the post-Covid environmental impacts in details across each of the segments—air, water, land, and noise.

5.5.2.1 Air Pollution

With over 91% of the global population living in places where poor air quality exceeds the permissible limits (Lal et al. 2020), air pollution is accountable for 8% of total deaths in the world. Out of these, the most affected countries in the world belong to Africa, Asia, and some parts of Europe (World Health Organisation 2020).

The most commonly assessed air pollutants are Nitrogen Oxides (NOx), Suspended Particulate Matters—PM2.5 and PM10, Carbon Monoxide (CO), Sulphur Dioxide (SO2), and Ozone (O3). CO in air is attributed to the incomplete burning of carbon-based fuels and is spread by wind circulation patterns throughout the lower atmosphere (Novelli et al. 1998). However, with the suspension of human activities and containment measures in response to the prevention of rapid spread of the infection impacted the environment all over the world, both positively and otherwise. A lot of research has been undertaken recently to assess the impacts on environment, with air pollution being the most widely focused segment.

Starting with the country from where this infection originated, China introduced perhaps the strictest lockdown measures of all. The country witnessed a significant improvement in the overall ambient air quality. If the monthly average of the suspended particles PM2.5 is to be compared with the same over the previous 3 years, a drop by 20–30% is observed over large parts of China (Lal et al. 2020; Zambrano-Monserrate et al. 2020). Furthermore, the NO2 levels also reduced by 22.8 $\mu g/m^3$ in Wuhan and 19.2 $\mu g/m^3$ overall in the country. Most interestingly, it was observed that in China alone, all of these quarantine-induced air quality improvements paved the way for human health benefits that has outnumbered the coronavirus-led deaths thus far (Chan et al. 2020).

Moving to the country where air pollution poses the most significant threat in the world, India remarkably became a hefty beneficiary of the infection-containing contingency measures. Sharma et al. (2020) monitored and assessed the improvements in both the suspended particle matters PM2.5, PM10, NO2, CO, and O3. The concentrations during the lockdown period were compared with the same time period over the last 4 years. Apart from O3 concentration, which increased by 17%, the rest of the pollutants observed a reduction of 43%, 31%, 10%, and 18% respectively. While the overall CO concentration observed slight decreases during the 11th and 16th week in 2020, Delhi in Northern India observed a significant reduction in CO concentration during the tenth to 16th week due to effective implementation of the shutdown of major industries (Lal et al. 2020). The positive results of this study will clearly present a confidence-boosting case to the regulatory bodies and help them device a strict air quality control plan in the future.

In Europe, Barcelona in Spain experienced an improvement in PM10, NO2, and Black Carbon concentrations by 31%, 51%, and 45% when the lockdown period was compared to a month before lockdown (Tobías et al. 2020). This significant reduction can be primarily attributed to the absence of traffic emissions. Ozone concentration, on the other hand, was found to increase by 33–57% after the lockdown. If NO2 alone is taken in to account, the air pollutant concentration improved by around 14–25% over Rome, Madrid, and Paris; the first three European cities that implemented strict quarantine measures (Lal et al. 2020).

Over the American Continents, Brazil showed a significant drop in CO levels (Between 30.3 and 48.5%), mostly owing to the restrictions in light duty vehicles and the corresponding vehicular emissions (Dantas et al. 2020). However, as the country started relaxing the restriction based on the consequence of lack of consensus about the importance and need of social distancing and lockdown, an increase in vehicular flux and movement of people was observed.

Researchers in Almaty, Kazakhstan also experienced a similar trend with improvements in PM2.5 (21%), CO (49%), and NO2 (35%) concentrations, with an increase in ozone concentration by 15%. This decrease is once again attributed to the restricted movement of vehicles in the city during the lockdown period.

5.5.2.2 Water Pollution

The coronavirus pandemic has introduced some significant changes in our water bodies. Unlike the case of air pollution, where the results are entirely positive, the infection-based responses have led to both positive and negative impacts on the water sources.

Numerous accounts of the crystal-clear canals, rivers, shorelines have been observed around the world. The canals of Venice have never displayed such clear water that even the sediment bed at the bottom is visible. With no tourists, motorboats, sediment churning, and other water pollutants have been dropped efficiently, residents got amazed by seeing the clear water and the fish could be seen once again in the canals (Saadat et al. 2020). In India, the Ganges has witnessed a massive improvement in term of visible clarity as viewed from the riverbanks in various parts of India along the course of the river (India Today 2020). Attributed as one of the most polluted in the world, Yamuna, in Delhi also witnessed a significant amount of visible improvement (TOI 2020).

However, one result of shut down of shops and restaurants is limited sell of food products, and of them all, the most pronounced effect towards water pollution comes from the discharge of dairy product into water bodies. In US alone, it is common to dump 3.7 million gallons of milk per day, according to estimates from Dairy Farmers of America, the country's largest cooperative (Jeffery 2020). In France, around 5000 tonnes of French cheese are at risk of going to waste amid the pandemic as demand dropped significantly by 60% (Michael Lacoste, president of National Council of

Appellations of Dairy Origin) (Snouwaert 2020). In the UK, similar records are observed. According to one farmer from Wiltshire, 17,000–18,000 l of milk is dumped down the drain every day (Fair 2020).

Milk could be 2300 times more harmful than treated sewage and even 350 times more harmful than untreated sewage if it enters the water courses (Slavov 2017). This is based on the high Biological Oxygen Demand (BOD) of milk and milk products.

5.5.2.3 Land Pollution

The quarantine policies, established in most countries, have led consumers to increase their demand for online shopping for home delivery. Consequently, organic waste generated by households has increased. Also, food purchased online is shipped packed, so inorganic waste has also increased. Medical waste is also on the rise. Hospitals in Wuhan produced an average of 240 metric tons of medical waste per day during the outbreak, compared to their previous average of fewer than 50 tons. In other countries such as the USA, there has been an increase in garbage from personal protective equipment such as masks and gloves (Calma 2020).

As a result of the pandemic, countries such as the USA have stopped recycling programs in some of their cities, as authorities have been concerned about the risk of COVID-19 spreading in recycling centres. In particularly affected European countries, waste management has been restricted. For example, Italy has prohibited infected residents from sorting their waste. Also, the industry has seized the opportunity to repeal disposable bag bans, even though single-use plastic can still harbour viruses and bacteria (Bir 2020).

5.5.2.4 Noise Pollution

Environmental noise is sourced from anthropogenic activities including industrial and commercial activities, along with vehicular sounds and music at high volumes. Environmental noise contributes as a source of discomfort towards both population and environment, causing health problems and altering the natural conditions of the ecosystems (Zambrano-Monserrate and Ruano 2019). The imposition of quarantine measures by most governments has caused people to stay at home. With this, the use of private and public transportation has decreased significantly (Zambrano-Monserrate et al. 2020). Also, commercial activities have stopped almost entirely. All these changes have caused the noise level to drop considerably in most cities in the world.

5.6 Conclusions

A review of economic impacts of previous pandemics shows that COVID-19 pandemic has been similar to them in many aspects

- All the pandemics had spread through trade and travel via land, air or shipping except for Spanish flu which spread through military routes. Higher the extent of global trade quicker is the transmission as in case of swine flu.
- There has been increase in businesses in healthcare services due to increased demand for healthcare products. This has led to higher prices and strains on healthcare expenditure.
- There were higher prices during Ebola due to scarcity of goods
- In case of some pandemics like Black death and Spanish flu (where death rate has been very high) have led to shortage of labour and higher wages. But, in some cases like COVID-19 impact on wages is ambiguous. Lower demand of labour has resulted into higher unemployment and lower wages (as is more likely happening in COVID-19)
- There have been disruptions in work, closure of schools, businesses and therefore, loss of production
- In service sector, hotels and entertainment industry had been adversely affected. There had been decline in tourism and travel revenue in more recent pandemics (2000 onwards). But, impact on productivity is ambiguous. In case of Spanish flu it was found that the flu lead to higher capital per worker and hence, higher productivity. But other study has found low productivity in long run due to lost skills (arising from unemployment) and adverse effect on productivity of kids from mothers pregnant during the pandemic. There was loss in productivity and labour force participation in Ebola.
- There has been adverse effect on GDP, investment and consumer confidence. The impact on GDP was greater for those countries which are highly dependent on tourism.
- The trade impacts have not been significant for pandemics before twenty-first century and for countries Guinea, Liberia and Sierra Leone which had a very low share in trade
- The news of the pandemic creates radical impacts on the mood and anxiety of investors in markets.
- It can be conclude that the concept of behavioural finance is applicable on the behaviour of the investors which is temporary and once the news of the pandemic gets absorbed in the minds of the investors, their anxiety level reduces and the market stabilizes by reverting back to their pre-pandemic levels or even surpassing those levels with a positive returns within a period of 6 months.
- Even the short term impact is sector specific. It has been observed that the travel related stocks showed a steep decline but at the same time, the industries whose demand rose with the news of the pandemic improved manifold.
- On the face of it, though it may seem that social impacts are increasing the short term stress levels and creating challenges in terms of women's abuse, child labour

and domestic violence among others—these can be far reaching and can turn into the long term impacts which can deteriorate the conditions for the upcoming future.

- It is really important to control these ill effects for the bright future of the economies as a significant amount of inter-dependencies exist between social aspects and economic, political, and financial aspects.
- Advancements in ice-core glaciology highlights the positive impact of pandemics in the past on air pollution reduction. Similar changes can be witnessed after Covid-19 induced-contingency measures limiting human activities.
- The positive effects of the Covid-19 pandemic towards the environment transcends the boundaries of air pollution and proves beneficial to other sectors such as water and noise pollution.
- Some negative aspects can however be attributed owing to increased waste production, lack of recycling, and dumping of waste food and dairy products to land and water bodies.

Overall, Covid-19 does differ in particular in terms of its international coverage and speed of transmission. "Preliminary medical studies find that Covid-19 is less deadly but more transmissible than SARS" (Baldwin and di Mauro 2020). It has engulfed many more countries than previous pandemics as the world now is much more integrated in terms of trade and travel. Since the pandemics in the past had spread mostly through trade and travel routes, it becomes clear that increased economic integration has made countries more vulnerable to negative economic impacts induced by such pandemics. Baldwin and di Mauro (2020) write that "The ten nations hit hardest by Covid-19 is almost identical to the list of the ten largest economies in the world (Iran and India are the exceptions). The US, China, Japan, Germany, Britain, France, and Italy are all in the top-ten most affected by the disease. Taking just the US, China, Japan, Germany, Britain, France, and Italy, they account for 60% of world supply and demand (GDP), 65% of world manufacturing and 41% of world manufacturing exports". These facts are indicative of the fact that a modern day pandemic like Covid-19 will have much significant economic consequences for the whole world than any other pandemic observed previously.

References

Baldwin R, di Mauro BW (2020) Economics in the time of COVID-19. CEPR Press, Washington, DC

Barnard H (2020) Coronavirus: what does it mean for people restricted by poverty? Joseph Rowntree Foundation. https://www.jrf.org.uk/blog/what-does-covid-19-mean-people-restricted-poverty

Bir B (2020) https://www.aa.com.tr/en/health/single-use-items-not-safest-option-amid-covid-19/1787067. Accessed 14 May 2020

Boerner L, Severgnini B (2011) Epidemic trade. SSRN. https://doi.org/10.2139/ssrn.1896603

Borse RH, Behravesh CB, Dumanovsky T, Zucker JR, Swerdlow D, Edelson P, Choe-Castillo J, Meltzer MI (2011) Closing schools in response to the 2009 pandemic influenza A H1N1 virus in New York City: economic impact on households. Clin Infect Dis 52(1):S168–S172. https://doi.org/10.1093/cid/ciq033

Britannica Encyclopedia (2020) Black death pandemic. Britannica Encyclopedia. https://www.britannica.com/event/Black-Death

Calma J (2020). https://www.theverge.com/2020/3/26/21194647/the-covid-19-pandemic-is-generating-tons-of-medical-waste. Accessed 14 May 2020

Cauchemez S, Ferguson N, Wachtel C, Tegnell A (2009) Closure of schools during an influenza pandemic. Lancet Infect Dis 9(8):473–481. https://doi.org/10.1016/S1473-3099(09)70176-8

Chan JFW, Yuan S, Kok K-H, To K-W, Chu H, Yang J, Xing F, Liu J (2020) A familial cluster of pneumonia associated with the 2019 novel coronavirus indicating person-to-person transmission: a study of a family cluster. Lancet 395:514–523. https://doi.org/10.1016/S0140-6736(20)30154-9

Chen W-C, Huang A, Chuang J-H, Chiu C-C (2011) Social and economic impact of school closure resulting from pandemic influenza A/H1N1. J Infect 62(3):200–203. https://doi.org/10.1016/j.jinf.2011.01.007

Dantas G, Siciliano B, França BB, da Silva CM, Arbilla G (2020) The impact of COVID-19 partial lockdown on the air quality of the city of Rio de Janeiro, Brazil. Sci Total Environ 729:139085. https://doi.org/10.1016/j.scitotenv.2020.139085

Fair J (2020) Pump and dump: could the Covid-19 pandemic create more pollution? Ends Report. https://www.endsreport.com/article/1681933/pump-dump-covid-19-pandemic-create-pollution

Fan EX (2003) SARS: economic impacts and implications. Policy Briefs – China. ISSN 1655–5260. https://www.adb.org/publications/sars-economic-impacts-and-implications. Accessed 18 May 2020

Garrett TA (2007) Economic effects of the 1918 influenza pandemic implications for a modern-day pandemic economic effects of the 1918 influenza pandemic implications for a modern-day pandemic (Issue November). https://www.stlouisfed.org/community/other_pubs.html

Girard MP, Tam JS, Assossou OM, Kieny MP (2010) The 2009 a (H1N1) influenza virus pandemic: a review. Vaccine 28(31):48995–44902. https://doi.org/10.1016/j.vaccine.2010.05.031

Harapan H, Itoh N, Yufika A, Winardi W, Keam S, Te H, Megawati D, Hayati Z, Wagner AL, Mudatsir M (2020) Coronavirus disease 2019 (COVID-19): a literature review. J Infect Public Health 13:667–673

Howard J (2019) Plague explained. National Geographic Society. https://www.nationalgeographic.com/science/health-and-human-body/human-diseases/the-plague/

Human Rights Watch (2020) COVID-19's devastating impact on children. Human Rights Watch. https://www.hrw.org/news/2020/04/09/covid-19s-devastating-impact-children

India Today (2020) India's coronavirus lockdown reveals fresh air, cleaner rivers. India Today. https://www.indiatoday.in/india/story/coronavirus-lockdown-india-fresh-air-clean-rivers-1669726-2020-04-22

Jeffery A (2020) Wasted milk, euthanized livestock: photos show how coronavirus has devastated US agriculture. CNBC. https://www.cnbc.com/2020/05/02/coronavirus-devastates-agriculture-dumped-milk-euthanized-livestock.html

Joo H, Maskery BA, Berro AD, Rotz LD, Lee Y-K, Brown CM (2019) Economic impact of the 2015 MERS outbreak on the Republic of Korea's tourism-related industries. Health Secur 17(2). https://doi.org/10.1089/hs.2018.0115

Jung H, Park M, Hong K, Hyun E (2016) The impact of an epidemic outbreak on consumer expenditures:an empirical assessment for MERS Korea. Sustainability 8(5):1–15

Lal P, Kumar A, Kumar S, Kumari S, Saikia P, Dayanandan A, Adhikari D, Khan ML (2020) The dark cloud with a silver lining: assessing the impact of the SARS COVID-19 pandemic on the global environment. Sci Total Environ. https://doi.org/10.1016/j.scitotenv.2020.139297

Liu H, Manzoor A, Wang C, Zhang L, Manzoor Z (2020) The COVID-19 Outbreak and Affected Countries Stock Markets Response. Int J Environ Res Public Health 17:2800

Long H, Feng W (2020) Research report on companies' survival and development strategy during a novel coronavirus epidemic. UIBE Press, Beijing

McConnell JR, Wilson AI, Stohl A, Arienzo MM, Chellman NJ, Eckhardt S, Thompson EM, Pollard AM, Steffensen JP (2018) Lead pollution recorded in Greenland ice indicates European emissions tracked plagues, wars, and imperial expansion during antiquity. PNAS 115 (22):5726–5731. https://doi.org/10.1073/pnas.1721818115

More AF, Spaulding NE, Bohleber P, Handley MJ, Hoffmann H, Korotkikh EV, Kurbatov AV, Loveluck CP, Sneed SB, McCormick M, Mayewski PA (2017) Next-generation ice core technology reveals true minimum natural levels of lead (Pb) in the atmosphere: Insights from the Black Death. GeoHealth 1(4):211–219. https://doi.org/10.1002/2017GH000064

Navarro JA, Kohl KS, Cetron MS, Markel H (2016) A tale of many cities: a contemporary historical study of the implementation of school closures during the 2009 pA(H1N1) influenza pandemic. J Health Polit Policy Law 41(3):393–421. https://doi.org/10.1215/03616878-3523958

Nippani S, Washer KM (2006) SARS: a non-event for affected countries' stock markets? Appl Financial Econ 14:1105–1110. https://doi.org/10.1080/0960310042000310579

Novelli PC, Masarie KA, Lang PM (1998) Distributions and recent changes of carbon monoxide in the lower troposphere. J Geophys Res Atmos 103(D15):19015–19033. https://doi.org/10.1029/98JD01366

Outlook (2020) Rejuvenate your relationship. Outlook. https://www.outlookindia.com/newsscroll/rejuvenate-your-relationship/1841427

Saadat S, Rawtani D, Mustansar C (2020) Environmental perspective of COVID-19. Sci Total Environ 728:138870. https://doi.org/10.1016/j.scitotenv.2020.138870

Saunders-Hastings PR, Krewski D (2016) Reviewing the history of pandemic influenza: understanding patterns of emergence and transmission. Pathogens 5(4):66. https://doi.org/10.3390/pathogens5040066

Shah I (2020) How pandemics impact financial markets. Towards Data Science Organisation. https://towardsdatascience.com/pandemics-impact-financial-markets-9a4feb6951f5

Sharma S, Zhang M, Anshika, Gao J, Zhang H, Kota SH (2020) Effect of restricted emissions during COVID-19 on air quality in India. Sci Total Environ 728:138878. https://doi.org/10.1016/j.scitotenv.2020.138878

Slavicek LC (2008) The black death. Chelsea House Publications, New York

Slavov AK (2017) General characteristics and treatment possibilities of dairy wastewater – a review. Food Technol Biotechnol 55(1):14–18. https://doi.org/10.17113/ftb.55.01.17.4520

Smith RD, Keogh-Brown MR, Barnett T, Tait J (2009) The economy-wide impact of pandemic influenza on the UK: a computable general equilibrium modelling experiment. BMJ. https://doi.org/10.1136/bmj.b4571

Snouwaert J (2020) 5,000 tons of French cheese are at risk of going to waste amid the coronavirus pandemic as demand sinks. Business Insider. https://www.businessinsider.in/retail/news/5000-tons-of-french-cheese-are-at-risk-of-going-to-waste-amid-the-coronavirus-pandemic-as-demand-sinks/articleshow/75583422.cms

Stevens J (2020) My food bank used to offer human contact, but coronavirus has ended that. The Guardian. https://www.theguardian.com/commentisfree/2020/apr/04/food-bank-coronavirus-human-contact-desperate-people-home

Tobías A, Carnerero C, Reche C, Massagué J, Via M, Minguillón MC, Alastuey A, Querol X (2020) Changes in air quality during the lockdown in Barcelona (Spain) one month into the SARS-CoV-2 epidemic. Sci Total Environ 726:138540. https://doi.org/10.1016/j.scitotenv.2020.138540

TOI (2020) Lockdown improves water quality of Ganga, Yamuna. The Times of India. https://timesofindia.indiatimes.com/india/lockdown-improves-water-quality-of-ganga-yamuna/articleshow/75456921.cms

Trading Economics (2020) Data source. Trading Economics. https://tradingeconomics.com/

UCW (2020) Child labour and the global crisis: an issues paper. https://www.unicef.org/protection/files/UCW_Summary_Financialcrisis_TCfinal.pdf

UN (2020) UN chief calls for domestic violence 'ceasefire' amid 'horrifying global surge.' United Nations - News. https://news.un.org/en/story/2020/04/1061052

UNDP (2015) Recovering from the Ebola crisis

UNESCO (2020) Education: From disruption to recovery. UNESCO. https://en.unesco.org/covid19/educationresponse

WHO (2019a) Lead poisoning and health. World Health Organisation. https://www.who.int/news-room/fact-sheets/detail/lead-poisoning-and-health#:~:text=The Institute for Health Metrics,-and middle-income countries

WHO (2019b) Middle East respiratory syndrome coronavirus (MERS-CoV). World Health Organisation. https://www.who.int/news-room/fact-sheets/detail/middle-east-respiratory-syndrome-coronavirus-(mers-cov)

WHO (2020a) Coronavirus disease (COVID-19) pandemic. World Health Organisation. https://www.who.int/emergencies/diseases/novel-coronavirus-2019

WHO (2020b) Ebola virus disease. World Health Organisation. https://www.who.int/news-room/fact-sheets/detail/ebola-virus-disease

WHO (2020c) International Travel and Health (SARS). World Health Organisation. https://www.who.int/ith/diseases/sars/en/

World Health Organisation (2020) Coronavirus disease 2019 (Covid-19): situation report, 81, WHO

Zambrano-Monserrate MA, Ruano MA (2019) Does environmental noise affect housing rental prices in developing countries? Evidence from Ecuador. Land Use Policy 87:104059. https://doi.org/10.1016/j.landusepol.2019.104059

Zambrano-Monserrate MA, Ruano MA, Sanchez-Alcalde L (2020) Indirect effects of COVID-19 on the environment. Sci Total Environ 728:138813. https://doi.org/10.1016/j.scitotenv.2020.138813

Chapter 6
Global Governance of COVID-19: Decline of Public Sphere and Transnational Democracy

Neeraj Mishra

Abstract This chapter provides a sociological analysis of the dynamic relationship between epidemics and human society in general, and the socio-political effects of CoViD-19 in particular. It first describes the fragility of human civilizations since ancient times due to the disease effects of the unprecedented concentration of crops, people, and livestock together with their attendant pathogens and parasites, which led to epidemics of one kind or another. It then looks at the origin of SARS-CoV-2 in similar circumstances in a wet seafood market in China and the disruption of lives and livelihoods that it caused in almost every country on the globe. Using Habermas's theory of public sphere and its democratic function, and Nancy Fraser's idea of 'transnational' public sphere, this chapter analyses how the growing weakness of public sphere and its critical democratic function, combined with the limited freedom of speech has resulted in the loss of the best early-warning system a world threatened by epidemics can have. With example of CoViD-19 pandemic and its governance in India and China, this chapter finally discusses the shortcomings in the functioning of a truly democratic society and its (in)ability to deal with such pandemics. It recommends the formation and strengthening of a 'global public sphere' beyond the nation state, to not only counter such epidemic threats to human life in the future, but to also build a resilient, just and deeply democratic world community.

6.1 Introduction

An unprecedented silence descended upon an otherwise unusually noisy country as 1.3 billion people of India were confined to their homes with the beginning of nationwide lockdown on 25 March 2020, 2 months after the Chinese government

N. Mishra (✉)
School of Humanities and Social Sciences, Indian Institute of Technology, Indore, Madhya Pradesh, India
e-mail: nmishra@iiti.ac.in

M. K. Goyal, A. K. Gupta (eds.), *Integrated Risk of Pandemic: Covid-19 Impacts, Resilience and Recommendations*, Disaster Resilience and Green Growth,
https://doi.org/10.1007/978-981-15-7679-9_6

imposed a complete lockdown of the Hubei province of China on 23 January 2020, in order to contain the outbreak of COVID-19 pandemic, caused by a family of coronaviruses called SARS-CoV-2. In our history of living together with microorganisms, humans have developed an immunity towards many of the commonly occurring viruses. However, every once in a while there comes a virus that either jumps from an animal host, or mutates in such a fashion that it takes the human immune system by surprise and infects large number of people, disrupting their normal lives and killing thousands, or even bringing an entire city, kingdom or a civilization to a sudden end. SARS-CoV-2 is one such coronavirus that has brought the entire world to a standstill, infecting and killing people of all nations mercilessly at its own discretion, irrespective of their caste, class, race or gender.

This chapter first traces the origin and evolution of pandemics from the prehistorical times till SARS-CoV-2 and discusses some of their important effects on human societies. The second part of the chapter focuses on CoViD-19 and enumerates the disruptions caused by this virus in people's lives in different locations as it moved from one country to the other till it circumambulated the entire globe. Along with the financial impacts on the global economy, the 10 week lockdown of countries such as India brought out a plethora of socio-political and psychological issues that lay hidden behind the statistical data on economic growth and 'demographic dividend'. The strictly differential and class-based impact of the national lockdown has not only brought into limelight the broader issues of social security, labor laws, social justice, and growing income inequality, but also the very nature of Indian economic organization as well as the democracy. At the same time, Chinese reaction to the initial reports of the outbreak and its curtailment, especially the first reports by Dr. Li Wenliang from Wuhan, forces us to think how freedom of speech in China could have drastically reduced the number of infections and deaths, if the world community could get a 3 week head start on the coronavirus SARS-CoV-2.

Building on Habermas's idea of 'public-sphere' as the realm of deliberation and discussion on collective issues, this chapter is theoretically anchored in the possibility and advantages of 'transnationalizing the public sphere' as described in the works of Nancy Fraser. From the limited realm of the nation state, the concept of public sphere must be expanded and 'scaled up' to correspond to existing realities of a globalized world. Although there is a good deal of interesting political theory now on how global governance must be democratized, the formation of the public sphere beyond the nation-state has received surprisingly little critical attention (Nash 2014: 3).

Such public sphere has the potential to act as the moral conscience of the world community and democratize policy making and information sharing in circumstances where the concerns are truly global. Realtime information about the detection of SARS-CoV-2 was one such concern, and a public bad that could have been efficiently contained and its negative effects drastically reduced if a vibrant global public sphere existed. The final section of this chapter ends with a brief discussion on some of the ways to prevent such pandemics in the future.

6.2 Creating Space for Zoonosis in Human Habitats: Prehistoric Conditions

This section argues that the growth and prosperity of human beings on this planet has always been fraught with dangers and uncertainties, geological as well as geographical, but most importantly biological. There are around 1.5 million viruses in wildlife that humans are not aware about and it is highly likely, and absolutely unpredictable as to when, where and how, any one of those would spillover into human population and may bring about the collapse of a civilization. Human civilization is a fragile interlude between ice ages, which needs favorable geological, geographical, economic and biological conditions for its growth and nourishment.

> The disappearance of these conditions—sometimes even one of them—may destroy a civilization. A geological cataclysm or a profound climatic change; an uncontrolled epidemic like that which wiped out half the population of the Roman Empire under the Antonines, or the "Black Death" that helped to end the Feudal Age...these are some of the ways in which a civilization may die (Durant 1954: 3–4).

Certainly it is possible, as Aristotle thought, that many civilizations came, made great inventions and luxuries, were destroyed, and lapsed from human memory. History, said Bacon, is the planks of a shipwreck; more of the past is lost than has been saved (Durant 1954: 107). From the collapse of ancient Rome to the fall of the Mayan empire, evidence from archaeology suggests that five factors have almost invariably been involved in the loss of civilizations: uncontrollable population movements, new epidemic diseases, failing states leading to increased warfare, collapse of trade routes leading to famine, and climate change (Hutt 2016). Most of these factors are observable today in the aftermath of the CoViD-19 pandemic, but it is hoped that our civilization does not come to an unceremonious end. This section explores some of the circumstances in which the human civilizations began in the first place, creating enough space for the zoonotic viruses to take root.

James Scott, in his search for the 'earliest states' explains that contrary to the popular belief that sedentism and cultivation led directly to state formation, it was only long after the fixed-field agriculture that the state began to appear (Scott 2017: 3). This basic fact is troublesome for popular version of human prehistory, which portrays it as a narrative of progress and civilization as codified by the great agrarian kingdoms. The dominant discourse in this regard has been an "ascent of man" story. Agriculture, it held, replaced the savage, wild, primitive, lawless, and violent world of hunter-gatherers and nomads. Contrary to this assumption, Scott argues that there is massive evidence of determined resistance by mobile peoples everywhere to permanent settlement, even under relatively favorable circumstances, and the implicit assumption of the standard narrative, namely that people couldn't wait to abandon mobility altogether and "settle down"—may also be mistaken (Scott 2017: 11).

Pastoralists and hunting-and-gathering populations have fought against permanent settlement, associating it, often correctly, with disease and state control. At the very least, we have no warrant at all for supposing that the sedentary "givens" of modern life can be read back into human history as a universal aspiration (Scott 2017: 08). In today's world, it is possible to compare them with the tribal people living in forests and secluded islands, which keeps them safe from any form of state control or spread of epidemic diseases. In his view, disease was a major factor in the fragility of the early states, which erupted mostly in congested living quarters. Its effects, however, are hard to document because many epidemic diseases left no obvious bone signature (Scott 2017: 7).

6.2.1 Sedentism and Birth of 'Multi-species Resettlement Camps'

This section shows that domestication of plants and animals for human consumption brought together many different species in close contact with each other, which increased the chances of mutation and formation of zoonotic viruses that were able to jump the species divide and infect the human population in homestead.

Domestication of plants and animals marked a paradigm shift in human history, which has also been referred to as "late-Neolithic multi-species resettlement camps" (Scott 2017: 18). Some authors define domestication as 'control over reproduction' of fire, plants, animals as well as slaves, state-subjects, and women in a patriarchal family. Domestication of fire, plants, and animals made concentration of food and people possible (Scott 2017: 12; Durant 1954: 8). It also changed the genetic makeup and morphology of both crops and animals around the homestead. The assemblage of plants, animals, and humans in agricultural settlements created a new and largely artificial environment, promoting new adaptations.

A second great and unanticipated outcome of agricultural lifestyle was the direct epidemiological effect of concentration, of people, livestock, crops, and the large suite of parasites and pathogens that followed them to the homestead or developed there. Such epidemics killed 99% of the population, especially those who were not exposed to domestication and living with animals in close proximity (Diamond 1997: 92). Diseases such as measles, mumps, diphtheria, and other community acquired infections, appeared for the first time in the early states. It seems almost certain that a great many of the earliest states collapsed as a result of epidemics analogous to the Antonine plague and the plague of Justinian in the first millennium CE or the Black Death of the fourteenth century in Europe (Scott 2017: 20–21). Hence it can be argued that the major killers of human beings in recent history are infectious diseases that evolved from diseases of animals.

When a virus jumps from an animal species to a human, it is called a zoonotic virus. Since the beginning of human life on this planet, these new kinds of viruses have been causing more and more outbreaks. The modern farming of animals for consumption has further increased the chances of zoonosis as the animals are reared in congested spaces and live in close proximity with the other species. SARS-CoV-2 is one such zoonotic virus that originated among the bats and is now spreading among the humans causing loss of lives and livelihoods worldwide.

6.3 Pandemics in Human History: A Review

This section shows how the human civilization has been faced with epidemic diseases since ancient times and how it has proved to be lethal on many occasions, decimating cities and towns each time it appeared.

The world's population in 10,000 BCE, according to one careful estimate, was roughly four million. A full 5000 years later, in 5000 BCE, it had risen only to five million. One likely explanation for this paradox of apparent human progress in subsistence techniques together with long period of demographic stagnation is that, epidemiologically, this was perhaps the most lethal period in human history. Epidemic disease is, I believe, the "loudest" silence in the Neolithic archaeological record (Scott 2017: 96–97). Epidemics affected domestic animals and crops that were also concentrated in the late-Neolithic multi-species resettlement camp. A population could as easily be devastated by a disease that swept through their flocks or their grain fields as by a plague that menaced them directly.

The Epic of Gilgamesh, an epic poem from ancient Mesopotamia provides perhaps the most powerful evidence when its hero claims that his fame will outlive death as he depicts a scene of bodies felled, probably by pestilence, floating down the Euphrates (Kruszelnicki 2007).

About 5000 years ago, an epidemic wiped out a prehistoric village in northeast China, leaving behind the remains of 97 human bodies, which were found stuffed into a small house. The site, whose modern-day name is '*Hamin Mangha*' is the largest and best-preserved prehistoric settlement site found to date in northeast China (Jarus 2015). In 430 BC, a plague struck the city of Athens, which was then under siege by Sparta during the Peloponnesian War (431–404 BC). In the next 3 years, most of the population was infected, and perhaps as many as 75,000 to 100,000 people, 25% of the city's population, died. It spared no segment of the population, including the statesman Pericles, who died of plague. The epidemic broke in early May 430 BC, with another wave in the summer of 428 BC and in the winter of 427–426 BC, and lasted 4.5–5 years (Littman 2009: 456).

Between 161 and 180 AD, in the reign of Marcus Aurelius Antoninus, an ancient pandemic was brought to the Roman Empire by troops returning from campaigns in Near East. Also known as the 'Antonine Plague' or the 'Plague of Galen', it claimed roughly 2000 lives per day in Rome, killing at least one-fourth of the people who were affected. More than five million people died, which reduced the overall population of the Empire by one-third and devastated the Roman army.

Another pandemic that afflicted the Roman Empire between 249 and 262 AD was called the 'Plague of Cyprian' named after the Bishop of Carthage. The city's population had declined by about 62% (from something like 500,000–190,000). An Athenian historian claimed that 5000 died each day. "The human race is wasted by the desolation of pestilence." Pandemic influenza might indeed account for the horrifying experience of the Plague of Cyprian (Harper 2017).

The 'Plague of Justinian' or the bubonic plague struck in Byzantium in the sixth century (541–542 AD) and killed a substantial part of the population in Constantinople as well as the port cities around the Mediterranean Sea, as merchant ships harbored rats that carried fleas infected with plague (Littman 2009: 457). Some historians believe the plague of Justinian was one of the deadliest pandemics in history, resulting in the deaths of an estimated 25–100 million people during two centuries of recurrence, a death toll equivalent to as much as half of Europe's population at the time of the first outbreak (Rosen 2007).

Between 1347 and 1351, another pandemic, popularly known as the 'Black Death' ravaged Europe and North Africa taking a proportionately greater toll of life than any other known epidemic or war up to that time. The Black Death probably arose in 1334 in the Chinese province of Hubei. It appears to have spread to Europe along the Mongol trading routes, arriving in Constantinople in 1347. By 1348, the Black Death had spread to Genoa and Venice, England, France, Spain and Portugal. The Black Death also devastated the Middle East, reaching Antioch in 1348–1349, Mecca and Mosul in 1349, and Yemen in 1351 (Kruszelnicki 2007). In China, some two-thirds of the population died in eight geographically separate incidents from 1353 to 1354, almost 90% of the people in Hubei had died in the initial outbreak. In Europe, some 70% of the people had died from the Black Plague by 1400, dropping the population from seven million to two million.

Some of the other epidemics resulting in massive loss of lives include 'Cocoliztli[1] epidemic' of 1545–1548 that killed more than 15 million inhabitants of Mexico and Central America. These illnesses, including smallpox, contributed to the collapse of the Inca and Aztec civilizations. Some estimates suggest that 90% of the indigenous population in the Western Hemisphere was killed off due to these diseases brought by the Spanish conquerers.

The 'Great Plague' reappeared in London in 1665–1666, killing more than 100,000 people and decimating 15% of the population of London. From London the plague moved to Marseille in France in the early eighteenth century (1720–1723) in which another 100,000 people died. Other countries that suffered from plague in the eighteenth century included Russia (1770–1772), where about 50,000–100,000 people died in Moscow alone. The Philadelphia 'yellow fever epidemic' of 1793 claimed 5000 lives before it subsided with the onset of winter.

Even nineteenth century had its share of epidemics and pestilence causing millions of deaths in different parts of the world. Most notable among them was the 1889–1890 flu pandemic, also called 'Asiatic Flu' or 'Russian Flu' that killed one million people worldwide.

As the World War-I came to an end in 1918, it had already killed nine million soldiers and 13 million civilians as a direct result of the war. But the worst was yet to come in the form of an influenza pandemic caused by the A/H1N1 virus, which came

[1]"Cocoliztli" is the Aztec word for pest.

to be known as the 'Spanish Flu'. The flu's spread and lethality was enhanced by the cramped conditions of soldiers and poor wartime nutrition that many people were experiencing during World War I (Jarus 2020). And the soldiers returning home from the War in different parts of the world, helped in its rapid global spread. Lasting for more than 12 months, it infected 500 million people and killed an estimated 50 million people worldwide, including an estimated 675,000 people in USA alone (Jordan et al. 2019). The scientists were able to isolate and recreate the 1918 virus in 2005 and reported that the constellation of all eight genes together make an exceptionally virulent virus, a uniquely deadly product of nature, evolution, and the intermingling of people and animals (Tumpey et al. 2005).

Since 1918, the world has experienced four additional pandemics, in 1957, 1968, 2009 and most recently in 2019. The 1957 H2N2 pandemic and the 1968 H3N2 pandemic each resulted in an estimated 1.1 million global deaths, with 116,000 occurring in USA.

The 2009 H1N1 Swine Flu originated in Mexico before spreading to the rest of the world. In 1 year, the pandemic infected as many as 1.4 billion people across the globe and killed between 151,700 and 575,400 people, according to the Center for Disease Protection and Control, USA. This perhaps begs the question of whether a high severity pandemic on the scale of 1918 could occur in modern times (Jester et al. 2018). Enter COVID-19.

Besides some of the major pandemics mentioned above, the humans in different parts of the world have been faced with pestilence of one kind or another at regular intervals, too many to be enumerated here. A complete list of such epidemics, from eleventh century BC to 2007 has been provided in *The Encyclopedia of Plague and Pestilence* which has more than 700 epidemics in its list, spread over time and space throughout the world (Kohn 2008).

From the review above, it is clear that human society has always lived in a shadow of pestilence that has the potential to wipe out entire cities and communities. It has happened many times in our historical past and is highly likely to reoccur in the future. The risk of eradication due to viruses and pandemics is a constant threat that would accompany our existence on this planet, and we as humans may not have the means to respond appropriately if we do not prepare ourselves well in advance.

6.4 CoViD-19: Origin of a Pandemic

The COVID-19 virus started spontaneously in Hunan Seafood Market in Wuhan, the capital city of Hubei province in central China, as per the official reports of the Chinese government. In a paper published by the doctors of Jin Yin-tan Hospital, Wuhan, it was reported that first patient with symptoms of COVID-19 was already found on 1 December 2019, and had no relation to the Hunan Seafood Market. Also, there were no epidemiological links between the first patient and later cases. On 10 December, three more cases were reported to the hospital, of which two patients had no link to the Hunan Seafood Market. Of the 41 cases studied by the authors,

27 (66%) had been exposed to Hunan Seafood Market, while 14 patients had no contact to the said market (Huang et al. 2020). If first cases were reported on 1 December 2019, the infections must have already happened in November, as there is an incubation period between the infection and its visible symptoms, argued Daniel Lucey, an infectious disease specialist at Georgetown University. This means that the "virus came into that marketplace, before it came out of that marketplace," Lucey asserts. Bin Cao of Capital Medical University, the corresponding author of *The Lancet* article clarified that "seafood market is not the only origin of the virus. But to be honest, we still do not know where the virus came from now" (Cohen 2020).

On 30 December 2019, Wuhan Health Commission issued an internal notice clarifying that "there has been continuous occurrence of pneumonia cases of unknown cause at Hunan Seafood Market". The next day, on 31st December, The Wuhan Health Commission issued a public notice saying that "some medical institutions found a link between the pneumonia cases and the Hunan Seafood Market. However, there was no evidence of obvious human-to-human transmission and no infection among medical personnel". On first January 2020, the Hunan Seafood Market was closed down for disinfection and cleaning.

On 26 January 2020, the Institute of Virology, China's Center for Disease Control and Prevention (CDC) announced that only 33 of the 585 environmental samples from the Hunan Seafood Market were found to contain the novel COVID-19 nucleic acid. This suggested that the virus originated from the wild animals sold at the Hunan Seafood Market, as reported. An argument that questions the origin of COVID-19 from seafood market is based on the fact that no one sold bats or any type of bat meat in the Hunan Seafood Market or the nearby areas. Hence some researchers are of the opinion that the official position of the Chinese government that COVID-19 originated spontaneously from Hunan Seafood Market, may be a coverup to avoid enquiry into the actual source of its origin, which maybe somewhere else (The Epoch Times 2020).

Another study revealed that the CoViD-19 virus was most closely related (89.1% nucleotide similarity) to a group of SARS-like coronaviruses that had previously been found in bats in China. This outbreak highlights the ongoing ability of viral spill-over from animals to cause severe disease in humans (Wu et al. 2020). A Chinese virologist and expert on bat viruses, at the Wuhan Institute of Virology, Zengli Shi, also called the 'bat woman of China', through her research on Coronavirus, was able to understand how these viruses can overcome cross species barriers and attack human bodies. In a study published in 2010 and then in 2015, Shi discussed her work on synthetic viruses, S-Protein, and the possibilities of mutation that coronaviruses can undergo, and how such viruses can be transmitted across species, especially humans, sometimes even without an intermediary host.

6.4.1 Reactions and Initial Responses to SARS-CoV-2

In the first few weeks of the outbreak, Chinese government as well as the other world leaders downplayed the possible impacts of the new virus, identified as SARS-CoV-2, that was sweeping the globe, arguing that the situation was absolutely under control. But the experts working in the field of infectious disease had already issued warnings that a global pandemic with massive destructive potential was coming soon.

In the last 6 months of its spread, CoV-19 has already infected 7,085,740 people and caused 405,272 deaths and many more are predicted. The SARS was a similar coronavirus that erupted in 2002 killing hundreds of people around the world. Carlos Urbani, an Italian physician working for the World Health Organization (WHO) in Hanoi, Vietnam, treated the first SARS cases there and notified the WHO of the severity of this disease. Unfortunately, Urbani contracted the disease and died of it in March 2003, thereafter the WHO put the entire world on alert (Wang and Cowled 2015: 30).

MERS in 2012 was also a coronavirus that caused similar damage and panic around the world. These were enough grounds for the experts to predict with extreme certainty that the next coronavirus was just around the corner.

What is so special about SARS-CoV-2 that has become a global problem like never before? The most devious aspect of CoViD-19 is that one can be infected and be without any symptoms for quite some time, or they can be mistaken for a flu. Due to this lack of any pronounced symptoms, the carrier continues to infect people in his/her vicinity, who further infect others, and the virus grows at an exponential rate.

The SARS-CoV-2 is the youngest in a whole family of 7 coronaviruses known to infect humans, now famous as SARS-CoV and MERS-CoV, which have already killed lots of people. But the other four are more successful viruses—HCoV-OC43, HCoV-229E, HCoV-NL63, and HCoV-HKU1, are causing up to a third of common colds everyday (Coronavirus Explained 2020).

Live animal markets such as those in China and other countries give animal viruses all kinds of opportunities to mix and mutate and jump to humans. Just like this zoonotic SARS-CoV-2, SARS also emerged in a live animal market in Guangdong province in China.

This prompted Chinese researchers to look for more such coronaviruses among the bats, especially those that were similar to the ones that had already affected the humans. They were able to find something called ‘bat coronavirus RaTG13’ which met the criteria for low risk virus. When the scientists sequenced the genome of the virus behind CoV-19, they found that 96% of it was identical to that bat virus, which led to the conclusion that bat virus, in all likelihood, evolved into this new virus that could infect humans. It was believed that this new virus might have mutated in another bat, or it may have jumped to another species, like a pangolin or a snake, before jumping to us.

The difficult part of containing such viruses is the absence of any vaccines or medicines that can counter their spread. This leaves us with only one ‘old school’ method of slowing down the spread of the virus, i.e. quarantine and maintaining 'spatial distance' from other human beings. This 'flattens the curve' killing lesser number of people over a relatively longer period of time. In the absence of such measures, a large number of people get very sick in a short span of time that causes more deaths. As it became known from the example of Italy, it was much harder for hospitals to save lives if too many people get infected at once. At the same time, countries such as South Korea were able to contain the virus even without a lockdown, by testing widely and retracing the steps of the people who came back positive.

A report was published by Global Preparedness Monitoring Board, WHO in September 2019 titled '*A World at Risk- Annual Report on Global Preparedness for Health Emergencies*' under the leadership of Dr. Brundtland (WHO 2019). It clearly mentioned that 'there is a very real threat of a rapidly moving, highly lethal pandemic of a respiratory pathogen killing 50–80 million people and wiping out nearly 5% of the world's economy. A global pandemic on that scale would be catastrophic, creating widespread havoc, instability and insecurity. The world is not prepared.'

Deforestation and growth of urban townships in such places where there were once thick forests, has brought wild animals in closer contact with human habitats. The factory farming is pushing animals closer together, giving their viruses more opportunities to combine into one that could infect humans. In a nutshell, if the human habitats and food habits overlap with, and include wild animals, there is a very real threat of zoonotic viruses moving into the human households and causing infectious diseases such as CoViD-19.

The Chinese authorities imposed a lockdown in Hubei on 23 January 2020 in an effort to quarantine the epicenter of COVID-19 preventing all movement to, and from, Wuhan. In the next two weeks, many more cities in China enforced a similar lockdown and restricted the movement of people completely. After China went into a lockdown, in February Italy became the next epicenter of the virus, as it remained open. While Italy locked down in March, USA did not, becoming the next epicenter. And soon the cases were being reported from all parts of the world, including the developing countries, where the lockdown is much harder to effect and the healthcare systems are almost non-existent.

On 30 January 2020, the Italian Prime Minister Giuseppe Conte informed the press that in 'Italy, the situation was fully under control'. On 25 February 2020, the President of Iran, Hassan Rouhani called COVID-19 a 'plot by the enemies to bring Iran to a complete shutdown'. Italian Prime Minister perhaps made a hasty judgement when he announced that in Italy the situation was fully under control, unaware of the fact that it was to become next epicenter of the virus. Soon Italy began the lockdown process towards the end of February, and complete shutdown of the country by early March. By 25 March 2020, more than 213 countries all over the world were in some form of lockdown, local or national, affecting more than three billion people.

As per the data provided by worldometers.info, a total of 213 countries around the world have reported a total of 7,085,740 confirmed cases of the coronavirus COVID-19 and a death toll of **405,272** people on 8 June 2020. From this website, it is clear that USA and Europe have been the worst hit regions in this pandemic. UK, Italy, Spain, France, Brazil, Belgium, Germany, India among others nations would need a long term strategy to revive their economies and socio-psychological lives. Coincidently, top 25 economies of the world were also the worst hit countries, reporting maximum number of deaths and therefore forced to lockdown their nations as well as their economic activities.

This chapter uses the theoretical lens of 'public sphere' along with the idea of 'global public sphere' and Fraser's concept of transnational democracy to account for some of the reasons that have led to the failure in containing the CoViD-19 and its spread from one part of the globe to the other.

6.5 Theoretical Approach: Public Sphere, Its 'Scaling-Up' and Global Democracy

This section first defines the concept of public sphere in the works of Habermas, and then introduces the idea of 'scaling up' of this concept to 'transnational public sphere' by Nancy Fraser. It argues that public sphere theory is in principle an important critical-conceptual resource that could be reconfigured and applied usefully to the modern globalized society to avert such pandemics in future by the formation of a critical public sphere beyond the nation state. It is argued that the development of an empowered 'global public sphere' could act as an effective deterrent and an 'early-warning system' in the fight against global emergencies such as CoViD-19 before they become a global hazard.

For Habermas, public sphere or '*Öffentlichkeit*' designates a sphere of open (public) spaces and communication where a public discourse on matters of common concern can take place and lead to the formation of an opinion on part of the public of citizens that in turn may influence political decision making. He defined the public sphere as a virtual or imaginary community which does not necessarily exist in any identifiable space. In its ideal form, the public sphere is "made up of private people gathered together as a public and articulating the needs of society with the state" (Habermas 1991). Through acts of assembly and dialogue, the public sphere generates opinions and attitudes which serve to affirm or challenge—therefore, to guide-the affairs of state. In ideal terms, the public sphere is the source of public opinion needed to "legitimate authority in any functioning democracy" (Rutherford 2000: 18).

The *normative legitimacy* and *political efficacy* of public opinion are essential to the concept of the public sphere in critical theory. Without them, the concept loses its critical force and its political point (Fraser 2014: 9). Fraser argues that these two features are not easily associated with the discursive arenas that we today call 'transnational public spheres'. In her view, the notion of a 'transnational public sphere' is indispensable in the current 'postnational constellation'. After Habermas proposed the theory of public sphere in 1962, it has remained within the frame of a bounded political community with its own territorial state. Tacitly, Habermas's account of the public sphere took for granted the Westphalian framing of political space, and the public sphere was conceptualized as a key institutional component of national democracy.

However, in the post Cold-War geopolitical reconfiguration, and increased international movements associated with globalization, it has become possible, and necessary to rethink public sphere theory in a transnational frame (Fraser 2014: 10). The present reconfiguration of the public sphere transcends its earlier confinement to the nation state. This understanding problematizes the critical function of the public sphere, its legitimacy and political efficacy as it is no more addressed to a sovereign state, but has a transnational character.

From global warming to CoViD-19, the geographical boundaries of the nation states have become redundant as far as such problems are concerned. And the people fighting against these global problems are not confined to any one nation either. They communicate and raise awareness throughout the planet and move as the problems move, from one country to another. Chinese doctors fighting CoViD-19 in Italy, French doctors fighting Ebola in Africa, Greenpeace warriors fighting climate change in different parts of the world, 'Doctor Without Borders' serving the poor in many developing countries etc. are some of the examples of this phenomenon.

6.5.1 Fraser's Six-Point Criticism: Challenges of Building a 'Global Public Sphere'

Nancy Fraser specifically enumerates six limitations of the Habermasian concept of public sphere that needs to be addressed before a *legitimate* and *efficacious* 'global public sphere' can be envisioned and established.

First of all, almost all nations of the world today depend on international institutions for many of their key governance functions such as environmental regulation, monitoring of the cyberspace, cross-border terrorism, inter-national river governance, etc. This situation leads to the inadequacy of a usable critical theory of the public sphere, and its confinement to the nation state does not hold (Fraser 2014: 21).

The second assumption that a public coincides with national citizenry, which formulates its common interest as the general will of a bounded political community, is belied by such phenomenon as migrations, diasporas, dual-citizenships, multiple residency etc. 'Often the interlocutors are neither co-nationals nor fellow citizens, and the opinion they generate, therefore, represents neither the common interest nor the general will of any *demos*' (Calhoun 2002). Far from institutionalizing debate among citizens who share a common status as political equals, post-Westphalian publicity appears in the eyes of many observers to empower transnational elites, who possess the material and symbolic prerequisites for global networking (Calhoun

2002). This leads to another problem regarding the criteria of efficacy and legitimacy of the meaningful application of the transnational public opinion in a post-Westphalian world.

The third presupposition of a national economy and its regulation by the territorial public-sphere does not hold in a global marketplace where transnational corporations, outsourcing, global financial markets, currency valuation etc. are all accountable to international forces of capital and labor movement and not to any public. Hence the sovereign state has little control over its economy, which implies that the national public opinion cannot be an effective force in articulating and directing the economy in the general interest of the people (Fraser 2014: 23).

The fourth assumption that public opinion is conveyed through a national communications infrastructure, centered on print and broadcasting, has come under severe criticism in the last couple of years. The free and independent character of the media has been sacrificed to the logic of capital and finance, as in most cases the media is owned by large capitalist houses. We can also observe the emergence of internet based communication channels such as twitter or facebook, which have the ability to bypass the state borders, signaling the denationalization of communication infrastructure. Granted, we see some new opportunities for critical public opinion formation, but these go along with the disaggregation and complexification of communicative flows. Given a field divided between corporate global media, restricted niche media, and decentered internet networks, how could critical public opinion possibly be generated on a large scale and mobilized as a political force (Fraser 2014: 24)? In the absence of any common citizenship or platform, it remains a challenge for the transnational media audiences to deliberate together as peers driven by a common goal.

The sixth assumption that a public sphere rests on a national vernacular literature, which supplies the shared social imaginary needed to underpin solidarity, too, is today counterfactual. The increasing hybridization of the world population, the rise of world literature, global mass entertainment aired by internet, popularity of Hollywood movies worldwide, among other things, have led to a relative decline of the shared national social imaginary, and it is difficult to recognize the sort of (national) literary cultural formation seen by Habermas as underpinning the subjective stance of public-sphere interlocutors.

From the above discussion it can be observed that the public-sphere theorized by Habermas as operating within the confines of a national territory has spilled over into the international arena, which has led to the decline of the critical function of public-sphere as a legitimate and efficacious force. As described by Fraser, public sphere theory has become insufficient to accommodate the socio-political changes that have come with globalization, at the same time the necessary modifications that would make it usable for the global public sphere remain unclear.

Nancy Fraser advocates a powerful case for post-national constellations which can protect all human beings from the adverse effects of globalization (Linklater 2007: 31). Fraser reflects on how the public sphere can be liberated from the traditional preoccupation with democratizing the 'Westphalian nation-state', arguing for the possibility (and necessity) of a global civilizing process which links citizens and aliens as co-legislators in a transnational public sphere, as all persons have the right to be consulted about decisions that affect them (Linklater 2007: 32).

6.6 Analysis of CoViD-19: Decline of Public Sphere and Global Democratic Governance

While one can partially agree with Fraser's criticism of 'disaggregated sovereignty' and the inadequacy of a usable critical theory of the public sphere, it can also be observed that international institutions involved in the partial governance of the affairs of a nation may not always be powerful enough to exert their will for the larger good of the global community. For e.g. WHO, which is supposed to lead the global response to such pandemics, is actually a very small organization, very dependent on voluntary contributions, without many resources or budget to actively pursue research and development (Gates 2019). Even if some states do not fully control their own territories, they are still in a position to control the people within their territories and confine the critical public opinion from escaping the national boundaries. The confinement of Dr. Li Wenliang and his early findings on CoViD-19 within the nation is an example of this phenomenon.

The second criticism of globally dispersed national citizens who do not amount to any specific 'demos' may be problematic from the perspective of legitimacy argument. But if the role of public-sphere is to act as the 'critical eye' on governmental policies, the dispersed citizens can still be critically active and build pressure on their home governments to act in the interests of national as well as the global community, based on Horkheimer's idea of shared vulnerabilities to suffering.

With CoViD-19, the entire globe is in an extremely vulnerable situation and the Chinese diaspora living in other countries such as USA had very explicitly spoken about the lack of freedom of speech in China and how it was leading to such disasters. But the Chinese government was under no obligation to pay any attention to their opinion as they were not the citizens of the country. And those who were actually the citizens, would not be able to speak, precisely because of this lack of freedom.

The international public opinion can still act as a potent force in directing some of the policies of the concerned state, which are evidently detrimental to the welfare of the world community. It may not be efficacious all the time, but its very presence in the international public sphere is the proof of its legitimacy. As said by one of the leading journalists in India that 'their opinion may not be efficacious enough to force the states to act in a certain manner, but it certainly is enough to show the world community, as a historical fact, that a group of international citizens opposed this act when it was happening'.

The other criticisms about the availability of a platform to build the global public opinion, the presence of a language that is understood worldwide, and a shared social imagery needed to underpin solidarity may emerge from some of the socio-political changes that we need to undertake as a world-community in our search for a truly just society. Hence it is important to deliberate upon the types of changes (institutional, economic, cultural, and communicative) that would now be required even to imagine a genuinely *critical* and democratizing role for transnational public spheres under current conditions.

Fraser responds to this challenge by defending the Kantian project of creating a cosmopolitan ethic, and the argument that a condition of vulnerability can be a substitute for shared nationality as a foundation for global social and political integration. The aim is to support a cosmopolitan ethic in an era of growing interconnectedness, and to stress the vulnerabilities to mental and physical suffering that form the most accessible points of solidarity between strangers (Linklater 2010: 24). The broadest understanding of vulnerability refers to the general fragility of human life, which means that we are all vulnerable (Nussbaum 1992). This theme can be linked with Horkheimer's defense of a critical-theoretical ethic which maintains that shared vulnerabilities to suffering can underpin global relations of solidarity. The recognition that human beings are 'finite beings whose community consists of fear of death and suffering' and 'of the struggle to improve and lengthen the life of all' was the only basis, he argued, for 'correct solidarity' (Horkheimer, c.f. Stirk 1992: 178).

As noted earlier, it may be that the transition to a post-Westphalian world is more likely to occur if human societies can agree on the harms that should be avoided rather than pursue the fruitless quest for a consensus on some universalizable vision of the good life (Linklater 2007: 34). For Linklater it is our mutual vulnerability to harm and suffering that grounds universal human solidarity and he is confident that there is a possibility of growth of such solidarity in the world today (Linklater 2010). Habermas contends that most political issues related to globalization are transnational, and public spheres that enable citizens and governments to learn to become less concerned with defending national interests would be sufficient to negotiate the making of law and policy to regulate cross-border affairs (Habermas 2001, 2009). Hence, the global public sphere can be erected on the humanitarian principle that the world is a community constituted by a form of solidarity grounded in our shared vulnerability to harm and suffering and in our mutual concern for each other as such vulnerable beings.

6.6.1 Weakening of Public Sphere as an Institution

If China valued free speech, there would be no coronavirus crisis, wrote Yu, Verna on 8 February 2020 after the death of Chinese doctor Li Wenliang due to CoViD-19. Three weeks before China began any containment measures against CoV-19, the 33 year old doctor at Wuhan Central Hospital, Wenliang, confirmed the presence of coronavirus in the city, indicating an outbreak. Rather than taking his findings as a starting point and investigating further, the Wuhan Police made him sign a letter that warned him to discontinue his 'illegal activity' immediately or he 'would receive the full sanction of the law, if he stubbornly persisted in his opinions'. The number of cases could have been reduced by 95% if the Chinese government took the advice of Dr. Li and acted upon it, rather than shutting him down.

Human rights in China may appear to have little to do with the rest of the world but as we have seen in this crisis, disaster could occur when China thwarts the freedoms of its citizens. Surely it is time the international community takes this issue more seriously (Yu 2020). One lesson that the world community learned from this (mis)management of disaster and death of Dr. Li Wenliang was that in a globalized world, suppression of freedom of speech in one nation would result in death and destruction worldwide, especially when the suppressed speech was actually an early-warning and caution about the outbreak of a pandemic. In the globalized world, what we do not have anymore is the luxury to turn away from injustice and unfreedom experienced by the common people in other countries, which we could have ignored earlier as a localized 'national' issue, territorially confined to a particular sovereign state. This probably is the another effect of globalization that it has taken away our freedom to be indifferent to the plight of oppressed groups in other countries, if we care about the safety of our own lives.

As the unlocking of India and other countries have started from the 1st week of June, there seems to be an unsaid agreement, building around the inevitability of living with the virus. Such expression, 'learning to live with the virus', in an immediate sense, seeks to individualize the responsibility involving safety from the virus. The very affirmative language of promises and prospects that is circulated by the central government through its periodically rhetorical announcements of economic package, however, fails to acknowledge the errors in the government's lockdown decisions and their implementation (EPW 2020).

6.7 Conclusions and Recommendations: Building a Pandemic Resilient World

> If you think of anything that could come along that would kill millions of people, the pandemic is our greatest risk. In terms of the death toll, a pandemic would rival even the gigantic wars of the past. The economy will shut down, the cost to humanity will be unbelievable, and no country will be immune from the problem this will create (Bill Gates, May 2019).

The statement made by Bill Gates a year ago, seems almost like a prophesy that has come true with the spread of CoViD-19. Along with the costs in terms of lives lost and financial transactions aborted, the pandemic also unearthed a series of socio-political problems in different countries that had become invisible, or were not considered important enough to merit any attention. The gender-based violence emerged as a serious problem worldwide, the exploitation of working classes and laborers in a capitalist economy came to the fore, the tattered condition of democracy in many countries was unmasked, and above all the denial of fundamental human rights such as 'freedom of speech' to a large section of the world community was observed. The surreal atmosphere of the COVID-19 pandemic has also exposed fault lines in trust among human beings, among countries, between citizens and governments, and it is pushing us to raise big questions about ourselves, our social relationships, and life in general.

Socio-culturally, it is pertinent that the spread of this pandemic is seen holistically, in its subtle interconnections with our economic behavior, food habits, and an overall disrespect for other forms of life and their habitats so that a holistic solution, which involves a change in our consumption patterns and lifestyle itself, can be effected. It needs to be questioned if wet seafood markets such as those in Hubei are essential for human survival, or the factory production of animals for human consumption is an absolute necessity that we cannot do without. If so, extensive research must be directed towards such 'factories' from a biological as well as ethical points of view, so that zoonosis is prevented. As Beck has argued, most modern risks are self-induced (Beck 1992), the continuation of wet markets and animal factories is also a step in this direction, of exposing humans to the dangers of such zoonotic viruses.

The lockdown is also showing us that the nature has begun to heal itself slowly. The rivers have become cleaner, the wildlife has taken to the streets in motor-free cities, the ozone layer has been healed, and the air quality has improved manifold reducing the risk of asthma and other lung diseases. It is evident that humans have strained the border between nature and culture. Our relationship to the earth has to be radically reinvented. The sustainability of our life on this planet is unthinkable without acknowledging our co-dependency with other human and non-human species. The ecological question has to be intertwined with the social question, especially in such times when cohabitation is being seen as a factor of contamination, while we are enticed to consider our neighbors as potential disease spreaders, and our social interaction is greatly reduced (Billaud 2020). Epidemics prompt the question: how and through what mechanisms can we continue to live together (Kelly et al. 2019).

The current post-national constellation demands something of the nature of 'transnational public sphere'. For e.g. global citizens like Bill Gates are important figures in the 'transnational public sphere' and when they speak about the imminent threat of a pandemic that could kill millions of people, they speak not for a country or a race but for the entire mankind and citizens of the world community. The need for a critical global public sphere has never been more urgent and necessary.

The 'silent summer' that followed the lockdown of India (China other countries of the world) was a silence not just of the chaotic traffic and honking noises, which has now become a signature of the Indian roads worldwide, but it was also a dictated silence of the critical voices of the people who were in a position to save the world from the CoViD-19 pandemic by issuing an early-warning. It is hoped that this crisis would enable the world community to reassess the implications of such unfreedom and initiate a genuine search for ways to ascertain the growth of a vibrant, critical, and efficacious public sphere worldwide, which could replace the existing simulative democracies with the establishment of a just global society.

References

Beck U (1992) Risk society: towards a new modernity. Sage, London
Calhoun C (2002) Imagining solidarity: cosmopolitanism, constitutional patriotism, and the public sphere. Publ Cult 14(1):147–171
Diamond J (1997) Guns, germs and steel: the fate of human societies. W.W. Norton & Company, New York
Durant W (1954) Our Oriental heritage. Simon and Schuster, New York
Economic and Political Weekly (EPW) (2020) Editorial. Living with CoViD-19. 55(20). https://www.epw.in/journal/2020/20/editorials/living-covid-19.html. Accessed 22 May 2020
Fraser N (2014) Transnationalizing the public sphere: on the legitimacy and efficacy of public opinion in a post-Westphalian world. In: Nash K (ed) Transnationalizing the public sphere. Polity Press, Cambridge
Gates B (2019) Coronavirus explained. 2020. 'The Pandemic'. Season-1, episode-1. Netflix Series
Habermas J (1991) The structural transformation of the public sphere: an inquiry into a category of Bourgeois Society (trans: Burger T, Lawrence F). MIT Press, Cambridge
Habermas J (2001) Why Europe needs a constitution. New Left Rev 11:5–6
Habermas J (2009) Europe: the faltering project. Polity, Cambridge
Harper K (2017) The fate of Rome: climate, disease, and the end of an empire. Princeton University Press, Princeton
Huang C, Wang Y, Li W et al (2020) Clinical features of patients infected with 2019 novel coronavirus in Wuhan, China. Lancet 395:497–506
Jester B et al (2018) Readiness for responding to a severe pandemic 100 years after 1918. Am J Epidemiol 187(12):2596–2602
Kelly AH, Keck F, Lynteris C (2019) The anthropology of epidemics. Routledge, London
Kohn GC (ed) (2008) Encyclopedia of plague and pestilence: from ancient times to the present. Facts-on-File Publishing, New York
Linklater A (2007) Public spheres and civilizing processes. Theory Cult Soc 24(4):31–37
Linklater A (2010) Moral progress and world history: ethics and global interconnectedness. In: Van Hooft S, Vandekerckhove W (eds) Questioning cosmopolitanism. Springer, New York
Littman RJ (2009) The plague of Athens: epidemiology and paleopathology. Mount Sinai School Med 76:456–467
Nash K (ed) (2014) Transnationalizing the public sphere. Polity Press, Cambridge
Nussbaum MC (1992) Human functioning and social justice: in defense of Aristotelian essentialism. Polit Theory 20(2):202–246
Rosen W (2007) Justinian's flea: plague, empire and the birth of Europe. Viking Books, New York
Rutherford P (2000) Endless propaganda: the advertising of public goods. University of Toronto Press, Toronto
Scott J (2017) Against the grain: a deep history of the earliest states. Yale University Press, London
Stirk PMR (1992) Max Horkheimer—a new interpretation. Barnes and Noble Books, Lanham
Tumpey TM, Baster CF, Aguilar PV et al (2005) Characterization of the reconstructed 1918 Spanish influenza pandemic virus. Science 310(5745):77–80
Wang LF, Cowled C (eds) (2015) Bats and viruses: a new frontier of emerging infectious diseases. Wiley Blackwell, New Jersey
WHO (2019) A world at risk—annual report on global preparedness for health emergencies. Global Preparedness Monitoring Board, Geneva
Wu F, Zhao S, Yu B et al (2020) A new coronavirus associated with human respiratory disease in China. Nature 579(7798):265–269

Websites

Billaud J (2020) Inventing a breathing planet after CoViD-19. https://graduateinstitute.ch/coronavirus-information-our-community-and-visitors/what-our-professors-have-say. Accessed 14 May 2020

Cohen J (2020) Wuhan seafood market may not be source of novel virus spreading globally. https://www.sciencemag.org/news/2020/01/wuhan-seafood-market-may-not-be-source-novel-virus-spreading-globally#. Accessed 13 May 2020

Coronavirus Explained (2020) The Pandemic. Season-1, episode-1. Netflix Series

Hutt R (2016) Why do civilizations collapse? https://www.weforum.org/agenda/2016/03/why-do-civilizations-collapse/. Accessed 10 May 2020

Jarus O (2015) Gruesome find: 100 bodies stuffed into ancient house. https://www.livescience.com/51662-100-bodies-found-prehistoric-house.html. Accessed 11 May 2020

Jarus O (2020) 20 of the worst epidemics and pandemics in history. https://www.livescience.com/worst-epidemics-and-pandemics-in-history.html. Accessed 11 May 2020

Jordan D, Tumpey T, Jester B (2019) The deadliest flu: the complete story of the discovery and reconstruction of the 1918 pandemic virus. https://www.cdc.gov/flu/pandemic-resources/reconstruction-1918-virus.html. Accessed 12 May 2020

Kruszelnicki KS (2007) Black death. http://www.abc.net.au/science/articles/2007/09/13/2031252.htm?site=science/greatestmomentsi. Accessed 12 May 2020

The Epoch Times (2020) Tracking down the origins of Wuhan coronavirus. https://www.youtube.com/watch?v=3bXWGxhd7ic. Accessed 13 May 2020

Yu V (2020) The Guardian. https://www.theguardian.com/world/2020/feb/08/if-china-valued-free-speech-there-would-be-no-coronavirus-crisis

Chapter 7
Psychological Impacts of COVID-19

Shivam Singh, Saket Dubey, Nikhil Kumar, Manish Kumar Goyal, and Indrajit Pal

Abstract The 2019 coronavirus disease (COVID-19) has high infectivity as well as high fatality rates and has haunted the whole world by creating psychosocial disturbances including mass hysteria, economic crisis, and financial instability. The fear generated due to highly contagious COVID-19 termed as "coronaphobia" has produced an excess of adverse psychiatric appearances across a different section of the society. From the time, this highly infectious novel coronavirus has started spreading around the globe with unprecedented pace, have haunted the people of all sections with many psychological disorders such as depression, anxiety, fear, and hopelessness. This chapter has been compiled to have a clear idea of the psychological impacts of COVID-19 among the various class of people. The pandemic itself coupled with forced quarantine to fight against novel coronavirus (COVID-19) imposed nation-wide lockdowns as an early measure can produce anxiety, obsession, hoarding, depression, and post-traumatic stress disorders (PTSD) in long-term. These psychological disorders have been fueled by an "infodemic" spread via irresponsible media houses and netizens. Outbursts of racism, stigmatization, and xenophobia against a particular community along with the infection of coronavirus have been widely reported. Frontline health care persons are more susceptible to get infected with the disease yet as experiencing adverse psychological outcomes within the kind of anxiety, fear of transmitting the infection to their relatives, depression, and PTSD. For developing a far better response system in coping with these psychological disorders at different section of society, psychological crisis prevention and intervention models should be implemented jointly by the government, health care personnel and other stakeholders.

S. Singh (✉) · S. Dubey · N. Kumar
Indian Institute of Technology Indore, Indore, Madhya Pradesh, India

M. K. Goyal
Civil Engineering, Indian Institute of Technology, Indore, Madhya Pradesh, India
e-mail: mkgoyal@iiti.ac.in

I. Pal
Asian Institute of Technology, Bangkok, Thailand

M. K. Goyal, A. K. Gupta (eds.), *Integrated Risk of Pandemic: Covid-19 Impacts, Resilience and Recommendations*, Disaster Resilience and Green Growth,
https://doi.org/10.1007/978-981-15-7679-9_7

7.1 Introduction

Ever since this pandemic COVID-19 has started spreading around the globe, ever one in the world have got affected in some way or other, forcing isolation, uncertainty, fear, and, hopelessness and this impact has been getting more intensified with the coupled economic meltdown which has caused great psychological distress all to the people in general. This pandemic is proving a threat to human society by risking human life, economic crisis along with hidden psychological strain. As per the United Nations, this coronavirus pandemic has seeds of all major mental health issues. This COVID-19 disease outbreak has been declared as global health emergency by World Health Organization (WHO) on January 30th, 2020 as the disease started haunting all countries throughout the world, starting from China in Dec 2019 with more severe consequence collapsing their health infrastructures and still widening its adverse impact in many of the countries (Jones 2020). Observing the spreading trend of this pandemic, nationwide lockdowns and home confinement strategies have been implemented in almost all the countries to control the further transmission of the disease (Pulla 2020; Rubin and Wessely 2020). The unpredictably faster spreading of infections of COVID-19 has been causing fear, anxiety, and distress in human communities, all of which by WHO are natural psychological responses to this critical situation (Kluge 2020). As a consequence of this rapidly worsening conditions, adverse psychosomatic responses from people are expected to increase significantly due to pandemic itself as well as with the incessant flow of information regarding COVID-19 being circulated in social media and news channels. Further the situations may get even worsen due to rapid expansion of mass hysteria and panic regarding COVID-19, which could eventually prove more detrimental than the virus itself (Depoux et al. 2020). From previous studies also, it is evident that mental well-being has been severely affected in these types of pandemic (Sim and Chua 2004). Therefore, it is important to assess the possible ways in which this pandemic may have an impact on the mental health of the people (Lima et al. 2020; Shigemura et al. 2020; Zandifar and Badrfam 2020).

COVID-19 has generated immense psychological crises in all sections of society. As the people of having less immunity (children and olds) or having some chronic diseases are more vulnerable to get infected by the virus and these are people who need most emotional support even in smooth going life. The extent, these people may face mental health challenges under these circumstances, is highly troubling. The WHO data says more than 264 million people of all ages suffer from depression around the globe which eventually may lead to committing suicide. Health care workers fighting with the pandemic at frontline face compounding stressors: more and longer working shifts, limited supplies of personal protective equipment (PPE), fear of bringing infection at home, quality of PPE kits, witnessing co-staff getting ill, and making tough allocation decisions about scarce, life-saving resources like oxygen cylinders and ventilators. Exposed people may have to face many challenging circumstances such as self-isolation to protect their near and dear ones. During this incubation phase, people living with uncertainty and limiting physical contact

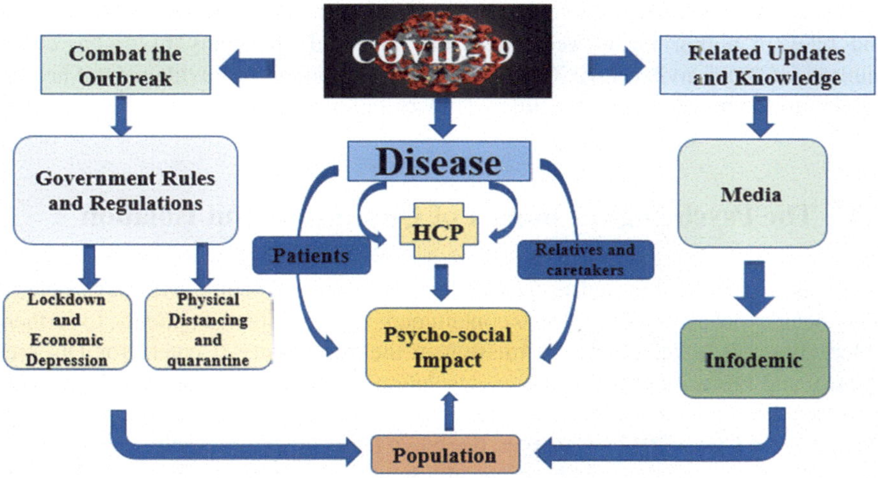

Fig. 7.1 Intricate psychosocial relationship between the disease, health care providers, government and population (Dubey et al. 2020)

(social-distancing) to avoid its opportunity for further spread of infection with derailing income-generating activities, losing source for income, and unmet obligations to the family may further fuel the stress. Feelings of being ignored and negligence by neighbors and relatives after getting exposed to the virus may further lead the anxiety, fear, and distress in the people during quarantine periods (Fig. 7.1).

7.2 COVID-19 Positive: A Stigma

From the outbreak of novel severe acute respiratory syndrome (SARS) in 2003, it was noticed that disease-associated stigmatization was remarkably present with the exposed people even after many years passed (Lee et al. 2005; Siu 2008). It was found that the people who worked in health care units and general medical practitioners were more prone to stigmatization (Verma et al. 2004). Similarly, this COVID-19 outbreak is also expected to leave stigmatization factors like fear of loneliness, discrimination, and marginalization with the exposed people (Siu 2008). In stigmatized communities, people have been found with the tendencies to seek medical care lately and hiding important medical history, especially related to travel history. This kind of public response increases more risk to community transmission. The WHO has also released specific psychosocial considerations for counteracting the rising stigma of novel coronavirus (WHO). Health crime has also been observed due to fear of being COVID-19 positive in some of the countries (The New Indian Express 2020). Various steps have been taken by the governing bodies of different countries to provide psychological support to the people. Especially in India, People have been asked to do regular yoga at their homes by the prime minister of the nation and relevant authorities to keep them mentally and physically fit. Even, chief minister of Delhi invited ISKCON a spiritual organization to provide mental health support to the people through online webinars and meetings.

7.3 The Psychological Impact of Quarantine and Isolation

Quarantine is separating and restricting the movement of people who have potentially been exposed to a COVID-19 containment zone or infected patients, to further reduce the risk of infection transmission in the community (Centers for Disease Control and Prevention 2017), whereas isolation is defined as separating the people who have been diagnosed with the infectious disease from those are not infected yet (Manuell and Cukor 2011). However, these two terms, quarantine and isolation have been often used interchangeably, especially in general communication. Most of the countries were forced to exercise quarantine as an early measure to have control over the unimaginably fast-spreading COVID-19 infection throughout the globe. Countries that could not implement quarantine and isolation at the initial step, have faced many challenges to make their health infrastructure function properly while combating with this pandemic. The quarantine have so deep psychological impact that it gives many folds mental miseries than physical sufferings to the exposed ones. Nationwide lockdown programs have imposed mass quarantine which has the potential to produce mass hysteria, fear, anxiety, hopelessness, and distress. This can be even more intensified with people having inadequate availability of essentials, economical crises and increased perception of risk. This even gets worst by improper and vague communications through social media and news channels at early stages of this pandemic (Brooks et al. 2020; Hawryluck et al. 2004; Maunder et al. 2003). The previous records of such outbreaks reveal that the psychological impact of quarantine can be immediate or can carry long-lasting impressions of mental disorders. The impact may vary from immediate effects, like fear of getting affected or spreading the infection to the family members, frustration, anxiety, depression, insomnia, to severe consequences leading to suicide (Barbisch et al. 2015; Jeong et al. 2016; Robertson et al. 2004). In one study, it was found that even 3 years later, hospital employees were showing posttraumatic stress symptoms due to the long-lasting effect of being quarantined (Wu et al. 2009). A study based on equine influenza (horse flu) outbreak in Australia, approximately 34% (938 out of 2760) of horse owners who were kept quarantined for several weeks, have shown high psychological disorders (Taylor et al. 2008). Suspected quarantined people may face anxiety due to uncertainty related to their heath updates and develop obsessive-compulsive symptoms, like repeated temperature checks and sterilization (Li et al. 2020b). The symptoms of these posttraumatic stress disorders were found positively associated with the duration of quarantine (Hawryluck et al. 2004; Reynolds et al. 2008). A study based on a comparative assessment of post-traumatic symptoms in parents and children revealed that mean post-traumatic stress scores were four times higher in the case of children who were quarantined than those who were not. In the case of parents who were quarantined, 28% (27 of 98) were found sufficient symptoms to warrant a diagnosis of a trauma-related mental health disorder in comparison with 6% of those who were not quarantined (Sprang and Silman 2013). In another study related to the impact of quarantine on health care workers revealed that even 3 years after being quarantined, they have been witnessing symptoms of depression and 9% (48 of 549) of the whole sample reported with high depressive symptoms in which only 60% (29 of 48) were quarantined (Liu et al. 2012).

It has been observed during the pandemic that forced home quarantine is often being violated because of social disobedience, irresponsible behavior, and low social perception (The Economic Times 2020). Therefore, it must be taken care of by the authorities at various levels to control the transmission of infection in the community. There is a need to further explore the psyche and mental status of people at a finer resolution for policymaking and implementation. Humane behavior towards self-isolation and voluntary quarantine should be appreciated and promoted through mass media to lessen anxiety and long-term impacts of imposed mass-quarantine (Brooks et al. 2020).

7.3.1 Stressors During Quarantine

- **Duration of quarantine**
  From studies based on the impacts of quarantine, it can be concluded that longer duration has a poorer impact on the mental health of the people (Marjanovic et al. 2007). Studies suggest that quarantine duration of more than 10 days have higher post-traumatic stress than that of less than 10 days (Hawryluck et al. 2004).
- **Fear of infection**
  People become fearful due to either own health concern or to infect others, especially their family members (Cava et al. 2005). They also get worried after experiencing any physical symptoms of infections and fear that the symptoms may get reflected positively which may lead one into a psychological trauma in the near future.
- **Frustration and boredom**
  In the quarantine period, due to confinement, loss of usual routine, and reduced physical and social connectivity with other members of society often resulted in boredom and frustration which is distressing for the people in general (Blendon et al. 2004; Robertson et al. 2004).
- **Inadequate supplies**
  Inadequate supplies of basic essentials (e.g. food, milk, clothes, etc.) lead to frustration associated with anxiety and anger (Blendon et al. 2004). Unavailability of proper access to regular healthcare and prescriptions also causes psychological problems to the people.
- **Inadequate information**

Poor information from health authorities and government officials act as a stressor, reporting insufficient clear guidelines about actions to implement and confusion about the purpose of quarantine (DiGiovanni et al. 2004).

7.3.2 *Stressors Post Quarantine*

- **Finances**
  In a situation like nationwide lockdown, people getting quarantined, the economy is going to be badly affected. Financial loss as a result of quarantine creates socioeconomic distress and found to be a risk factor for psychological disorders such as anger and anxiety in the long term.
- **Stigma**
  Quarantined health-care workers were more likely to face stigmatization and rejection from neighbors in comparison with non-quarantined. It suggests that there exists a stigma specifically surrounding people who had been quarantined. Reports reveal that people were treating them differently: avoiding them, treating with fear and suspicion, withdrawing social invitation, and making critical comments (Lee et al. 2005; Wester and Giesecke 2018).

7.3.3 *Mitigating the Consequences of Quarantine*

Quarantine has been widely accepted, appreciated, and an effective early step to combat the outbreak of such highly infectious diseases. However, it has been observed that quarantine is most often associated with a negative psychological effect. Various studies have been conducted to quantify the psychological influence of quarantine on various sections of people such as old, children, health care workers. Seeing the adverse consequences of quarantines, it is important to ensure the effective mitigation measures are put in place as part of the quarantine planning process. Some of the mitigation processes are as follows.

- Shorter duration of quarantine period
- Spreading awareness through correct information
- Availability of essential services
- Reduce the boredom and improve communication
- Distinct attention to health care personals
- Encouraging altruism than compulsion

7.4 Internet Users, Social Media and COVID-19

Nowadays, the whole world has been interconnected through online social networks and high-speed internet connections which can be potentially used to generate real-time maps which can be proved as an important tool for tracking a pandemic and for making interventional campaigns when required. But social media can have disastrous effects on the control and handling of an infectious disease pandemic (Al-garadi et al. 2016). The irresponsible behavior of social media creates panic among the people by a relentless plethora of fake information and this kind of negatively skewed misinformation spreads all around even faster than coronavirus itself (Depoux et al. 2020; Shimizu 2020). In this connection, it has been referred to as "coronavirus infodemic" by the director-general of WHO which is spreading fright and panic among people through mind blogging rumors without any factual verification just for their news propaganda and sensationalism (Zarocostas 2020). Seeing the emerging trend of COVID-19 in social media, people through their blogs, Facebook account, Instagram, twitter handle started publicizing relevant content to make profit off COVID-19 through many impetuous and volatile courses of action (Merchant and Lurie 2020). Since sensationally charged contents attract more attention in social media, many users feigned COVID-19 symptoms to become popular easily and thus tenaciously spread mass confusion and panic among the people (Sokolov 2020). As a result of this irresponsible attitude, these give invitation for mental stress in form of anxiety, depression, phobia, obsession, and irritability (Asmundson and Taylor 2020; Ho et al. 2020). Health-care seekers get perplexed and worried about COVID-19 symptoms through this false propaganda based upon massive misinformation on social media. On the other dimension, another group of social media may mislead people by not taking the pandemic as a serious matter of concern. This may result in a violating basic rule of pandemic that may further lead to social disobedience and other oppositional activities. After being influence by misdirecting social media messages, therapeutic misadventures tried by common people are increasingly being noticed (Soltaninejad 2020). Thus irresponsible social media users and mainstream media-houses plays a significant role in boosting stressful environment causing long-term stigma and mental health problems in the public (Zhai and Du 2020). Media should play a responsible role along with health care persons by providing correct information and creating effective communication with the citizens in combating these types of pandemic.

7.5 Psychological Impacts of COVID-19 on Health Care and Frontline Workers

The impact of the pandemic on healthcare workers (HCWs) and frontline workers in terms of their psychological health is a significant question. However, it remains partially understood due to its complexity and insufficient research. Past studies from

Canada, Hong Kong, and Taiwan during SARS 2003 outbreak (Chong et al. 2004; Sim and Chua 2004) illustrate that HCWs faced enormous psychological challenges in their battle against the epidemic. And, in Taiwan during this outbreak, HCWs associated with curbing the epidemic encountered more depressive and symptoms (Su et al. 2007). Moreover, even after the epidemic was over, a substantial number of HCWs in China experienced Post-traumatic stress disorder (PTSD) (Wu et al. 2009).

Likewise, psychological issues among HCWs and frontline workers (policemen, bankers, and others) can shoot up during the COVID-19 outbreak. The mental wellbeing of HCWs can be negatively impacted, due to exposure to COVID-19 patients, quarantine period, loss of family/friends from COVID-19, and perception of an overemphasized risk of lethality of virus (Greenberg et al. 2020; Wu et al. 2009). In China, health care professionals faced high-performance pressure and other psychological challenges due to a sudden surge in work, insufficient protection gear, and isolation (Kang et al. 2020; Lai et al. 2020). And, many psychological challenges associated with 'vicarious traumatization' among the nursing staff have been reported in China (Li et al. 2020c). In third world countries such as India, the health care system is overloaded beforehand, therefore the current upsurge of COVID-19 cases might induce anxiety, depression, frustration, and stress among HCWs. Among HCWs, this can further be augmented owing to insufficient supply of personal protective equipment (PPE) and hygiene tools, as they bear the highest risk of transmission (Biswas and Chatterjee 2014; Chen et al. 2020). Li et al. (2020c) in its nationwide study found that during the COVID-19 outbreak, 14% of physicians and around 16% of nurses experienced moderate/severe clinical depression. And, in another study Maunder et al. (2003) found the heightened episodes of the anxiety of HCWs while catering to the needs of fellow ill colleagues, which might further affect their performance and skills at work. Moreover, it is cumbersome to address the psychological aspects such as anxiety, depression, etc. of critical COVID-19 patients by HCWs due to scarce formal training on mental health interventions and standard infection-control measures during a pandemic(Xiang et al. 2020). This might lead to apathy for COVID-19 patients and induce a sense of vulnerability among HCWs due to an incomplete understanding of virus (incubation period, asymptomatic transmission, preventive vaccines, definitive therapy) and psychological aspects of the pandemic, which can result in a perception of an overemphasized risk about the virus. And, incidents such as assaults on HCWs after fatalities from COVID-19 (Independent 2020; Indiatoday 2020a, b). This can impact the performance and competency of HCWs. Therefore, mental well-being of HCWs is of utmost importance at this vulnerable hour.

7.6 Responsibilities for Mental Health-Care Workers

A novel and new psychosocial crisis avoidance strategy model with the sole purpose of dealing with unprecedented and unique issues of different sectors and divisions of the population during the times of global pandemic, COVID-19, should be established with the use of digital technologies and internet resources(Qiu et al. 2020; Zhou et al. 2020), the main reason for which being the to form an amalgam of the various health institutions, mental and psychiatric health consultants, psychologists, psychiatrists, mental and tertiary health departments, medical practitioners and also the earlier involvements if any with rehabilitation centers and organizations (Liu et al. 2020; Zhang et al. 2020). The pandemic has made most of the divisions of the population vulnerable to mental health illnesses and therefore calls for the relevant and respective authorities and organizations to recognize and categorize the psychologically morbid and affected groups by proper screening, timely referral and early involvement in a directed and targeted way (Jung and Jun 2020). This vulnerability observed in health care providers, of age groups, quarantined population, older population, children, and side-lined communities of daily wage and migrant workers and homeless populations of the society and people with previous histories of mental health illnesses need greater attention and intervention.

Mental health care workers and professionals can be assigned work in other healthcare sectors so that help may be provided in case severe manpower concerns arise, but such allocation of help would lead to the deterioration of physical and psychological health during a pandemic of this nature like COVID-19 (Cullen et al. 2020). The frontline healthcare providers tending to the population affected by COVID-19 would not be able to cater to the problems faced by the patients who are psychologically distressed because of the dearth of training for dealing with psychological illnesses. Strict measures for the control of infections discourage the entry of psychologists, mental health workers, and psychiatrists into quarantine wards and confinement units (Duan and Zhu 2020; Lima et al. 2020). An experienced and professional team with leading mental health physicians can be prepared for proper directions and actions to be taken in case of greater psychological distress faced during the pandemic. For healthy interactions and daily sessions for the mental exhaustion in people recovered from the effect of the virus, the psychiatric departments and organizations can be involved and considered (Liu et al. 2020). To lessen the stigma around COVID-19, governmental organizations, political front-runners, and healthcare establishments have to play a central role in keeping up the interracial harmony pre and post the pandemic (Liu et al. 2020).

7.7 Impacts on Various Sections of Society

The unpredictable pandemic and its psychological involvement of the masses because of its uncertain nature is giving a rise to greater behavioral changes and emotional peaks in the common people (Taha et al. 2014). Many studies have been conducted which give a sense of how on an intense and broad level the general public is facing the wrath of mental disorders and illnesses. The perceived risks of the pandemic and getting infected are enough to generate very unusual and unexpected anxiety levels among the general public. The mass fear of COVID-19 referred to as "coronaphobia" (Asmundson and Taylor 2020) is due to the perceived risks mentioned earlier. The early stages of the outbreak of this deadly novel coronavirus where the identification of people affected and the surge in the number of cases made rounds in the media were enough to let the dreads of the masses peak. This as an immediate reaction demanded the common people seek help and this help-seeking and overthinking in the anxious population lead to greater instability. The untimely and unexpected lockdown and the stockpiling and storing of essential supplies further deteriorated the mental health and the social functioning of many and hindered daily schedules which later led to more psychological distress. All sections of the society suffered in one way or the other leading to a rise in the rate of recurrence of anxiety, hopelessness, vague phobias, cognitive-behavioral changes, averting, and compulsive behavioral patterns.

Developmental psychology research founded that experiences from early childhood shape the behavior of an individual, as it is a critical period for cognitive, psychological, and emotional development. The psychological impact of the COVID-19 outbreak on children is a very crucial aspect, however, it has been largely overlooked (Ghosh et al. 2020; Wang et al. 2020). Disruptions in children's usual routine (such as the closing of playgrounds, schools, parks, home confinement, etc.) and exposure to media associated with potential risks of virus can induce boredom, anxiety related to educational development, irritability, development issues and fear of infection (Remmerswaal and Muris 2011; Sprang and Silman 2013). To address this issue, some interventions such as proper parenting, online classes, online study material, maintenance of sleep cycle, proper physical exercise schedule, and education regarding proper hygiene measures can be undertaken to curb this issue. And, as the elderly are most vulnerable to the virus, it might lead to irritability, excess fear, anger, and anxiety (Li et al. 2020a; Muris and Field 2010). Additionally, difficulty in access to medicines due to imposed travel restrictions can also trigger psychological issues. Some interventions such as exercise at home, online counseling sessions, online regular physical health check-ups, and reliable drug delivery systems can help the elderly fight against these psychological issues (Jiménez-Pavón et al. 2020; Liu et al. 2020; Qiu et al. 2020). On the other hand, disadvantaged groups of our society (daily wage workers, slum dwellers, prisoners, migrants, etc.) are the most vulnerable to exclusion, stigma, and discrimination during an infectious pandemic (Liem et al. 2020; UNHCR n.d.). Also, COVID-19 being a highly infectious Such marginalized groups are at the highest risk of contracting and transmitting the disease (UN n.d.). This situation might cause depression, stress, the stigma of discrimination, and financial insecurity among marginalized groups. Interventions for the protection of basic human rights (Tsai and Wilson 2020), adequate accommodation, food and water supply, health care, proper social distance education can be carried out to curb this issue.

7.8 The Psychological Effect on People with Pre-existing Illness

Overwhelming news related to COVID-19 caused public anxiety and fear, this fear is increased due to limited social interaction and loneliness (Gao et al. 2020), additionally, people with a pre-existing illness such as diabetes and asthma are under tremendous stress as this disease strongly affects those with a weak immune response (Zhu et al. 2020). COVID-19 has also caused various psychological problems among the public such as depression, panic, and anxiety (Ho et al. 2020). Studies have shown that patients with mental instability are substantially more prone to getting infectious diseases due to poor awareness level, impaired perception of risk, and lack of personal hygiene (Dubey et al. 2020). In addition to this, social discrimination with people having a mental illness makes the management of COVID-19 more challenging when it coexists. Patients with mental illness are also prone to deteriorate their pre-existing signs and symptoms during a pandemic, such as people with pre-existing obsessive-compulsive disorder (OCD) tends to self-check their temperature frequently to look for the signs of disease at the same time they also swallow saliva frequently to check for any throat pain, similarly patients with washing OCD tend to wash their hands more frequently. Strictly applied lockdowns also affect their scheduled therapeutic sessions and impose difficulties on the availability of medicines. Unfortunately, people with high health anxiety, are likely to perceive harmless bodily symptoms and feelings as evidence of COVID-19 (for example perceiving normal body pain or cough as COVID-19). This will increase their anxiety and distress, and unnecessarily increasing the burden on health care system. Dependents of the people with such mental disorders are also highly affected during the pandemics, as these people are subjected not only to the fear of virus but at the same time are subjected to observe the deteriorating mental conditions of their loved once. As these pandemics affect a large population, and no one has idea about how long it is going to last, the psychological effect on the population is unquantifiable (Tan et al. 2020).

7.9 COVID-19 and Homeless Human Beings

The homeless people will be the section of human society that will be greatly affected by this pandemic, despite already they lack access to basic human needs (Tsai and Wilson 2020). In India, this section is predominant in almost every state of the country and contributes to nearly 0.15% of the country's population (Singh et al. 2018). This will pose a challenge and concern for the country to fight against COVID-19. The homeless themselves face the challenge to meet the basic regulations of social distancing, isolation at home, and keeping one-self clean at all times. They are also the victims of a lack of awareness and understanding of how the infection spreads and how to avoid and care for in any situations (Ahmed et al.

2020). If anyone of them is infected, they are most likely to get scared, panic, and run away from being treated. They may unknowingly be the silent carriers of the contagious infection and go on for many days while spreading it to others. Many of the homeless people may be suffering from pre-existing health and mental health issues that worsened the situation for them as it puts them at high risk of losing their precious lives due to the infection. Their poor health and poor living conditions make it almost impossible for them to survive or cope during this pandemic. Unless special attention, care and help is issued by the government or NGOs for them it's unimaginable to hope for their well-being. At the same time the primary source of survival for the homeless is by begging, which is highly affected during the pandemic as people are scared of social interaction, which interrupts the source of food supply. The poor hygiene of homeless due to their limited resources subjects them to social discrimination due the fear of viral spread.

7.10 Ways Forward

Recent days have shown a significant rise in patients with mental illness (Torrey and Miller 2001), thus psychological preparedness carries high importance especially during the times of global pandemics. Stakeholders must understand the psychological morbidities of the pandemic to assess the stress, fatalities, and associated consequences. Blames of spreading the virus on specific communities may hinder trade, finance, and social relationships (Casciano 2020). Racisms and religious propagandas during the times of pandemic must be intervened with a regular discussion of equality, love, and kindness. Mental health organizations should be specifically trained for the pandemics, these institutions should be instructed to organize awareness programs at both personal and community level. Toll-free phone numbers and well-structured websites must be developed to reduce the social distress and strict regulation by the government on the spread of fake or misleading information should be issued. The recent virus outbreak has shown us that we are still not equipped to deal with such crises and the greatest assets to humankind are health, love, solidarity, peace, and knowledge.

References

Ahmed F, Ahmed N, Pissarides C, Stiglitz J (2020) Why inequality could spread COVID-19. Lancet Public Heal 5:e240

Al-garadi MA, Khan MS, Varathan KD, Mujtaba G, Al-Kabsi AM (2016) Using online social networks to track a pandemic: a systematic review. J Biomed Inform 62:1–11. https://doi.org/10.1016/j.jbi.2016.05.005

Asmundson GJG, Taylor S (2020) Coronaphobia: fear and the 2019-nCoV outbreak. J Anxiety Disord 70:102196. https://doi.org/10.1016/j.janxdis.2020.102196

Barbisch D, Koenig KL, Shih F-Y (2015) Is there a case for quarantine? Perspectives from SARS to Ebola. Disaster Med Public Health Prep 9:547–553. https://doi.org/10.1017/dmp.2015.38
Biswas P, Chatterjee S (2014) Hand hygiene compliance among doctors in a tertiary care hospital of India. Indian J Pediatr 81:967–968
Blendon RJ, Benson JM, DesRoches CM, Raleigh E, Taylor-Clark K (2004) The public's response to severe acute respiratory syndrome in Toronto and the United States. Clin Infect Dis 38:925–931. https://doi.org/10.1086/382355
Brooks SK, Webster RK, Smith LE, Woodland L, Wessely S, Greenberg N, Rubin GJ (2020) The psychological impact of quarantine and how to reduce it: rapid review of the evidence. Lancet (London, England) 395:912–920. https://doi.org/10.1016/S0140-6736(20)30460-8
Casciano D (2020) COVID-19, discipline and blame. J Extrem Anthropol 4:E18–E24
Cava MA, Fay KE, Beanlands HJ, McCay EA, Wignall R (2005) The experience of quarantine for individuals affected by SARS in Toronto. Public Health Nurs 22:398–406. https://doi.org/10.1111/j.0737-1209.2005.220504.x
Centers for Disease Control and Prevention (2017) Quarantine and isolation. [WWW Document]. https://www.cdc.gov/quarantine/index.html. Accessed 17 Jun 20
Chen Q, Liang M, Li Y, Guo J, Fei D, Wang L, He L, Sheng C, Cai Y, Li X (2020) Mental health care for medical staff in China during the COVID-19 outbreak. Lancet Psychiatry 7:e15–e16
Chong M-Y, Wang W-C, Hsieh W-C, Lee C-Y, Chiu N-M, Yeh W-C, Huang T-L, Wen J-K, Chen C-L (2004) Psychological impact of severe acute respiratory syndrome on health workers in a tertiary hospital. Br J Psychiatry 185:127–133
Cullen W, Gulati G, Kelly BD (2020) Mental health in the COVID-19 pandemic. QJM An Int J Med 113:311–312
Depoux A, Martin S, Karafillakis E, Preet R, Wilder-Smith A, Larson H (2020) The pandemic of social media panic travels faster than the COVID-19 outbreak. J Travel Med 27(3):taaa031. https://doi.org/10.1093/jtm/taaa031
DiGiovanni C, Conley J, Chiu D, Zaborski J (2004) Factors influencing compliance with quarantine in Toronto during the 2003 SARS outbreak. Biosecur Bioterror 2:265–272. https://doi.org/10.1089/bsp.2004.2.265
Duan L, Zhu G (2020) Psychological interventions for people affected by the COVID-19 epidemic. Lancet Psychiatry 7:300–302
Dubey S, Biswas P, Ghosh R, Chatterjee S, Dubey MJ, Chatterjee S, Lahiri D, Lavie CJ (2020) Psychosocial impact of COVID-19. Diabetes Metab Syndr Clin Res Rev 14(5):779–788
Gao J, Zheng P, Jia Y, Chen H, Mao Y, Chen S, Wang Y, Fu H, Dai J (2020) Mental health problems and social media exposure during COVID-19 outbreak. PLoS One 15:e0231924
Ghosh R, Dubey MJ, Chatterjee S, Dubey S (2020) Impact of COVID-19 on children: special focus on psychosocial aspect. Education 31:34
Greenberg N, Docherty M, Gnanapragasam S, Wessely S (2020) Managing mental health challenges faced by healthcare workers during COVID-19 pandemic. BMJ 368:m1211
Hawryluck L, Gold WL, Robinson S, Pogorski S, Galea S, Styra R (2004) SARS control and psychological effects of quarantine, Toronto, Canada. Emerg Infect Dis 10:1206–1212. https://doi.org/10.3201/eid1007.030703
Ho CSH, Chee CY, Ho RC (2020) Mental health strategies to combat the psychological impact of COVID-19 beyond paranoia and panic. Ann Acad Med Singap 49:1–3
Independent (2020) Coronavirus: NHS doctor kicked out by landlord because of COVID-19 fears | The Independent [WWW Document]. https://www.independent.co.uk/news/health/coronavirus-latest-nhs-doctor-evicted-COVID-19-oxford-a9425166.html
Indiatoday (2020a) Kolkata landlords evict medical professionals, Resident Doctor's Association steps in—India News [WWW Document]. 2020. https://www.indiatoday.in/india/story/bengal-kolkata-landlords-evict-medical-professionals-coronavirus-1659333-2020-03-25
Indiatoday (2020b) 49-year old man becomes Telangana's latest COVID-19 fatality—Coronavirus Outbreak News [WWW Document]. https://www.indiatoday.in/coronavirus-outbreak/story/49-year-old-man-becomes-telangana-latest-COVID-19-fatality-1662363-2020-04-02

Jeong H, Yim HW, Song Y-J, Ki M, Min J-A, Cho J, Chae J-H (2016) Mental health status of people isolated due to Middle East Respiratory Syndrome. Epidemiol Heal 38:e2016048. https://doi.org/10.4178/epih.e2016048

Jiménez-Pavón D, Carbonell-Baeza A, Lavie CJ (2020) Physical exercise as therapy to fight against the mental and physical consequences of COVID-19 quarantine: special focus in older people. Prog Cardiovasc Dis 63(3):386–388. https://doi.org/10.1016/j.pcad.2020.03.009

Jones DS (2020) History in a crisis—lessons for COVID-19. N Engl J Med 382:1681–1683. https://doi.org/10.1056/NEJMp2004361

Jung SJ, Jun JY (2020) Mental health and psychological intervention amid COVID-19 outbreak: perspectives from South Korea. Yonsei Med J 61:271–272

Kang L, Li Y, Hu S, Chen M, Yang C, Yang BX, Wang Y, Hu J, Lai J, Ma X (2020) The mental health of medical workers in Wuhan, China dealing with the 2019 novel coronavirus. Lancet Psychiatry 7:e14

Kluge HHP (2020) Statement to the press by Dr Hans Henri P. Kluge. WHO Regional Director for EuropeTitle

Lai J, Ma S, Wang Y, Cai Z, Hu J, Wei N, Wu J, Du H, Chen T, Li R (2020) Factors associated with mental health outcomes among health care workers exposed to coronavirus disease 2019. JAMA Netw Open 3:e203976

Lee S, Chan LYY, Chau AMY, Kwok KPS, Kleinman A (2005) The experience of SARS-related stigma at Amoy Gardens. Soc Sci Med 61:2038–2046. https://doi.org/10.1016/j.socscimed.2005.04.010

Li Q, Guan X, Wu P, Wang X, Zhou L, Tong Y, Ren R, Leung KSM, Lau EHY, Wong JY (2020a) Early transmission dynamics in Wuhan, China, of novel coronavirus–infected pneumonia. N Engl J Med 382(13):1199–1207

Li W, Yang Y, Liu Z-H, Zhao Y-J, Zhang Q, Zhang L, Cheung T, Xiang Y-T (2020b) Progression of Mental Health Services during the COVID-19 outbreak in China. Int J Biol Sci 16:1732–1738. https://doi.org/10.7150/ijbs.45120

Li Z, Ge J, Yang M, Feng J, Qiao M, Jiang R, Bi J, Zhan G, Xu X, Wang L, Zhou Q, Zhou C, Pan Y, Liu S, Zhang H, Yang J, Zhu B, Hu Y, Hashimoto K, Jia Y, Wang H, Wang R, Liu C, Yang C (2020c) Vicarious traumatization in the general public, members, and non-members of medical teams aiding in COVID-19 control. Brain Behav Immun 88:916–919. https://doi.org/10.1016/j.bbi.2020.03.007

Liem A, Wang C, Wariyanti Y, Latkin CA, Hall BJ (2020) The neglected health of international migrant workers in the COVID-19 epidemic. Lancet Psychiatry 7:e20

Lima CKT, Carvalho PMM, Lima IAAS, Nunes JVAO, Saraiva JS, de Souza RI, da Silva CGL, Neto MLR (2020) The emotional impact of coronavirus 2019-nCoV (new coronavirus disease). Psychiatry Res 287:112915. https://doi.org/10.1016/j.psychres.2020.112915

Liu X, Kakade M, Fuller CJ, Fan B, Fang Y, Kong J, Guan Z, Wu P (2012) Depression after exposure to stressful events: lessons learned from the severe acute respiratory syndrome epidemic. Compr Psychiatry 53:15–23. https://doi.org/10.1016/j.comppsych.2011.02.003

Liu S, Yang L, Zhang C, Xiang Y-T, Liu Z, Hu S, Zhang B (2020) Online mental health services in China during the COVID-19 outbreak. Lancet Psychiatry 7:e17–e18

Manuell M-E, Cukor J (2011) Mother nature versus human nature: public compliance with evacuation and quarantine. Disasters 35:417–442. https://doi.org/10.1111/j.1467-7717.2010.01219.x

Marjanovic Z, Greenglass ER, Coffey S (2007) The relevance of psychosocial variables and working conditions in predicting nurses' coping strategies during the SARS crisis: an online questionnaire survey. Int J Nurs Stud 44:991–998. https://doi.org/10.1016/j.ijnurstu.2006.02.012

Maunder R, Hunter J, Vincent L, Bennett J, Peladeau N, Leszcz M, Sadavoy J, Verhaeghe LM, Steinberg R, Mazzulli T (2003) The immediate psychological and occupational impact of the 2003 SARS outbreak in a teaching hospital. CMAJ 168:1245–1251

Merchant RM, Lurie N (2020) Social media and emergency preparedness in response to novel coronavirus. JAMA 323:2011–2012. https://doi.org/10.1001/jama.2020.4469

Muris P, Field AP (2010) The role of verbal threat information in the development of childhood fear. "Beware the Jabberwock!". Clin Child Fam Psychol Rev 13:129–150

Pulla P (2020) COVID-19: India imposes lockdown for 21 days and cases rise. BMJ 368:m1251. https://doi.org/10.1136/bmj.m1251

Qiu J, Shen B, Zhao M, Wang Z, Xie B, Xu Y (2020) A nationwide survey of psychological distress among Chinese people in the COVID-19 epidemic: implications and policy recommendations. Gen psychiatry 33(2):e100213

Remmerswaal D, Muris P (2011) Children's fear reactions to the 2009 Swine Flu pandemic: the role of threat information as provided by parents. J Anxiety Disord 25:444–449

Reynolds DL, Garay JR, Deamond SL, Moran MK, Gold W, Styra R (2008) Understanding, compliance and psychological impact of the SARS quarantine experience. Epidemiol Infect 136:997–1007. https://doi.org/10.1017/S0950268807009156

Robertson E, Hershenfield K, Grace SL, Stewart DE (2004) The psychosocial effects of being quarantined following exposure to SARS: a qualitative study of Toronto Health Care Workers. Can J Psychiatr 49:403–407. https://doi.org/10.1177/070674370404900612

Rubin GJ, Wessely S (2020) The psychological effects of quarantining a city. BMJ 368:m313. https://doi.org/10.1136/bmj.m313

Shigemura J, Ursano RJ, Morganstein JC, Kurosawa M, Benedek DM (2020) Public responses to the novel 2019 coronavirus (2019-nCoV) in Japan: mental health consequences and target populations. Psychiatry Clin Neurosci 74:281–282. https://doi.org/10.1111/pcn.12988

Shimizu K (2020) 2019-nCoV, fake news, and racism. Lancet 395:685–686. https://doi.org/10.1016/S0140-6736(20)30357-3

Sim K, Chua HC (2004) The psychological impact of SARS: a matter of heart and mind. CMAJ 170:811–812

Singh N, Koiri P, Shukla SK (2018) Signposting invisibles: a study of the homeless population in India. Chin Soc Dialogue 3:179–196

Siu JY (2008) The SARS-associated stigma of SARS victims in the post-SARS era of Hong Kong. Qual Health Res 18:729–738. https://doi.org/10.1177/1049732308318372

Sokolov M (2020) The pandemic infodemic: how social media helps (and hurts) during the coronavirus outbreak [WWW Document]. Drum. https://www.thedrum.com/opinion/2020/03/03/the-pandemic-infodemic-how-social-media-helps-and-hurts-during-the-coronavirus

Soltaninejad K (2020) Methanol mass poisoning outbreak: a consequence of COVID-19 pandemic and misleading messages on social media. Int J Occup Environ Med 11:1–3. https://doi.org/10.34172/ijoem.2020.1983

Sprang G, Silman M (2013) Posttraumatic stress disorder in parents and youth after health-related disasters. Disaster Med Public Health Prep 7:105–110

Su T-P, Lien T-C, Yang C-Y, Su YL, Wang J-H, Tsai S-L, Yin J-C (2007) Prevalence of psychiatric morbidity and psychological adaptation of the nurses in a structured SARS caring unit during outbreak: a prospective and periodic assessment study in Taiwan. J Psychiatr Res 41:119–130

Taha S, Matheson K, Cronin T, Anisman H (2014) Intolerance of uncertainty, appraisals, coping, and anxiety: the case of the 2009 H 1 N 1 pandemic. Br J Health Psychol 19:592–605

Tan W, Hao F, McIntyre RS, Jiang L, Jiang X, Zhang L, Zhao X, Zou Y, Hu Y, Luo X (2020) Is returning to work during the COVID-19 pandemic stressful? A study on immediate mental health status and psychoneuroimmunity prevention measures of Chinese workforce. Brain Behav Immun 87:84–92

Taylor MR, Agho KE, Stevens GJ, Raphael B (2008) Factors influencing psychological distress during a disease epidemic: data from Australia's first outbreak of equine influenza. BMC Public Health 8:347. https://doi.org/10.1186/1471-2458-8-347

The Economic Times (2020) 11 Coronavirus suspects flee from a hospital in Maharashtra

The New Indian Express (2020) Bihar man beaten to death for informing COVID-19 medical help center about arrival of two people from Maharashtra

Torrey EF, Miller J (2001) The invisible plague: the rise of mental illness from 1750 to the present. Rutgers University Press, New Brunswick

Tsai J, Wilson M (2020) COVID-19: a potential public health problem for homeless populations. Lancet Public Heal 5:e186–e187

UN (n.d.) Water access critical to beating back COVID-19 spread in slum areas | | UN News [WWW Document]. https://news.un.org/en/story/2020/03/1060042

UNHCR (n.d.) UNHCR—The rights and health of refugees, migrants and stateless must be protected in COVID-19 response [WWW Document]. https://www.unhcr.org/news/press/2020/3/5e836f164/rights-health-refugees-migrants-stateless-must-protected-COVID-19-response.html

Verma S, Mythily S, Chan YH, Deslypere JP, Teo EK, Chong SA (2004) Post-SARS psychological morbidity and stigma among general practitioners and traditional Chinese medicine practitioners in Singapore. Ann Acad Med Singap 33:743–748

Wang G, Zhang Y, Zhao J, Zhang J, Jiang F (2020) Mitigate the effects of home confinement on children during the COVID-19 outbreak. Lancet 395:945–947

Wester M, Giesecke J (2018) Ebola and healthcare worker stigma. Scand J Public Health 47:99–104. https://doi.org/10.1177/1403494817753450

World Health Organization (2020) Mental health and psychosocial considerations during the COVID-19 outbreak

Wu P, Fang Y, Guan Z, Fan B, Kong J, Yao Z, Liu X, Fuller CJ, Susser E, Lu J, Hoven CW (2009) The psychological impact of the SARS epidemic on hospital employees in China: exposure, risk perception, and altruistic acceptance of risk. Can J Psychiatr 54:302–311. https://doi.org/10.1177/070674370905400504

Xiang Y-T, Yang Y, Li W, Zhang L, Zhang Q, Cheung T, Ng CH (2020) Timely mental health care for the 2019 novel coronavirus outbreak is urgently needed. Lancet Psychiatry 7:228–229

Zandifar A, Badrfam R (2020) Iranian mental health during the COVID-19 epidemic. Asian J Psychiatr 51:101990. https://doi.org/10.1016/j.ajp.2020.101990

Zarocostas J (2020) How to fight an infodemic. Lancet 395:676. https://doi.org/10.1016/S0140-6736(20)30461-X

Zhai Y, Du X (2020) Mental health care for international Chinese students affected by the COVID-19 outbreak. Lancet Psychiatry 7:e22. https://doi.org/10.1016/S2215-0366(20)30089-4

Zhang J, Wu W, Zhao X, Zhang W (2020) Recommended psychological crisis intervention response to the 2019 novel coronavirus pneumonia outbreak in China: a model of West China Hospital. Precis Clin Med 3:3–8

Zhou X, Snoswell CL, Harding LE, Bambling M, Edirippulige S, Bai X, Smith AC (2020) The role of telehealth in reducing the mental health burden from COVID-19. Telemed e-Health 26:377–379

Zhu L, She Z-G, Cheng X, Qin J-J, Zhang X-J, Cai J, Lei F, Wang H, Xie J, Wang W (2020) Association of blood glucose control and outcomes in patients with COVID-19 and pre-existing type 2 diabetes. Cell Metab 31(6):1068–1077.e3

Chapter 8
Environment Impact Assessment of COVID-19

Bhaskar Jyoti Deka, Varsha Bohra, Wazir Alam, Suchitra Sanasam, Jiaxin Guo, Lalit Borana, and Alicia Kyoungjin An

Abstract Novel infectious coronavirus epidemic (COVID-19) with vulnerable transmission among human was detected initially in late 2019 at Wuhan, China. Later, it transforms to global pandemic within the first quarter of 2020. COVID-19 has a significant impact on our environment specially in land, water and atmosphere. During handling of COVID-19, huge surplus quantity (~ 600% rise) of waste were generated majorly from healthcare sectors, which caused tremendous load to pre-exist solid waste management facilities. Also, the presence of infectious personal protective equipment (PPE) such as masks, gloves, face shield, gown and head cover were reported in municipal waste sectors. These wastes have become a great threat to our society and has ability for community outbreak. Additionally, the pandemic makes us realize about the ability of our accessible treatment infrastructure and further requirement of superior management initiatives to deal with such emergencies. Many countries have imposed different types of restrictions and even lockdown to stop the spread of pandemic. Despite devastation, the lockdown due to COVID-19 has significantly brought down the pollution level in all forms of natural environment, specifically water resources (rivers, lakes) and air environment.

B. J. Deka · J. Guo · A. K. An (✉)
School of Energy and Environment, City University of Hong Kong, Kowloon, Hong Kong SAR
e-mail: bdeka2-c@my.cityu.edu.hk; jxguo4-c@my.cityu.edu.hk; alicia.kjan@cityu.edu.hk

V. Bohra
Hong Kong Baptist University, Kowloon, Hong Kong SAR

W. Alam
Department of Forestry and Environmental Science, Manipur University, Imphal, Manipur, India
e-mail: wazir@manipuruniv.ac.in

S. Sanasam
Biochemistry, Pravabati College, Manipur University, Mayang Imphal, India

L. Borana
Civil Engineering Department, IIT Indore, Indore, India
e-mail: lalitborana@iiti.ac.in

M. K. Goyal, A. K. Gupta (eds.), *Integrated Risk of Pandemic: Covid-19 Impacts, Resilience and Recommendations*, Disaster Resilience and Green Growth,
https://doi.org/10.1007/978-981-15-7679-9_8

Keywords Environmental pollution · Lockdown · Plastic waste · Pandemic · Virus · Waste management · Water quality · Wastewater

8.1 Introduction

The novel coronavirus disease 2019 (COVID-19) outbreak has spread profusely across the globe posing massive health, environmental, economic and social challenges to the entire humankind. It is considered as the greatest challenge the human population have faced after the second world war (WW-II) bringing health care units of the world's most developed countries on the brink of collapse (Wang et al. 2020; Nghiem et al. 2020). Following the severe acute respiratory syndrome (SARS) in 2002–2003 and middle east respiratory syndrome (MERS) in 2012, it is the third hefty zoonotic coronavirus disease outbreak in last 20 years (Bruinen de Bruin et al. 2020). COVID-19 as shown in Fig. 8.1 is a kind of lower respiratory tract contagion which may lead to serious and probably fatal atypical novel coronavirus infected pneumonia (NCIP) in human beings (Sohrabi et al. 2020). The disease was initially reported on 31 December 2019 by Chinese Health Officials as atypical pneumonia of unknown cause in several peoples at Wuhan, city of central Hubei Province, China (Dantas et al. 2020; He et al. 2020; Giordano et al. 2020), which was followed by a series of local risk mitigation measures. For the remaining world, business sustained as usual with minimal risk perception, until, 13 January 2020, wherein the first case of COVID-19 was detected outside China in a patient from Thailand (Bruinen de Bruin et al. 2020). Meanwhile, the virus dispersed all around the world: in Asia, Africa, Europe, Oceania and America (Hopkins 2020). By March 11, 2020, the number of cases identified surpassed 118,000 cases across 114 countries and World Health Organization (WHO) declared COVID-19 outbreak as a pandemic and global public health emergency owing to its growing danger to global health (WHO 2020a). As of June 3, 2020, the outbreak spanned across 216 countries or territories resulting in 6,287,771 infected cases including 379,941 deaths globally with maximum mortalities recorded in the United States of America, Italy and Spain as depicted

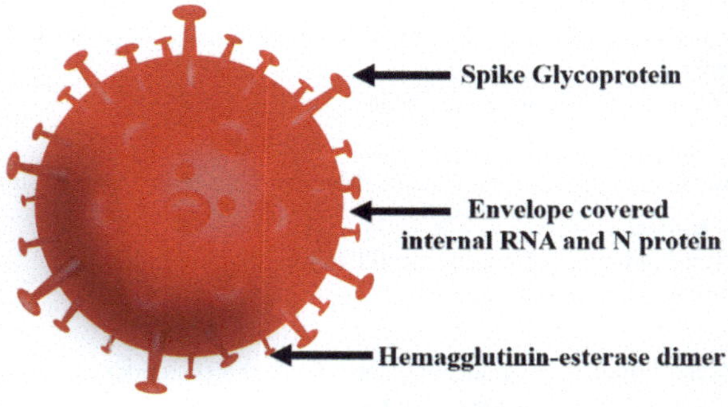

Fig. 8.1 General structure of Covid-19

in Figs. 8.2 and 8.3 (WHO 2020b) The causative agent of COVID-19 has been recognized as an extremely contagious novel coronavirus which was named as "severe acute respiratory syndrome coronavirus 2 (SARS-CoV-2)" by the International Committee on Taxonomy of Viruses on January 7, 2020 (Liu et al. 2020) and on 12 January 2020, genomic sequence of SARS-CoV-2 was reported by China (Bruinen de Bruin et al. 2020). The virus shows genetic resemblance to SARS coronavirus and in a similar way presumed to have spanned the species blockade from animal to human (Gorbalenya et al. 2020). However, its precise origin still remained undetermined, bat coronavirus is considered as the prospective ancestor (Chin et al. 2020). SARS-CoV-2 is a β-coronavirus from subgenus sarbecovirus. It is a small spherical shaped virion of diameter ≈100 nm comprising positive single stranded RNA genome enveloped in a delicate lipid membrane (Zhu et al. 2020). SARS-CoV-2 has spike proteins (CoV-2-S) on its external surface which precisely targets host cells (cells in lung, heart, artery, intestine and kidney) containing angiotensin-converting enzyme 2 (ACE2) protein receptors on its membrane, that helps in viral entry into healthy cells. As soon the CoV-2-S protein binds with ACE2 protein, the S protein cleaves into two subunits; S1 and S2. The S1 subunit is comprised of receptor binding domain which helps it in binding to the peptidase domain of ACE2 protein. After binding, S2 protein assist virus to fuse with the host cell membrane in order to discharge its nucleic acid content (Paital et al. 2020). After infection, the virus multiplies inside the host cell exhausting all the resources and escapes from the cell in an act called shedding. Reports suggests that shedding starts substantially early during the course of disease, may be before or within few days since the inception of symptoms (WHO 2020c). Viral shedding has been observed in respiratory secretions, sputum, blood, and faeces from symptomatic patients (Wang et al. 2020). Though, it is expected that viral count is relatively low in asymptomatic carriers, additional investigation is required to establish the same (Nicola et al. 2020a). The reports from WHO indicates extremely infectious attribute of SARS-CoV-2, the transmission pathways of which includes: (1) inhaling liquid droplets comprising virus (2) close contact with infected individual (3) contact with SARS-CoV-2 infected surfaces (4) aerosol transmission and (5) faecal-oral route (Liu et al. 2020; Yuen et al. 2020; Gormley et al. 2020). COVID-19 is detected primarily with symptoms like fever, dry cough, cold and breathing problem caused due to fibrosis and cough deposit in air sacs of lungs (Huang et al. 2020a), which may on severe cases lead to death of the patient. The disease morbidity rate is perceived to be higher in elderly peoples and people suffering from co-morbidity like diabetes, asthma, cardiac diseases and cancer possibly because of their subordinate immunity (Paital et al. 2020). A currently used method for unveiling the presence of SARS-CoV-2 in an individual is testing for the presence of viral nucleic acid by means of real time polymerase chain reaction (RT-PCR) technology.

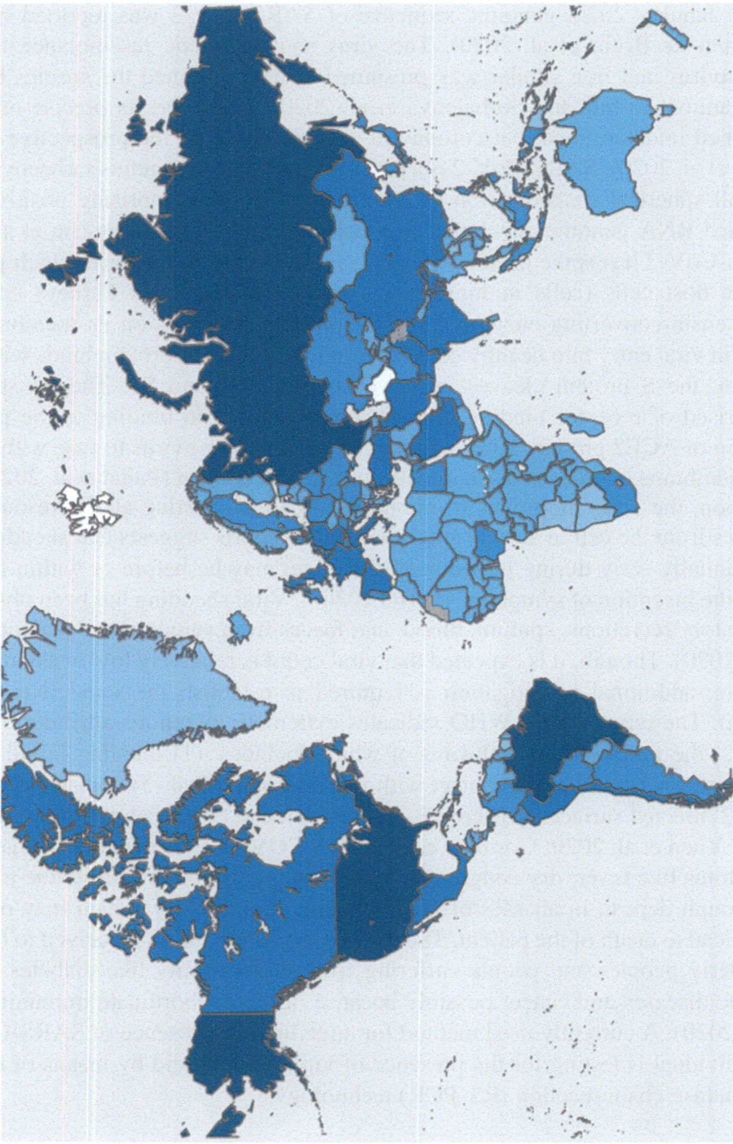

Fig. 8.2 Distribution of COVID-19 in various countries. (Source: WHO, 2020, Accessed date: 3 June 2020)

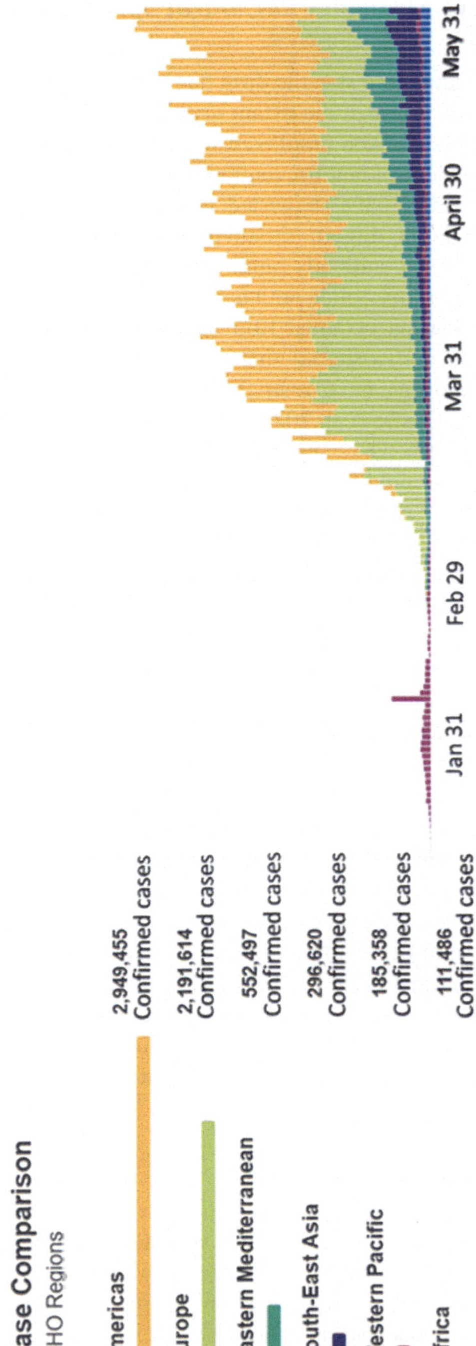

Fig. 8.3 COVID-19 case comparison in WHO regions. (Source: WHO, 2020, Accessed date: 3 June 2020)

8.1.1 Preventive Strategies to Control Spreading of Pandemic

Owing to its novel nature, no specific drug has been discovered for the treatment of COVID-19. However, some of the extant drug types like antibacterial, antimalarial and anti-HIV drugs have been employed to treat COVID-19. Establishments like Centres for Disease Control and Prevention (CDC), U.S. Food and Drug Administration (FDA) and WHO have suggested few drugs namely hydroxychloroquine, chloroquine (antimalarial drugs); ritonavir, lopinavir (HIV protease inhibitors); tocilizumab (monoclonal antibody inhibiting interleukin-6 and its receptor the binding); azithromycin (macrolide antibacterial drug); convalescent plasma therapy (infusion of blood plasma obtained from people recovered from COVID-19 into active patients), and corticosteroid therapy, but not yet approved any of them. However, all these drugs have restrictions and risk associated with them and symptom dependent (Gostic et al. 2020). As no antiviral drug or vaccine is available for the treatment of COVID-19, preventive strategies are eventually considered as the only intrusion that could be practiced to control the spread of COVID-19 (Lai et al. 2020). COVID-19 pandemic has emerged as a global threat which necessitates global retort encircling all the countries. Governments ought to be accountable for disseminating precise information in order to restrict the spread of virus and help public fight this novel infection. The unprecedented strategies presently followed to prevent COVID-19 disease are discussed as follow.

8.1.1.1 Movement Restriction

Mobility restriction refers to limit people's movement to prevent or curtail the spread of virus. This leads to reduction in the number of people who might get exposed and infected by symptomatic and non-symptomatic cases (Gostic et al. 2020; Hellewell et al. 2020). Mobility restriction includes closing/restraining public transport (road and rail transport), air transport, water transport and sometimes even transient closure/restricted use of personal transportation including bicycles, cars and outside activities like walking and running, specifically in groups as guided by the government. Though, mobility restriction allows for the minimal and emergency movement of general public, unnecessary use of personal transportation must be circumvented and provisions to work from home should be made.

8.1.1.2 Socio-Economic Restrictions

The objective of this restriction is to prevent two individuals from distinct families or neighboring residents to get in close proximity with each other. These restriction targets social and economic activities where people assemble for sportive, recreational, educational or occupational purposes. It comprises the closing/restricted opening of restaurants, hotels, swimming pools, museums, outdoor sport facilities, schools, nurseries, and elderly homes. This cluster also incorporates restriction on retail trade, saving of necessary goods (medicines), limiting/terminating production in factories along with financial, banking and insurance services. Emergency services like medical care, medicine supply and food security are usually permitted, however, these services may also be closed in case of emergency and the duration of lockdown is for 1 or 2 days. In order to avoid any kind of gatherings, people may use remote technologies and online services to contact friends, family, healthcare professionals and other critical services (Nicola et al. 2020b).

8.1.1.3 Physical Distancing

Individual's behavior is critical in regulating the outbreak of COVID-19. Physical distancing is considered as the most suitable measure to avoid the spread of virus by avoiding contacts including (asymptomatic) infected persons. All the individuals, especially those over 70 years of age should follow proper distancing (presently guided between 1.5 and 2 m) from other individuals (Nicola et al. 2020b). Also, prohibition on large groups, regulating count of individual present per square meter in indoor spaces like home, offices, and commercial spaces must be well regulated. On identification of symptomatic case in the household, patient must be kept in isolation for 14 days in order to diminish the number of contacts outside the house. Also, the whole family should remain in voluntary home quarantine for 14 days and avoid any form of social contact (Nicola et al. 2020b).

8.1.1.4 Sanitation Measures

Transmission of virus through direct or indirect contamination can be restricted by following proper sanitary measures. These cluster comprises of washing hand at regular intervals for at least 20 s, avoid touching environmental surfaces, using e-payment methods, wearing personal protective equipment like masks, gloves, eye protectors. Usage of high-filtration masks such as N95, gowns and goggles by health care professionals working in close proximity of infected patients. Patient should be instructed to cover their mouth and nose, maintain hand hygiene on contact with respiratory exudations and avoid contaminating environmental surfaces. Proper disposal of household and clinical waste must be practiced to prevent exposure of waste collectors and waste processing operators with contaminated infectious materials (Semple and Cherrie 2020). Placement of protective glass barrier in places where physical distancing is not feasible (example, between cashier and customer) and, symptomatic screening such as obligatory temperature check at various public places (malls, railway stations, airports) should be implemented.

8.1.1.5 Information Communication

This cluster is thought to play most important role in risk mitigation measures. Necessary information regarding nature of the disease, its transmission mechanism, risk perception, precautions and government policies should be propagated among the public. Every update in context of pandemic should be disseminated globally. Social networking, mass media, internet, television, radio, newspapers can play crucial role in spreading the right information throughout the globe. A WHO 'WhatsApp' group was formed on March 2020, for providing information and risk mitigation measures, globally (Bruinen de Bruin et al. 2020). Countries have likewise launched national communication strategies in the form of news and information websites for encouraging understanding, awareness and amenability towards restrictions and for combatting false news.

8.1.2 Impact on Environment

In order to fulfil the needs of rapidly expanding population, human beings started destructing nature in several ways. Urbanization and industrialization became unavoidable whose obvious consequence in the form of environment pollution has become major problem of the present world. However, the novel coronavirus disease pandemic has hit the world severely, halting all sorts of industries, transportation and human activity, for the only time in human history. As increased industrialization and anthropogenic activities have long been envisaged as among the major causes of pollution in various environmental spheres, it is presumed that pollution loads on environment will decrease due to shut down of industrial activities and public movements for months (Yunus et al. 2020). For years, industrialization and rapid urbanization has severely polluted the Hydrosphere, the major component of environment. However, lockdown implemented to prevent the spread of pandemic have restricted or completely shut down the major sources (industrial wastewater disposal, plastics and crude oil) of aquatic pollution, thus helping water bodies to rejuvenate (Häder et al. 2020). For instance, Canal in Italy have turned clear along with recurrences of several aquatic species (Saadat et al. 2020a). In India, one of the extremely polluted river the Ganges, get clear at certain locations (Mani 2020) surface water quality of the longest freshwater Vembanad Lake get enhanced (Yunus et al. 2020) as compared to the pre lockdown period. On a similar note, drastic reduction in air pollution has been experienced in many cities of India. Prime cities such as Delhi, Mumbai, Pune and Ahmedabad have reported 40–50% reduction in average nitrogen dioxide level by March 2020 in comparison to March 2019 (Wright 2020). The reason was accredited to substantially reduced usage of conventional energy sources or fossil fuels in sectors such as transportation and industries. Residents from several big cities can visualize clear sky for the first time ever. The depleted ozone layer has been found to have rejuvenated to some extent. The extent of pollution in tourist places like sea beaches, forests, hill areas and others are also getting reduced considerably.

On a contrary note, the diseases outbreak has brought about an abrupt rise in the demand of plastic products. Consumption of personal protective equipment (gloves and masks), routine plastic supplies (syringes), disposable plastic part of life support equipage and respirators to be used as protective gear for general public, patients, health care and service workers has increased dramatically. Lockdown measures have also resulted in an increased usage of plastic packaging materials for the distribution of foodstuffs and groceries to the dwellings. These alterations may intensify environmental issues associated with plastics, befitting management of which has always been considered as one of the major environmental issue, even before the outbreak of the COVID-19 pandemic. Although the current upsurge is inevitable, environmental safety efforts needs to be sustained. Comprehensive management of plastic waste can minimize the unpredicted effects on environment or human health. Overall, the pandemic exhibited a very favourable impact on the world environment along with the worldwide destruction of human civilization.

8.2 COVID-19 Pandemic and Environmental Pollution

The impact of any form of global pandemic situation causes potentially significant changes in many features of the physical, biological and human (socio-economic) environment. In case of physical and biological environment, there may be direct or indirect impacts that can be either favorable or harmful to the environment. However, in most of the cases, human environment faces severe detrimental impacts due to halt in all forms of developmental activities (Madurai Elavarasan and Pugazhendhi 2020). The process of environmental impact assessment (EIA) for any form of pandemic can facilitate the systematic identification, quantification and appropriate interpretation of significant impacts for future preparedness to such crisis. Human environment are associated with all forms of developmental activities related to financial sectors, infrastructure development, industrial and manufacturing sectors, public service facilities such as power supply, communication facilities etc., that have tremendous impacts on our day to day environmental (air, water, soil, noise, solid waste) quality. Sudden halt in all these developmental processes have significant short and long term positive impacts on overall environmental quality due to decline in all forms of pollutant load.

8.2.1 *Assessment of the Impact Due to COVID-19 on Society and Global Environment*

All forms of developmental activities are linked with every attribute of socio-economic features, problems and solutions. It is apparent that socio-economic features of any state or nation have control over environmental quality and *vice-versa*. Present pandemic situation due to COVID-19 have reflected the similar state regarding the environmental quality as well as its impacts on the society. Due to lockdown, all mode of communications, industrial functioning and other developmental activities have come to a halt, leading to a great impact on the physical, biological and socio-economical environmental quality (Muhammad et al. 2020). As a result of lockdown in various parts of world, the drop in air travel have reached lowest (96%) since last 75 years and it is also expected that there will be around 10% fall in global oil demand from 99.9 to 90.3 million barrel per day (bpd) during 2019–2020 (Fernandes 2020). Moreover, decline in industrial, manufacturing and other developmental sectors have resulted in significant drop of global FDI, which is further predicted to shrink by another 5–15%, due to lockdown scenario worldwide (Fernandes 2020). The G-20 (19 countries and European Union) contributes 66% of global population and shares 85% global economy with 80% of primary electricity consumption till 2019, which was projected to increase by 5.35% in 2020 (Fernandes 2020). However, due to COVID-19 crisis, the consumption scenario has completely changed leading to lesser environmental stress and pollution. The pandemic have not only affected 215 countries across the globe but also have affected all sectors of the society, economy and health care systems of countries worldwide (WHO 2020d). The COVID-19 pandemic have significantly changed the priorities of the society and quality of environmental media (air, water soil, noise) due to slow down of developmental activities across the globe (Saadat et al. 2020b).

8.2.2 *Socio-Economic Environmental Aspects*

Due to outbreak of COVID-19 pandemic, an evitable consequence of socio-economic environment has become prior concern of the present world situation. The countries like China, Italy, USA, UK, France, Spain, Turkey, Iran, Germany, Taiwan, South Korea, India, Australia and many more are in the grip of partial or complete lockdown situation since the outbreak of pandemic and uncertainty in lifting of lockdown have caused tribulations in socio-economic features of all the 215 affected countries across the world (WHO 2020d). As per the Interim Economic Assessment report, the annual Gross Domestic Product (GDP) is projected to drop by 2.4% in the first quarter of 2020 (WHO 2020e). As per the report of Asian Development Bank (ADB), the global economy is expected to suffer around 6.4–9.7% losses as GDP, under 3 months and 6 months containment scenario, respectively (Park et al. 2020). As the situation have emerged unexpectedly, no country was prepared to cope up with such situation of health and medical emergencies, economic crisis, transportation and communication sectors, public services, agricultural sectors, manufacturing and production sector (Chakraborty and Maity 2020). Government mechanisms of all the affected countries have tried their best to restrict widespread spreading and community transmission of COVID-19 by controlling free movement, social and commercial activities and maintenance of social distancing of people. As a result, many religious, cultural, social, sports, scientific and political mass gathering have been banned and dealt with strict enforcement of laws against the violators. Some of the world events such as Hajj, all forms of pilgrimages, Olympic sports, events of entertainment industries, scientific and research gatherings, all forms of global tourism activities have come to a halt across the world impacting a huge population associated with these sectors (Gu et al. 2020). Global tourism sector alone has faced loss of US$ 1.2 trillion revenues resulting in liability of 100–120 million jobs in global travel and tourism sector (UNWTO 2020). Moreover, due to the uncertainty regarding the future consequences of the pandemic and restoration of normal situation, there is a prevailing phobia among the people across the globe leading to changes in community cohesion, attitude and behavior (Díaz de León-Martínez et al. 2020). Although, many countries are responding to the crisis with pro-social behavior, many social sectors including migrant workers, part-time and contract labors, low-income families, socially weaker sections, farmers and linguistically disadvantage groups etc. are facing serious problems and adversities (Brouard et al. 2020). The education system across the globe have faced remarkable challenges due to closedown of schools, colleges and universities affecting around 1,198,530,172 learners across 153 countries, for maintenance of social distancing to curb the spread of COVID-19 (Huang et al. 2020b). Apart from severe health emergencies, COVID-19 pandemic have caused intense impacts on global economy, social ramification and disruption of public services of all affected countries across the world (Nicola et al. 2020a). The cumulative socio-economic challenges due to the pandemic phenomenon will have more widespread and multifaceted global impacts in all sectors, till the world community successfully fight backs the pandemic and restore the current crisis.

8.2.3 Pollution Assessment During COVID-19

As the world moves ahead with COVID-19 pandemic situation, the countries will face more complex challenges for green and inclusive recovery with already existing looming environmental challenges such as combating climate change, loss of biodiversity, environmental pollution management, indiscriminate use of natural resources etc. (Chakraborty and Maity 2020). Global emerging environmental problems are mostly associated with human centric developmental activities. Due to the pandemic, most of the developmental activities have been put to a halt from the beginning of the year. As a result, many incidents of improvement of environmental quality have been reported across the world recently (Muhammad et al.

2020). The pandemic has lockdown all forms of anthropogenic activities leading to natural revival of air, water and other environmental quality. The environmental assessment on footprints of water, air, solid waste, and energy due to COVID-19 pandemic have shown significant affirmative changes in a short span of few months (Saadat et al. 2020b). Air pollution status in all major cities/urban centres of USA, Europe and Asian countries were found to be improving due to reduction of emission from transport mobility (Nicola et al. 2020a). Ministry of Ecology and Environment, Republic of China has reported around 11.4% rise in good air quality across 337 cities of the country (Yunus et al. 2020). Similarly, drop in the NO_2 emission were observed in Spain, Italy and UK (Ficetola and Rubolini 2020). After the outbreak of COVID-19 pandemic, lockdown of major developmental activities in manufacturing sectors, transportation sectors, industrial activities and infrastructure projects etc. have ceased and resulted in reduction of effluent significantly. Consequently, various instances of natural recovery of environmental qualities were observed after the outbreak of the pandemic. For example, the water quality of Grand Canal of Italy (Yunus et al. 2020), river Ganges, Yamuna River in Delhi (Kapil 2020) and many urban wetlands, water bodies and aquatic ecosystem (Häder et al. 2020) have been found with significant changes in water quality. Similarly, traffic noise pollution levels of major cities across the globe have drastically fallen due to lockdown of all forms of commercial activities and traffic mobility (Zambrano-Monserrate et al. 2020). Generations of municipal solid waste and hazardous biomedical waste in the large and medium cities of China have reduced by 30% during the COVID-19 pandemic according to State Council's Joint Prevention and Control Mechanism (Klemeš et al. 2020a).

8.3 Prolong and Shorten Impacts of Pandemic

The COVID-19 pandemic have multifaceted impacts of varied dimensions including both short and long term impacts on various sectors of the environment. These impacts may manifest as both beneficial and detrimental, in term of environmental quality are concerned. According to ADB report, it predicted that 7.0 billion people would be affected with 40 billion deaths without the lockdown, social distancing and necessary interventions taken by the countries worldwide (Park et al. 2020). Moreover, it is believed that the most vulnerable section of the society to be affected by the pandemic are elderly people and infants, due to weaker and immature immune system, respectively (WHO 2020d). As per the present global trend of impacts due to pandemic, it is expected that the countries across the globe will have to face sustained challenges from COVID-19 in upcoming days. The world will need to fight the multifaceted impacts of this pandemic as part of new normal to human society and civilization.

8.3.1 *Evaluation of Temporary and Permanent Effects on Environment*

The environmental and socio-economic impact assessment due to COVID-19 will apparently show significant impacts in wide-range of dimensions. The unprecedent global health crisis and emergencies have already created various direct and indirect effect on socio-economic and environmental sectors which are predicted to be short-term, long-term and cumulative in nature as presented in Table 8.1. Countries under lockdown which restrict social gathering, commercial, and economic activities have already started exhibiting ongoing rippling impacts in various dimensions. These impacts are expected to have profound long-term cumulative negative affect in the

Table 8.1 The nature of associated impacts due to COVID-19 on various sectors

	Sector	Direct	Indirect
COVID-19 PENDEMIC	Economy (−)	Livelihood and Employment (−)	Gross domestic Product GDP (−)
		Manufacturing Industries (−)	Human Development Index (HDI) (−)
		Petroleum and Oil Sector (−)	Quality of Life (QOL) Index (−)
		Agricultural sector (−)	
		Transportation and Communication (−)	
		Tourism sector (−)	
		Entertainment Sector (−)	
	Environment (±)	Air Quality (+)	Climate Variability and Change (+)
		Water Quality (±)	Biodiversity (+)
		Solid Waste Management (±)	Aesthetic Environmental Quality (+)
		Noise Environment (+)	Visual Pollution (±)
		Biodiversity(+)	Forestry and Wildlife (±)
			Landuse/Land cover (+)
	Society (±)	Public Health (−)	Community Attitudes (±)
		Education (−)	Social Cohesion (±)
		Research & Development activities (±)	Social Bonding (±)
		Sanitation and Hygiene (−)	Public Services (−)
		Religion and Festivals (−)	Standard of living (−)
		Global Sports (−)	Household violence (−)
		Recreational activities (−)	Psychological wellbeing (−)
			Behavioral responses (−)

(+) indicates positive impacts, (−) indicates negative impacts, (±) indicates both positive and negative impacts depending various factors

form of livelihood, employment opportunities, agriculture productivity, transportation, manufacturing, local and global trade and other commercial activities. The primary sectors (agriculture, petroleum, oil etc.), secondary sectors (manufacturing industry) and tertiary sectors (education, research, tourism and other public services) may face numerous challenges. Gradually, with time things may get normalize and restore with human effort but ongoing lockdown has improved the quality environmental media and also positive shift in climatic variables due to reduction in emissions from anthropogenic activities (Muhammad et al. 2020).

8.3.2 Case Studies

The river Yamuna is the largest tributary of River Ganga and considered to be one of the important river systems of India. The river originates from Yamunotri glacier near lower Himalayan range of Bannder Punch at elevation of 6320 m asl. The studies were carried out at 36 locations of the river Yamuna by Central Pollution Control Board (CPCB) of India before and week after the lockdown due to COVID-19 (Kapil 2020; Shrangi 2020). The study shows significant changes in water quality of Yamuna river compared to Ganga river. The Ganga river flows through UP, Bihar, WB and finally flows into Bay of Bengal in Bangladesh. There are total 97 urban establishments/towns situated in the bank of Ganga river at various locations where industrial and sewage disposal are the major causes of water pollution. The Ganga river receives 35,000 million litre per day (MLD) of sewage, out of which 68.6% are untreated and around 9% are from untreated Industrial wastes (Yunus et al. 2020). In Yamuna, CPCB claimed more noticeable improvement in water quality with respect to dissolved oxygen (DO), biological oxygen demand (BOD) and chemical oxygen demand (COD) in three locations in Delhi. These changes are noticeable due to closedown of 42 industrial units that are situated in the bank of Yamuna river and in its stretch within the city of Delhi (Mahato et al. 2020). According to Yamuna Monitoring Committee, around 80% of Yamuna river water pollution is contributed by the continuous discharge of domestic sewage. As per the Delhi Pollution Control Board (DPCC), the assessment of water quality of Yamuna river in Delhi was improved by 33%. The Table 8.2 explains the pH, DO levels, BOD and COD in three major locations Yamuna river in the capital city of Delhi *i.e.* Palla (the entry

Table 8.2 Comparison of few water quality parameters of Yamuna River during pre-lockdown and post-lockdown period

	pH		DO (mg/L)		BOD (mg/L)		COD (mg/L)	
Locations	BL	AL	BL	AL	BL	AL	BL	AL
Palla	8.7	7.8	8.3	17.1	7.9	2.0	28	6
Nizamuddin bridge	7.3	7.2	–	–	5.6	0.57	90	16
Okhla barrage upstream	7.2	7.1	–	1.2	6.1	0.27	95	18

BL (before lockdown) and AL (After lockdown). (Source: Shrangi 2020)

point of Yamuna river in Delhi), Nizamuddin bridge (between Wazirabad and Okhla barrages where 14 drains discharges) and Okhla barrage upstream (receive discharges from 16 drains) collected before lockdown (March 2020) and after lock down (April 2020). At Nizamuddin, BOD and COD values were found to be reduced by 77.4% and 81.1%, respectively, signifying no industrial effluents were released in the stretch of 7.5 km after Nizamuddin Bridge. Similar observations were noticed at Najafgarh and Shahdara where there are 30–45% decreases in BOD levels in post-lockdown scenario due to reduction in various anthropocentric effluents. As per the report of committee consisting of CPCB and DPCB, significant changes of Yamuna river quality were attributes to dilution of river water by fresh and clean water of around 5000 cumec during lockdown period (Shrangi 2020).

8.4 Role of Lockdown on Overall Environment

8.4.1 Water Pollution in Ganga River Before and After Lockdown

The subsequent lockdown in the wake of COVID-19 pandemic has created massive devastation in every sector of human life and also has taught us several lessons in respect of the hydrology of river, maintaining the ecological balance, various environmental pollution and the role of the community at large. The decrease in the rate of industrial production and commercial use, besides lowering in the irrigation system have also contributed to the change.

The river Ganga which has been a lifeline for those living along its course of flow originates from the Gangotri Glacier, along the Gangetic plains and finally merged into the Bay of Bengal. The Hindus considered it as a holy river and worshiped it. Since the last few decades due to rapid industrialization and urbanization along with increased anthropogenic activities, the river severely polluted. Many worshippers carried out various religious rituals and bathing round the year along with laundry activities, fishing, ferry and freight services resulting to the contamination of the river water with many organic and inorganic wastes like heavy metals, surfactant and polycyclic aromatic hydrocarbons (PAHs) (Goswami and Mazumdar 2016; Rakshit and Sarkar 2018). The additional point sources of coliform bacteria, heavy metals and hydrocarbons are from various industries located at the banks and untreated sewage discharges (Choudhury et al. 2019). The untreated sewage discharges is received by the river leads to fecal contamination (Strauss 1997). As the Ganges is home to several species of fishes, reptiles and mammals, crocodiles and the South Asian river dolphins, these pollutants not only threaten the humans but also the other aquatic lives. The overall quality of river water influence significantly on the microbial flora of the river as the pollutants and the water serves as the nutrients for the microbial activity and proliferation.

The levels of faecal coliform bacteria found in the river are much higher than the government permissible limit. The central government of India has invested several crores of rupees through Ganga Action Plan since 1986 for cleaning the river ganga, but the success is yet far because of several reasons mainly lack of awareness, environmental planning and poor technical expertise.

The main sources of pollution of Ganga river are domestic wastewater from the towns situated near the river, and various industrial wastewater. The pollution level was assessed by the CPCB, India a week before lockdown and weeks after lockdown at 36 locations in Uttar Pradesh and West Bengal. Out of the 36 monitoring units, water quality around 27 points placed at various points of the river was found to be fit for bath and proliferation of wildlife (The Tribune 2020). The DO (>6 mg/L), BOD (<2 mg/L), total coliform levels (5000 per 100 mL) and pH (6.5–8.5) were the different parameters that the monitoring stations monitor online to assess the health of the river. Other than few places of Uttarakhand and Uttar Pradesh, the water in Ganga river was not suitable before the lockdown for bathing till it merged into the Bay of Bengal.

Though the pandemic as brought massive lost in several sectors, it has become a boon for the environmental sectors due to shutting down of the point and nonpoint sources of pollution (Mitra et al. 2020). In a recent study (Sengupta et al. 2020), it was found that the load of both nitrate and phosphate has sharply declined due to the imposed lockdown at three selected sites viz.: Botanical Garden, Babughat and second Hoogly Bridge in Kolkata, West Bengal, as a result of decline in the anthropogenic industrial wastes. Significant changes were experienced for Babughat (dissolved nitrate: 62.56 μgmat/L and phosphate: 14.59 μgmat/L) and second Hooghly bridge (dissolved nitrate: 49.18 μgmat/L and phosphate: 11.27 μgmat/L). These changes were mainly because of the sudden decrease in waste generation from highly congested Babughat area, which comprises of public bus terminals, markets, lodges and restaurants. Without any treatment of the released nutrients from these point sources, the adjacent water in Ganga river become richer in nutrients compared to other two sites. During the lockdown period, these units remains shut, as a result, the load of nutrients has gradually decreased as per weekly data of April 2020. The previous study (Tiwari et al. 2016) revealed occurrence of eutrophication in these areas, which reduce the level of dissolved oxygen and pose threat to aquatic lives. Low nutrients in the aquatic phase (observed during the COVID-19 lockdown phase) is pleasing for aquatic ecosystem thus preventing eutrophication, leading to better water quality. Decline in anthropogenic pollution may reduce the pathogenic microbes and coliforms, attributed to cleaner and healthier water.

There are several reports of microbial multiplication due to increase pollution around the world (https://www.coronavirus.gov/). A comparative study of the total coliform load was carried out at two sites of the Kolkata metropolis before and during the lockdown (Mukherjee et al. 2020). The experimental findings presented about the severe decrease in total coliform load which may be due to the shutting down of industries, traffic, tourism sector and decline in performing various religious rituals and bathing along the river banks during lock down. The results of their study show the advantageous of reduced human intervention on the environment (Ganges here) during the lockdown which improved the microbiological quality of the river drastically.

Since the enforcement of the lockdown as reported by experts, the water quality of Ganga river, especially around the industrial clusters has improved remarkably as humans stay in quarantine, keeping the Ganga Ghats deserted. Prof PK Mishra of the department of chemical engineering at the IIT Banaras Hindu University reported that the total effluent dumped into the Ganga is around 6500–6700 MLD (in) UP stretch and onwards of which 30% of the total BOD load is due to industries along the river Ganga and amounting to 130–150 tons per day. He further reported that around 10% of toxic load from industries (approximately 700 MLD) which usually dumped in the river is not entering the river at present as all the major grossly polluting industries are closed due to the lockdown. With the factories discharging toxic industrial waste into the river were closed, the stretches of the holy river have become cleaner. The water is now clean and the number of fishes in the river has increase due to the rise in water level. The DO has increased amid lockdown indicating that aquatic life is getting better. The river has finally shown signs of much improvement and rejuvenation on almost all parameters. And none of it is due to any government-sponsored programme or project. The bacteriophages in the river has a self-cleaning mechanism and it relentlessly work to clean the river. But the unprecedented pollutants in it, far outnumbered the bacteriophages, rendering it a stagnant decaying body of drain water. If one monitors the activities of the industrial effluents and make sure that the raw sewage does not go in the river Ganga, it will become clean for sure. The flow has now been resumed and the parameters of river water purity have shown signs of significant improvement. And this is mainly due to the abstinence of human intervention in the flow of the river. Today we are battling coronavirus and survival has become critical. This is indeed a learning for us. This COVID-19 pandemic has taught us how badly we treat our environment.

For years, the governments at central and respective states have not seriously and sincerely worked for revival of rivers in India. Despite plans and funds, the number of polluted river stretches have increased from 302 in 2016 to 351 as per 2018 assessment of CPCB. Another report of Union Environment Ministry states that the total sewage generation in the country is 61,754 MLD while the treatment capacity is 22,963 MLD and the remaining 38,791 MLD sewage is flowing into river without treatment. What 'the Centre's flagship programme Namami Gange', to clean the Ganga, could not achieve in several years, nature has done in 2 months' time.

The National Mission for Clean Ganga launched IDEAthon on May 22,020 to focus on "The Future of River Management". It will basically explore the concepts how COVID-19 crisis will change the river management strategies. Some of the mitigating measures that could be undertaken to clean the River Ganges water and make it free from contamination includes controlled of various industrial waste disposal, proper treatment of the sewage prior to their discharge in the river, controlled of anthropogenic ventures, periodic monitoring of water quality and microbial analysis (coliform load), prohibition of community bathing and controlled of routine religious rituals.

8.4.2 *Municipal, Industrial and Hospital Waste*

The new coronavirus (COVID-19) has generated an unprecedented impact throughout the World and India is no exception to it. Hence, handling of municipal solid waste (MSW) and other hazardous medical waste is a major challenge now. One of the major environmental problem is the recycling of waste as reported by Liu et al. (2020) though it is a common and effective way of preventing pollution and conserving natural resources besides saving energy (Varotto and Spagnolli 2017). Because of the current pandemic, the usual normal treatment facilities loaded with sudden huge quantities of waste, which paralyzed the overall process. Scientific modelling and design are also essential for the system to manage the evolving nature of the pandemic. In countries such as USA, the authorities concerned have stopped recycling programs due to the risk of COVID-19 spreading in the recycling centers. In some of the affected European countries like Italy as a precautionary measure infected resident have been prohibited from sorting their waste. The lifespan activity of coronaviruses on various surfaces are shown in the Fig. 8.4 below as per the study of van Doremalen et al. (2020). Similar reports were corroborated in another study (Kampf et al. 2020; van Doremalen et al. 2020).

The waste materials that originates from households and other quarantine centres with suspected or positive cases of COVID-19 patients could be a source of infection as it may contain viable COVID-19 and could survive on hard surfaces and plastics for few days. During the initial stages of this outbreak, the waste collection procedures, have not been monitored properly to address the potential threat of COVID-19 for the larger community. Hence, while collecting waste materials from affected households and quarantine centres, strict standard operating protocols should be maintained. The lockdown policies that have been established in most countries, have led to the demand for home delivery system of various essential items through online shopping, thereby increasing the inorganic and organic waste generated by households.

The sudden jump in the requirement and utilisation of plastic products for protecting the general masses, patients, paramedics and those working in the frontlines is one of the major impact of this pandemic. The trend in demand for different

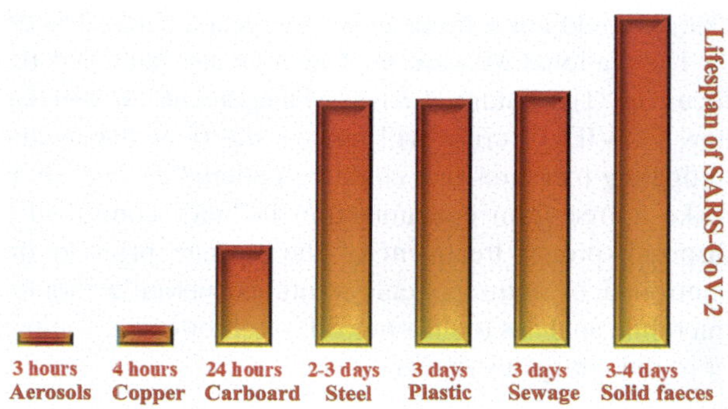

Fig. 8.4 Lifespan of COVID-19 in the environment

plastic made products, such as PPE including syringes are required to meet international pandemic curve. Even before the pandemic, plastic waste management have been a key problem due to the increasing concerns about environment (Rajmohan et al. 2019).

This sudden increased in the amount of waste endangers to devastate current waste management systems as does the healthcare capacity. As per the State Council's Joint Prevention and Control Mechanism in China (www.gov.cn/xinwen/gwylflkjz53/index.htm. Accessed 7 April 2020) during the outbreak, the amount of MSW was reduced by 30% though the medical waste generated in Hubei Province alone increased sharply (+370%) with a high percentage of plastics. During the period from 20 January to 31 March, the medical waste accumulated was estimated as 207 kt. In Wuhan, it increased from 40 tons/day (normal level) to a peak of about 240 tons/day, which far exceeds the maximum incineration capacity (49 tons/day). China has estimated the incineration cost for medical waste as 281.7–422.6 USD/tons, which is relatively much higher than incineration of MSW (14.1 USD/tons) (Lai et al. 2020).

As coronavirus is rapidly spreading at a fast rate all over the world, there is a sharp increase in the need for healthcare products and packaging, thereby raising the waste management load. It has now become a crucial issue for coronavirus decontamination process though many waste managements companies have already taken up proper steps to find solutions at the earliest. There is also a change in both the quantity and quality of plastic waste due to various kinds of suppression measures taken up by different nations. As a safety alternative, consumer now opt for single-use plastics. The lifespan of corona virus (COVID-19) on different surfaces were studied by Van Doremalen et al. (2020). Similar reports were corroborated in another study (Kampf et al. 2020). Consumers prioritising on hygiene opt for disposability as an advantage though plastics are no less better than other material in terms of retention of viruses and thus increase use of disposal of plastic products, even for non-medical applications despite a decline in other sectors such as in automotive and aviation applications. An increased trend in the use of packaged products for delivery of foods and other essential items to different household is on the rise. Although this increase trend is unavoidable, environmental protection efforts should be sustained.

The pandemic has created disruptions in both the upstream supply chain and downstream waste disposal problems due to the widespread use of protective equipment worldwide. Used plastic products ought to be handled as hazardous waste as they are frequently pathogen contaminated. And moreover medical waste from hospitals also need to be destroyed for any residual pathogens as reported by Windfeld and Brooks (2015).

The collection of waste from households and quarantine facilities with suspected or positive COVID-19 patients is a very critical control point given the potential role of the environment in the spread of COVID-19 (Qu et al. 2020). Hence implementation of monitoring or control measures especially at those locations along the wastewater handling and treatment points, will be suitable to prevent the spread of the disease. Cleaners, trash collectors and those people who are often exposed to

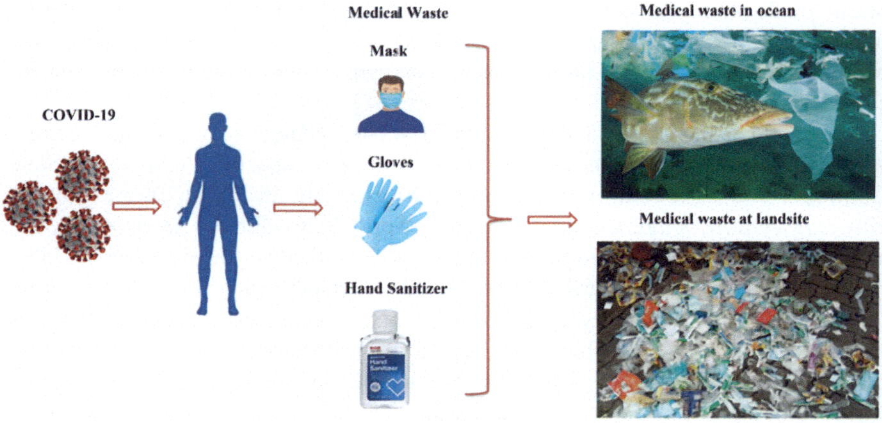

Fig. 8.5 Waste generation from healthcare sectors during pandemic

public places are at greater risk and can have adverse health effects due to contact with medical wastes as well. The respective governments of different countries have closed all the educational institutions, and employees are advised to work from home or attend offices on rotation basis except for those who are under the essential services, making them the most vulnerable and highly susceptible groups to the virus. The surgical masks that are being used are advised not to be worn more than a single day. So, disposing the single used—along with tissue papers and empty bottles of hand sanitizer are getting up piled up in huge quantity in the environment. The sea beach of Soko islands, Hong Kong was observed to have large quantities of discarded single-used masks as per the survey of environmental NGO Ocean Asia. According to the data of a survey conducted by an environmental NGO Ocean Asia in Soko Due to the pandemic, it was noticed all along the high tide line and also in seashore. Alarming amount of trash is generated when every individual suddenly starts utilizing one time use gloves, disposable masks and hand sanitizers. The outcome of the impact of such medical wastes are going to be profound. As reported (Hellewell et al. 2020), when medical wastes remained discarded in an animal's natural habitat in both land and ocean, these could lead to the death of animals as they can mistakenly eat it as food. The above diagram illustrates the problems of environmental pollution by medical wastes during COVID-19 pandemic (Fig. 8.5) (Saadat et al. 2020b).

8.4.3 Strategies to Reduce, Reuse and Decompose of Waste

We human beings often forget that we are dependent on Mother Nature and become quite ignorant towards taking care of it. The lockdown imposed throughout the world due to Covid-19 pandemic has struck a chord in every one of us and has made us think how nature is so important and indispensable for the mankind to survive.

A separate waste collection facility was arranged in Europe specially for COVID-19 infected houses to ensure the safety of waste collectors as per waste management norms (Fig. 8.6). It is also recommended to delay waste collection time for 72 h to overcome the life span of COVID-19. Addition to above, the collected materials are directly transported directly without any segregation to landfills or waste incinerators (Nghiem et al. 2020).

During the pandemic, the excellent management systems should re-emphasize the limiting exposure of worker who are handling potentially contaminated waste, infected PPE and equipment.

Significant hype by 600% (40–240 tons/day) in medical waste generation was reported in Hubei province of China during COVID-19, which paralysed the existing waste management facilities. These similar challenges will be faced by other countries. It is thus essential to develop and have pre-exiting system to deal with the waste generated during pandemic to avoid spreading of diseases.

Various types of medical hazardous waste like infected masks, gloves and other protective equipment along with non-infected items of the same nature are generated during the pandemic. Use of mask is recommended in public places after a report of airborne transmission came out (Bourouiba 2020). Biomedical and healthcare waste can be effectively detected, collected, separated, stocked, transferred, treated, disposed, and sterilized. The approaches for waste treatment during the pandemic is summarized in the Fig. 8.7 below (Klemeš et al. 2020b). The source of contamination of waste was not limited to hospitals.

The rapid increase in the number of infected cases have become insufficient to cope even with advanced healthcare facilities. Hence a considerable structural change in the management of waste is required starting from waste collection to treatment, and as those patients at home isolation with mild symptoms generate contaminated MSW, safety protocol must be followed by the waste collectors. In the EU, the contaminated waste like used gloves, masks and tissues are required to put in double-bagged. The biodegradable containers of food are considered as potential threat in Germany for spreading disease (Municipal Waste Management and COVID-19 2020). Those households with suspected or positive COVID-19 cases are advised to collect the waste separately. Though these practices serve as precautionary measure, it increased plastic usages and more mixed waste are generated. Other economic and developmental issues are put on hold in the coronavirus crisis.

Under the non-hazardous solid waste regulation, incineration and steam sterilization (90 min, 120 °C) is the common thermal treatment for hazardous medical waste and its residues can be safely handled, however WHO recommended between 900 and 1200 °C for healthcare waste. Whether the capacity of MSW incineration for medical waste is to be utilized or not, remains an open question as the amount of waste surging due to COVID-19 and fear of exceeding the treatment capacity by huge margin remains a major challenge now. In Spain, when necessary cement plants can co-incinerate waste. To cope with the surge of medical waste, Norway

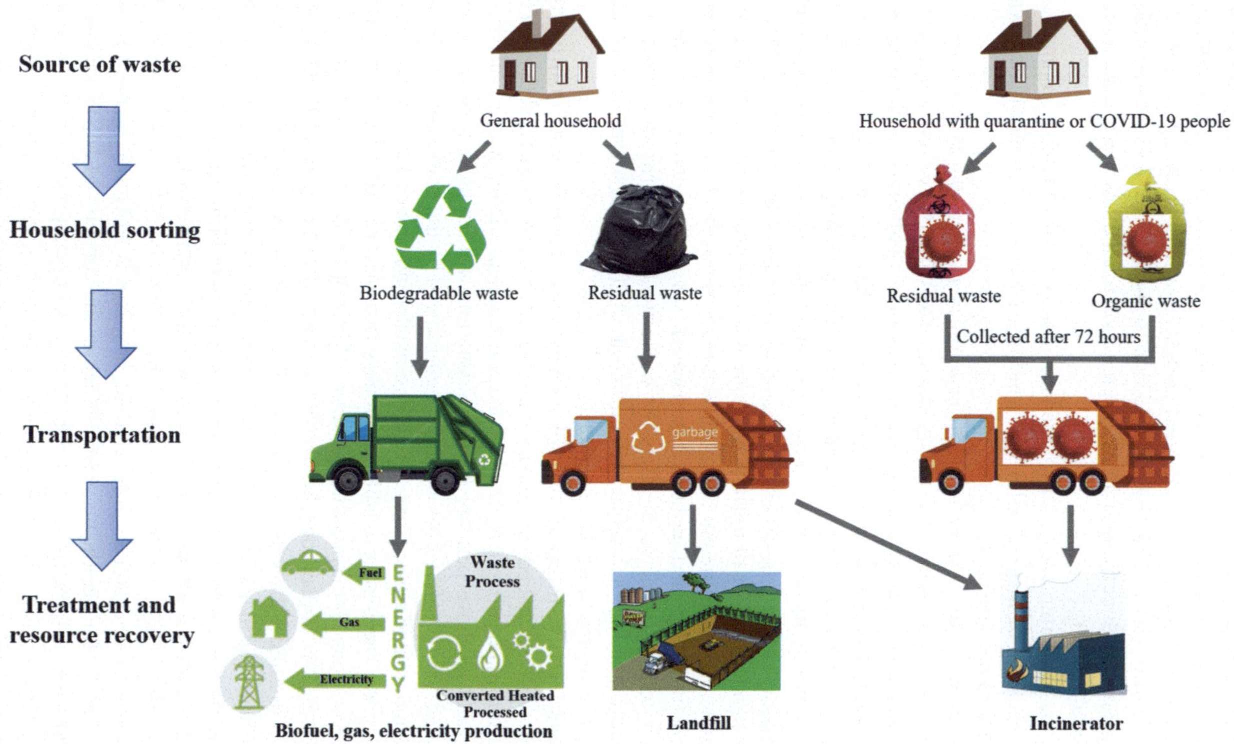

Fig. 8.6 Municipal waste management during the COVID-19 crisis

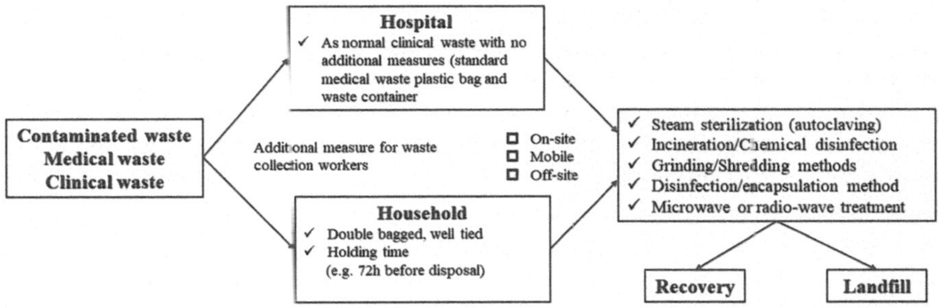

Fig. 8.7 The main contaminated waste handling approaches during COVID-19

permits temporary change in landfill and to carry waste elsewhere. Whether to treat on-site mobile or off-site dealing with this unexpected crisis is a big current debate.

Though recycling of plastic is now a possible option, it has some disadvantages. Zheng and Suh (2019) stated that fossil-based energy can be replaced by renewable energy to significantly reduce the environmental impact of plastic, especially greenhouse gases. Some characteristics of plastics that are important for applications in COVID-19 management. However, before the current pandemic, due to the government regulations and public perception, use of plastic minimized, also the impact of the plastics on environment (e.g., microplastics pollution) are not due to the inherent properties of plastics but are consequences of behavioral patterns of humans (e.g., improper disposal).

To support the planning of waste management, there is need for optimization and decision-making tools such as treatment strategies, infrastructure, protection, and governing aspects link to the bio disaster response etc. (Klemeš et al. 2020b). Taking motivation from the present pandemic, proper treatment and management facilities may be created to deal with medical waste and used masks for uncertainties of future.

During each pandemic crisis, the main priority is given to public health. The existing waste treatment and disposal facilities far exceeds its total amount, so if proper waste management is not done there is high risk of secondary contagion. We should now learn a lesson from the impacts induced by this pandemic for building a different and better society free of pollutions. The emergency measures that are put in place to deal with the current pandemic crisis to translate into long-term waste management options is an important concern for us.

8.5 Conclusion

The chapter presented an extended discussion on environmental impact on waste management and natural water resources during COVID-19 pandemic. Incapable handling of waste generated specially from healthcare sectors has become a big challenge to our present treatment facilities. To avoid the transmission of disease among society, the creation of additional infrastructure facilities is essential. Due to COVID-19 crisis, the consumption scenario has completely changed leading to lesser environmental stress and pollution. The pandemic has not only affected 215 countries across the globe but also have affected all sectors of the society, economy and health care systems of countries worldwide. Case studies and comparative analysis of different water quality parameters before and during the pandemic situation has presented substantial improvement. Additionally, certain measures and strategies are necessary to reduce, reuse and decompose of waste.

References

Bourouiba L (2020) Turbulent gas clouds and respiratory pathogen emissions: potential implications for reducing transmission of COVID-19. JAMA 323:1837–1838. https://doi.org/10.1001/jama.2020.4756

Brouard S, Vasilopoulos P, Becher M (2020) Sociodemographic and psychological correlates of compliance with the Covid-19 public health measures in France. Can J Polit Sci:1–6. https://doi.org/10.1017/S0008423920000335

Bruinen de Bruin Y, Lequarre A, Mccourt J et al (2020) Initial impacts of global risk mitigation measures taken during the combatting of the COVID-19 pandemic. Saf Sci 128:104773. https://doi.org/10.1016/j.ssci.2020.104773

Chakraborty I, Maity P (2020) COVID-19 outbreak: migration, effects on society, global environment and prevention. Sci Total Environ 728:138882. https://doi.org/10.1016/j.scitotenv.2020.138882

Chin AWH, Chu JTS, Perera MRA et al (2020) Stability of SARS-CoV-2 in different environmental conditions. Lancet Microbe 1:e10. https://doi.org/10.1016/S2666-5247(20)30003-3

Choudhury M, Fazli P, Pramanick P et al (2019) Sensitivity of the Indian Sundarban mangrove ecosystem to local level climate change. Parana J Sci Educ 5:24–28

Dantas G, Siciliano B, Boscaro B et al (2020) The impact of COVID-19 partial lockdown on the air quality of the city of Rio de Janeiro, Brazil. Sci Total Environ 729:139085

Díaz de León-Martínez L, Rodríguez-Aguilar M, Wong-Arce A et al (2020) Evaluation of acute and chronic exposure to aflatoxin B1 in indigenous women of the Huasteca Potosina, Mexico. Environ Sci Pollut Res. https://doi.org/10.1007/s11356-020-09361-4

Fernandes N (2020) Economic effects of coronavirus outbreak (COVID-19) on the world economy. SSRN Electron J. https://doi.org/10.2139/ssrn.3557504

Ficetola GF, Rubolini D (2020) Climate affects global patterns of COVID-19 early outbreak dynamics. medRxiv. https://doi.org/10.1101/2020.03.23.20040501

Giordano G, Blanchini F, Bruno R, Colaneri P, Filippo AD, Matteo AD, Colaneri M (2020) Modelling the COVID-19 epidemic and implementation of population-wide interventions in Italy. Nat Med. https://doi.org/10.1038/s41591-020-0883-7

Gorbalenya AE, Baker SC, Baric RS, et al (2020) The species and its viruses—a statement of the Coronavirus Study Group. Biorxiv (Cold Spring Harb Lab) 1–15. https://doi.org/10.1101/2020.02.07.937862

Gormley M, Aspray TJ, Kelly DA (2020) COVID-19: mitigating transmission via wastewater plumbing systems. Lancet Glob Health 8:e643. https://doi.org/10.1016/S2214-109X(20)30112-1

Gostic KM, Gomez ACR, Mummah RO et al (2020) Estimated effectiveness of symptom and risk screening to prevent the spread of COVID-19. elife 9:1–18. https://doi.org/10.7554/eLife.55570

Goswami K, Mazumdar I (2016) How idol immersion is polluting the Ganga river in Kolkata, West Bengal: an overview. Indian J Appl Res 6:460–462

Gu Y, Lu J, Yang Z (2020) Pilgrimage and COVID-19: the risk among returnees from Muslim countries. Int J Infect Dis 95:457–458. https://doi.org/10.1016/j.ijid.2020.04.066

Häder DP, Banaszak AT, Villafañe VE et al (2020) Anthropogenic pollution of aquatic ecosystems: emerging problems with global implications. Sci Total Environ 713:136586. https://doi.org/10.1016/j.scitotenv.2020.136586

He FJ, Tan M, Song J, MacGregor GA (2020) Emergence of a novel human coronavirus threatening. Nat Med 26:313–314

Hellewell J, Abbott S, Gimma A et al (2020) Feasibility of controlling COVID-19 outbreaks by isolation of cases and contacts. Lancet Glob Health 8:e488–e496. https://doi.org/10.1016/S2214-109X(20)30074-7

Hopkins (2020) Coronavirus Research Center. John Hopkins Univ. Med

Huang C, Wang Y, Li X et al (2020a) Clinical features of patients infected with 2019 novel coronavirus in Wuhan, China. Lancet 395:497–506. https://doi.org/10.1016/S0140-6736(20)30183-5

Huang RH, Liu DJ, Tlili A et al (2020b) Handbook on facilitating flexible learning during educational disruption: the Chinese experience in maintaining undisrupted learning in COVID-19 outbreak. Smart Learning Institute of Beijing Normal University, Beijing

Kampf G, Todt D, Pfaender S, Steinmann E (2020) Persistence of coronaviruses on inanimate surfaces and their inactivation with biocidal agents. J Hosp Infect 104:246–251. https://doi.org/10.1016/j.jhin.2020.01.022

Kapil S (2020) Has COVID-19 lockdown really helped Yamuna health in Delhi? DownToEarth

Klemeš JJ, Van Fan Y, Tan RR, Jiang P (2020a) Minimising the present and future plastic waste, energy and environmental footprints related to COVID-19. Renew Sustain Energy Rev 127:109883. https://doi.org/10.1016/j.rser.2020.109883

Klemeš JJ, Van Fan Y, Tan RR, Jiang P (2020b) Minimising the present and future plastic waste, energy and environmental footprints related to COVID-19. Renew Sust Energ Rev 127:109883. https://doi.org/10.1016/j.rser.2020.109883

Lai CC, Shih TP, Ko WC et al (2020) Severe acute respiratory syndrome coronavirus 2 (SARS-CoV-2) and coronavirus disease-2019 (COVID-19): the epidemic and the challenges. Int J Antimicrob Agents 55:105924. https://doi.org/10.1016/j.ijantimicag.2020.105924

Liu Y, Ning Z, Chen Y et al (2020) Aerodynamic analysis of SARS-CoV-2 in two Wuhan hospitals. Nature 582(7813):557–560. https://doi.org/10.1038/s41586-020-2271-3

Madurai Elavarasan R, Pugazhendhi R (2020) Restructured society and environment: a review on potential technological strategies to control the COVID-19 pandemic. Sci Total Environ 725:138858. https://doi.org/10.1016/j.scitotenv.2020.138858

Mahato S, Pal S, Ghosh KG (2020) Effect of lockdown amid COVID-19 pandemic on air quality of the megacity Delhi, India. Sci Total Environ 730:139086. https://doi.org/10.1016/j.scitotenv.2020.139086

Mani KAS (2020) The lockdown cleaned the Ganga more than 'Namami Gange' ever did

Mitra A, Chaudhuri TR, Mitra A et al (2020) Impact of COVID-19 related shutdown on atmospheric carbon dioxide level in the city of Kolkata. Parana J Sci Educ 6:84–92

Muhammad S, Long X, Salman M (2020) COVID-19 pandemic and environmental pollution: a blessing in disguise? Sci Total Environ 728:138820. https://doi.org/10.1016/j.scitotenv.2020.138820

Mukherjee P, Pramanick P, Zaman S, Mitra A (2020) Eco-restoration of River Ganga water quality during COVID-19 lockdown period using Total Coliform (TC) as proxy. NUJS J Regul Stud 67–74. ISSN: 2456-4605 (O)

Municipal Waste Management and COVID-19 (2020) Association of Cities and Regions for Sustainable Resource Management. www.acrplus.org/en/municipal-waste-management-covid-19

Nghiem LD, Morgan B, Donner E, Short MD (2020) The COVID-19 pandemic considerations for the waste and wastewater services sector. Case Stud Chem Environ Eng 1:100006. https://doi.org/10.1016/j.cscee.2020.100006

Nicola M, O'Neill N, Sohrabi C et al (2020a) Evidence based management guideline for the COVID-19 pandemic—review article. Int J Surg 77:206–216. https://doi.org/10.1016/j.ijsu.2020.04.001

Nicola M, Neill NO, Sohrabi C et al (2020b) Evidence based management guideline for the COVID-19 pandemic—review. Int J Surg 77:206–216. https://doi.org/10.1016/j.ijsu.2020.04.001

Paital B, Das K, Kumar S (2020) Inter nation social lockdown versus medical care against COVID-19, a mild environmental insight with special reference to India. Sci Total Environ 728:138914. https://doi.org/10.1016/j.scitotenv.2020.138914

Park C-Y, Villafuerte J, Abiad A, et al (2020) An updated assessment of the economic impact of COVID-19. Manila, Philippines

Qu G, Li X, Hu L, Jiang G (2020) An imperative need for research on the role of environmental factors in transmission of novel coronavirus (COVID-19). Environ Sci Technol 54:3730–3732. https://doi.org/10.1021/acs.est.0c01102

Rajmohan KVS, Ramya C, Raja Viswanathan M, Varjani S (2019) Plastic pollutants: effective waste management for pollution control and abatement. Curr Opin Environ Sci Health 12:72–84. https://doi.org/10.1016/j.coesh.2019.08.006

Rakshit D, Sarkar SK (2018) Idol immersion and its adverse impact on water quality and plankton community in Hooghly (Ganges) River Estuary, India: implications for conservation management. Indian J Geo-Marine Sci 47:1870–1879

Saadat S, Rawtani D, Mustansar C (2020a) Environmental perspective of COVID-19. Sci Total Environ 728:138870. https://doi.org/10.1016/j.scitotenv.2020.138870

Saadat S, Rawtani D, Hussain CM (2020b) Environmental perspective of COVID-19. Sci Total Environ 728:138870. https://doi.org/10.1016/j.scitotenv.2020.138870

Semple S, Cherrie JW (2020) Covid-19: protecting worker health. Ann Work Expo Health 64 (5):461–464. https://doi.org/10.1093/annweh/wxaa033

Sengupta T, Pramanick P, Mitra A (2020) Nutrient load in the River Ganges during the COVID-19 lockdown phase: a ground zero observation. NUJS J Regul Stued 79–83. ISSN (O):2456-4605

Shrangi V (2020) Yamuna water improves during lockdown period, CPCB analysis shows. Hindustan Times

Sohrabi C, Alsafi Z, O'Neill N et al (2020) World Health Organization declares global emergency: a review of the 2019 novel coronavirus (COVID-19). Int J Surg 76:71–76. https://doi.org/10.1016/j.ijsu.2020.02.034

Strauss M (1997) Health (pathogen) considerations regarding the use of human waste in aquaculture. Environ Res Forum 5(6):83–98

The Tribune (2020) Lockdown helps improve Health of Ganga. Trib

Tiwari A, Dwivedi A, Mayank P (2016) Time scale changes in the water quality of the Ganga River, India and estimation of suitability for exotic and hardy fishes. J Waste Water Treat Anal 7:1000254. https://doi.org/10.4172/2157-7587.1000254

UNWTO (2020) Impact assessment of the COVID-19 outbreak on international tourism. United Nations World Tourism Organization, Madrid. https://www.unwto.org/impact-assessment-of-the-covid-19-outbreak-on-international-tourism

van Doremalen N, Bushmaker T, Morris DH et al (2020) Aerosol and surface stability of SARS-CoV-2 as compared with SARS-CoV-1. N Engl J Med 382:1564–1567. https://doi.org/10.1056/NEJMc2004973

Varotto A, Spagnolli A (2017) Psychological strategies to promote household recycling. A systematic review with meta-analysis of validated field interventions. J Environ Psychol 51:168–188. https://doi.org/10.1016/j.jenvp.2017.03.011

Wang D, Hu B, Hu C et al (2020) Clinical characteristics of 138 hospitalized patients with 2019 novel coronavirus-infected pneumonia in Wuhan, China. J Am Med Assoc 323:1061–1069. https://doi.org/10.1001/jama.2020.1585

WHO (2020a) WHO Director-General's opening remarks at the media briefing on COVID-19—11 March 2020

WHO (2020b) Coronavirus disease (COVID-19) pandemic
WHO (2020c) Report of the WHO-China Joint Mission on Coronavirus Disease 2019 (COVID-19). WHO, Geneva
WHO (2020d) Water, sanitation, hygiene and waste management for the COVID-19 virus. World Health Organization, Geneva
WHO (2020e) Rolling updates on coronavirus disease (COVID-19). World Health Organization, Geneva. https://www.who.int/emergencies/diseases/novel-coronavirus-2019/events-as-they-happen
Windfeld ES, Brooks MSL (2015) Medical waste management—a review. J Environ Manag 163:98–108. https://doi.org/10.1016/j.jenvman.2015.08.013
Wright R (2020) The world's largest coronavirus lockdown is having a dramatic impact on pollution in India
Yuen KS, Ye ZW, Fung SY et al (2020) SARS-CoV-2 and COVID-19: the most important research questions. Cell Biosci 10:1–5. https://doi.org/10.1186/s13578-020-00404-4
Yunus AP, Masago Y, Hijioka Y (2020) COVID-19 and surface water quality: Improved lake water quality during the lockdown. Sci Total Environ 731:139012. https://doi.org/10.1016/j.scitotenv.2020.139012
Zambrano-Monserrate MA, Ruano MA, Sanchez-Alcalde L (2020) Indirect effects of COVID-19 on the environment. Sci Total Environ 728:138813. https://doi.org/10.1016/j.scitotenv.2020.138813
Zheng J, Suh S (2019) Strategies to reduce the global carbon footprint of plastics. Nat Clim Chang 9:374–378
Zhu N, Zhang D, Wang W et al (2020) A novel coronavirus from patients with pneumonia in China, 2019. N Engl J Med 382:727–733. https://doi.org/10.1056/NEJMoa2001017

Chapter 9
Impact of COVID in Agriculture

Laura Şmuleac, Daniel Botănoiu, Florin Imbrea, Cosmin A. Popescu, Tiberiu Iancu, Adrian Şmuleac, Tabita Adamov, Laurenţiu Gavra, and Raul Paşcalău

Abstract Current pandemic turned into a worldwide health crisis that having a devastating impact on both human society and the world economy. COVID-19 has had repercussions in most economic sectors and implicitly in the food and agricultural sector. While food supplies have been kept in more states, measures implemented to stop virus spreading have disrupted the ability to provide supplies of agri-food good and products for markets and consumers. COVID-19 affected all elements of the food system, from primary supply, to processing, to trade, as well as national and international logistics systems, to intermediate and final demand. Thus, COVID 19 started threatening all the sectors in the poorest countries, where agricultural production systems consume labor above the average, and where the ability to resist to a macro economic shock is very week. Demand for goods is expected to fall in the coming months, and prices will fall, and this will have a negative impact on farmers and the agricultural sector. Policy makers and managerial staff from all over the globe must pay attention and make the most suitable steps so that they do not transform the current health crisis into a completely regrettable one related to goods and products from all sectors. While the pandemic presents some serious shortterm challenges for the agrifood system, it represents in the same time a suitable moment to speed up changes in the agricultural area in order to rethink the strategy to combat the effects of climate change on this area of activity.

Keywords COVID-19 pandemic · Agriculture · Food markets · Recovery measures

---

L. Şmuleac · F. Imbrea · C. A. Popescu · T. Iancu · A. Şmuleac · T. Adamov · R. Paşcalău (✉)
Banat University of Agricultural Sciences and Veterinary Medicine "King Michael I of Romania", Timişoara, Romania

D. Botănoiu
Ministry of Agriculture and Rural Development, Bucharest, Romania

L. Gavra
European Parliament, Brussels, Belgium

M. K. Goyal, A. K. Gupta (eds.), *Integrated Risk of Pandemic: Covid-19 Impacts, Resilience and Recommendations*, Disaster Resilience and Green Growth,
https://doi.org/10.1007/978-981-15-7679-9_9

9.1 The Impact of COVID-19 on the Agri-Economic World

Straight impact of the contagion over basic agriculture would be better restraint, as the infection has not an influence over natural resources, which represent the basis of production.

Figure 9.1 shows a degree of damage to agriculture and services resulting from the COVID-19 pandemic, as moderate. However, the COVID-19 pandemic affects the entire food system. Exercising a symmetrical but asynchronous shock on the global food system. It has affected all elements of the food system, from primary supply, to processing, to trade, as well as national and international logistics systems, to intermediate and final demand. It also affects factor markets, namely labor and capital input of production (Schmidhuber et al. 2020). The virus started threatening all the sectors in the poorest countries, where agricultural production systems consume labor above the average, and where the ability to resist to a macro economic shock is very weak.

Demand shouldn't be so influenced and damaged by this critical situation than demand for different goods and services, registering important changes in demand's framework.

The occurrence of the coronavirus in China in 2019, and its rapid spread has led to the freezing of world markets, especially agricultural ones, because China is the country that imports the most agricultural goods. Analysing the situation before the pandemic, we notice that the markets were incorporated and interconnected, with a Chinese economy contributing 16% to the worldwide growth. Consequently, any impact affecting China now involves bigger repercussions for the global economy.

For the agricultural sectors in Europe, the effect of the coronavirus is felt with each passing day, the situation being able to worsen without some concrete measures to avoid its spread. China, with a population of more than 1.4 billion people, who need to be fed, needs imports in addition to domestic production. As with all crises, consumption affects the supply chain.

The many existing transport restrictions in the world lead to a decrease in the possibility of trade. The products that show a high degree of concern, in the short and medium term, are dairy products and pork. The price of these products decreased throughout February 2020.

Coronavirus has spread rapidly in over 50 countries around the world: on all continents.

The European beef chain has also been affected by the presence of coronavirus in Italy. As in all sectors, agrifood production can be massively affected by the absence of labor, the virus being considered a case of force majeure which allows the establishment of aid for suffering companies, companies can also resort to partial unemployment.

The many existing transport restrictions in the world lead to a decrease in the possibility of trade. The products that show a high degree of concern, in the short and medium term, are dairy products and pork. The price of these products decreased throughout February 2020.

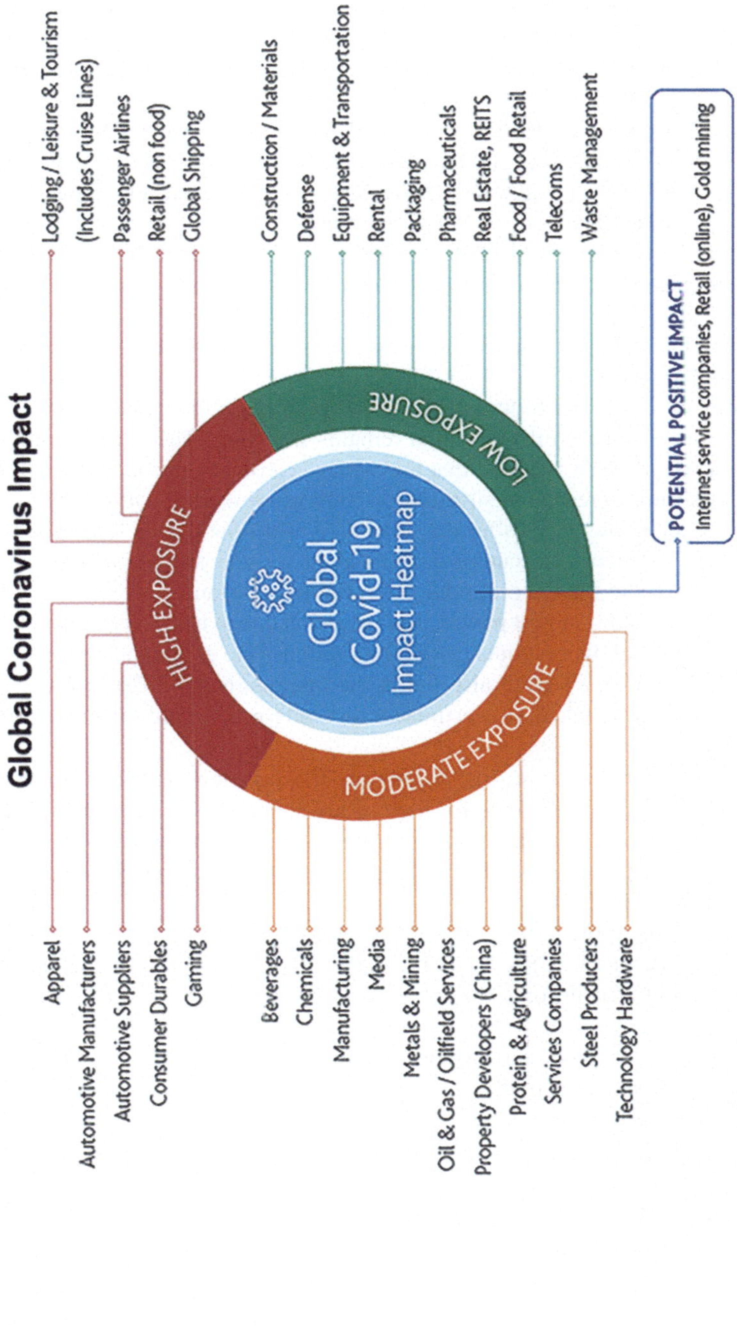

Source: Moody's Investors Service

Fig. 9.1 The degree of impact of Coronavirus on different economic sectors (https://www.moodys.com/)

Coronavirus has spread rapidly in over 50 countries around the world: on all continents.

The European beef chain has also been affected by the presence of coronavirus in Italy. As in all sectors, agrifood production can be massively affected by the absence of labor, the virus being considered a case of force majeure which allows the establishment of aid for suffering companies, companies can also resort to partial unemployment.

Markets have reacted well for weeks to the spread of the virus in China, but its rapid spread to South Korea, Iran and Italy has raised concerns. World stock markets have fallen, the volatility of agricultural products has increased and interest has migrated which: gold, US sovereign bonds, the Swiss franc, etc.

China's contribution to world GDP in 2002 was 4.2% and in 2018 it reached 15.8%, which means that it contributes a quarter to the growth of the world economy.

Coronavirus has a major impact on the world economy, with signs that it is leading to a global recession.

The opening of China and India to a market economy has increased the share of world GDP, but the situation in Argentina remains worrying. The world economy may enter a recession and it is hoped that the damage caused by coronavirus will be mitigated in order not to reach the situation of 2009 when world GDP fell by 0.1% being the worst economic result after the second. World War II. It is known that if world GDP falls for two consecutive quarters then we can talk about recession, already two countries in the G8 group have entered a recession.

Japan, the world's third largest power, is in recession due to negative results in the last quarter of 2019 and the first quarter of 2020. Italy is also experiencing a second consecutive decline. China's GDP also fell in the first quarter of 2020 and with all the measures that will be taken, it is estimated that only in the third quarter can the economy recover.

In March, the Organization for Economic Cooperation and Development (OECD) lowered all the predictions for the worldwide economic growth in 2020, from 2.9 interest to 24 interest, representing the weaker degree in the economic down turn, 10 years back, pointing out that a longer and intense outbreak could even diminish it to 1.5% (Fig. 9.2).

The estimate of the growth of the world's economies is more pessimistic for this year: China 3.5%, India 4%, USA 1.1%, EU 0.5%, this counter performance makes the forecast of world growth for 2020 to be only 1.6% which is a lower figure than in 2009.

At the moment there is no certainty of the annual decline in GDP in 2020, but, unfortunately, the blockade of trade, travel, international trade in general can lead to a more severe recession than in 2009.

Following the introduction of COVID-19 containment measures worldwide, real gross domestic product (GDP) in the OECD area fell by 1.8% in the first quarter of 2020, the largest decrease since a contraction of 2.3% in the first quarter of 2009 at the height of the financial crisis, according to provisional estimates.

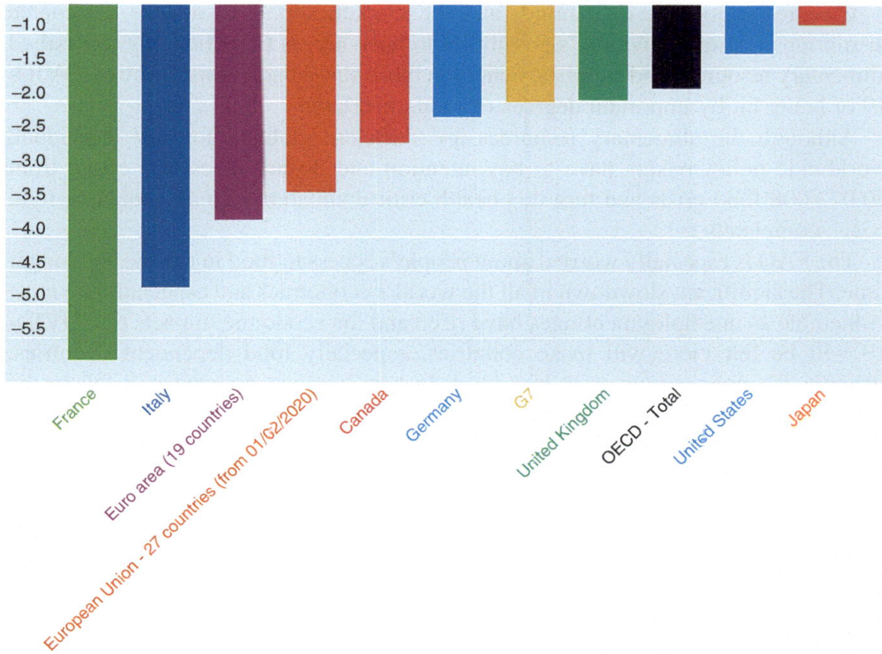

Fig. 9.2 GDP in the first quarter of 2020 (https://www.oecd.org)

Among the seven major powers, GDP fell significantly in France and Italy, where the austerity measures were the strictest and were implemented the earliest (−5.8% and 4.7%, respectively, compared to −0.1% and −0.3% in the previous quarter).

GDP also fell sharply in Canada, Germany and the United Kingdom (−2.6%, −2.2% and 2.0%, respectively, compared with 0.1%, −0.1% and 0%, 0% in the initial semester).

In the United States, where many states introduced "stayathome" measures at the end of March, the contraction in GDP was less dramatic (−1.2% compared to 0.5% in the initial semester).

In Japan, where containment measures were less stringent, GDP contracted by −0.9% in the first quarter of 2020, likened to −1.9% in the previous semester.

In the euro area and the European Union, GDP fell by 3.8% and 3.3%, respectively, compared with an increase of 0.1% and 0.2% in the previous quarter.

Annual GDP growth for the OECD zone fell to −0.8% in the first semester of 2020, after rising by 1.6% in the previous semester Among the seven major economies, America registered the largest annual increase (0.3%), while France recorded the strongest annual decrease (−5.4%).

The Food and Agriculture Organization of the United Nations (FAO) warns that we risk a growing food crisis if measures of prevention and protection are not adopted for the exposed ones, preserving worldwide supply chains active and diminish in the same time the impact of the pandemic on the food system.

Closure of borders, quarantines and the market, supply chain and commerce disruptions restrict individual's possibility to have access to sufficiently diversified alimentary resources and supplies, mainly in those states badly impacted by COVID-19 or being hit by important degrees of food uncertainty.

Although the alimentary resources are sufficient globally for any individual, worldwide policymakers have to pay attention and do not repeat the errors from 20,072,008 food crisis and turn this health critical situation into an avoidable food crisis, completely.

The FAO is especially worried about people's access to food in middle and longer time. The significant slowdown in all the world's economies and especially the most vulnerable as unemployment rates have risen and the economic impacts of COVID-19 will be felt more will make countries, especially food dependent countries, struggle. To need resources to buy food. In turn, as food demand will fall in the coming months, prices are expected to fall in 2020, and this will register a damaging effect on farmers and the agricultural sector.

From now on, interruptions have been insignificant, as food supply has been adequate and markets have been stable so far. Still, we have seen constraints regarding logistical obstacles (unable transfer goods and food from A to B), which were largely resolved in midApril; and fewer highvalue foods (i.e. fruits and vegetables) are likely to be placed on the market.

We can still expect disruptions in food supply chains, especially highvalue goods (fruits, vegetables, meat, fish, milk, etc.). Consequently: transfer limitations, along with main repellent manners at the hand of workers can prevent actors in the alimentary area and processors, dealing with the vast area of all type of goods from processing.

Deficiency of manure, veterinary medicines and different inputs could influence farming, globally.

Eating places being shut down, automatically, less alimentary products, have diminished demand for fresh products and fishery products, influencing both growers and manufacturers but also the providers.

The agriculture, fisheries and aquaculture sectors were mainly hit by limitation on tourism, the closure of eating places, and cafes and the suspension of school meals.

In any eventuality, the exposed ones, segments of the population (migrants, included, moved individuals and those affected by the conflict) will be the most affected. Countries in protracted crises also suffer from underinvestment in public health, which increases the impact of the pandemic.

9.2 The Impact of COVID in the Agricultural, Forestry and Fisheries Sector

The supply of agrifood products remained stable, even in the conditions of the crisis situation due to COVID-19. Initially, uncoordinated measures taken by states to combat COVID-19 led to disruptions in the single market which affected the

Fig. 9.3 Market situation (http://www.amisoutlook.org)

	From previous forecast	From previous season
Wheat	N/A	▲
Maize	▲	▲
Rice	▬	▬
Soybeans	▲	▼

▲ Easing ▬ Neutral ▼ Tightening

operation of food supply chains. Farmers and their agricultural cooperatives have managed to maintain a stable supply of high quality, safe and affordable food for their citizens.

Support should come from outside the agricultural budget. In this context, the use of the crisis reserve under the agricultural budget heading will not alleviate the market problems currently facing farmers and their cooperatives. In a letter from European Commissioner Janusz Wojciechowski to Member States' agriculture ministers on 8 April, he encouraged the use, as far as possible, of the funds available for rural development (EC 2020a, b).

Even if at a worldwide level, the aliments will still be plentiful, the impact generated by COVID-19 began to affect food markets in April. The historic plunge in oil prices, the sudden decline in ethyl alcohol production and the slowing in alimentation demands have conducted to important declines in the costs for the maize. Meanwhile, protectionist exchanges strategies, although not permanent, have increased worries about global wheat and rice flows.

Given that worldwide diminution in 2020 becomes weaker than the Great Recession of 2009, low demand in the face of generous supply is foreseen to maintain markets with a powerful constraint and coercion (Fig. 9.3). Former critical situations have shown international cooperation being essential for supporting maintenance of a balance within food markets and to support extended economic disturbance on the most exposed ones (Amis Market Monitor 2020).

9.2.1 *Sector Analysis*

Wine sector The situation is deteriorating every day. The crisis will have a longterm impact on the sector. The decrease in domestic consumption, as well as the decrease in exports (expected to decrease by 14% compared to last year) are not offset at all by retail and online sales. Due to the closure of the HoReCa channel, 30% of the market share was lost, which represents 50% of the market value. Stocks are accumulating and, in particular, high value products are seriously affected. Some proposals, such as crisis distillation, private storage and organic harvesting, are insufficient from a budgetary point of view. It is impossible for operators to implement such measures with the remaining funds. A strong budget is needed, from outside the agricultural funds.

Fruit and vegetables As far as the market is concerned, the prices of seasonal products tend to fall due to seasonal circumstances. In terms of trade, exports remain problematic due to high air transport prices. Transport costs, the implementation of social distance measures and the availability of labor remain major issues. There is a greater demand from retailers for more packaged products.

Cereals The automatic import duty calculation mechanism was launched on 27 April, setting the import duty on maize, sorghum and rye at EUR 5.27 per tonne. The new calculation has led to a revised duty of EUR 10.40 per tonne since 5 May. The collapse in oil prices has led to lower demand for bioethanol fuel in the US, as well as lower transportation costs. In addition, a record global corn harvest is expected for the 2020/2021 marketing year. The combination of these factors led to a US cif market price for maize in the port of Rotterdam of EUR 146.63 per tonne.

According to AMIS Market Monitor, grain productivity in the current year, 2020, will continue to stay close to the levels of the last year (the second largest registered); the likelihood of recoveries in Australia and Kazakhstan, hand in hand with a larger production in the Russian Federation and in other states from Asia, may setoff diminutions in all European Union, but also in northern African states, Ukraine and America. The use of this production in 2020/2021 to stagnate the price, as an increase in alimentation consuming is mostly set off by lower expectations for food demand and manufacturing usage.

Business prognosis for 2020/2021 (07/06), indicates a minor raise because of UK commerce with the countries from European Union from 2020/2021; furthermore, global commerce is declining slightly, even if there are registration of good progress regarding Australian and Canadian exports. Resources (terminated by 2021) will raise, the increase being a projection of subsequent accumulations in China; with the exception of China, global grain resources are directing towards a fall of nearly 5% to their minimal degree since 2013, with expectations waiting in the America and northern African countries.

Without any certainty regarding COVID-19's impact on the worldwide grain business in the past month, and the focus was largely put on output borders in the Black Sea zone, meanwhile Russia's export quota for all cereals in the last few months of the period, carried out by the end of March, was consumed by the end of April. It has been considered potential by shifting some late applications to different sources.

Many notable importers preserved operative in the market, like Egypt, which persisted to provide freight for 2019/2020 transport, in spite of the beginning of domestic harvest purchases. Worldwide cost assistance has resulted from just as much as optimal conditions for 2020/2021 crops in different regions, especially in European countries and the Black Sea area, where precipitations have been necessary to ensure and keep output ability. Global price feeling has been cushioned by

Table 9.1 International Grains Advice (IGC) Commodity prices (http://www.amisoutlook.org)

		GOI	Wheat	Maize	Rice	Soybeans
		(January 2000 = 100)				
2019	April	**182.9**	180.7	172.7	161.3	167.9
	May	**181.6**	177.0	177.9	160.7	166.1
	June	**193.0**	185.4	196.0	161.4	179.7
	July	**190.2**	178.7	196.3	162.5	177.4
	August	**184.7**	171.1	171.5	165.3	177.0
	September	**183.4**	173.8	163.0	161.8	176.5
	October	**190.9**	184.1	173.6	161.8	182.9
	November	**190.3**	181.4	177.7	161.0	182.1
	December	**193.6**	186.9	181.9	161.6	184.4
2020	January	**197.0**	194.9	189.1	166.7	182.8
	February	**192.5**	191.2	186.6	169.6	175.9
	March	**188.2**	186.5	179.5	178.0	170.1
	April	**189.2**	191.5	170.5	195.1	168.4

GOI Grains and Oilseeds Index

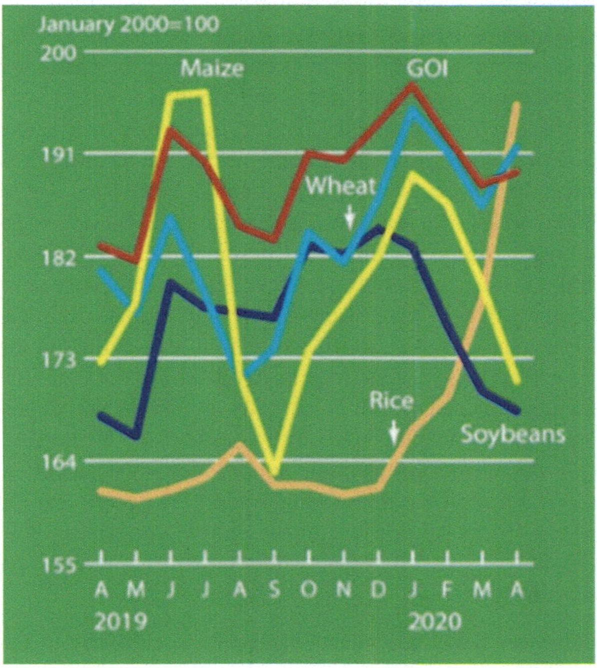

Fig. 9.4 Evolution of prices (http://www.amisoutlook.org)

concerns about longterm worldwide requests projections, in terms of lower business prospects, the ability of extended restrictions on COVID-19 development and declining oil income have been taken into account as potential to reduce purchases. In some key markets (Table 9.1, Fig. 9.4).

Corn production in 2019 was almost unchanged, with only a slight revision in Mexico, surpassing 2018 production by about 22 million tons (2%). Forecasts for 2019/2020 use fell this month, largely due to significant corrections on the decline regarding America and China, countries where the impact of COVID-19 on the energy economics and trading is considered to bring an important decline in ethanol, amidine and also, to a diminished level, upon the need and request for food.

Commerce for 2019/2020 (July/June) is foreseen to be around 167,000,000 to, representing a peripheral enlargement from 2018/2019 standards, backed by broad output availability. Resources (reaching and end by 2020) have risen significantly (22.7 million to) only from previous month, while decreases in industry and food sectors are foreseen to rise more than expected storage in the US and China.

The IGC GOI subindex for corn (Table 9.1) decreased for the third consecutive month, mainly due to problems with a decrease in coronavirus related consuming. Prices in America were particularly low as a number of processors of ethyl alcohol closed or decreased power due to low demands for petrol. Registering more factories specialised in good packing, especially meat, which have been closed or closed for the moment, important quantities have been redirected to output channels, adding additional constraints for FOB rates and quotation. Another fall happened in Argentina, where values went down quickly, based on shortterm or seasonal supply growth, and traders sought remaining competitive with America. On the contrary, medium rates in Ukraine have become largely steady and firm, although registering weaker tone in the last part of the month.

Productivity of rice in 2019 declined, mainly due to decreased rates and estimations for Nigeria and diminished production results in Pakistan; these intensifications exceeded especially in Cambodia and Mali. Rice use will be lower in 2019/2020, based on low estimations of alimentation consumption in Nigeria but also because of industrial usage in China. (a) Commerce in 2020 has continued to register a decrease and it is now close to the low level of 2019. (b) Rice stocks have grown fractionally, bigger as they have been previously estimated to be in Cambodia, China and Myanmar, offsetting small production in Nigeria and the Philippines.

Rice prices rose sharply in April on the back of COVID-19 as some important traders imposed output constraints on efforts to maintain local prices, including in Vietnam, registering deliveries for April to be limited by governmental rates or quota. Logistical constraints amid wider isolation actions have also restricted shipments from different important traders, like India and Pakistan, Thailand. However, prices fell in the last part of the month, confirming that exportation from Vietnam will be at its best by May, while logistical restriction have also begun to register a certain ease.

Oilseeds Rapeseed suffers because it is currently unable to compete with soybean meal. However, rapeseed prices have finally risen slightly in recent days due to the sharp rise in crude oil prices. The EU is gradually easing blockade measures, and in light of this, global fuel consumption could increase rapidly, benefiting all raw materials used in the manufacture of biofuels. Biofuels obtained from European arable crops improve the supply of protein rich feed for animal husbandry.

Forecasts for *soybean production* 2019/2020 have further declined in unfavourable cropping conditions in some regions in Argentina, Brazil and Uruguay, conducting to an important decline in overall productivity in the former short period of times. Use in 2019/2020 has decreased in European countries and also in South America ones, showing smaller harvests and increasing needs. In the current season, the extension of worldwide use has fallen below 1%. (a) Trade forecasts for 2019/2020 have increased, with upward revisions for Chinese imports, which exceed the diminished importation foreseen for countries from South Asia, mainly related to the pandemic of COVID-19. (b) Inventories (achievement of 2019/2020) were reduced, mainly reflecting the renewal of potential stocks in China and the accumulation of inventory in the USA; In some countries, as Brazil, for example, revised transfer resources bring a contribution to a bigger and more important than it has been foreseen stocktaking especially in season endings.

Reflecting lower prices, the IGC GOI subindex decreased with 1%. After less early weather support than the ideal for South American crops, the market has shrunk to growing concern related to COVID-19 impact upon worldwide trading and economics activity (Fig. 9.5). In spite of the recent acquisitions from China, pressure on U.S. stocks has been linked to generally thin international demand, as evidenced by a series of disappointing weekly sales values, while soybean and crude oil losses have also been affected. Key. Even if deliveries registered an important in terms of higher shipments to China, rates from Brazil were considered by rising supply of new crops and lighter internal cash.

Flowers and ornamental plants Recognition that the sector has been hit hard was welcome. In our view, the measures, including an exception to the competition rules for a period of 6 months, will not provide any tangible or immediate exemption for entrepreneurs in the flower sector who face major financial problems including cash flow and high cost of destruction.. Although easing the blockade measures in some Member States could have positive effects, it is still too early for a realistic forecast.

Dairy and meat sector In general, the situation remains similar to that of recent weeks. However, the impact of the COVID-19 crisis on pork and poultry prices is clear (EUR 20/100 kg in the last 7 weeks). Closing key channels (e.g. HoReCa) puts our internal market under increasing pressure. These imports without an alternative outlet accumulate in the freezer. Their placing on the market after the lifting of the restrictions has the potential to cause major disruptions. In addition to COVID-19, the livestock sector has been hit just as hard by the drought. The pandemic has an impact on the livestock sector because of the diminished approach to animal feed and low slaughterhouse ability (because of logistical issues and labour insufficiencies) in a identic context with the one from China. The dairy market remains under pressure due to the COVID-19 outbreak. Market measures have recently been introduced, which help to improve market sentiment.

Olive oil There has been a substantial decrease in consumption due to the closure of the HoReCa channel in both domestic and international markets. Some third countries have imposed restrictions, significantly affecting exports, which are expected to

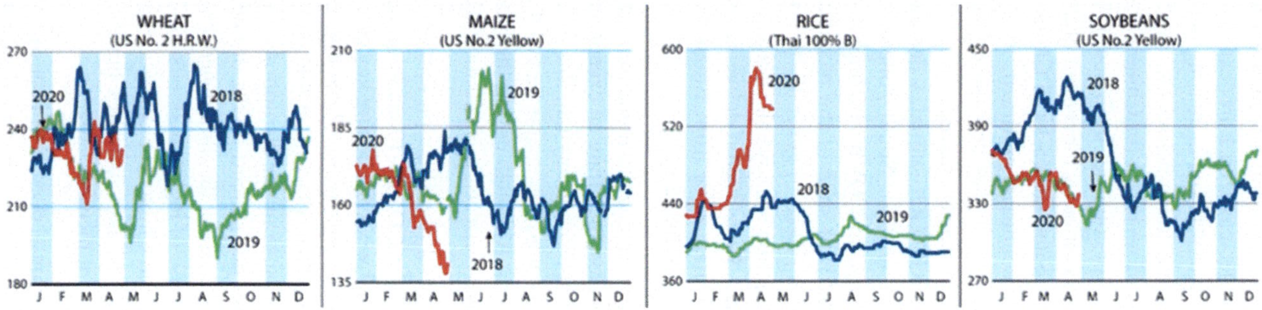

Fig. 9.5 Daily quotations of selected export prices (USD/ton, 20,182,020) (http://www.amisoutlook.org/)

fall by 8% compared to last year. Stopping in tourism is also a major concern. WP is finalizing a letter with a wide range of proposals, including market measures and tools to support demand and cope with the COVID-19 crisis.

Sugar The sugar market experts, who met on 6 May, provided evidence of the downward trend in EU sugar consumption. The discussion on the sugar balance confirmed an increase in depleted stocks compared to the previous estimate. According to analysts, spot market prices in the EU fell further in the first week of May. The persistent weakness of the Brazilian real against the dollar is a major influencing factor for world sugar prices. Real fell to a new record high of 5.8744 Real/USD on May 7th. The weaker real encourages the export sales of sugar producers in Brazil. In fact, while sugar prices in the Brazilian currency have remained fairly stable, they have fallen in dollars and euros. World sugar prices remain under pressure due to the latest forecasts from Conab, the official forecasting agency of the Brazilian government. On May 5, Conab announced that Brazil's sugar production 2020/2021 will rise +18.5% year/year to 35.3 MMT, while milking diverts more cane juice in sugar production. This is due to the fact that the outlook for ethanol is bleak due to declining consumption and prices.

Conab predicts that Brazilian factories will divert 42.4% of cane juice in sugar production in 2020/2021, which is up from 34.9% in 2019/2020. The WTO CXL erga omnes TRQ sugar volume with a reduced duty of € 98/ton has started to enter the EU market, reflecting the relative attractiveness of the EU market and the very low level of sugar import prices. According to the latest figures from the European Commission, by mid April, total imports for 2019/2020 were 25% higher than last year in the same period, and total exports for 2019/2020 were 56% lower than last year, at the same time.

Organic Agriculture Although the impact of COVID-19 on the organic sector depends on the country and the commodity concerned, there has been an increase in retail sales in several countries. For example, in Germany, retail sales of organic products increased by 25% in the first quarter of 2020 compared to the previous year. In recent weeks, organic farmers in several Member States, such as Sweden, Belgium, the United Kingdom and Italy, have begun to face some difficulties in purchasing organic feed, especially for nonruminants. These difficulties are largely due to the disturbances caused by COVID-19 in the flow from major organic food suppliers such as China and India. In some regions, the price for organic feed has risen by more than 25%.

Tobacco Due to the lack of access to seasonal workers, in some countries farmers could not plant all the seedlings on their farms. For example, in Germany, tobacco growth is estimated to be 20% lower in 2020. Exports are also blocked because other products have priority in customs clearance.

Aquaculture The situation of freshwater aquaculture is still alarming in several countries due to the closure of HoReCa and export channels. Fresh seafood products, short lived ones, being necessary to be commercialized, processed or preserved for a shorter periods of time, pose a particular risk. Transport restrictions and quarantine measures have prevented farmers and fishermen from gaining entrance to trading places, limiting their constructive capacity and preventing them from trading all their products.

However, for marine aquaculture, local shell and fir sales have increased in Italy, France and Spain. The implications in this area can vary and have a bigger complexity. Regarding wild fish fishing, the impossibility of the specific ships to be operational (because of market limitation or collapse, as well as unsafe sanitary measures within the ship) may have a domino effect on the entire value chain in terms providing goods from all specific varieties. Additionally, for wild catch fishing and other aquaculture activities, logistical issues associated with restricted transport, border closures and reduced demand for restaurants and hotels can lead to significant market changes affecting prices.

Illegal Trading Practices (UTPs) Reports from CopaCogeca members indicate a perceived increase in the use of UTPs. Most of the reported situations include unilateral changes to terms, prices and/or conditions, especially for perishable and shortlived products. The different situations reported in previous weeks remain the same (EC 2020a, b).

9.3 Reaction of Agricultural Markets to the Impact of COVID-19

COVID-19 and its effects have led to important changes in the structure of demand, with important crash in the area of food distribution and consumption, hotels and catering, the closure of open markets and an increase in demand in supermarkets. There is a clear evidence of business adaptation, within the food distribution products and consumption areas to current requirements and demands, for example by changing the main manufacturing lines, delivery to individuals and companies, developing this way new horizons: use of temporary staff, switch on to on line activity and platforms, managing larger stocks,

The biggest issues within this area, come from the measures needed to prevent the extension of COVID-19; the imperative adjustments within the sector to respect although it may increase costs; and the necessity to go for diversity in markets for impacted products as people modify their consumption habits in response to COVID-19.

Prices are declining compared to the same period last year (Fig. 9.6).

Whoever thinks about the impact of coronavirus on agricultural product markets, in fact, thinks about the impact on agricultural holdings.

Indeed, after the first coronavirus case detected in China, there is a key word regarding the evolution of the price of agricultural raw materials, namely: volatility.

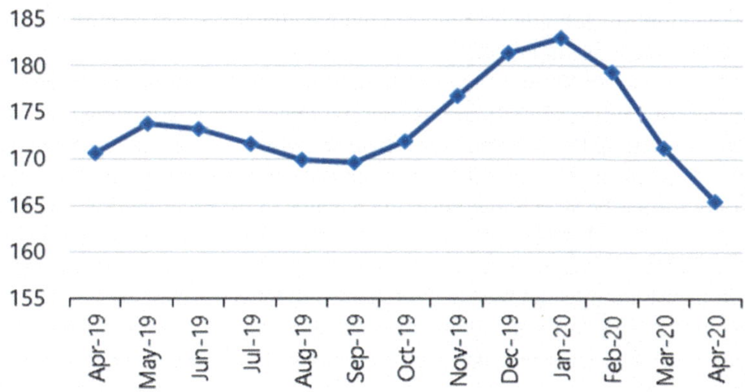

Fig. 9.6 Food Price Index (http://www.fao.org/)

The price exploded with the beginning of the pandemic, increasing by 25 €/ton in the case of wheat and reaching 80 €/ton increase in the case of rape.

At this point, it is very difficult to see clearly what is happening in the markets. One thing is certain: coronavirus has contributed to the collapse of oil prices and has had serious consequences for agricultural markets.

Wheat is facing a possible decline in demand in the medium term while inducing a panic in case of purchases, for the time being.

At the beginning of the coronavirus crisis, the slowdown in the "engines" of the world economy led to lower wheat consumption and as a result the price fell. Importing countries such as Egypt, Saudi Arabia, Algeria or Indonesia have sold their oil and bought wheat, but there is some uncertainty about whether to maintain purchases in the coming months. The lack of liquidity in case of the sale of oil makes the purchases of wheat decrease.

For a few days now, however, the price of wheat has begun to rise. While consumers have emptied their shelves, producers and mills, large importing countries are stepping up their precautionary purchases to build reserves. At this moment, the logistics necessary for the supply are maintained, but there are fears of stopping it, which would block the inflow of goods to the processing units.

China gradually emerged from the "lethargy" of acquisitions and returned to imports of wheat, corn and sorghum from the US and Europe. China has so far imported 1.5 million tons of wheat from France, a large amount compared to what happened years ago.

The Russian economy is largely based on oil (25%), and the fact that its price has plummeted has affected the country's economy, causing a devaluation of the ruble. Russian wheat has automatically gained in competitiveness on the international market, which negatively influences EU wheat exports. However, the decrease of the ruble led to the increase of the price of wheat on the domestic market, which determined the government to limit the export and to ensure the necessary on the domestic market at a reasonable price with the consumers' incomes.

Beer consumption at home is important but the closure of bars and restaurants has led to a drastic decrease in consumption.

The decrease in consumption led to a reduction in the quantities produced, which led to smaller quantities of malt and as such less beer barley consumed.

Sowing conditions last autumn were delicate, which could lead to an increase in spring barley areas in Western Europe, a measure that would increase production. We must not forget the Brexit that upset the balance of different markets. The British want to sell all the barley beer before the start of the 2020 harvest campaign, even if the price drops. This creates an imbalance between increasing supply and decreasing consumption, which means losses for all actors in the supply chain.

The decrease of the oil price led to a decrease in ethanol consumption, which led to a decrease in the price and production of ethanol. In the US, ethanol is the lowest price ever recorded.

The decrease in the profitability of ethanol has led to the decrease of interest for the purchase of corn, it is known that in the USA 40% of the quantity of corn produced is used in the ethanol producing industry.

This shows why the price of corn on the Chicago Board of Trade is at the lowest level in the last 10 years of market. European corn has a limited "weight" on the international market compared to the 360 million tons produced by the USA. We are dependent on what is happening to US corn at the moment.

As the quantities of rapeseed are reduced, now the problem is represented by srot and oil, two markets that today are extremely antagonistic.

The restrictions imposed as a result of the crisis caused by COVID-19 have led to a decrease in biofuel consumption. The price of oil has reached the lowest level in the last 18 years, so all the "green energy" is in full swing. Today it is easier for "gas stations" to pay the fine for not introducing the share of biofuels in gasoline or diesel than to comply with their obligations. Biodiesel processors have reduced production and demand for rape is declining. With 70% of the amount of rapeseed produced in Europe consumed for biodiesel, it is a real crisis to find a market in which this industry goes into "sleep".

Rapeseed largely depends on what is happening on the palm oil market. India has decided to quarantine, for several weeks, as such, the prices of vegetable oils can only fall. At the same time, there may be hope for the rape complex. In Europe, the areas cultivated with rapeseed are smaller than in the previous year, the processing is also decreasing and the quantities of srot will be smaller.

Argentina and Brazil, the world's largest exporters of soybean meal, imposed restrictions on South American ports, leading to rising prices for soybeans.

The decrease in the price of oil led to the decrease in the price of biofuels, this being one of the rare positive effects of the COVID-19 action.

Lack of labor could disrupt food production and processing, especially for labor intensive industries (e.g. high value crops, meat and fish).

Also, interruptions downstream of the farm gate cause, in some cases, the accumulation of surpluses, putting effort on storage facilities and, for high perishability, increase food losses. For some products, supply disruptions are exacerbated by demand cuts (especially foods commonly consumed at home and luxury items see below). In combination, these effects put a strain on farm income. Moreover, these loss of income of agricultural households can be aggravated by a low income from the farm.

The COVID-19 pandemic may also affect the availability of key interim contributions for farmers. For the time being, there do not appear to be shortcomings in the producing regions of developed countries, although farmers may face additional difficulties in terms of input due to additional restrictions on the movement of people and goods. However, in the People's Republic of China (hereinafter 'China'), for example, pesticide production fell sharply and gradually resumed only after production plants were closed following the outbreak.

Low availability and/or high input prices, such as pesticides, could skew crop yields and production in 2020 and 2021, especially in developing countries. Closing borders or slowing down the cross-border movement of seeds could impede seed supply chains and the timely delivery of seeds with a negative impact on agriculture, feed and food production in the next season and beyond. The closure of restaurants and street food stores removes a key market for many producers and processors who may temporarily close or trigger reductions in upstream production, as can be seen in the fish and meat sectors. In some developing countries, urban supply and demand for fresh produce are declining due to restrictions and aversive behavior by traders and consumers.

Developing countries pose a particular risk, as COVID-19 can lead to a reduction in the workforce and affect incomes and livelihoods, as well as labor intensive forms of production (agriculture, fishing/aquaculture). Of particular concern is sub-Saharan Africa, where most of the countries facing food crises are located and where the pandemic is spreading at crucial times for both farmers and pastoralists when people need access to seeds and other inputs and their farms to plant, and when nomadic shepherds have to move with their animals in search of pastures and water sources, as they dry up more and more before the start of the rainy season (http://www.fao.org/).

The need to modernize international standards on hygiene, working conditions and living facilities for agricultural activities and fishing vessels on board, as well as the entire fish value chain, must be taken into account.

9.4 Measures Imposed to Restore the Agricultural System

Policy makers are facing uncertainties about the impact of COVID-19 global food sectors and identify the most suitable measures to reinforce the idea that the current epidemic will not turn on to a food crisis.

Statements and regulations to address, are saying these disruptions may exacerbate market situations and their impact, as was the case in the global food price crisis of 20,072,008 (Demeke et al. 2009). Although the magnitude of the COVID-19 pandemic is different from any other crisis in recent history, the policy responses available to governments against real or registered in agrifood markets are similar to those taken during previous crises.

These include the 20,072,008 food price crisis and the Ebola epidemics (West Africa, 2014), severe acute respiratory syndrome (SARS) (East Asia, 2003), HIV/AIDS (Africa, 1990s, 2000), plague (South Asia, 1994) and cholera (Latin America, 1991). Based on these experiences, appropriate policy measures must be applied in order to support political decision making in this difficult time of COVID-19 (Boubaker BenBelhassen et al. 2020).

In the context of the COVID-19 pandemic, with a severe state of public health emergency for citizens and societies, there has been a major shock to global economies. Under these conditions, a coordinated economic response is crucial to mitigate these negative repercussions.

This shock affects the economy through various channels, with an impact on supply resulting from the disruption of supply chains and a shock on demand caused by lower consumer demand with a negative effect of uncertainty on investment plans and the impact of liquidity constraints on farmers, and processors of agricultural products.

The various measures taken by the affected States, such as social distancing measures, travel restrictions, quarantines and blockages, are intended to ensure that the shock is as short and limited as possible. These measures have an immediate impact on both supply and demand and affect businesses and employees, including in the agricultural sector, and the impact is also felt on the financial markets, especially with regard to liquidity constraints.

In the current exceptional circumstances created by the COVID-19 outbreak, farmers and processors face a severe lack of liquidity. Also, insolvent or less solvent companies may face a sudden lack or even unavailability of liquidity. SMEs pose a particular risk. Therefore, this can seriously affect the economic situation of many healthy businesses and their employees in the short and medium term, while having long-term effects, endangering their survival.

Businesses not only face insufficient liquidity, but can also suffer significant damage due to the COVID-19 outbreak. Even solid undertakings, well prepared for the risks inherent in the normal course of business, may suffer in these exceptional circumstances, to the extent that their viability may be undermined.

Taking into account those from the above, decisive action with a high impact is required to ensure the liquidity of enterprises in the agricultural sector during the period when the effects of COVID-19 are manifested.

Considering the need to adopt tools to protect farmers and processors as well as the economic system to inject not only money into the economy, but also trust and mobilizing spirit, contractual loyalty and willingness to adapt business relationships to severe challenges and difficulties which the whole nation is going through.

As a result of the current crisis, regarding the outbreak of COVID-19, crossed worldwide, a significant number of domestic agricultural producers face real difficulties both in accessing and, especially, in reimbursing the exposures/credits carried out, as well as due to the diminishing guarantees value for these loans, guarantees that have greatly diminished their values compared to the initial assessments.

An immediate priority is the support given to producers and processors of agricultural products, their support through specific financial programs and economic projects, created and supported at national level, by the elaboration, and the creation of a package to support them.

Coronavirus has hit Europe and the world to its core, testing our systems, our societies and economies, and our way of living, interacting, and working together. In order to protect lives and individuals, repair the single market, and build a sustainable and prosperous recovery, the European Commission aims to realize the full potential of the EU budget. The next generation of the € 750 billion EU, as well as targeted long-term consolidation of the EU budget for 20,212,027, will bring the total financial strength of the EU budget to € 1.85 trillion. European Commission President Ursula von der Leyen said: "The recovery plan turns the immense challenge we face into an opportunity, not only by supporting the recovery but also by investing in our future: the European Green Deal and digitalization will boost jobs and growth, the resilience of our societies and the health of our environment. This is Europe's moment. Our willingness to act must live up to the challenges we are all facing. With Next Generation EU we are providing an ambitious answer" (EC 2020a, b).

Under this budget, € 15 billion is allocated to the European Agricultural Fund for Rural Development to support rural areas in making the necessary structural changes under the European Green Deal and achieving ambitious targets under the new Biodiversity and Farm to Fork strategies.

The G20 recent statement given by governmental authorities in the field of agriculture provides justifications to hope that instructions have been learned from 20,072,008. In that statement, they reaffirmed their agreement not to impose export restrictions or extraordinary taxes on food and agricultural products purchased for non-commercial humanitarian purposes by the World Food Program (WFP) and other humanitarian agencies.

Additional actions need to be identified to reduce the impact of COVID-19 on food safety and nutrition; share best practices and lessons learned, such as resolving barriers in supply chains; promoting evidence based information and combating misinformation; provide capacity building and technical assistance; and to promote research, responsible investment, innovation and reform that will improve the sustainability and resilience of agricultural and food systems.

Additional actions need to be identified to reduce the impact of COVID-19 on food safety and nutrition; share best practices and lessons learned, such as resolving barriers in supply chains; promoting evidence based information and combating misinformation; provide capacity building and technical assistance; and to promote research, responsible investment, innovation and reform that will improve the sustainability and resilience of agricultural and food systems. The statement states the relevance of work to secure the continuous circulation of aliments and food, essential goods and contributions to agriculture and cross-border alimentary production in spite of the challenges facing COVID-19. These types of activities are meant to give the possibility to worldwide trading to become an important agent in order to avoid feed deficiencies but also to mitigate for an inevitable crash on all economic levels, resulting from a potential pandemic (https://g20.org).

References

AMIS Market Monitor (2020) No 78. http://www.amisoutlook.org/amismonitoring

BenBelhassen B, Mermigkas G, Gadhok I, Avesani C (2020) Agrifood markets and trade policy in the time of COVID-191 brief policy. Trade and Markets Division, Economic and Social Development Department, FAO, Rome

Demeke M, Maetz M, Pangrazio G (2009) Country responses to the food security crisis: nature and preliminary implications of the polices. Initiative on mooring food prices. FAO, Rome

European Commission (2020a) Europe's moment: repair and prepare for the next generation, Brussels, 27 May 2020

European Commission (2020b) Coronavirus: package of measures to further support the agrifood sector, Brussels, 4 May 2020

Food and Agriculture Organization of United Nations (2020) Novel coronavirus (COVID-19). http://www.fao.org/

Food Security Information Network (FSIN) (2020) Global report on food criss. Joint analysis for better decisions. https://www.fsinplatform.org/

G20 Extraordinary Agriculture Ministers Meeting (n.d.) Ministerial statement on COVID-19, virtual meeting—April 21, 2020. https://g20.org/

Moody's (2020). https://www.moodys.com/

Organisation for Economic Cooperation and Development (OECD) (2020). https://www.oecd.org

Schmidhuber J, Pound J, Qiao B (2020) COVID-19: channels of transmission to food and agriculture. FAO, Rome. https://doi.org/10.4060/ca8430en

Chapter 10
Impacts of COVID-19 on Air Pollution

Shubham Sharma and Sri Harsha Kota

10.1 Introduction

10.1.1 Air Pollution

Atmospheric pollution and climate change is a major problem being faced by all nations of the world (Annadanam and Kota 2019; Lou et al. 2019). It has now become a serious issue of concern and many countries are formulating action plans to deal with it. The technological revolution and ever-increasing urbanization and industrialization have contributed to the deterioration of the environment.

In general, the sources of air pollution include residential, transportation, industrial, windblown dust and power generation sources. The major regulated pollutants that affect the air quality and health include particulate matter (PM), ozone (O_3), sulfur dioxide (SO_2), carbon monoxide (CO) and oxides of nitrogen. Air pollution can lead to many diseases in humans and is said to even cause death.

Air pollution is a menace that is affecting people all over the world. The most affected are the developing nations undergoing rapid expansion of infrastructure and cities, industrialization, in the past two decades. With a large population and an evolving economy India is one of the leading developing countries which is facing the problem of high air pollution in majority of its cities (Lou et al. 2019; Garaga et al. 2020; Sahu et al. 2020). Air pollution is known to cause a large number of premature deaths in India. According to a study by Guo et al. (2017), nearly one million people died in India due to $PM_{2.5}$ in the year 2015. Pollution levels have been

S. Sharma · S. H. Kota (✉)
Department of Civil Engineering, IIT Delhi, Delhi, India
e-mail: harshakota@iitd.ac.in

M. K. Goyal, A. K. Gupta (eds.), *Integrated Risk of Pandemic: Covid-19 Impacts, Resilience and Recommendations*, Disaster Resilience and Green Growth,
https://doi.org/10.1007/978-981-15-7679-9_10

found to exceed the Central Pollution Control Board (CPCB) and WHO air quality standards (Garaga et al. 2018; Kota et al. 2018).

Vehicular, residential, windblown dust, energy and industrial emissions contribute to the particulate matter, which is the major pollutant in India (Guo et al. 2017, 2019a). Many initiatives aimed at reducing the air pollution in India. The National Clean Air Programme (NCAP) is one such 5 year programme that was rolled out in 2019 and aimed at a 30% reduction in the PM concentrations nationwide (MoEFC 2019).

However, the success of these strategies generally focused on the reduction of emissions from the sources. However, the meteorology also plays an important role in the formation and the transport of the pollutants. The role of meteorology has been emphasized in many studies in the past. Aggressive measures taken in the form of 5 year clean air action plans in China starting 2013, have been found to have reduced pollutant concentrations significantly (Sun et al. 2019). However, unfavorable meteorological conditions during episodes in winters lead to high peak $PM_{2.5}$ concentrations (Wang et al. 2019). Even during the COVID-19 lockdown in China during January and February 2020, no improvement in the air quality was observed and meteorology was observed to be responsible for the high air pollution (Wang et al. 2020).

10.2 COVID-19 Spread in India

The COVID-19 pandemic is one of the major global health crisis humanity has ever faced. China alerted the WHO about cases of unusual pneumonia in Wuhan on December 31. The COVID-19 outbreak that emanated out of Wuhan in China was declared as a Public Health Emergency of International Concern by WHO on 30 January 2020. The unprecedented COVID-19 pandemic spread rapidly across nations which then responded with lockdowns, making people globally remain at home.

India confirmed its first case of COVID-19 on January 30th, 2020. The COVID-19 cases starting sharply increasing in India (https://www.mohfw.gov.in/).

Following this, the Indian Government imposed restrictions on travel to countries Italy, China, Iran, Korea and Japan and issued travel advisories on March 11th, 2020. Starting March 10th, Kerala, which was first state in India hit, imposed restrictions on mass gatherings. Educational institutions, shopping malls and theatres were also asked to remain closed across the country from March 16th. A 14 h nationwide lockdown was observed after the call by Prime Minister Narendra Modi on March 22nd and was given the name 'Janta Curfew'. Following this a 21 days lockdown was announced on March 24th and which would end on April 14th but was further extended. Restrictions on the movement of people during this lockdown were imposed during the lockdown and as people were confined to their homes, a reduction in the emissions form industrial and transportation sector is expected. The variations in air pollution levels during this lockdown period of 21 days can help in exploring and estimating the amount of reduction in the air pollutant concentrations. This will also help regulators devise new strategies and improve the existing action plans.

This chapter discusses the change in the pollutant concentrations during 1st March to 14th April for the years 2019 and 2020. The change in meteorology has also been analyzed to understand how meteorology affects the concentrations of pollutants across India.

10.3 Methodology

More details about the methodology is available in Sharma et al. (2020) and only briefly described below. The main difference between Sharma et al. (2020) and this study is the time period used for analysis.

10.3.1 Data Sources

The online portal for air quality data dissemination of CPCB was used to retrieve the hourly pollutant concentration ($PM_{2.5}$, PM_{10}, SO_2, O_3, NO_2, and CO) and meteorological data (wind speed, wind direction) for the study period of March 1st to April 14th for the years 2019 to 2020 collected for 22 Indian cities. The cities included seven north, five in south, two in center, four in the west and five in the east India. More details about the cities is available.

10.3.2 Air Quality Index (AQI) and Risk Calculations

The weighted concentrations of different pollutants is used to AQI (CPCB 2014; Sahu and Kota 2017). AQI was calculated to understand and estimate the improvement in air quality across India during the period of the lockdown.

AQI is calculated using the concentrations of at least three of the pollutants out of PM_{10}, $PM_{2.5}$, CO, SO_2, NO_2, O_3, NH_3 and Pb. $PM_{2.5}$ or PM_{10} should compulsorily be one of the minimums of three pollutants considered in the AQI calculation. For the AQI calculation, firstly the pollutant concentrations are first converted to a sub-index ranging from 0 to 500. These sub-indices (AQI_i) are calculated for all available pollutant(i) using Eq. (10.1).

$$AQI_i = \frac{IN_{HI} - IN_{LO}}{B_{HI} - B_{LO}} \times (C_i - B_{LO}) + IN_{LO} \tag{10.1}$$

Table 10.1 β values for different pollutants (Sharma et al. 2020)

Pollutant	β
$PM_{2.5}$	0.038
PM_{10}	0.032
SO_2	0.081
NO_2	0.13
O_3	0.048
CO	3.7

Note: Except CO, all other pollutants are for increment of 10 μg/m^3

In this equation, C, B and IN represents ambient, breakpoint concentrations and index values. 'i' is the pollutant; LO and HI are low and high values. The maximum of all the sub-indices is overall AQI.

Relative risk (RR) has been used to estimate the overall health benefits that can be achieved due to the lowered concentrations of the pollutants during the lockdown across the 22 Indian cities. RR greater than 1 means there is increased risk of a disease on being exposed to a certain substance and vice versa (Guo et al. 2019b). Relative risks were then calculated using the Eq. (10.2)

$$RR_i = \exp\left[\beta_i (C_i - C_{i,0})\right], C_i > C_{i,0} \tag{10.2}$$

In Eq. (10.2), β is the exposure-response coefficient that indicates added health risk that can be caused by the exposure to a unit amount of the pollutant i, if it exceeds a certain threshold concentration ($C_{i,0}$). The β values for different pollutants have been listed in Table 10.1. Threshold concentration ($C_{i,0}$) is the concentration of a pollutant at or below which the pollutant poses no excess health risk.

The Excess Relative Risk (ER) is then calculated, which is equal to the percentage change in the health risk in one group compared to a reference group (2010).

$$ER_{total} = \sum_{i=1}^{n} ER_i = \sum_{i=1}^{n} (RR_i - 1) \tag{10.3}$$

Higher the value of ER_{total}, higher is the health risk of the disease. The change in the ER_{total} for the year 2019 and 2020 were analyzed to calculate the health benefits.

10.4 Results and Discussion

10.4.1 Changes in Meteorology

The wind rose plots for the study period in India have been presented in Fig. 10.1. From the figure, it is evident that the wind pattern remains quite similar for both years 2019 and 2020 apart from the Central India region. In central India, the wind speeds were found to be higher in the year 2019 and the dominant wind direction in

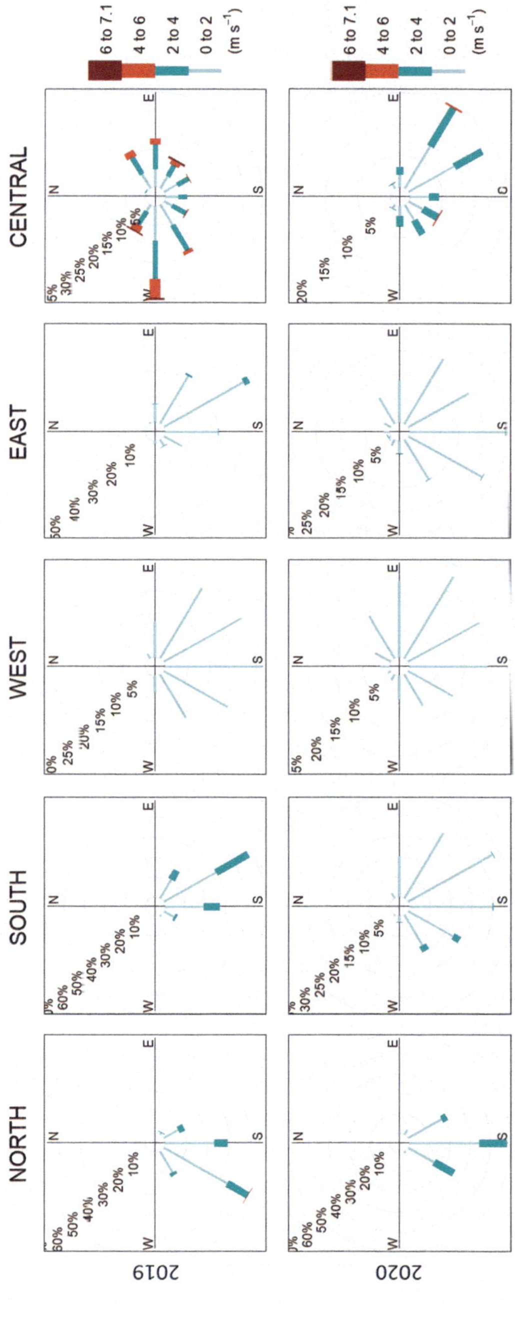

Fig. 10.1 Variations in wind directions and wind speeds across different regions in India in the year 2019 and 2020

2020 was found to be southeast when compared to the west in the year 2019. In North India the observed average wind speed was ~1.5 ms^{-1} and the dominant wind direction being south and southwest in the year 2020 and 2019 respectively. A similar trend in the wind directions was observed in the east, south, and west India where the dominant wind directions were south and southwest, and the average wind speed was found to lie between 0.8 and 1 ms^{-1}. Little to no variations were observed in the Temperature and RH for the 2 years. For the year 2019 and 2020 the average temperature in North India was 30 and 28.94 °C respectively. The RH values were also very similar in both the years. Overall, for the 2 years the meteorology was found to be very similar.

10.4.2 Variation in Pollutant Concentrations

The variations in the pollutant concentrations have been represented in Fig. 10.2 as the percentage changes in the average concentrations of pollutants considered in this study across different regions of India in the years 2019 to 2020. Overall, the $PM_{2.5}$, PM_{10} and NO_2 concentrations decreased by an amount of 28.62, 27.48 and 24.96% during lockdown period in 2020 as compared to 2019. However, an average increase of 35.76 and 20.62% was observed in the concentrations of CO in South, west and central India, and SO_2 across entire India, during the lockdown period as compared to that in 2019. The O_3 concentrations in central and east India in 2020 were found to be almost thrice and twice of the concentrations in 2019, however they were still under the permissible limits prescribed by CPCB. The higher decrease in NOx concentrations when compared to concentrations of VOCs can be the reason for this increase in O_3 as most cities in India are VOC limited (Sharma et al. 2016). The decrease in the $PM_{2.5}$ concentrations across India could also be another reason as it results in more photochemical activities by allowing more sunlight to pass through and thus encouraging the O_3 production (Dang and Liao 2019; Li et al. 2019).

North India witnessed notable decreases in the concentrations of $PM_{2.5}$, PM_{10} and NO_2 in 2020 as compared to 2019. For example, a ~35% decrease in the concentrations of $PM_{2.5}$, thereby emphasizing the effect of the reduced emissions during the lockdown. A similar trend of decrease in the concentrations of $PM_{2.5}$ and PM_{10} was observed in other regions of India as well.

SO_2 concentration was found to have increased during the lockdown in 2020 from 2019. The use of coal generated electricity and no restrictions on the power plants can be one of the reasons for the same.

In east India an increase in the concentrations of the gaseous pollutants except CO was noticed in 2020 to those in 2019. O_3, NO, NO_2 and SO_2 had 77, 11, 4 and 3% increase in the concentrations respectively when compared to 2019.

For south India, the concentrations of NO, NO_2 and O_3 were found to have decreased during the lockdown in 2020 when compared to 2019. However, an increase of 12.20% in the concentrations of CO were observed. In central India,

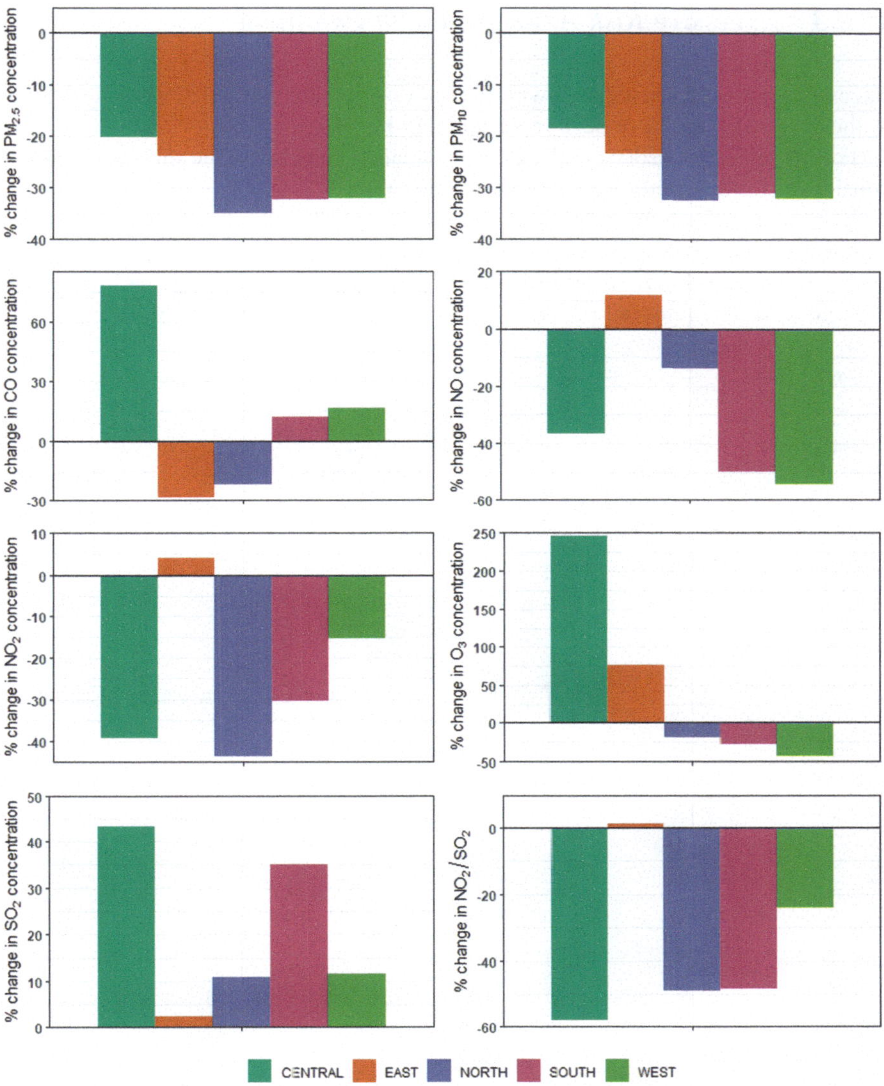

Fig. 10.2 Percentage change in the average $PM_{2.5}$, PM_{10}, CO, NO, NO_2, NO_x and O_3 concentrations for the time period of March 1 to April 14 for the years 2019 and 2020 in different regions are shown. Positive values show an increase in the concentrations in 2020 as compared to 2019 and vice-versa

decreases of 39 and 36% in the concentration of NO_2 and NO concentrations were observed.

Overall $PM_{2.5}$ and PM_{10} concentrations were observed to have reduced significantly across the country.

10.4.3 Excessive Risk Associated with Pollutants

Excessive risks (ER) associated with the various pollutants in 2020 were compared to those in 2019 and have been represented in Fig. 10.3.

Following WHO (2005) and CPCB air quality guidelines, the threshold values considered for this study have been listed in Table 10.2.

Even after reduction on the concentrations of pollutants during the lockdown, significant excess risks of diseases were estimated. An average reduction of ~53 and ~44% in the excess risk due to $PM_{2.5}$ and PM_{10} was observed. However, an overall average increase of around ~32% in the ER due to SO_2 was observed.

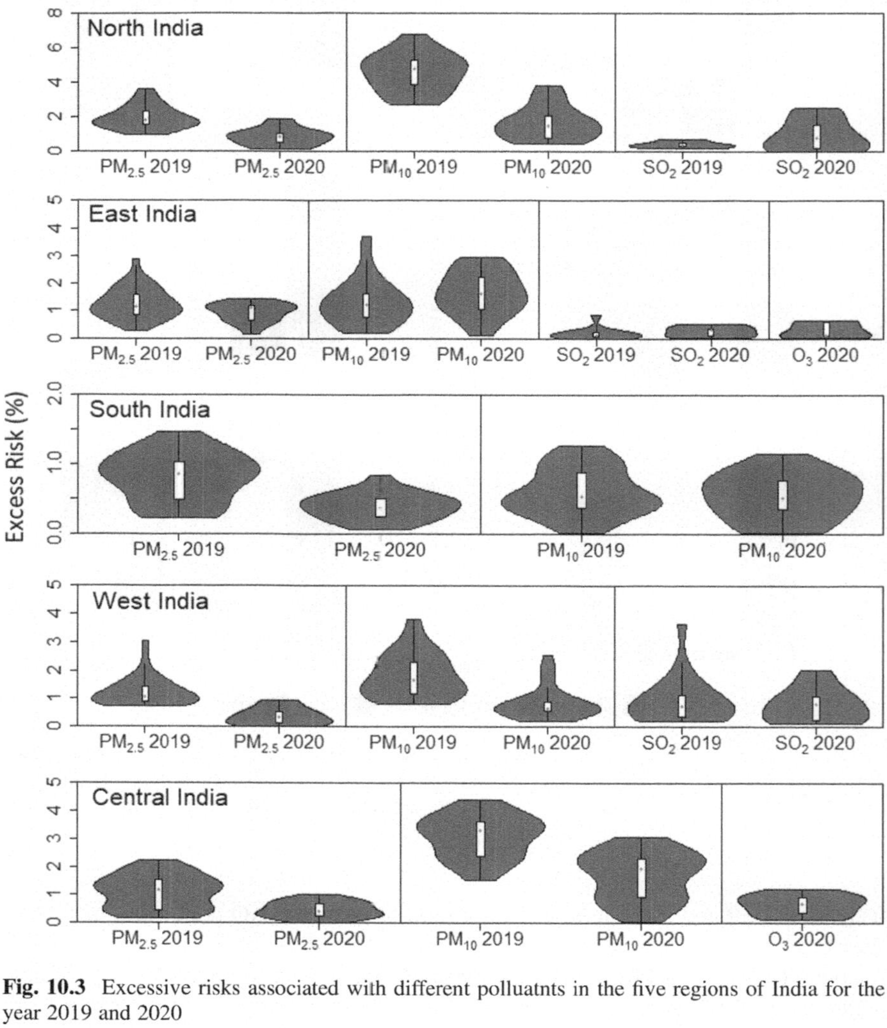

Fig. 10.3 Excessive risks associated with different polluatnts in the five regions of India for the year 2019 and 2020

Table 10.2 Threshold concentrations for six pollutants considered in this study

Pollutant	Threshold concentration	Averaging period (h)
$PM_{2.5}$	25 μg/m^3	24
PM_{10}	50 μg/m^3	24
O_3	100 μg/m^3	8
NO_2	200 μg/m^3	1
SO_2	20 μg/m^3	24
CO	4 mg/m^3	1

Excess risk for nearly all pollutants reduced in the year 2020 as compared to that in 2019 except that for SO_2 which increased in northern and eastern regions of India. From our mortality estimates, approximately 6.5 lakh deaths in India could be saved in a year from the reductions in concentrations observed during this lockdown.

10.4.4 Variations in AQI

Figure 10.4 represents the variations in AQI and dominant pollutants during the study period in 22 Indian cities in the year 2019 and 2020. Overall, a major improvement in the overall AQI was observed across the country in 2020 during the lock down when compared to 2019. An improvement of 31.51% was observed in the AQI during the lockdown in 2020 compared to the 2019.

About 50, 18, 29, 58 and 8% improvement in the AQI were observed across northern, southern, eastern, central and western regions respectively. Highest improvement on the AQI was observed for Delhi where the AQI improved by ~66% from the previous year. Along with the change on AQI, the dominant pollutants for different cities also changes in 2020. The dominant pollutant changed to O_3 from PM_{10} in the four cities (Gaya, Kolkata, Kanpur and Nagpur) in 2020 as compared to 2019. In Agra and Patna, it changed from $PM_{2.5}$ to NO_2. As maximum reductions were observed for $PM_{2.5}$ and PM_{10}, the dominant pollutant changing to O_3 in these cities is justified.

Figure 10.5 shows the correlation of AQI values among cities in different regions in India for the study period in 2020 and 2019. The correlation improved in 2020 as compared to 2019, specially in north and east India. For example, the correlation among Delhi and other cities improved on an average by about a factor of 2.35. The highest correlation (0.78) in all the regions was observed among Bhopal and Dewas in Central India during 2020. This emphasizes the major contributions to the pollutant concentrations by regionally transported pollutants during the lockdown period.

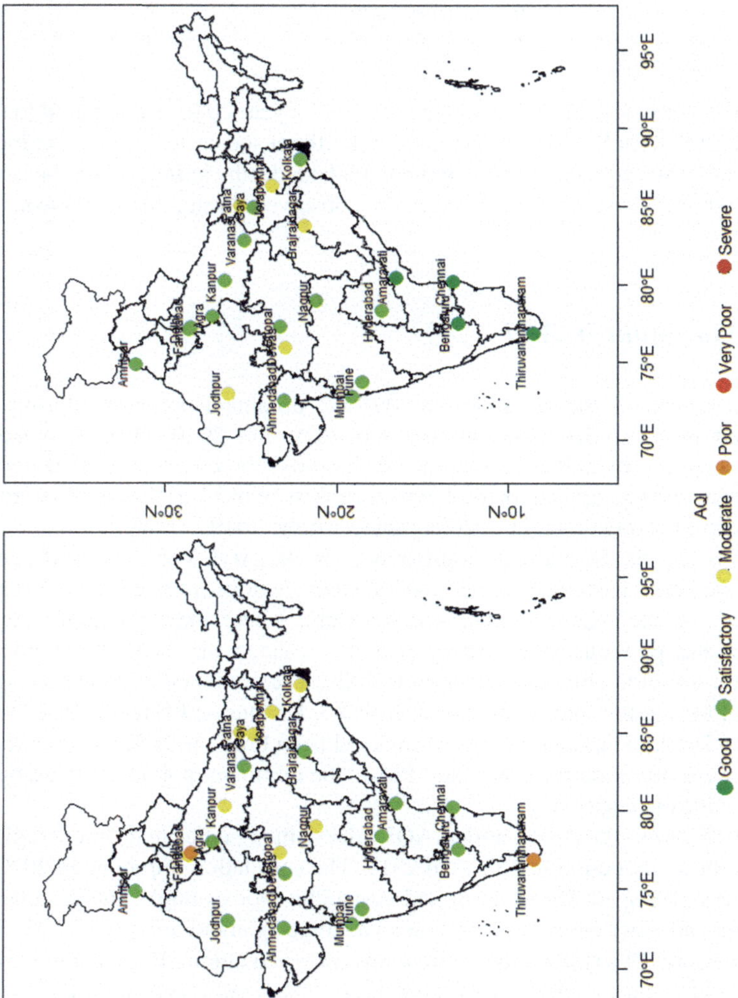

Fig. 10.4 Change in AQI across Indian cities from March 1st to April 14th of the years 2019 and 2020. Different symbols are used to denote the dominant pollutant in a city

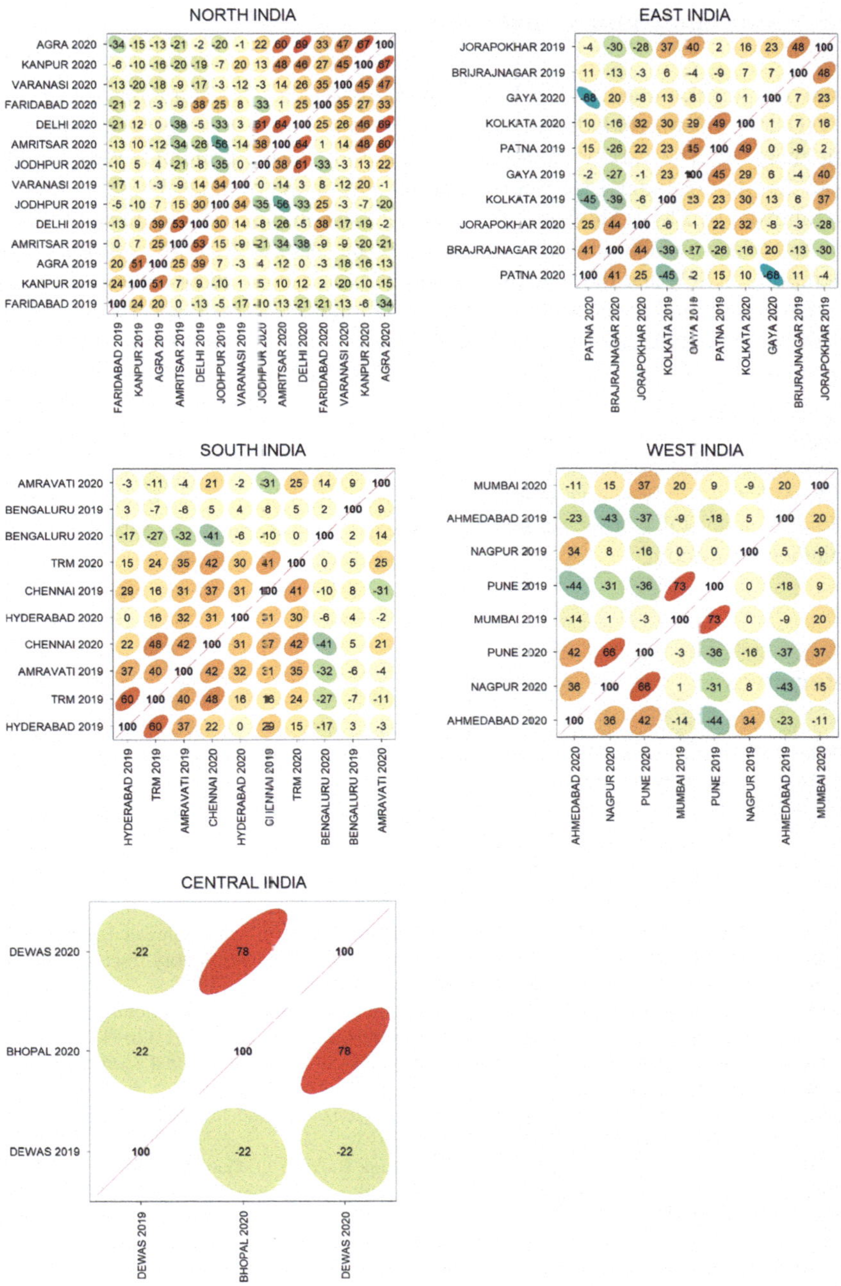

Fig. 10.5 Regional correlation among AQI values among cities during the study period in 2019 and 2020

10.5 Summary and Conclusions

In this chapter the impact of reduced emissions on the concentrations during the COVID-19 pandemic in India for the study period of March 1 to April 14, 2020 have been analyzed. The variations in the concentrations of the pollutants were studied across 22 cities of India. Maximum reduction was observed in the $PM_{2.5}$ concentration followed by PM_{10}. However, SO_2 and O_3 concentrations increased in almost all regions of India which could be attributed to the increased use of coal for power generation for SO_2 and the reduction in the NOx and $PM_{2.5}$ concentrations for O_3. The reduction in the concentrations of these pollutants also resulted in a decrease in the ER by about a factor of 4. The AQI also improved in the lockdown period when compared to 2019. But as the dominant pollutants changed in the year 2020 for many cities, strategies must be made to reduce the secondary pollutants.

Significant regional transport was observed as the correlations in 2020 improved among cities when compared to 2019.

Overall, this study shows that significant improvement in the concentrations of pollutants and AQI can be achieved if inclusive action plans are formulated and implemented strictly.

References

Annadanam SK, Kota SH (2019) Emission of greenhouse gases and criteria pollutants from railways in India estimated using a modified top-down approach. J Clean Prod 213:610–617

CPCB (2014) National air quality index report. Central Pollution Control Board, New Delhi

Dang R, Liao H (2019) Radiative forcing and health impact of aerosols and ozone in China as the consequence of clean air actions over 2012–2017. Geophys Res Lett 46(21):12511–12519

Garaga R, Sahu SK, Kota SH (2018) A review of air quality modeling studies in India: local and regional scale. Curr Pollut Rep 4(2):59–73

Garaga R, Gokhale S, Kota SH (2020) Source apportionment of size-segregated atmospheric particles and the influence of particles deposition in the human respiratory tract in rural and urban locations of north-east India. Chemosphere 255:126980

Guo H, Kota SH, Sahu SK, Hu J, Ying Q, Gao A, Zhang H (2017) Source apportionment of PM2.5 in north India using source-oriented air quality models. Environ Pollut 231:426–436

Guo H, Kota SH, Sahu SK, Zhang H (2019a) Contributions of local and regional sources to PM2.5 and its health effects in north India. Atmos Environ 214:116867

Guo H, Sahu SK, Kota SH, Zhang H (2019b) Characterization and health risks of criteria air pollutants in Delhi, 2017. Chemosphere 225:27–34

Kota SH, Guo H, Myllyvirta L, Hu J, Sahu SK, Garaga R, Ying Q, Gao A, Dahiya S, Wang Y, Zhang H (2018) Year-long simulation of gaseous and particulate air pollutants in India. Atmos Environ 180:244–255

Li K, Jacob DJ, Liao H, Shen L, Zhang Q, Bates KH (2019) Anthropogenic drivers of 2013–2017 trends in summer surface ozone in China. Proc Natl Acad Sci 116(2):422–427

Lou J, Wu Y, Liu P, Kota SH, Huang L (2019) Health effects of climate change through temperature and air pollution. Curr Pollut Rep 5(3):144–158

MoEFC (2019) NCAP National Clean Air Programme. In: Sundaray SNK, Bharadwaj DSR (eds) Ministry of environment forest and climate change, government of India 122. MoEFCC, New Delhi

Sahu SK, Kota SH (2017) Significance of PM2.5 air quality at the Indian capital. Aerosol Air Qual Res 17:588–597

Sahu SK, Sharma S, Zhang H, Chejarla V, Guo H, Hu J, Ying Q, Xing J, Kota SH (2020) Estimating ground level PM2.5 concentrations and associated health risk in India using satellite based AOD and WRF predicted meteorological parameters. Chemosphere 255:126969

Sharma S, Chatani S, Mahtta R, Goel A, Kumar A (2016) Sensitivity analysis of ground level ozone in India using WRF-CMAQ models. Atmos Environ 131:29–40

Sharma S, Zhang M, Gao J, Zhang H, Kota SH (2020) Effect of restricted emissions during COVID-19 on air quality in India. Sci Total Environ 728:138878

Sun J, Liang M, Shi Z, Shen F, Li J, Huang L, Ge X, Chen Q, Sun Y, Zhang Y, Chang Y, Ji D, Ying Q, Zhang H, Kota SH, Hu J (2019) Investigating the PM2.5 mass concentration growth processes during 2013–2016 in Beijing and Shanghai. Chemosphere 221:452–463

Wang P, Guo H, Hu J, Kota SH, Ying Q, Zhang H (2019) Responses of PM2.5 and O3 concentrations to changes of meteorology and emissions in China. Sci Total Environ 662:297–306

Wang P, Chen K, Zhu S, Wang P, Zhang H (2020) Severe air pollution events not avoided by reduced anthropogenic activities during COVID-19 outbreak. Resour Conserv Recycl 158:104814

WHO (2005) Air Quality Guidelines. Global update 2005. WHO Regional Office for Europe, Copenhagen

Chapter 11
Economic Impact of Covid-19 and Challenges of Recovery

Santosh Kumar

Abstract Covid-19 came as a surprise to the global community. More than 203 countries got affected. Different countries took different measures at different points of time for fighting the virus. One common action was taken by all and that was the lockdown for protecting the lives of the people and slowing down the spread of the pandemic. Lockdown by the individual country led to global lockdown. Pandemic killed more than 300 thousand people with huge global economic loss (until May 31st 2020). This pandemic turned out to be the costliest disaster in the history of mankind. COVID-19 led to slowing down of the GDP, mass unemployment, bankruptcy, migration, closure of manufacturing sectors, a complete halt of travel, trade, tourism and economy as a whole. Its recovery is going to be a real challenge. There is a huge problem of poverty, unemployment, supply chain and the absence of effective demand. Pandemic disaster is a global issue and hence solutions should also be worked out collectively by the global community. There is a need to evolve or innovate other strategies than the lockdown for a fighting pandemic by the global community.

Keywords Public Health · COVID · Biological disasters · Pandemic · Global recession · GDP · Growth gap · Resiliency · Recovery

11.1 The Context: Global Recession

Before the pandemic global economy was facing an economic crisis. The economy was slowing down and it was a recession type of situation. GDP of the countries was showing a declining trend. Depression or slowing down of the economy has been a

S. Kumar (✉)
National Institute of Disaster Management, Ministry of Home-Affairs, Government of India, New Delhi, India

M. K. Goyal, A. K. Gupta (eds.), *Integrated Risk of Pandemic: Covid-19 Impacts, Resilience and Recommendations*, Disaster Resilience and Green Growth,
https://doi.org/10.1007/978-981-15-7679-9_11

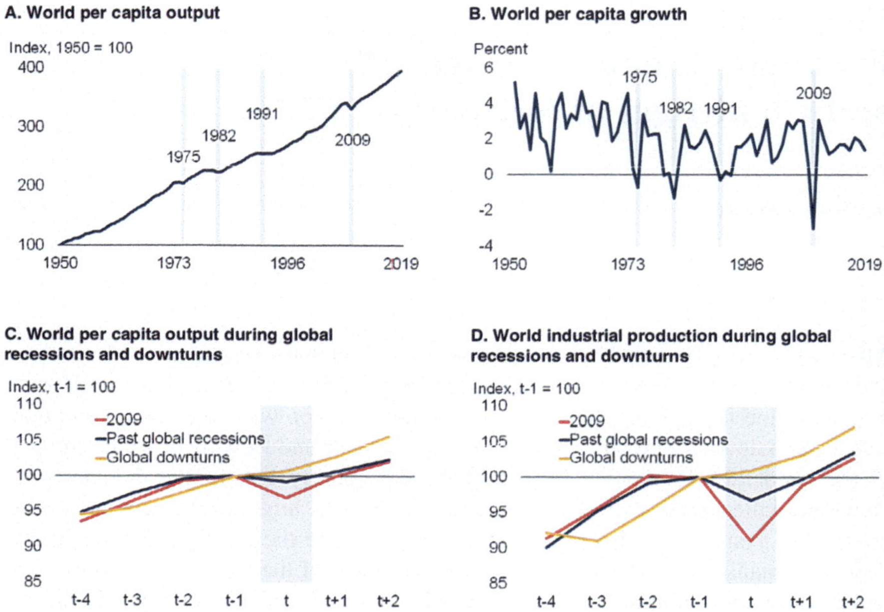

Fig. 11.1 Showing the global trend of economic growth. (Source: A Decade after the global recession, The World Bank 2019)

part of economic development. After the great depression of the 1930s, there have been almost six economic phases at different time intervals where there was a serious economic downturn which the global economy has faced. The years 1975, 1982, 1991, 2009 are the bad phases for growth of GDP, Industrial output, infrastructure growth, employment, capital flows, international trade.

As per the World Bank report 2019, the global downturn could be seen in the world per capita output, per capita growth, per capita output and world industrial production in Fig. 11.1.

As per the World Bank report, "Global recessions are highly synchronized events internationally, with many economies sliding into recessions simultaneously. Remarkably, the proportion of economies in recession during successive global recessions has increased over time: it was close to 40% in the 1975 episode and about 61% in the 2009 global recession. The proportion of countries in recession typically starts rising ahead of the recession year. The degree of synchronicity in the last global recession was the highest in the past 70 years, possibly reflecting the depth of the global financial crisis and stronger international trade and financial linkages compared to the prior decade".

Suddenly, The World is witnessing a global pandemic. In this slowing down of the economy, an additional monster like a burden gives jitters which will lead to a serious downturn as the economics of the World is not in very healthy conditions. This is the time of uncertainty. No one knows how big the pandemic will be. How long it is going to stay. The public health expenditure is bound to escalate. The uncertainty is making economic predictions difficult. In this situation, there would be a serious issue of supply chain and also reviving the demand. There is no trade, no production, national economies of the globe will be in stress, the decline in remittances etc.

As IMF, 2020 "Many countries face a multi-layered crisis comprising a health shock, domestic economic disruptions, plummeting external demand, capital flow reversals, and a collapse in commodity prices. Risks of a worse outcome predominate."

11.2 The Pandemic (COVID-19) and Economics

The ongoing global pandemic COVID-19 has shaken the World. This has happened after almost 100 years (Spanish flu of 1918) where millions of people got killed. People use to fight by confining themselves, maintaining social distancing and personal hygiene practice. The WHO guidelines in 2020 for fighting pandemic are almost on similar lines in 2020. There is no vaccine for fighting coronavirus. The World is puzzled with its uncertainty and with the confusing signals of the WHO for mitigating the impact of a pandemic. Global communities are practising almost similar methods of fighting with it. Lockdown of the country has been the most popular method of pandemic management. Testing, social distancing, quarantine and personal hygiene are the other methods. The locking down the country became one of the most important actions for flattening of the curve. Its enforcement was the big challenge. The diversity of the population, different economic conditions, poverty, daily wage earners, labourers, political affiliations were making it difficult to execute. However, most of the countries adopted the same. Developed economies—China, the US, Japan, Germany, France, Italy, Spain had a different time frame for lockdown whereas the emerging Market economy—India, Brazil, Iran, Saudi Arabia, North Korea, Singapore, Cambodia, Vietnam.

11.3 The Characteristics of Pandemic

Pandemic is not the new phenomenon and hence many thinkers do not consider it as a Black swain event. Global concern about the pandemic threat escalated in 2005–2008. But then it dropped rapidly, especially when the 2009 H1N1 flu pandemic, turned out to be less disruptive by making quick efforts than many flu seasons. The drop in concern reinforced four pandemic myths, creating a challenge when policy planners, governance and people accepted the myths, the mitigation of impact became much more challenging.

World Bank report on Pandemic risk, 2016 mentioned that the reactions by governments, businesses and consumers to the 2003 SARS outbreak (which was

	Plausible…	*… but:*
Myth 1	Nothing can be done to prevent pandemics. It's just nature.	Human activity helps or hinders the onset of contagion. Most pandemics start in animals, so human-managed **livestock health makes a big difference.** Robust public veterinary and human health systems can work to **stop contagion early**, so it does not become a pandemic. The science is clear; tools and effective disease control methods are well known in most cases.
Myth 2	We can deal with it when it comes. There is no risk. Do you want to scare people?	Severity of pandemic impact would depend on human reactions, on the resilience of communities, and on whole-of-society preparedness, including for official communications. Preparedness for household, community, country, and international responses will reduce costs. **Risk awareness** is a necessary first step.
Myth 3	Health authorities will protect us from pandemics.	Pandemic risk reduction is **not a priority for the health sector**, which is mainly concerned with existing patients, not potential ones. Since prevention brings few observable recognized rewards, it is often grossly neglected. Vaccines are no panacea and will not prevent a pandemic with current technologies. They may be available with delay after a pandemic starts, in limited quantities, and have low effectiveness. There is still no AIDS vaccine, decades after the onset of the AIDS pandemic. Vaccines and antiviral medicines could protect the health of a fortunate minority with access to them, while possibly harming social cohesion if scarce vaccine is allocated arbitrarily. Mitigation of pandemic impacts requires **measures in all sectors**, not only health interventions.
Myth 4	Pandemic risk is not a development concern.	This is mistaken, on at least three counts. First, a severe pandemic would impose widespread health, economic and social costs, **setting back progress by years**. The poor and vulnerable could be most affected. Second, pandemic prevention hinges on **robust public health systems (veterinary and human) that collaborate** to detect contagion early, respond rapidly, and stop it. These same systems are needed to prevent and control other diseases as well, especially the human diseases that are caused by 2.3 billion infections of animal origin every year. Third, pandemic preparedness, which supports **resilience across the whole-of-society and business continuity in key sectors**, will help developing countries reduce the impacts of other disasters, especially complex ones.

Fig. 11.2 Explains the four myths. (Source: Jonas 2013)

arrested after 8000 cases and 800 fatalities) gave rise to economic costs of $54 billion, confirming that impacts of contagion outside the health sector predominate, possibly by a very wide margin. Some 60% of the impact would be due to demand and supply shifts driven by people's avoidance reactions. The other major driver of costs would be lost production due to high worker absenteeism, amounting to about 28% of the total cost (Fig. 11.2).

The report of the World Bank on pandemic risk, that If a disease is not controlled at its animal source and a sudden-onset pandemic occurs, people in all countries may experience both a contagion of disease and a sharp, possibly catastrophic, economic downturn associated with shifts in demand, supply shocks, and economic and social disruptions. Because countries are connected by, and depend on, travel, trade and capital movements, the shocks would propagate across interconnected economic and financial systems worldwide, possibly ahead of the contagion itself.

These risks were evident in the SARS outbreak in 2003 (Fig. 11.3) and should be anticipated, based on connectedness characteristics of each country (some will be more vulnerable than others). This was truly found in the case of COVID-19. The pandemic started from the Wuhan city of China travelling to Europe and then spread to all across the globe. From a few, everyone got affected. It spread over to more than 203 countries of the World. Ted the travel and tourism the most. People love

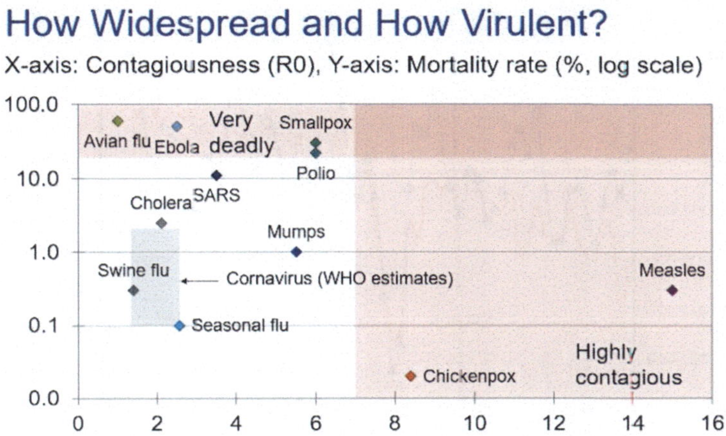

Fig. 11.3 Viral widespread and mortality. (Source: The World Bank Report 2016 on pandemic risk)

travelling in the time of globalisation, lockdown affected the trade and tourism sector severely.

11.4 Pandemic and Flattening of the Curve

Figure 11.4 explains how the impact of SARS 2005 on travel and tourism of China and Thailand get affected. This impact is higher than the natural disaster such as the Tsunami. The dip in case of SARS is much higher than the Tsunami.

The pandemic based economic loss will go only when the pandemic is treated well or the health delivery system is robust. A study conducted by the New York Times that for saving the economy it is important to focus on the flattening of the curve and that is possible only when ex-ante preparedness is done. Dew Harries has analysed the phenomenon and concluded that if proactive preparedness is done the outbreak is bound to happen and then it is difficult to contain. Studies conducted by the ADB, World Bank, UNESCAP and other individual countries show that once it goes out of hand, it will have a huge economic loss. The containment becomes uncertain and so the loss to the economy. Although there are no direct damages (loss to stock happening to the property as it happens in the case of natural disasters but the indirect loss (revenue loss – loss of flow) becomes fatal and will continue until the pandemic is contained (Fig. 11.5).

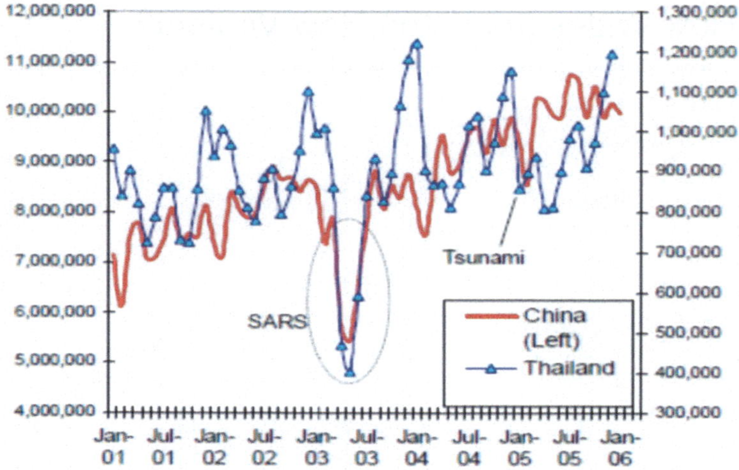

Fig. 11.4 Initial response to SARS: sudden collapses in tourist arrivals. (Source: The World Bank 2016)

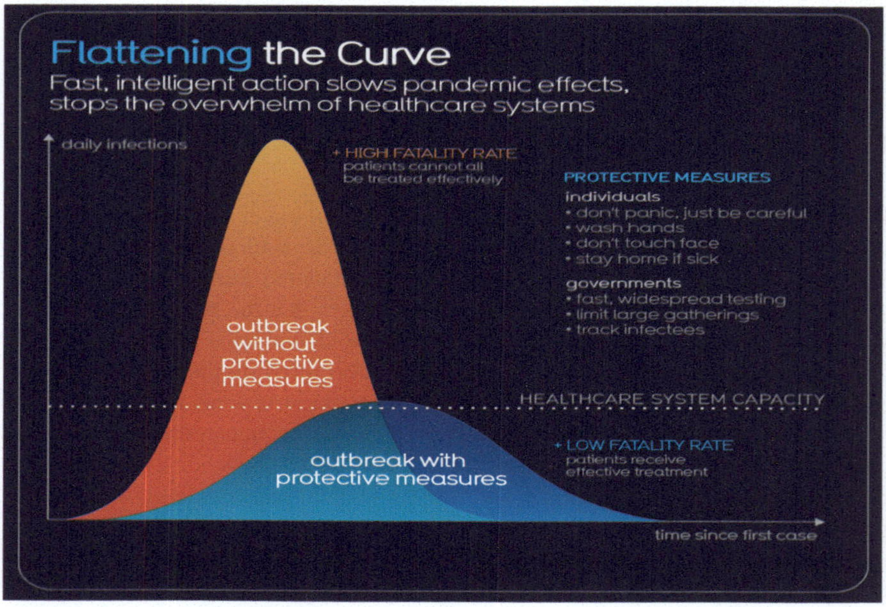

Fig. 11.5 A self-explanatory explanation about the effective management of pandemic. Pre-planning and investment in public health services are the keys of containing the pandemic. (Source: Drew Harris, New York Times)

11.5 Economic Impacts of Other Disasters

Pandemic is one but on the other hand, losses occurring due to other disasters are also huge and due to climate change it is going to be even much sharper. As per the Global Assessment report 2019 on disaster management, "economic losses from disasters such as earthquakes, tsunamis, cyclones and flooding are now reaching an average of US$250 billion to US$300 billion each year. Future losses (expected annual losses) are now estimated at US$314 billion in the built environment alone. This is the amount that countries should set aside each year to cover future disaster loses".

The IMF, 2020 has predicted a cumulative global output loss of around US$ 9 trillion from the COVID-19 crisis and a huge shrink in per capita income for over 170 countries. The global economy was already displaying signs of a downward steeping slope entering a recession. The duration and the intensity of the impact of COVID-19 in different countries would depend on the span of the health crisis in each country, the proactive policy measures put forth by the governments, and their diplomatic, geopolitical and economic relations with three important international commitments for making the World a better place to other countries across the globe. In 2015, global communities have adopted and committed lives. Sustainable Development Goals, Sendai Framework for Disaster Risk Reduction and Paris Agreement on Climate Change Adaptation are fundamentals for the global communities for resilient development.

11.6 Each Disaster Is Adding Colossal Damage to Property and Lives

In the time of Covid-19 India, US and Philippines and Cambodia faced hurricanes. India faced two consecutive typhoons during the lockdown—one on its west coast and another was on the East coast. People and the governance of the country is fighting for COVID and stretched out and at the same point of time they have to fight with the losses that occurred due to other disasters. Cyclone UPPAM of India has been estimated at a loss of US $1000 billion (Govt of West Bengal, Rapid Assessment-memorandum 2020). Every such disaster creates further the challenge of increased poverty, unemployment, distress, trauma slowing down the growth etc. Hence, countries which are exposed to natural disasters will have a double effect of slowing GDP. Their challenge of recoveries would be different than others who are less exposed.

11.7 Growth Projections

As per IMF Report in April, 2020, as a result of the pandemic, the global economy is projected to contract sharply by 3% in 2020, much worse than during the 2008–2009 financial crisis. In a baseline scenario, which assumes that the pandemic fades in the second half of 2020 and containment efforts can be gradually unwound, the global economy is projected to grow by 5.8% in 2021 as economic activity normalizes, helped by policy support. Revising its estimates, IMF on 25th May, 2020 announced that the decline in global growth due to the pandemic would be worse than its first reckoned and indicated a shrink in world's GDP by 4.9% (The Economist, 25th June 2020).

As the WTO Director-General defines the crisis, Covid-19 is first and foremost a health crisis which has forced governments to take unprecedented measures to protect people's lives. The unavoidable declines in trade and output will have painful consequences for households and businesses, on top of the human suffering caused by the disease itself.

Growth projection, quarterly, of the advance economy and emerging market economy have been studied and projected by the IMF study on great lockdown. The projected and actual impacts for both global and emerging market economies are quite big. Recovery of the projected path is not easy (Fig. 11.6) of the IMF. It will again be depending on the flattening and wiping out of the pandemic, type and volume of stimulants offered by each country, global economic policy on trade, transport and other activities. As currently, lockdowns have been relaxed in many countries but still the economy has to be picked up. We still have to wait for another two quarters as along with the supply side, effective demand and type of demand is also to be looked out. This may go under changes as we are seeing anything depending on how the e-businesses are doing better but others are still to pick. Many countries are now struggling with a cash crunch. The monetary and fiscal policy of the countries are to be monitored on a regular basis.

The study conducted by IMF indicates that total global downturn is going to be a negative growth whereas it is showing a recovery in 2021 by almost close to 8.5%. The improvement may be attributed to the combined recovery of advanced economies and emerging market and developing economies. It's an interesting outcome that emerging market and developing economies have the least economic impact of covid-19 than the advanced economies (−6.1%). The economic stimulus allocated for the countries by the multilateral agencies and respective governments (Fig. 11.7).

This pandemic has created uncertainty all over. There cannot be any forecast that how long it is going to continue and keep affecting the growth of the economy. Countries are evolving different economic measures and changing their monetary and fiscal policies to address the economic impact of coronavirus. There is extreme uncertainty around the global growth forecast. IMF report 2020, the economic fallout depends on factors that interact in ways that are hard to predict, including the pathway of the pandemic, the intensity and efficacy of containment efforts, the extent of supply disruptions, the repercussions of the dramatic tightening in global

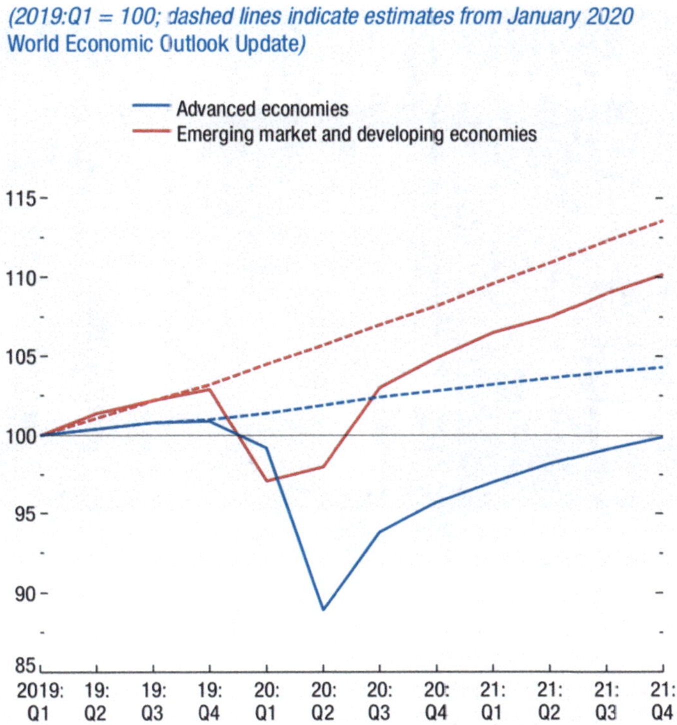

Fig. 11.6 IMF, The great lockdown 2020. (Source: IMF staff estimates)

financial market conditions, shifts in spending patterns, behavioural changes (such as people avoiding shopping malls and public transportation), confidence effects, and volatile commodity prices. Consumer demand patterns would also not be the same as it used to be in pre Covid time. The office is also more banking on the work from home. More digital divide may also become a future challenge.

Many countries face a multi-layered crisis comprising a health shock, domestic economic disruptions, plummeting external demand, capital flow reversals, and a collapse in commodity prices. Risks of a worse outcome predominate (Table 11.1).

In the above analysis, it is clear that the pandemic has put the economy to complete halt. The implications are huge. Countries gradually trying to unlock their economies as the negative economic impact is having a cascading effect of poverty, unemployment, trauma, dispersion, anxiety, violence, trafficking, reverse migration etc. Most of the Asian economies also followed the lockdown, social distancing, personal hygiene care, sanitization, mask etc. for slowing down the spread of the pandemic. They are also spending more in strengthening public health facilities. The capacity of public health services (despite no vaccine is available as yet) are getting enhanced by equipping hospitals, increasing the facilities of the ventilator, PPE, personal hygiene etc. Countries small and medium enterprises are

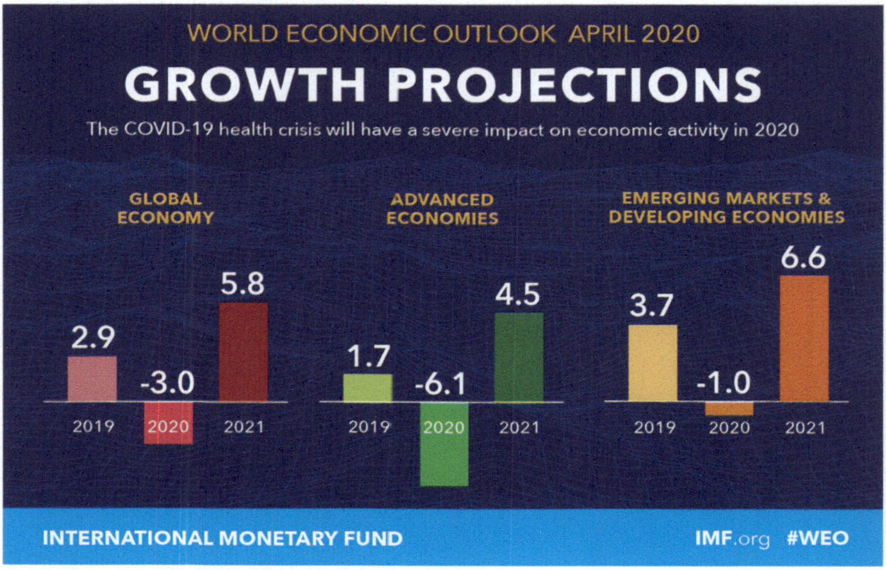

Fig. 11.7 Explaining the growth projection during the recovery from Covid-19. (Source: IMF 2020)

being encouraged to produce PPE, mask, sanitizer etc. Incentives are being provided by the government to quickly enhance capacity for fighting Covid-19. Once it is strengthened, the confidence of the people for participation in the economic activities will increase. Studies are showing revival in 2020, certainly not the V shape recovery but it is indicating towards a better story.

11.8 Mitigation and Economic Stimulants

Various activities are undertaken by the countries but amongst all few became almost universal when it comes to the percentage of implementation. UNESCAP in its study has quoted IF study where analysis has been made for the Asian nations. And with this, an estimation is being drawn of how these would work once it is opened in economic recovery (Fig. 11.8).

Countries after realising the economic losses and social distress started thinking of an alternative of opening the economy and removing border sealing for restoring trade and also travel and tour. In between countries announced different economic stimulants for protecting the economy. The monetary and fiscal policies of the countries got revised differently. Many countries did not opt for the option of adopting economic stimulants. Hence the revival of the regional economy would be different and challenging. Countries adopted their stimulant packages as per their need and priority. Not all countries in the region have announced economic

Table 11.1 The World Economic Forum has outlined the top 10 areas of loss due to COVID-19

(Percent change, unless noted otherwise)

		Projections		Difference from January 2020 WEO update[a]		Difference from October 2019 WEO[a]	
	2019	2020	2021	2020	2021	2020	2021
World Output	**2.9**	**−3.0**	**5.8**	**−6.3**	**2.4**	**−6.4**	**2.2**
Advanced Economies	**1.7**	**−6.1**	**4.5**	**−7.7**	**2.9**	**−7.8**	**2.9**
United States	2.3	−5.9	4.7	−7.9	3.0	−8.0	3.0
Euro Area	1.2	−7.5	4.7	−8.8	3.3	−8.9	3.3
Germany	0.6	−7.0	5.2	−8.1	3.8	−8.2	3.8
France	1.3	−7.2	4.5	−8.5	3.2	−8.5	3.2
Italy	0.3	−9.1	4.8	−9.6	4.1	−9.6	4.0
Spain	2.0	−8.0	4.3	−9.6	2.7	−9.8	2.6
Japan	0.7	−5.2	3.0	−5.9	2.5	−5.7	2.5
United Kingdom	1.4	−6.5	4.0	−7.9	2.5	−7.9	2.5
Canada	1.6	−6.2	4.2	−8.0	2.4	−8.0	2.4
Other Advanced Ecoonomies[b]	1.7	−4.6	4.5	−6.5	2.1	−6.6	2.2
Emerging Market and Developing Economies	**3.7**	**−1.0**	**6.6**	**−5.4**	**2.0**	**−5.6**	**1.8**
Emerging and Developing Asia	5.5	1.0	8.5	−4.8	2.6	−5.0	2.3
China	6.1	1.2	9.2	−4.8	3.4	−4.6	3.3
India[c]	4.2	1.9	7.4	−3.9	0.9	−5.1	0.0
ASEAN-5[d]	4.8	−0.6	7.8	−5.4	2.7	−5.5	2.6
Emerging and Developing Europe	2.1	−5.2	4.2	−7.8	1.7	−7.7	1.7
Russia	1.3	−5.5	3.5	−7.4	1.5	−7.4	1.5
Latin America and the Caribbean	0.1	−5.2	3.4	−6.8	1.1	−7.0	1.0
Brazil	1.1	−5.3	2.9	−7.5	0.6	−7.3	0.5
Mexico	−0.1	−6.6	3.0	−7.6	1.4	−7.9	1.1
Middle East and Central Asia	1.2	−2.8	4.0	−5.6	0.8	−5.7	0.8
Saudi Arabia	0.3	−2.3	2.9	−4.2	0.7	−4.5	0.7
Sub-Saharan Africa	3.1	−1.6	4.1	−5.1	0.6	−5.2	0.4
Nigeria	2.2	−3.4	2.4	−5.9	−0.1	−5.9	−0.1
South Africa	0.2	−5.8	4.0	−6.6	3.0	−6.9	2.6
Memorandum							
European Union[e]	1.7	−7.1	4.8	−8.7	3.1	−8.8	3.1
Low-Income Developing Countries	5.1	0.4	5.6	−4.7	0.5	−4.7	0.4
Middle East and North Africa	0.3	−3.3	4.2	−5.9	1.2	−6.0	1.2
World Growth Based on Market Exchange Rates	2.4	−4.2	5.4	−6.9	2.6	−6.9	2.6
World Trade Volume (goods and services)	**0.9**	**−11.0**	**8.4**	**−13.9**	**4.7**	**−14.2**	**4.6**
Imports							
Advanced Economies	1.5	−11.5	7.5	−13.8	4.3	−14.2	4.2

(continued)

Table 11.1 (continued)

(Percent change, unless noted otherwise)

		Projections		Difference from January 2020 WEO update[a]		Difference from October 2019 WEO[a]	
	2019	2020	2021	2020	2021	2020	2021
Emerging Market and Developing Economies	−0.8	−8.2	9.1	−12.5	4.0	−12.5	4.0
Exports							
Advanced Economies	1.2	−12.8	7.4	−14.9	4.4	−15.3	4.3
Emerging Market and Developing Economies	0.3	−9.6	11.0	−13.7	6.8	−13.7	6.6
Commodity Prices (US dollars)							
Oil[f]	−10.2	−42.0	6.3	−37.7	11.0	−35.8	10.9
Nonfuel (average based on world commodity import weights)	0.8	−1.1	−0.6	−2.8	−1.2	−2.8	−1.9
Consumer Prices							
Advanced Economies	1.4	0.5	1.5	−1.2	−0.4	−1.3	−0.3
Emerging Market and Developing Economies[g]	5.0	4.6	4.5	0.0	0.0	−0.2	0.0
London Interbank Offered Rates (percent)							
On US Dollar Deposits (6 month)	2.3	0.7	0.6	−1.2	−1.3	−1.3	−1.5
On Euro Deposits (3 month)	−0.4	−0.4	−0.4	0.0	0.0	0.2	0.2
On Japanese Yen Deposits (6 month)	0.0	−0.1	−0.1	0.0	−0.1	0.0	0.1

Source: IMF Staff

Note: Real effective exchange rates are assumed to remain constant at the levels prevailing during February 17–March 16, 2020. Economies are listed on the basis of economic size. The aggregated quarterly data are seasonally adjusted. *WEO World Economic Outlook*

Source: IMF (2020)

[a]Difference based on rounded figures for the current, January 2020 *WEO Update*, and October 2019 WEO forecasts

[b]Excludes the Group of Seven (Canada, France, Germany, Italy, Japan, United Kingdom, United States) and euro area countries

[c]For India, data and forecasts are presented on a fiscal year basis, and GDP from 2011 onward is based on GDP at marked prices with fiscal year 2011/2012 as a base year

stimulants. But the majority have done. Multilateral agencies are also allocating resources for supporting countries in designing the economic packages for reviving the economy. In the globalized world each country is connected to another. Trade and travel is one which is significantly linking economy and the people. In the lockdown, the maximum brunt came on supply chain disruptions. Trade and travel has to quickly revive for achieving a high rate of recovery (Fig. 11.9).

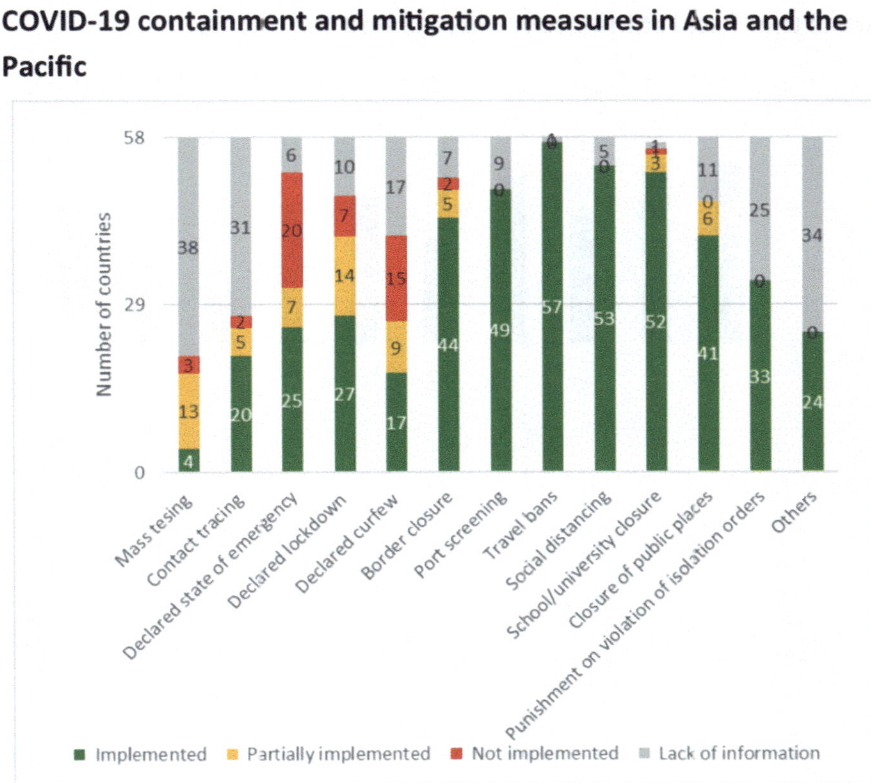

Fig. 11.8 Steps taken for mitigation of Covid UNESCAP report 2020. (Source: ESCAP, based on IMF Policy Responses to COVID-19, ILO COVID-19 Country Policy Responses, Oxford COVID-19 Government Response Tracker, and various national sources.) *Note:* The figure is based on information available up to 1 May 2020. Implemented = Implemented nationwide; Partially implemented = implemented in selected locations

11.9 India: A Case Study

India has been under a severe attack of coronavirus pandemic since January 2020. As a country, several steps have been taken to be in terms of screening the immigrants, foreigners at the airport or door to door in certain areas, social distancing, complete lockdown, etc. Almost all the states are reeling under the impact of this virus. The country as a whole including citizens are fighting this virus as one unit following the directions being provided from time to time by Government agencies and also Doctors. This has not only brought the country to a standstill but also taken the economy to a back seat. Livelihood, corporate sector, industries etc. have been badly hit.

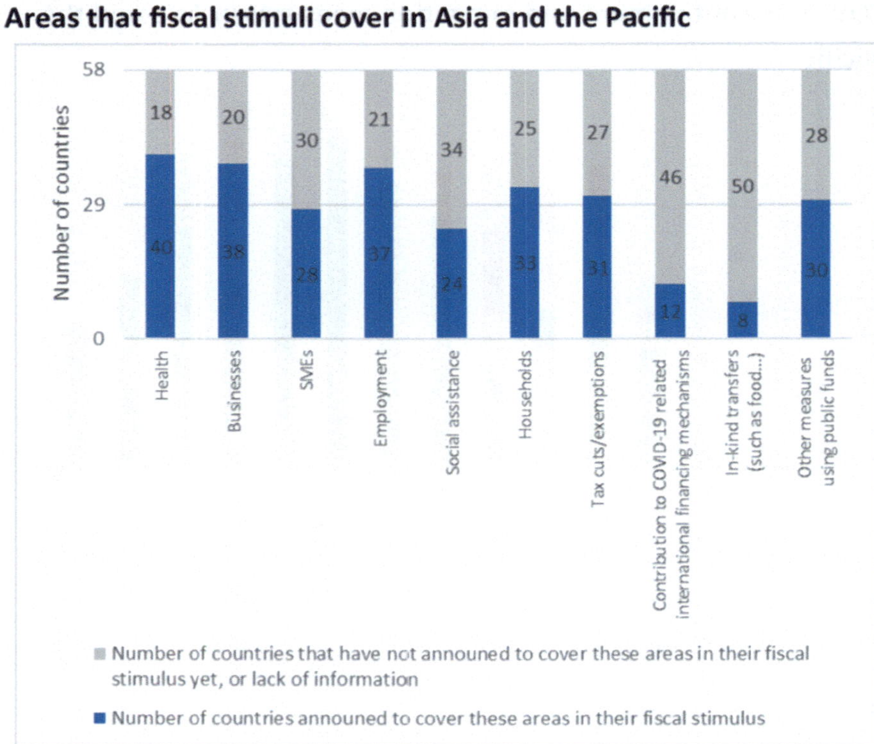

Fig. 11.9 Sector wise Fiscal stimulants, UNESCAP 2020. (Source: ESCAP, based on information available up to 1 May 2020 from IMF Policy Responses to COVID-19, ILO COVID-19 Country Policy Responses, OECD Country Policy Tracker, Oxford COVID-19 Government Response Tracker, and various national sources)

India as a whole has about 61 lakhs of the urban migrated population for work/employment out of which 32.51% have gone beyond the boundaries of their own state/UT. Out of this 61 lakhs population, 6 lakhs is illiterate population and the percentage of this population which has migrated outside the state/UT is about 32.27%. Similarly, when we talk about the rural population the data comes out to be 1 crore and 11 lakhs approximately as total out of which 34.78% has migrated outside and 37.07% of the illiterate population has migrated outside the border.

11.9.1 Impact on Indian Economy and Stimulants

The Indian economy is likely to lose over ₹ 32,000 crore (US$4.5 billion) every day during a lockdown. Up to 53% of businesses in the country will be significantly affected. Supply chains have been disrupted Those in the informal sectors and daily

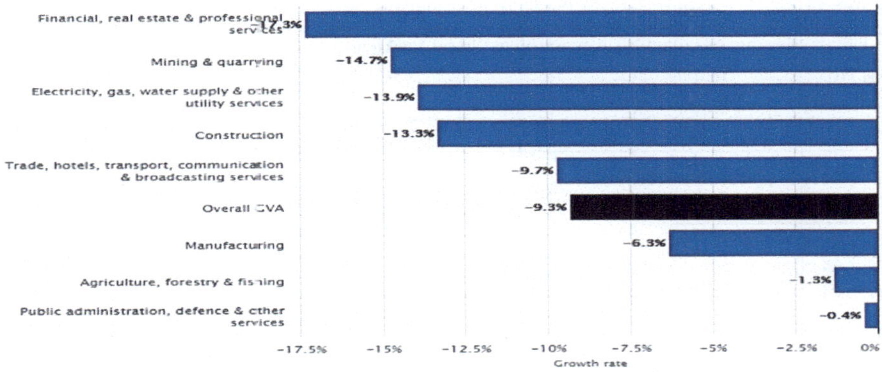

Fig. 11.10 Economic impact of Covid-19 on India. (Source: MOSPI (2020))

wage groups and living below the poverty line are vulnerable. A large number of farmers around the country are also facing uncertainty about the selling of their products. Tourism, small traders, retailers, transport (OLA/Uber) E-rickshaw, construction workers, restaurants, hotels etc. are going to be deeply affected. The analysis is done by the GVA (Fig. 11.10) on how the Indian economy is getting impacted by the pandemic COVID-19.

11.9.2 Estimated Economic Impact from COVID-19 in India from period April–June 2020

The Government of India has announced a variety of proactive measures to tackle the situation. On 27th March 2020, the Reserve Bank of India also announced a number of measures which would make available ₹ 374,000 crore (US$52 billion) to the country's financial system. On 29th March, with partial opening, the government allowed the movement of all essential as well as non-essential goods. The World Bank, on 1st April, has approved US $1 billion in support to India to tackle the coronavirus pandemic. Further to this, on 3rd April, the central government released more funds to the states for tackling the coronavirus totalling to ₹ 28,379 crore (US $4.0 billion). One day salary for the month of March and April of all the Govt employees have been cut for the support to the newly created Prime Minister Care Fund. Industrial houses and others, all supported this initiative by strengthening the fund. Prime Minister also took initiative for supporting SAARC nations by creating SAARC fund for fighting COVID-19. Govt of India also took the lead for fighting COVID-19 in a collaborative approach with all the affected nations. Accordingly, the initiative for the G-20 meeting was called. Govt of India also used Insurance for covering pandemic under the old health Insurance scheme for the treatment of

Table 11.2 Allocation of 10% of the GDP as stimulants, MoF, GoI

Overall stimulus provided by Atmanirbhar Bharat Package			
S. No.	Item		(Rs. Cr.)
1	Part 1		5,94,550
2	Part 2		3,10,000
3	Part 3		1,50,000
4	Parts 4 and 5		48,100
		Sub-total	11,02,650
5	Earlier measures incl. PMGKP	(earlier slide)	1,92,800
6	RBI measures (actual)		8,01,603
		Sub-total	9,94,403
		GRAND TOTAL	20,97,053

Covid-19. This was implemented after the direction issued by the central government. This became very handy for the victims.

Now there is a shift from Jaan hai toh Jahaan hai to Jaan bhi aur Jahan Bhi. If we need life, we need an economy too. Hon'ble PM has announced while extending the lockdown I to Lockdown 2.0 with a promise to open the economy for infusing growth. RBI has already eased monetary policy. Now, it is up to the central and state governments to put their levers of fiscal policy in the highest gears. Public spending will revive the economic activities, create demand for goods and services and put the money into the hands of working people resulting in increased aggregate demand. Public spending on the consumption of goods or services, subsidies, and welfare schemes are also capable of raising the aggregate demand. But this route is susceptible to inflation resulting from supply constraint and higher liquidity in the economy. However, if public spending is done on investment goods and services, it will create long term capacity to have a healthier Indian economy. Monetary and Fiscal measures of the government with more focus on small and medium-term enterprises may lead to supply chain management. It is also expected that trade would be opened and the supply chain would be restored. All speculation will be determined when the COVID-19 is taken care of well by enhanced spending on the public health services. India's total stimulant package is equivalent to the 10% the GDP which is the highest ever allocation for post-disaster recovery with the slogan of Atma Nirbhar Bharat (Table 11.2). While Unlocking the economy, the table below explains about the five stages stimulants announced by the Prime Minister for the recovery. This has given hope that in the year 2021 India may do better. Details of allocation are given in Annexure.

As per the damage assessment report, India, also recently got hit by two severe hurricanes which led to the total loss of Rs. 1.5 lac crore as estimated by the respective affected states governments. There is an additional funding requirement for recovery. Hence, during the Covid-19 planning for disaster management cannot be ignored and vice versa.

11.10 The Key Recommendations

11.10.1 Integrated Framework of DRR with Pandemic

Currently, the situation is not so simple. Countries are in complex situations. For V shape recovery on a sustainable basis, something different has to be innovated. Global community facing the problem of resilient development, Climate change and now pandemic. The new normal cannot be handled by business as usual. It has to be flexible and ready to accommodate change fast. A three-dimensional development model would work. Global communities are striving for achieving sustainable development goals, climate change and Paris agreement and Disaster Risk Reduction. The pandemic challenges, repeated after a hundred years, have reinforced the need for strengthening it. The framework now should include all the conventions together (Fig. 11.11).

This could be a possible framework which is recommended for the recovery of the economy with building resilience. To achieve the integrated framework would be important for considering development planning. People's risk profile has changed entirely now. In 2015 global communities have signed an agreement for achieving all the targets and goals of SDG, CCA and DRR by 2030. But in the time of pandemic especially its impact on the economy have pushed back and delayed the time and cost of achieving the costs. UNESCAP has also conducted the study and suggestions for the additional allocation of resources. The Framework is explained further below.

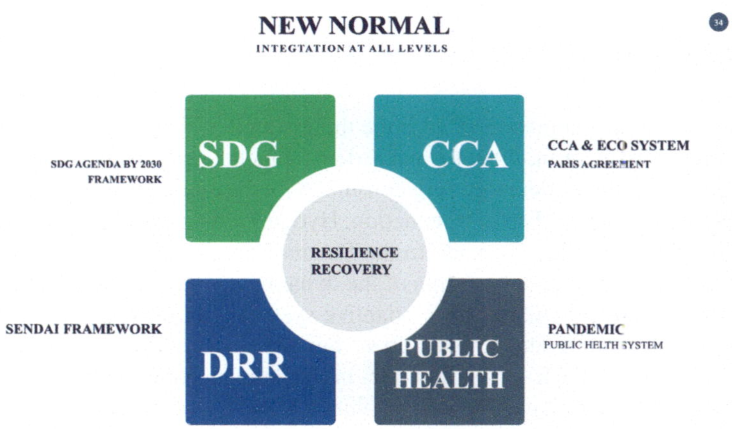

Fig. 11.11 An integrated framework for SDG, CCA, DRR and Public Health

11.10.2 Sustainable Development Goals (SDG)

The sustainable development goals have been adopted by the UN and the member countries after the Millennium Development Goals 2000–2015. The sustainable development goal is the commitment of the government for fighting against poverty, development, inclusion, health, education and many more. It was adopted in the year 2015 by the global community. As per the UN, "the 17 Goals are all interconnected, and to leave no one behind, we must achieve them all by 2030" (Fig. 11.12).

11.10.3 Climate Change Adaptation (CCA)

(UNFCCC 2015), describes it, "The Paris Agreement central aim is to strengthen the global response to the threat of climate change by keeping a global temperature rise this century well below 2 °C above pre-industrial levels and to pursue efforts to limit the temperature increase even further to 1.5 °C. Additionally, the agreement aims to strengthen the ability of countries to deal with the impacts of climate change. To reach these ambitious goals, appropriate financial flows, a new technology framework and an enhanced capacity-building framework will be put in place, thus supporting action by developing countries and the most vulnerable countries, in line with their own national objectives."

11.10.4 Disaster Risk Reduction (DRR)

Risk is not a static phenomenon. This is the thing which is changing every day. The global community is struggling to make development resilient as there is huge development investment that is getting consumed by disasters. Yokohama Strategy, an International decade of Disaster reduction, Hyogo Framework of Action and now in the year 2015 commitment made to the Sandia Framework for Risk Reduction 2015–2030 "with a new understanding of the dynamic nature of systemic risks, new structures to govern risk in complex, adaptive systems and develop new tools for risk-informed decision-making that allows human societies to live in and with uncertainty. The UNDRR 2019 and the institutions dealing with disaster risk management all have to have integrated solutions. Pandemic is nowhere chosen as the rule of business of disaster management. It goes separately with the Ministry of health where the scale and its management needed regular support from the disaster management. India imposed national lockdown by sighting the provision of the national disaster management act 2005." (Fig. 11.13).

Fig. 11.12 Sustainable Development Goals, UNDP

Fig. 11.13 Innovative Disaster Risk Reduction, Global Assessment Report 2019. (Source: UNDRR 2019)

11.10.5 Public Health

Integration and Revamping Public Health: After the outbreak of a pandemic, the world order has changed drastically, The living style, consumer demands, supply chain etc. are being defined as New Normal. It has created enormous economic and cascading losses which has not ended even today. Each economy is grappling with its reputational risk. Low public health capacity will remain a challenge for the emerging market economies. Hence Asian and south Asian nations should plan a long-term investment in building the capacity of public health. The current expenditure is not enough for fighting against odds and striving for resilient sustainable development. ESCAP report also enforcing and recommending countries to make more investment in the public health system.

11.10.6 Alternative to Lockdown

Lockdown was applied in 1918 Spanish Flu almost 100 years ago but that time the economy was not connected with each other. Each country had their own economic zone and they used it to maximise their economic output accordingly. In the recent time of globalised economy lockdown turned out to be the most expensive solution. However, it has contributed immensely in the reduction of death, spreading out of the disease and flattening of the curve. Business continuity plans should go as mandatory with the inclusion of pandemic. But still countries have to sit together and find out best solutions for the future as an alternative to lockdown. As long as it is not there, the country will have no choice but to impose lockdown.

11.10.6.1 Economic Stimulants

Other than the supply side, we also need to focus on the demand side. Demand is understood as people will make rational choices. The definition of rational choice to be re looked in the light of Covid-19. The economic stimulants have to be designed and packaged for increasing domestic demand. Many countries might have fallen into the trap of cash crunch. Hence, how the cash is being planned whether by reducing the interest rate-cheap money policy, designing incentive, moratorium on EMI payment of the Bank etc. Monetary policy may also be monitored strictly and there is a need it may be altered (making flexible). But at the same time efforts for strengthening health infrastructure should go hand in hand. Govt

11.10.6.2 Formation of Global Consortium

In the close globalised economy, high interdependence on each other cannot be ignored. For fast revival countries collectively have to think to draft a common policy and stimulants for the restoration of trade and travel. A global consortium could be constituted for wiping out the pandemic from everywhere, promoting research and development for finding vaccines and speeding up of the protocol for its approval for the launch in the market and economic stimulus be supported to the countries which cannot afford the provisions of stimulus due the losses occurred . Global consortium may be facilitated by the regional intergovernmental bodies and multilateral agencies and United nations. This consortium may also help the countries who are willing to implement an integrated framework for building resilience.

11.10.7 Enhanced Allocations to Public Health

Most of the European countries have a robust health system. Also, south-east Asian nations—Japan, Singapore, Korea, Vietnam, Cambodia, Singapore recovery would be better in those countries. The public health system and pandemic management have a symbiotic relationship with each other. The robust public health system will determine the shape of the recovery. Economies are uncertain about the shape of recovery but it is being anticipated and projected that South Asian nations would recover strongly.

11.11 Disaster Risk Financing, Insurance And Risk Transfer

Understood and addressed. It is clearly not the optimal time to initiate the process while the fire is burning, the ground shaking or the water rising and area is inundated. Through funding and expertise, countries could develop and implement tailored financial protection strategies that increase the ability of governments, homeowners, businesses, agricultural producers and low-income populations to respond more quickly and resiliently to disasters. The focus shifts clearly on developing ex-ante financial solutions that can effectively redress the shock for the affected people and seamlessly take over the burden of disaster from the government. Disaster risk financing through insurance has been a time-honoured mechanism that has helped countries to ensure that their people are financially protected in the event of a disaster (Fig. 11.14).

11.12 Conclusion

Covid-19 pandemic has jolted most of the countries of the World. Starting from China to Europe, America, West Asia, Asia, South East Asia, South Asia etc. It has been a devastating experience for everyone. This has exposed the human's capacity to fight against the virus. Still, a hundred years old methodology of lockdown,

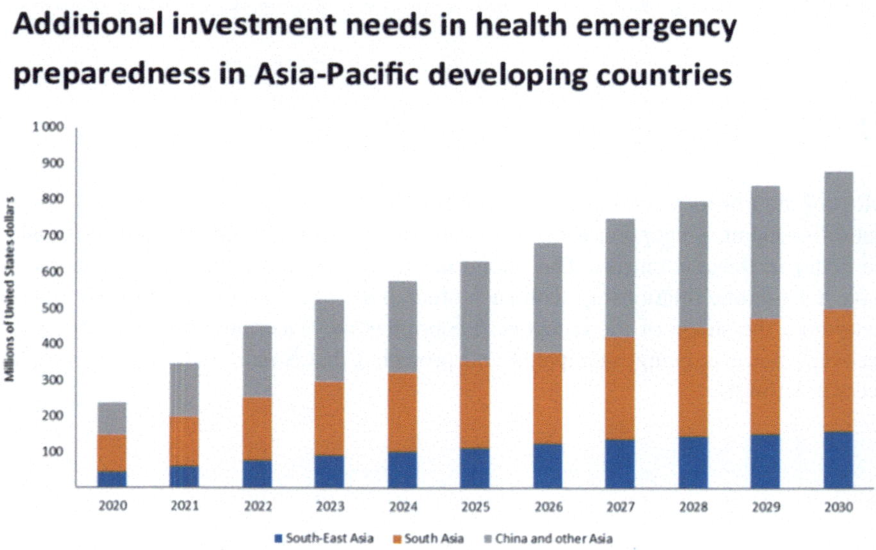

Fig. 11.14 Allocation of health stimulants by the Asia-pacific nations, UNESCAP 2020. (Source: ESCAP estimates, based on WHO SDG Health Price Tag)

personal hygiene, social distancing are being adopted by the affected countries. Countries are losing trillions of dollars. The economy has already taken a big downturn. Countries are desperate for a recovery. But this recovery is not simple unless they start parallel strengthening their health services. The key revival also has to be sought in the same way. We need to revisit lockdown as a tool for public health crisis management for the future. Other than the pandemic, Climate-induced disasters and other disasters (natural/man-made) keep hiring the economy and hence an isolated approach will not work. With such disruptions, achieving the goals and targets of Sendai = Framework of Disaster Risk reduction along with Sustainable Resilient development will remain as pipes dream. An integrated framework—DRR, CCA, SDG and Pandemic is the need of the hour with ex-ante strengthening capacity of the nations. For Pandemic management, a global consortium could be established for solving it together. COVID-19 has been the global problem and hence it has to be handled globally too with expertise, research, resources, co-operation etc. For restoring the economy COVID-19 must be wiped out completely from the globe with each other's support—local, national, sub-regional, regional and global. Story getting repeated will take no time. We have seen how fast it travelled across the globe from the Wuhan city of China. If the trace of the Covid-19 remains (if there is no vaccine) it may burst again. This has to be kept in mind while moving ahead for building resilience. There is a need to be watchful on increasing, poverty, unemployment , gender based violence, trafficking and other social distress while focusing on economic recovery. The issues related to new normal has to be dealt with new strategies and innovations. Business as usual has to be revisited and redrafted for addressing new challenges.

Annexure

Stimulus provided by announcements in Part-1

S. No.	Item	(Rs. Cr.)
1.	Emergency W/C Facility for Businesses, incl MSMEs	3,00,000
2.	Subordinate Debt for Stressed MSMEs	20,000
3.	Fund of Funds for MSME	50,000
4.	EPF Support for Business & Workers	2800
5.	Reduction in EPF rates	6750
6.	Special liquidity Scheme for NBFC/HFC/MFIs	30,000
7.	Partial credit guarantee Scheme 2.0 for Liabilities of NBFCs/MFIs	45,000
8.	Liquidity Injection for DISCOMs	90,000
9.	Reduction in TDS/TCS rates	50,000
Sub total		5,94,550

Stimulus provided by announcements in Part-2

S. No.	Item	(Rs. Cr.)
1.	Free Food grain Supply to Migrant Workers for 2 months	3500
2.	Interest Subvention for MUDRA Shishu Loans	1500
3	Special Credit Facility to Street Vendors	5000
4	Housing CLSS-MIG	70,000
5	Additional Emergency Working Capital through NABARD	30,000
6	Additional credit through KCC	2,00,000
Sub-total		3,10,000

Stimulus provided by announcements in Part-3

S. No.	Item	(Rs. Cr.)
1.	Food Micro enterprises	10,000
2.	Pradhan Mantri Matsya Sampada Yojana	20,000
3.	TOP to TOTAL: Operation Greens	500
4.	Agri Infrastructure Fund	1,00,000
5.	Animal Husbandry Infrastructure Development Fund	15,000
6	Promotion of Herbal Cultivation	4000
7	Beekeeping Initiative	500
Sub-total		1,50,000

Stimulus provided by unnouncements in Part-4 and Part-5

S. No.	Item	(Rs. Cr.)
1	Viability Gap Funding	8,100
2	Additional MGNREGS allocation	40,000
Sub-total		48,100

References

Huang Z, Saxena SC (2020) Combating covid-19, Asia and pacific. UNESCAP, Bangkok. https://www.unescap.org/resources/policy-responses-covid-19-combating-covid-19-asia-and-pacific-measures-lessons-and-way

IMF (2020) World Economic Outlook—the great lockdown. IMF, Washington, DC

Jonas OB (2013) World development report, pandemic risk. World Bank, Washington, DC

Kose MA, Ohnsorge F (2019) A decade after the global recession: lesson & challenges for emergencies & developing economics

MOSPI (2020) Economist Intelligence Unit: Observer Research India. Economist Intelligence Unit: MOSPI (NSO) Foundation

National DM Act (2005) Govt of India

The Global Assessment Report (2019) GIDRR

UNDRR (2019) Global assessment report on disaster risk reduction. United Nations Office for Disaster Risk Reduction (UNDRR), Geneva

UNFCCC (2015) Paris Agreement

Chapter 12
Impact of COVID-19 on Agro-Food Industry and Transitions Towards Food Security

Vikas Poonia, Manish Kumar Goyal, Chandra A. Madramootoo, Anil Kumar Gupta, and Chitresh Saraswat

Abstract Global pandemics, epidemics, and disease outbreaks have plagued humanity for ages. However, the scale and spread of pandemics and epidemics increased drastically in recent history. Currently, COVID-19 pandemic (respiratory illness) caused by SARS-CoV-2 virus ravaging fright around the world. The infectious nature of the disease affected the global and local economies and societies. Nations are forced to implement precautionary measure such as restriction of mobility to stop the spread of disease, which vastly affecting the major economic sectors including food and agricultural industry. The precautionary measures such as travel restrictions disrupt the food production, distribution, and supply chain network. The impact is widely seen on livestock and aquaculture farming, which threatens the food security and calls for immediate policy interventions from the government. This chapter investigates the challenges posed by COVID-19 pandemic

V. Poonia (✉)
Discipline of Civil Engineering, Indian Institute of Technology, Indore, India

M. K. Goyal
Civil Engineering, Indian Institute of Technology, Indore, India
e-mail: mkgoyal@iiti.ac.in

C. A. Madramootoo
McGill University, Quebec, Canada

A. K. Gupta
Environment and Disaster Risk Management and Centre for Excellence on Climate Resilience, National Institute of Disaster, New Delhi, Delhi, India

C. Saraswat
The Fenner School of Environment & Society, Australian National University, Canberra, ACT, Australia

M. K. Goyal, A. K. Gupta (eds.), *Integrated Risk of Pandemic: Covid-19 Impacts, Resilience and Recommendations*, Disaster Resilience and Green Growth,
https://doi.org/10.1007/978-981-15-7679-9_12

on agricultural sector and food industry in India and determine the possible mitigation measures. The study presented the recommendation to call for government support in strengthen policies to boost agricultural sector activities to achieve transitions towards food security during and post-pandemic period.

Keywords Agriculture · Coronavirus · COVID-19 · Food security · Harvest · Travel restrictions · Mitigation · Pandemic · Social distancing

12.1 Introduction

Global pandemics and disease outbreaks have plagued humanity for ages. However, the scale and spread of pandemics and epidemics increased drastically in recent history, as there are several incidents of different scale of pandemics and epidemics such as Spanish Flu, HIV/AIDS, MERS etc. (Table 12.1). Each of the pandemics or epidemics has calamitous impacts on the economic growth and human development across the world (Hanashima and Tomobe 2012). Currently, the world is witnessing the COVID-19 pandemic caused by a novel coronavirus (SARS-Cov-2) (Wang et al. 2020), which is different from MERS-CoV and SARS-CoV. COVID-19 is respiratory illness caused by SARS-CoV-2 (new form of coronavirus) and is infectious disease that has affected human activities and social development on global scale. According to World Health Organization (WHO), the first confirmed case of new coronavirus was reported in Wuhan, China (the epicenter of COVID-19 outbreak) on December 31, 2019 (WHO 2020a). After the first few reported cases, the COVID-19 infection spreads around the world at an alarming rate due to the novelty of the virus. This is turning out to be one of the biggest global health crises in recent years. On March 11, 2020, the WHO declared the disease as a global pandemic (Cucinotta and Vanelli 2020). Figure 12.1 presents the globally confirmed cases and deaths due to the corona pandemic (data used till June 25, 2020). The organization of Economic Co-operation and Development (OCED) predicts that annual global GDP growth is projected to drop to 2.4% in 2020 and has warned that it may reach to 1.5% due to prolonging of the pandemic (OECD 2020). The International Labour Organisation

Table 12.1 Major global pandemics and epidemics during the twentieth and twenty-first century

Pandemic	Virus	Time period	Reference
Spanish flu	H1N1 strain of influenza A	1918–1919	Johnson and Mueller (2002)
Asian flu	H2N2 strain of influenza A	1957–1958	CDC (2020a)
Hong Kong flu	H3N2 strain of influenza A	1968–1970	CDC (2020b)
HIV/AIDS	Human immunodeficiency virus	1981-till now	WHO (2020b)
SARS	Corona virus (SARS-CoV)	2002–2003	Cherry and Krogstad (2004)
Swine flu	H1N1 strain of influenza A	2009-2010	Coburn et al. (2009)
MERS	Corona virus (MERS-CoV)	2015-till now	CDC (2015)
COVID-19	SARS-CoV-2	2019-till now	WHO (2020c)

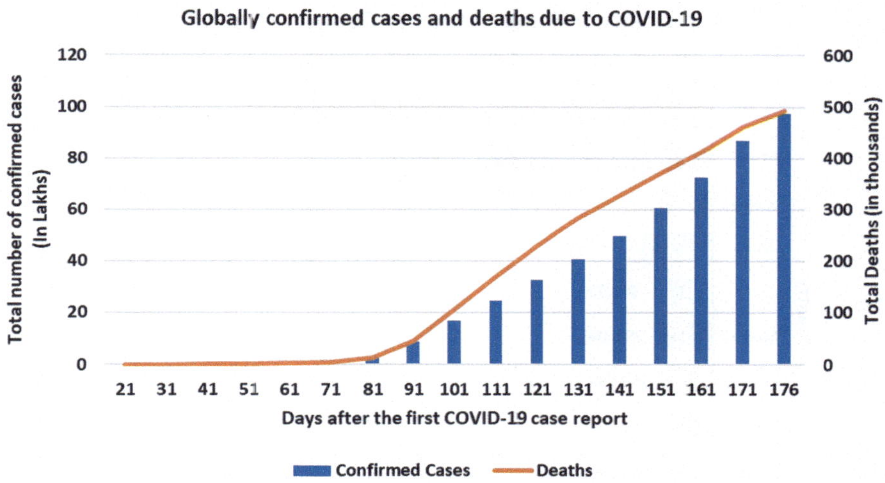

Fig. 12.1 Globally total COVID-19 infection cases and deaths (data used up to June 25, 2020). (Source: WHO (2020c))

stated that the COVID pandemic is the biggest crisis in terms of global unemployment since the Second World War (ILO 2020a). Many experts have already called COVID-19 pandemic a "black swan" for the world economy due to the lasting and substantial impact on the economy. Here, we will argue that due to frequent occurrence of pandemics, epidemics and outbreak in recent years (Table 12.1) provides us chance to predict the global scale distortion in economy and preparedness. So, the question arises how prepared are we?

To understand the preparedness, we explored developing country as a case study. India confronted the first confirmed case of the COVID-19 on January 30, 2020. After that, the number of cases has intensified significantly. In the initial phase, India recorded fewer cases compared to other developed countries like the United States, Brazil, Italy, Spain, etc To contain the spread of coronavirus, the government of India declared a three-week nationwide travel restriction from March 25 to April 14, 2020, which was then extended till May 3, 2020 and then further extended to May 31, 2020 to achieve satisfactory control of the coronavirus. All non-essential facilities such as educational institutions, religious places, government offices, transportation, and retail establishments across the country were closed. Currently, the WHO reported 9.73 million confirmed cases and 492,390deaths globally (data used up to June 25, 2020). Figure 12.2 represents the top ten country-wise total confirmed cases across the world. However, in the case of India, the total confirmed cases reached 4,75,099 and reported 15,301 deaths by June 25, 2020. Figure 12.3 presents the state-wise total number of confirmed cases in India. Ray et al. (2020) have claimed that by July, the number of coronavirus cases will rise beyond five hundred thousand in India. She based her facts on her study of the 5 days that is May

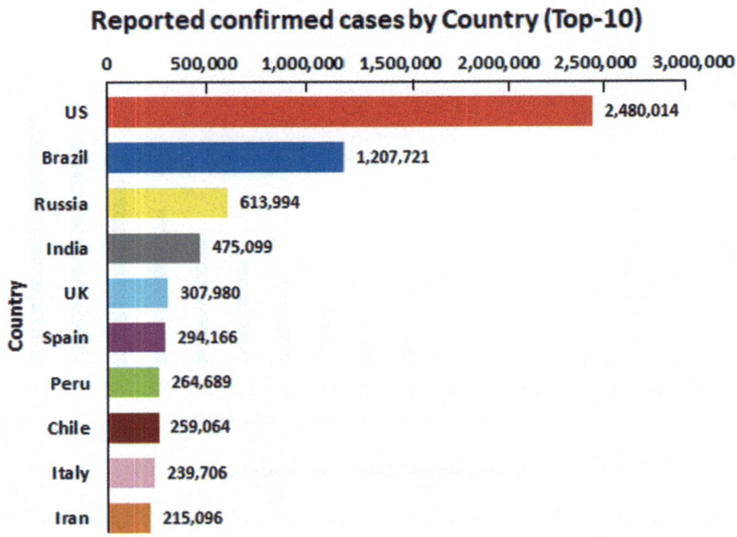

Fig. 12.2 Country-wise confirmed cases across the world (Top 10 countries: Data used up to June 25, 2020). (Data Source: WHO (2020c))

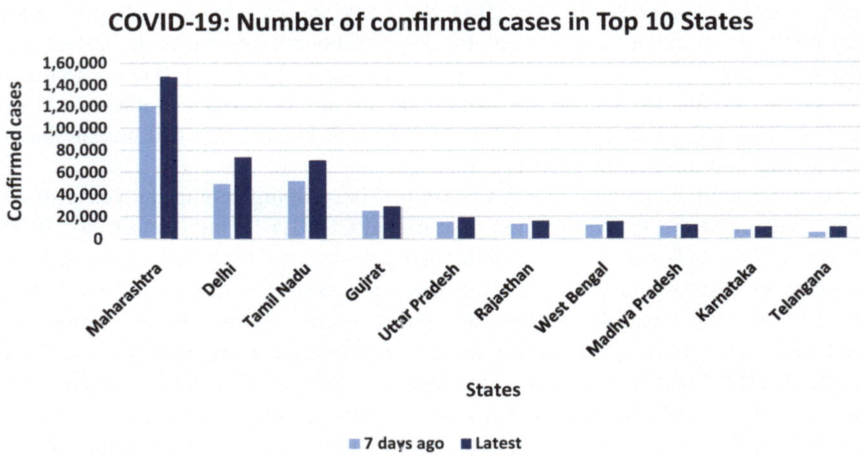

Fig. 12.3 State-wise confirmed cases in India (Top 10 states: Data used up to June 25, 2020). (Data Source: (Ministry of Health and Family Welfare))

20 to May 25, 2020. Moreover, she also said that the number would cross 2.1 million in the worsening situation.

From past pandemics, it has been found that panic and quarantine regulations significantly impacted the economic growth and human development (Arndt and Lewis 2001; Bermejo 2004). However, it also affects agricultural activities due to decreased availability of agricultural industry workers, decreased demands for agricultural exports, impacts of longer lead times in supply, limited transportation and logistics services, decreased supply capacity, stringent market protocols, reduced average annual precipitation (Goyal and OJha (2012); Goyal and Ojha (2014); Sharma and Goyal (2018)) and globally the increase biosecurity regulations. Burgui (2020) and Sar et al. (2010) found that an outbreak of contagious diseases will result in malnutrition and hunger due to significant impact on agricultural related activities.

Worldwide, agriculture is one of the crucial sectors in human growth and impact on food security of region and nations (Kogo et al. 2020; Lopez-Ridaura et al. 2019; Das et al. (2020)). FAO has projected that nearly 60% of the global population depends on the agriculture and food industry for their existence (Zavatta 2014). ILO claimed that out of the total employer population worldwide in 2019, 26.85% of were employed in the agricultural sector (WorldBank 2019). It is one of the largest sectors and backbone of the India's economy. The three major reasons to call the agriculture industry as the backbone of the Indian economy, first it plays a vital role in the country's gross domestic product (GDP); second, it helps to employ a major chunk of country's informal workers; and finally, the development of other industries depends on the agricultural industry. During the COVID-19 pandemic, it is important to examine the impact of pandemic on agriculture and related activities. It is vital to explore that how agricultural sectors will respond to the situations arise from pandemic on global and national scale. Also, government decision making to improve the condition of agriculture related sectors will affect the future. This chapter investigates the challenges posed by COVID-19 pandemic on agricultural sector in India and determine the possible mitigation measures for recovery. The presented recommendation in the chapter will strengthen policies responses to boost the agricultural sector activities to transitions towards food security during and the preparedness in post-pandemic period.

12.2 Impact of COVID-19 on Agricultural Sector

COVID-19 precautionary measures adopted by the Indian government has severely impacted the supply chain networks of agricultural sector, disrupted the value chain between agricultural producer and consumer. The FAO (2020a) claimed that the COVID-19 pandemic has severely disrupted the agricultural sector by impacting on food demand and supply during the period. These negative impacts of pandemic might lead to food insecurity (Siche 2020) in developing economies such as India.

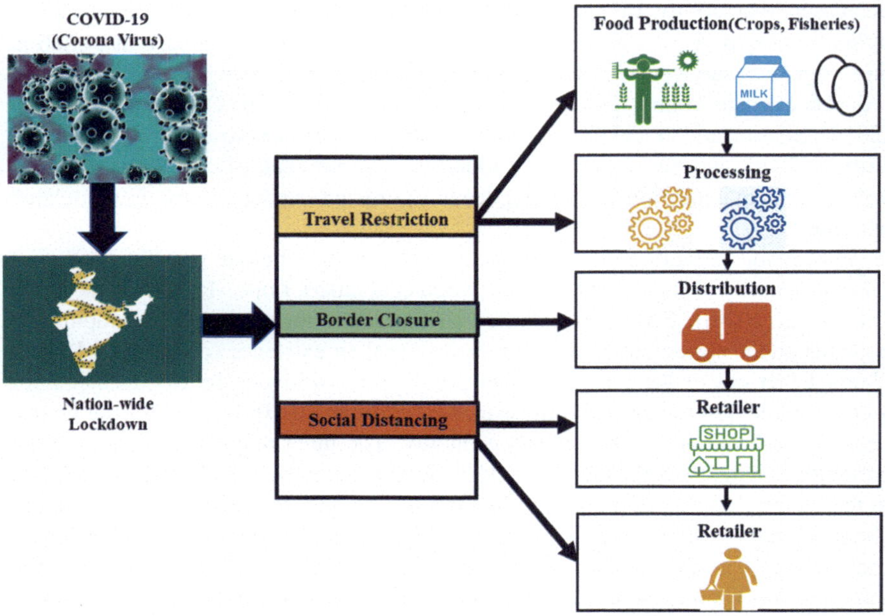

Fig. 12.4 COVID-19 affecting each phase of the food supply chain especially transport and distribution

12.2.1 Impact on Agricultural Production and Food Supply

Several countries have announced nation-wide travel restrictions by imposing several restrictions such as home quarantine, business closure, travel bans, etc. to contain the spread of the corona infection. These restrictions have affected agricultural activities especially on the food supply chain (Fig. 12.4). Moreover, FAO (2020a) also claimed that the restrictions on mobility cause labour shortages which may hamper food processing and farming activities. For example, *mandis* (food market) are closed in several parts of India during the travel restrictions, thereby disturbing the food supply chain from production to consumption centres (Narayanan 2020). World trade is believed to decline by 13–22% due to corona pandemic (WTO 2020). Various agricultural production sectors such as livestock, fisheries, and crops have been affected by COVID-19. For example, March–April is the best time for the sale of rabi crop, however, the harvesting process will be delayed due to the departure of migrant workers to their homes. In China, the pandemic has caused a negative impact on livestock farming due to lack of labour and inadequate access to animal feedstock (Zhang 2020). Travel restrictions and worker shortages have led to thousands of containers of frozen chicken, pork, and beef piling up at some of China's major ports. If such travel restrictions continue for a long time, there will be no hatching of eggs and breeding stock as claimed by the

International Poultry Council (Vorotnikov 2020). Fish is a vital source of energy and accounts for nearly 20% of animal protein for three million people (FAO 2020a). Labour shortage and social distancing have affected fishing activities in several parts of Europe, Asia, and Africa (FAO 2020b). Fish farmers witness's difficulty in aquaculture production due to the absence of fish seeds as well as they were unable to sell their harvest. In the case of agricultural products, farmers have to store their unsold perishable product for a long time which leads to diminishing the food quality, along with an increase in the cost of production (FAO 2020b). The supply of dairy and milk products has been extremely affected during the coronavirus pandemic. Closure of confectionery shops, hotels and restaurants, and tea shops has reduced milk sales during the travel restrictions travel restrictions, especially in India. The departure of migrant workers to the countryside might also decrease the demand for milk in urban areas. Hence, dairy farmers are forced to dump their milk due to remarkable reduction in milk requirements. Forstadt (2020) estimates that American dairy farmers are dumping approximately four million gallons of milk every day. Moreover, the situation is more drastic in Nepal such that dairy products worth two billion (NPR) has already been spoiled, while products in stock worth five billion (NPR) appear on the edge of deterioration (NepaliSansar 2020).

Due to the disruption of global trade, farmers suffer from lack of agricultural inputs such as fertilizer, seed, and pesticide. Worldwide, China is a major fertilizer producer as well as exporter, however, the shut down in China has severely impacted international fertilizer sales (Marlow 2020). For example, India requires approximately 25 million quintals of agricultural input (seed) in the Kharif season (Singh 2020). Therefore, the sowing of crops like spring wheat, maize, canola, barley, open field vegetables, maize can't be carried out due to the pandemic situation. If the pandemic continues for a longer time, the sowing of Kharif and Zaid season crops might be affected. Hence, import-dependent nations are most likely to be affected during COVID-19 (Cullen 2020). Due to the alarming rate of COVID infection, labour shortages and their health problems have been the major issue in the agricultural industry especially in under-developed and developing countries where agricultural workers are deprived of adequate social protection and health services. However, such workers were compelled to work in agricultural fields without protections during the pandemic. Hence, these agricultural workers are at high risk of infection (ILO 2020a).

Mobility restriction has created a shortage of seasonal and farm workers (Brelie and Petit 2020). The pandemic results in a quick increase in the unemployment rate across the world. ILO projected that pandemic has impacted the 2.7 billion workers (approximately 81% of the global workforce) due to part or full closure of the workplace (ILO 2020b). To overcome these drastic circumstances, governments and NGOs are playing an important role to maintain regular food supply chains. At present, FAO has taken the responsibility to maintain continuity of the food supply (FAO 2020c).

Despite that agriculture and the food supply chain are considered an essential service, they were largely affected during the initial phase of travel restrictions. However later, recovery is being observed to some extent. As a result, the prices of vegetables and cereal goes down which had primarily gone up during the period March–April. Thus, despite the restrictions that governments have imposed on the mobility of labour in agricultural systems, although with some problems, the supply of essential goods is normally assured.

12.2.2 *Impact on Food Demand*

Demand can be defined as the consumer's ability and willingness to buy goods in a given time (Gottheil 2013). The food demand has been affected during the pandemic due to the uncertainty of the future and the reduction in income and expenditure capacity. However, the situation could worsen if the COVID-19 pandemic lasts for a long time (FAO 2020c). The travel restrictions made some consumers panic and go to the market for food stocks and thus create a scarcity of basic commodities. This stockpiling has resulted in higher food prices, shortage of essentials and panic in consumer buying behaviour. Siche (2020) claimed that prices of basic goods and services are projected to be constant, however, an increase in the price of high-value commodities such as packaged meat and frozen desserts may occur. In the present condition, animal protein sector is severely hit during this pandemic due to misinformation that chickens' can be the carriers of COVID-19. Millions of poultry farmers especially from Orissa, Maharashtra, Andhra Pradesh, and Karnataka are compelled to dump their products. As China is a crucial market in global trade, their experience indicates an offtake in e-commerce and online demand in the beverage and food sector due to travel restrictions guidelines (FAO 2020a). During such travel restrictions, contactless food distribution services are favoured by customers, for example, the use of drones for food delivery in China (Yang and Reuter 2020) (Fig. 12.5). This would be extremely beneficial for global response to COVID-19. Moreover, it would increase social distancing, thus reducing the spread of COVID-19 and saving lives (Skorup and Haaland 2020).

Due to COVID-19 the decreased in food demand was identified from the consumer behaviour in multiple setting which created imbalance between supply surplus and deficit for certain foods. The reason behind this is the time taken in

Fig. 12.5 Non-contact food delivery using a drone

adaptation to the new consumer behaviour change during the pandemic period and uncertainty caused by enforced travel restrictions and quarantine regulations. The uncertainty caused a serious risk to food security as capacity of produce and distribute the food products, impacted buying power and intensified precautionary steps to ensure the food safety at different distributing levels. This uncertainty calls for the community managed agriculture that directly provides produce to consumer with less or no hassle of distribution channel and handling.

12.2.3 Impact on Food Security

Food demand denotes the accessibility and availability of an adequate amount of nutritious food in a constant manner. However, the above goal is extremely difficult under humanitarian crises such as disruption in food production, food supply, and international trade, hence, food insecurity may occur. According to FAO, fishers and small farmers might face difficulty in selling their goods that would result in a reduction in their income and buying capacity. COVID-19 increases food insecurity which will affect the most vulnerable (Fig. 12.6) and poorest sections of the society (FAO 2020a). Policies must concentrate on retaining the food supply network in continuity across the world and should mitigate the effects of COVID-19 across the agricultural sector. Social protection projects will be served as a solution that reduces the impact of short-term calamities.

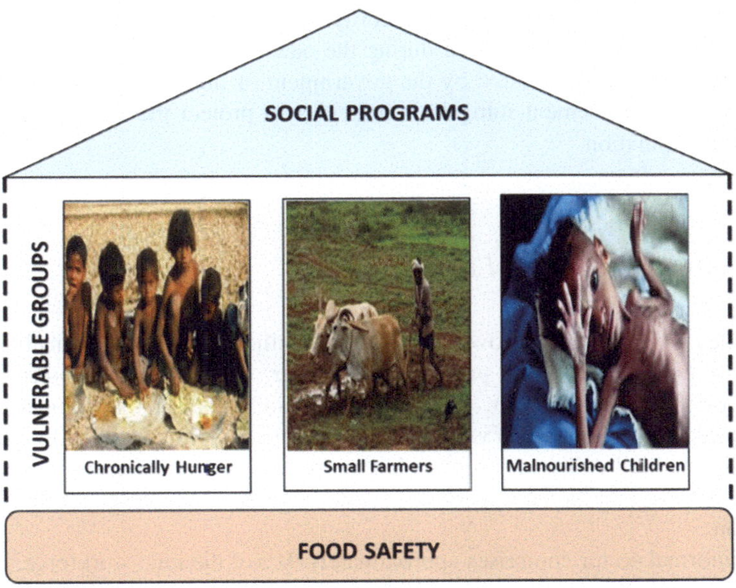

Fig. 12.6 Vulnerable groups to a food crisis

Three vulnerable groups of the population will be affected most during the pandemic. The first vulnerable group involves the people who do not consume proper caloric energy and experience chronic hunger. In the current scenario, 820 million people deal with chronic hunger while 113 million deal with severe insecurity (FAO et al. 2019). Therefore, any disruption in their food access and livelihoods during the pandemic affects these populations severely. If COVID-19 spreads over such groups with limited health capacity, the outcomes could be dangerous. Another group includes small farmers, those could be restricted from working on their field and reaching markets to buy seeds or trade their goods. The third group involves children from poor families (around ten million in number), who are mainly dependent on social programs for their nourishment; however, any interruption in these programs makes their food and nourishment at risk and reduces their power to deal with virus infection (FAO 2020d). Hence, each country needs to continue social food schemes, considering the appropriate protections to contain the spread of the coronavirus.

12.3 Strategies and Precautionary Measures: Indian Context

The government of second populous country of the world, India was forced to take immediate safety action to contain the spread of the pandemic among the urban and rural sprawls. As most of the population is categorized as vulnerable due to economic conditions, therefore, food insecurity was identified as a major problem prioritised over spread of infection during the pandemic. As soon as the complete travel restrictions were declared by the government of India, the authorities took a step forward to implement mitigation strategies to protect the livelihoods of the vulnerable population.

12.3.1 Impact of Travel Restrictions

The sudden travel restrictions over India with 1.3 billion people was a sensible move to control the spread of the pandemic. However, the local food systems were interrupted. The worst part of the travel restrictions was that it coincided with India's peak harvesting time of various seasonal crops. Summer fruits and vegetables were fully matured and ready to pick; barley, paddy, and wheat crops were set for harvest; however, all the hard work was wasted due to the unexpected shutdown of the nation.

The informal sector comprises approximately 91%of the total workforce, including migrant, agricultural, and several other workers. However, their livelihood is entirely based on daily wages. These workers and their families will be the most affected during the pandemic. After the travel restrictions announcement, the informal workers went hungry and began walking back home with their families without any means of transport. The survival of temporary workers in the city was implausible without the availability of food, shelter, and regular income. This mass migration of the workers to their villages brought the agricultural activities to a halt. Various preliminary reports demonstrate that complete shutdown has disrupted both harvesting and post-harvesting activities across the country. A few instances are highlighted below:

- The travel restrictions severely hamper the bumper harvest of wheat especially in Northern India due to lack of workers and transportation bottlenecks.
- Farmers in the western part of Pune have grown grapes in abundance, however, they had to search for student volunteers to harvest their crops due to the shortage of farmworkers. In the absence of inadequate storage facilities, the farmers were compelled to sell their farm products at a very low price.
- Zero business in the market leads to huge post-harvest damage of fish stocks and vegetables. Closure of borders and transportation facilities made it tougher for the trade to happen.
- Asia's leading onion trade market prevailing in Maharashtra found it difficult to transport the freshly harvested onion across the country due to the shortage of workers and drivers.
- The meat and poultry industry were badly hit due to the fake rumours that chickens are carriers of COVID-19. Millions of poultry farmers especially in Orissa, Maharashtra, Andhra Pradesh, and Karnataka were compelled to dump their products. The decreasing demand for animal protein resulted in an unbelievable reduction in their prices leading to a loss of INR 22.5 million for the industry.
- The bumper harvest of bell peppers and cucumbers in Haryana started to get rotten due to the unavailability of buyers and retailers in the market. During the initial phase of the shutdown, people were hesitant to buy foods and vegetables due to fear of coronavirus infection.
- Intercropped peppers are left untouched in most coffee estates due to a shortage of workers.
- India's leading coffee-producing state i.e. Karnataka was unable to sell coffee due to the unavailability of sellers and buyers in the market. Tons of cured coffee worth USD 52 million were kept in storehouses due to the disruption in the supply chain.
- Food prices climb sharply across the country due to the mobility restriction on transportation facilities. The major reasons behind the shortage of groceries in the market were hoarding and panic buying among the consumers.
- Shopkeepers took advantage of the shutdown situation and practiced black marketing by selling essential goods at an outrageous price on existing stocks.
- Processed foods like biscuits, noodles, snacks were in high demand during the travel restrictions. However, all food manufacturing activities suddenly stopped,

hence, companies like Parle, ITC, Britannia, and Nestle are running at low capacity.
- Markets become unstable due to the selling of non-branded goods at more price. Even the government gave assurance about food security, however, on the ground, mobs rushed to stockpile essential commodities.

Innovations such as mobile vegetable trucks were extensively recognized, especially in the state of Kerala, where they provide vegetables and fruits on household doorsteps. This step was taken by local vendors due to the unavailability of customers in the market. While agro e-commerce did not succeed in India during the coronavirus pandemic. This shows the digital illiteracy among customers and their disinterest to buy basic goods online.

12.3.2 Social Mitigation Strategies

The Kerala government was the first who took immediate action to diminish the risk of starvation and hunger of the poorest population. Several mitigation actions are discussed below:

- The government declared free ration including 35 kg rice to below poverty lines (BPL) families for 1 month. Approximately, 11.3% of the total population of the state took benefit from the scheme. In addition to this, 15 kg of rice is also made available to others through the public distribution system (PDS).
- Kerala government has also carried out the distribution of food kits comprising 17 essential commodities worth INR 1000 to every household starting from the first week of April 2020. The food kits were distributed through the ration shops under the PDS scheme. The scheme was sponsored by the Kerala Chief Minister Distress Relief Fund and has allocated around USD 45 million. Authorities have implemented this scheme during the initial phase of the pandemic and plan to continue the scheme according to the severity of the situation thereafter.
- Another important mitigation action was setting up community kitchens in various parts of the state to ensure food security. The Kerala government was the first to provide cooked food to the poorest population. District authorities were appointed to supervise the project while the local volunteers took responsibility in distribution.
- The government has instructed rural childcare centres (*Anganwadi*) to provide nutritious meals to children under the age of 6. According to the instructions, the *Anganwadi* staff is appointed to deliver raw materials to their families weekly.

Followed by Kerala's initiative, the central government announced 1.7 trillion (INR) relief package, to ensure food security to the vulnerable sections. The declaration involves the release of INR 2000 directly to the farmer's bank account under the PM-KISAN scheme. Moreover, the government also increased the pay for workers involved in the NREGS, world's largest wage guarantee scheme. Especially for the vulnerable groups, the special scheme called *Pradhan Mantri Garib Kalyan Yojana* has been announced. Further, grain allocations (5 kg wheat or rice per head, 1 kg of pulse per family in addition to free cooking gas) to listed beneficiaries were also declared by the Indian government for the next 3 months after the nationwide travel restrictions. Food and cash provision to migrant workers has been declared by using the separate fund called PM-CARES (Prime Minister Citizen Assistance and Relief in Emergency Situations) fund. The above models were also implemented by other states in India such as Haryana, Andhra Pradesh, Karnataka, and Tamil Nadu to cope with hunger and starvation.

In addition to the governmental mitigation projects, the food-based research laboratory in South India named as Defence Food Research Laboratory (DFRL) took the responsibility to feed the country under extreme conditions such as conflicts and disasters. During the pandemic, the laboratory has been working 24 h to feed the health workers across the country. Further, the food delivery application was also adopted by the Karnataka government to deliver one million meals to the poorest population in Bengaluru. Private sectors also came forward with government authorities to help the vulnerable section. Companies such as SRK and Wipro groups funded nearly 60,000 meals per day to the poor population through government canteens. Free meals were given in *Amma canteens*, Tamil Nadu and *Anna Purna Canteen*, Hyderabad for the migrant workers, and students who were stranded.

12.3.3 Mitigation Measures: Indian Context

As the agriculture and food industry was greatly disrupted, the Indian Council of Agricultural Research (ICAR) took the responsibility for controlling the agricultural activities under the Ministry of Agriculture and Farmers Welfare and executed the following measures:

- ICAR experts have delivered state-wise crop-specific advisories to the farmers regarding harvesting as well as post-harvesting operations.
- Assessment of all possible impact of the pandemic on agriculture and food sectors and advised mitigation measures to reduce its effects on the food system.
- Detailed guidelines were given for threshing and harvesting of rabi crops like barley and wheat.
- Local field agencies were advised to monitor the free movement of farm products and their related machinery.
- Experts also recommended farmers to delay wheat harvest which was about to give record harvest after the monsoon.

The government of India claimed that granaries of FCI are overflowing with a buffer stock of 71 million tons of wheat and rice to ensure food security. Other mitigation activities by the government are shown below:

- An official announcement was made to support the poultry farmers by providing Rs 100 per bird as monetary support as compensation for zero business.
- Food safety inspectors were advised to examine perishable goods like meat, fish products, and vegetables at local levels to avoid contamination and adulteration.

Moreover, the Reserve Bank of India (RBI) has also declared some mitigation plans to control the burden of debt servicing during the pandemic. Agricultural loans have been decided on a suspension/freeze for 3 months (up to 31 May) with a 3% concession on loans up to INR 300,000 for debtors.

12.4 Policy Recommendations

The travel restrictions result in the disruption of all economic activity. Especially in urban areas, there is extensive loss of jobs and incomes for the poor and informal workers. Centre for Monitoring Indian Economy estimates that the unemployment rate in India rose from 8.4% on March 15, 2020, to 23% on April 5, 2020. Moreover, in the case of urban areas, it will rise to 30.9% as of April 5. This travel restrictions have left informal workers to struggle to meet daily needs of food with limited access to food products. The best method to help these workers is to use social safety nets to stabilize their lives with money and food. The government of India announced 1.7 trillion (INR) relief package, aimed to provide safety nets to agricultural workers during enforced COVID-19 travel restrictions. However, the declared budget is inadequate compared to the enormous problem existing during shut down. According to Duflo and Banerjee (2020), the government should have been much bolder towards this social transfer scheme. This budget is only 0.85% of the country's GDP and much lower compared to packages delivered by some Asian and European countries. According to experts, India must spend a minimum of 4–5% of GDP to counteract the effect of travel restrictions.

The government and policymakers must take some mitigation measures to gear up to deal with the economic crisis, hence, a combined effort of both central and state governments is critical.

- ***Supply chain:*** Proper focus should be given on the smooth functioning of post-harvest activities, marketing of retail, production, storage, wholesale, and transport. Additionally, storage and godowns should be intensified. If travel restrictions are extended, home delivery and E-commerce should be promoted by the government.
- ***Proper implementation:*** The government has issued proper travel restrictions guidelines that have exempted farming operations. However, due to improper implementation, the information regarding guidelines have not reached the police personnel and local authorities. As a result, the food supply chain has been affected. There is a need for proper implementation of the guidelines.
- ***Procurement measures:*** There must be proper and continued market access to the farmers. This must be a combination of government procurement and private

markets. The government should help farmers with perishables products as they face problems such as inadequate storage, compelled to sell their goods at a lower price. Several states have already announced their procurement plans.

- ***Relaxation in APMC regulations:*** APMC regulations should be easy-going which forbid purchase and sale outside market jurisdiction. Farmers should be allowed to sell their perishable produce such as fruits, fish, meat, vegetables, milk etc. outside the regulated markets.
- ***Milk and poultry industry:*** The government must put more emphasis on dairy farmers and small poultry farmers as they need more help since their market access difficulties are crucial. For industry, restructuring or moratorium of loans may be needed.
- ***Safety of agricultural population:*** Proper testing facilities, social distancing during harvesting, marketing, procurement, and packaging, etc. must be ensured to protect agricultural workers and farmers from coronavirus infection. All measures will help to reduce spread of the disease.
- ***Avoid export bans:*** The crucial supply chain must be well functional to ensure food security. For example, approximately two tothree million deaths occurred in Bengal during 1943 famine due to improper food supply. However, it was not due to deficiency of food availability. The government should encourage trade and business by avoiding import and export restrictions. After the travel restrictions, exports of agricultural goods have to be continued.
- ***Food security:*** The government should include agricultural workers and farmers in their assistance package as well as in social protection programs during the crisis. Currently, the PM-KISAN scheme announced by the central government includes only landowners. However, the tenant farmers should also be included in this scheme. There should be an immediate expansion of the Pradhan Mantri Fasal Bima Yojana (PMFBY) to ensure compensation payments to farmers affected by the Covid-19 pandemic. The Government should consider waiving the interest costs of all outstanding crop loans and ensure a fresh flow of credit to small and marginal farmers for the kharif season of 2020.

Some additional mitigation measures need to be covered in the government's relief package:

- ***Cash Transfers*****:** The authorities must help informal workers through cash transfers directly to their account. The government had already provided INR 500 per month to 200 million women directly to their bank account through the *Jan Dhan* financial program. However, this is insufficient to fight against the pandemic. The government must provide a minimum of INR 3000 per month for the next 3 months as support. Khera (2020) suggested to use the NEFT system instead of using the Aadhar Payment Bridge system to avoid failed payments. Additionally, it is important to note that some of the vulnerable groups and informal workers do not have *Jan Dhan* accounts. Hence, it is important to have an optimal cash transfer program to figure out the targeted beneficiaries, amounts, and durations.

- ***Food and nutritional security:*** At present, the government storages are abundant with 71 million tons of wheat and rice. Therefore, it is the best time to declare that at least 10 kg free ration per head per month must be given for the next 3 months. Many state governments have already announced free basic rations. The nutrition level of poor labors was low even before the pandemic. However, it will drop further in the absence of jobs and incomes during shut down. Hence, there is an urgent need to ensure a diversified diet including pulses, jagger, oil, etc. for them. Government programs such as mid-day meals, *Anganwadis,* and Integrated Child Development Services must provide rations to beneficiaries at home. Several state governments have already started such innovative programs to support and assist informal workers. In Kerala, the government is distributing meals at the doorsteps of beneficiaries.
- ***Migrant workers:*** Migrant workers have faced extreme hardships during the travel restrictions and hence, there have been some suggestions to help them. The government must use government schools and colleges, *panchayat bhavans*, community halls, district headquarters, block offices, *anganwadis,* etc. to arrange free cooked food, sanitation, and medical care for these people. Benefits of social protection programs such as the *Ujjwala* scheme, Public Distribution System (PDS) are made available to them (Kapur and Subramanian 2020). Even after the travel restrictions, the government must arrange an orderly return of the migrant workers to their workplaces.

12.5 Conclusion

The chapter investigates the challenges posed by COVID-19 pandemic on the food insecurity in India and analysed the government response to tackle the situation. The investigation further elaborated that worldwide COVID-19 pandemic significantly impacted the social development and economic activities. The agricultural industry and food systems are severely affected by the pandemic and overall outlook for the industry are not very encouraging. In developing countries such as India, where the agriculture accounts for the major share of GDP, the self-isolation, travel restrictions, quarantine regulations and aversion behaviour results in low reserves of food staples and induced uncertainty in food security. The analysis of government response to the situation resulted in increase in food demand due to travel restrictions and decreased purchasing capacity of consumer. Most of the seasonal, migrant, and informal farm labors are losing their jobs that also affected the food demand. The supply chain has been greatly affected by the pandemic, which causes the food insecurity for the most vulnerable segment of population. The challenges faced by sector comprises of decreased availability of agricultural industry workers, decreased demands for agricultural exports, impacts of longer lead times in agro-food supply, limited transportation and logistics services, decreased supply capacity, stringent market protocols, and globally the increase biosecurity regulations. India government's possible mitigation measures are significant and useful but not suffice to tackle the situation. To achieve the comprehensive plan and strengthen the agricultural policies, we presented the recommendation to strengthening supply chain network, increasing market access via reducing tariffs and increasing budget allocation for the sector revival. The presented policy recommendation calls for transitions towards food security during and after COVID-19 pandemic.

References

Arndt C, Lewis JD (2001) The HIV/AIDS pandemic in South Africa: sectoral impacts and unemployment. J Int Dev 13(4):427–449. https://doi.org/10.1002/jid.796

Bermejo A (2004) HIV/AIDS in Africa: international responses to the pandemic. New Econ 11 (3):164–119

Brelie HVD, Petit G (2020) French farm producers suffer during the COVID-19 restrictions. Euronews. https://www.euronews.com/2020/04/17/french-farm-producers-suffer-during-the-covid-19-restrictions

Burgui D (2020) Coronavirus: how action against hunger is responding to the pandemic. https://www.actionagainsthunger.org/story/coronav%0Airus-how-action-against-hunger-respondingpandemic

CDC (2015) Interim infection prevention and control recommendations for hospitalized patients with Middle East Respiratory Syndrome Coronavirus (MERS-CoV). https://www.cdc.gov/coronavirus/mers/infection-prevention-control.html

CDC (2020a) 1957–1958 pandemic (H2N2 virus). https://www.cdc.gov/flu/pandemic-resources/1957-1958-pandemic.html

CDC (2020b) 1968 pandemic (H3N2 virus). https://www.cdc.gov/flu/pandemic-resources/1968-pandemic.html

Cherry JD, Krogstad P (2004) Sars: the first pandemic of the 21st century. Pediatr Res 56(1):1–5

Coburn BJ, Wagner BG, Blower S (2009) Modeling influenza epidemics and pandemics: insights into the future of swine flu (H1N1). BMC Med 7:30. https://doi.org/10.1186/1741-7015-7-30

Cucinotta D, Vanelli M (2020) WHO declares COVID-19 a pandemic. Acta Biomed 91:157–160. https://doi.org/10.23750/abm.v91i1.9397

Cullen MT (2020) Coronavirus food supply chain under strain what to do? 1–39

Das J, Poonia V, Jha S, Goyal MK (2020) Understanding the climate change impact on crop yield over Eastern Himalayan Region: ascertaining GCM and scenario uncertainty. Theor Appl Climatol, 2011. https://doi.org/10.1007/s00704-020-03332-y

Duflo E, Banerjee A (2020) A prescription for action: nine steps after the next 21 days. Indian Express

FAO (2020a) Covid-19 pandemic: impact on food and agriculture. http://www.fao.org/2019-ncov/q-and-a/en/

FAO (2020b) How is COVID-19 affecting the fisheries and aquaculture food systems

FAO (2020c) FAO Director-General urges G20 to ensure that food value chains are not disrupted during COVID-19 pandemic

FAO (2020d) FAO warns of the impact of COVID-19 on school feeding in Latin America and the Caribbean

FAO, IFAD, UNICEF (2019) The State of Food Security and Nutrition in the World 2019, safeguarding against economic slowdowns and downturns. FAO, Rome

Forstadt J (2020) Dairy industry upended by COVID-19. https://wskg.org/news/dairy-industry-upended-by-covid-19/

Gottheil FM (2013) Principles of microeconomics, 7th edn. Cengage Learning. EEUU, Belmont

Goyal MK, Ojha CSP (2012) Downscaling of precipitation on a lake basin: evaluation of rule and decision tree induction algorithms. Hydrol Res 43(3):215–230. https://doi.org/10.2166/nh.2012.040

Goyal MK, Ojha CSP (2014) Evaluation of rule and decision tree induction algorithms for generating climate change scenarios for temperature and pan evaporation on a lake basin. J Hydrol Eng 19(4):825–835. https://doi.org/10.1061/(ASCE)HE.1943-5584.0000795

Hanashima M, Tomobe K (2012) Urbanization, industrialization, and mortality in modern Japan: a spatio-temporal perspective. Ann GIS 18:57–70. https://doi.org/10.1080/19475683.2011.647078

ILO (ed) (2020a) ILO monitor: COVID-19 and the world of work, 2nd edn. ILO, Geneva

ILO (2020b) ILO: COVID-19 causes devastating losses in working hours and employment. https://www.ilo.org/global/about-the-ilo/newsroom/news/WCMS_740893/lang%2D%2Den/index.htm

Johnson N, Mueller J (2002) Updating the accounts: global mortality of the 1918–1920 "Spanish" influenza pandemic. Bull Hist Med 76:105–115

Kapur D, Subramanian A (2020) How coronavirus crisis can be converted to opportunity to fundamentally strengthen Indian economy. Indian Express

Khera R (2020) Covid-19: what can be done immediately to help vulnerable population. Ideas for India

Kogo BK, Kumar L, Koech R (2020) Climate change and variability in Kenya: a review of impacts on agriculture and food security. Environ Dev Sustain:1–21. https://doi.org/10.1007/s10668-020-00589-1

Lopez-Ridaura S, Barba-Escoto L, Reyna C, Hellin J, Gerard B, van Wijk M (2019) Food security and agriculture in the Western Highlands of Guatemala. Food Secur 11:817–833. https://doi.org/10.1007/s12571-019-00940-z

Marlow S (2020) COVID-19: effects on the fertilizer industry. IHS Markit. https://ihsmarkit.com/research-analysis/report-covid19-effects-on-the-fertilizer-industry.html

Narayanan S (2020) Food and agriculture during a pandemic: managing the consequence. https://www.ideasforindia.in/topics/agriculture/food-and-agriculture-during-a-pandemic-managing-theconsequences.%0Ahtml

NepaliSansar (2020) COVID-19 inflicts NPR 2 Bn loss on Nepal Dairy Industry! https://www.nepalisansar.com/business/covid-19-inflicts-npr-2-bn-loss-on-nepal-dairy-industry/

OECD (2020) Coronavirus: the world economy at risk. OECD Interim Econ Assess:1–15

Ray D, Salvatore M, Bhattacharyya R, Wang L, Mohammed S, Purkayastha S, Halder A, Rix A, Barker D, Kleinsasser M, Zhou Y, Song P, Bose D, Banerjee M, Baladandayuthapani V, Ghosh P, Mukherjee B (2020) Predictions, role of interventions and effects of a historic national lockdown in India's response to the COVID-19 pandemic: data science call to arms. medRxiv. https://doi.org/10.1101/2020.04.15.20067256

Sar TT, Aernan PT, Houmsou RS (2010) H1N1 influenza epidemic: public health implications for Nigeria. Int J Virol 6:1–6

Sharma A, Goyal MK (2018) District-level assessment of the ecohydrological resilience to hydroclimatic disturbances and its controlling factors in India. J Hydrol 564(July):1048–1057. https://doi.org/10.1016/j.jhydrol.2018.07.079

Siche R (2020) What is the impact of COVID-19 disease on agriculture? Sci Agropecu 11:3–6

Singh IS (2020) Agriculture in the time of Covid-19. The Hindu BusinessLine. https://www.thehindubusinessline.com/economy/agri-business/agriculture-in-the-time-of-covid-19/article31248717.ece

Skorup B, Haaland C (2020) How drones can help fight the coronavirus. SSRN Electron J. https://doi.org/10.2139/ssrn.3564671

Vorotnikov V (2020) Coronavirus could disrupt poultry production. Poult. World. https://www.poultryworld.net/Meat/Articles/2020/4/Coronavirus-could-disrupt-poultry-production-563092E/

Wang H, Wang Z, Dong Y, Chang R, Xu C, Yu X, Zhang S, Tsamlag L, Shang M, Huang J, Wang Y, Xu G, Shen T, Zhang X, Cai Y (2020) Phase-adjusted estimation of the number of coronavirus disease 2019 cases in Wuhan. China Cell Discov 6:4–11. https://doi.org/10.1038/s41421-020-0148-0

WHO—World Health Organization (2020a) Novel coronavirus (2019-nCoV). situation report-1. https://www.who.int/emergencies/diseases/novelcoronavirus-%0A2019/situation-reports/
WHO—World Health Organization (2020b) HIV/AIDS: global situation and trends. Global Health Observatory (GHO) data. https://www.who.int/gho/hiv/en/
WHO—World Health Organization (2020c) Coronavirus disease (COVID-2019) situation reports. https://www.who.int/emergencies/diseases/novelcoronavirus-2019/situation-reports
WorldBank (2019) Employment in agriculture (% of total employment) (modeled ILO estimate)
WTO (2020) WTO | Frequently asked questions: the WTO and COVID-19. https://www.wto.org/english/tratop_e/covid19_e/faqcovid19_e.htm
Yang J, Reuter T (2020) 3 ways China is using drones to fight coronavirus. World Economic Forum. https://www.weforum.org/agenda/2020/03/three-ways-china-is-using-drones-to-fight-coronavirus/
Zavatta G (2014) Agriculture remains central to the world economy. 60% of the population depends on agriculture for survival
Zhang X (2020) Chinese livestock farms struggle under COVID-19 restrictions. Research Post of International Food Policy Research Institute. https://www.ifpri.org/blog/chinese-livestock-farmsstruggle-%0Aunder-covid-19-restrictions

Chapter 13
Impact of COVID19 on MSME in Water Sector

Harinarayan Tiwari and Subash Prasad Rai

Abstract The novel Covid-19 virus has impacted the life of humans all across the world in many ways. The virus has brought India to a standstill with human beings inside their homes to supplement the governmental efforts of social distancing. This has led to economic crisis affecting all the major sectors like manufacturing, automobile, hospitality, tourism, aviation to name a few. Even in this time of crisis, the resource personnel involved in water supply, water treatment and water management have stood up and carried out a commendable job to ensure the water services were not disrupted. The Covid crisis can provide insights of the officials involved in water sector about the challenges faced by the water industry as well as the impacts on the water and water related industries, particularly the MSME's. The situation also underlines the way forward for the water industry keeping in view the sustainability and adaptability of the system. This has provided a push towards the digital management of water infrastructure, telecommunication systems for remote operations, etc.

Keywords Crisis · Water management · Challenges in MSME's · Sustainability · Digital management

13.1 Introduction

The Micro, Small and Medium Enterprises (MSME) sector has developed as an exceedingly lively and self-motivated sector of the Indian economy over the last five decades. It contributes significantly in the economic and social development of the country by fostering entrepreneurship and generating largest employment opportunities at comparatively lower capital cost, next only to agriculture. Contribution of MSME's in GDP of Indian Economy is 30–35% along with average 114 million

H. Tiwari (✉) · S. P. Rai
Floodkon Consultants LLP, Noida, India

M. K. Goyal, A. K. Gupta (eds.), *Integrated Risk of Pandemic: Covid-19 Impacts, Resilience and Recommendations*, Disaster Resilience and Green Growth,
https://doi.org/10.1007/978-981-15-7679-9_13

employment (Beal et al. 2016; Brears 2020; KPMG n.d.). In which sharing of micro enterprises are significantly higher (around 99%) than small and medium. Total Number of MSMEs registered on Udyog Aadhar Memorandum (UAM) as Micro—60,32,100, Small—7,28,516, Medium—28,611 (Ministry of Micro, Small & Medium Enterprises n.d.).

Classification of micro, small and medium enterprises are based on the total annual turnover as per The Medium, Small and Micro Enterprises Development (Amendment) Bill, 2018. Earlier, it was based on total investment in the plant, machinery or equipment. India was expected to emerge as one of the leading economies in the world over the next decade in the light of a positive political and economic scenario before COVID-19. The Micro, Small & Medium Enterprises (MSME) segment was also expected to play a significant role in the emergence of the Indian economy. COVID-19 and subsequent lockdown puts a rider in worldwide economy. This paper tries to evaluate the basic aspects like economic challenges, Consumer Behaviour & Other aspects of MSME related to water sector before and after COVID-19.

13.2 Change in Definition

Earlier, Manufacturing and service enterprises was classified separately. Earlier definition the criteria of categorization was only the investment; either it is Manufacturing or service enterprises. The definition of MSME has changed and now it is categorized in uniform manner for both sectors. Figure 13.1 depicts new definition of MSME (Ministry of Micro, Small & Medium Enterprises n.d.).

This change in definition bring the additional criteria of turnover into the place, which can attract the investor to invest in MSME. In a major initiative, Global tenders to be disallowed for Government tenders up to Rs. 200 crore to enable

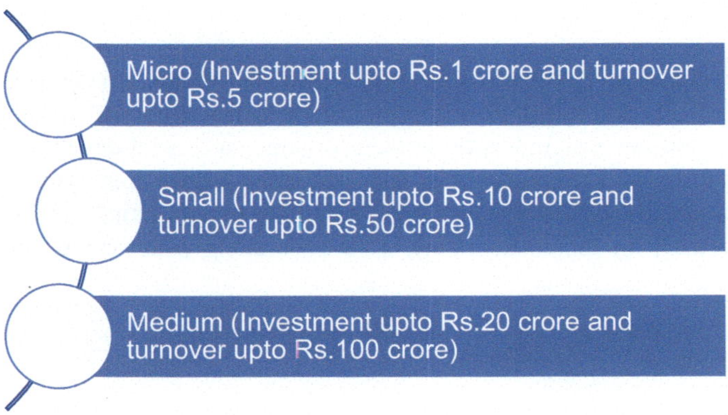

Fig. 13.1 Definition of MSME (Ministry of Micro, Small & Medium Enterprises n.d.)

MSMEs to participate in the Government procurement process and it will also attract young and energetic mind to come for new ventures.

13.3 Rivers in COVID

One of the major benefits of the lockdown has resulted in is a significant improvement in the river water quality. Over a month into the nationwide lockdown, air and water pollution levels have shrunk and the wildlife is free. The nationwide lockdown was imposed on March 25, 2020, and within 10 days, signs of improvement in water quality started surfacing. According to the real-time water monitoring data of the CPCB, out of the 36 monitoring units placed at various points of the Ganga river, the water quality around 27 points was found suitable for bathing and propagation of wildlife and fisheries.

More than 80% of pollution in the Ganga is due to domestic sewage from surrounding towns and villages. The rest is contributed by industrial waste. During the lockdown, domestic sewage would have increased owing to increased demand for water to maintain hand-washing hygiene. Industrial waste, however, stopped entering the Ganga. Other activities such as tourism, fairs, bathing and cloth washing near the ghats were curtailed. Experts said these observations reflected that domestic sewerage was not the only cause of concern. Zero industrial pollution increased quality of water in the Ganga.

It would be a challenge for the local bodies as well as authorities to maintain the water quality in the rivers after the lockdown opens and economic activities gather steam. It would also be critical to observe how the institutions involved in maintaining the water quality ensure the quality of the water as well as look up to any changes in the policies and guidelines for maintaining the quality standards.

13.4 Water Industry in and After COVID

As Covid-19 wreaks havoc on national healthcare systems and economies, water and wastewater providers have been thrust into uncharted territory. The municipal water sector is predicted to be affected “long beyond” the current healthcare crisis, with the short-term impacts catalysing new approaches to urban resilience. The area’s most likely to be impacted by the current pandemic include: capital and operating expenditures; workforce management; affordability; customer communications and resilience planning.

The Covid-19 pandemic is both a massive near-term threat, and a long-term opportunity for the water sector. In the near term, municipal utilities are suffering the cumulative impact of local economies freezing, demand patterns shifting massively overnight, and limited personnel to manage operations. This calls for enhanced remote management, emergency response, and restructuring of operating budgets.

For solution providers and engineering consultancies, there could be a major hit to corporate earnings, downsizings and even "bankruptcies for smaller, venture-dependent organizations on discretionary budgets".

Water, and other major utilities such as telecom and power workforces have "morphed overnight", with entire sectors shifting to telecommuting models. As a result, network operators and their emergency protocols, asset redundancy and remote management are automation efforts are being tested. Utilities that have already invested heavily in remote monitoring and digital asset management will likely see more immediate benefits from a resiliency perspective. Long term, the pandemic offers an opportunity to improve resilience through digitalisation, which will strengthen both emergency response capabilities, and day-to-day operations.

COVID-19 can be a watershed moment for digital transformation of water sector. It is difficult to overstate the challenges that COVID-19 poses for the water sector, from operational difficulties to acute financial shocks. But the pandemic also presents a long-term opportunity for the industry to rethink the status quo and embrace innovation. In particular, we see the crisis as a proving ground for digital water technologies, and a potential watershed moment for the digital transformation of the water sector.

Utilities that have already made digital investments are best positioned to cope with crisis. Digital technologies have already played an invaluable role in helping utility workers remain connected to their critical assets, customers, and co-workers amid widespread lockdowns. In many cases, utilities which made robust investments before the crisis in digital monitoring and control capabilities—e.g. automation and optimization platforms, sensor and telemetry equipment, remote workforce and customer management tools—are faring better than those that did not.

Meanwhile, technologies that can be deployed rapidly to fill existing operational gaps have seen a boost in demand. According to a recent American Water Works Association survey, 74% of U.S. and Canadian utilities have implemented telework policies for non-field employees in response to the pandemic. Some utilities have asked as many as four-fifths of their staff to stay at home, necessitating significant investment in remote communication and productivity (e.g. laptops, teleconferencing platforms), online billing, and cyber security.

13.5 Key Challenges in Water Sector

13.5.1 Wastewater

Wastewater sector is getting challenge in the Post-Covid world. Generation of higher municipal wastewater due to frequent hand washing, maintaining hygienic environment around us. Chain of wastewater systems is also disrupted due to long lockdown period. Cost of maintenance will increase due to need of higher protective environments. In the same time, contactless measurement technologies will get higher opportunities in the sector. Post-Covid world will also lead fear for waste water workers, so the new technologies for the safety will take place in future.

13.5.2 Finance

In the United States, the water industry is facing substantial financial losses. According to a recent report prepared for the AWWA and the AMWA, the combined water and wastewater sector impact of the COVID-19 crisis is calculated to be more than $27 billion. According to the Statista 2020, water supply including utility services is losing more than 13% revenue between April and June 2020 in India (Statista n.d.).

13.5.3 Development of Remotely Sensed Water Management System

Covid-19 leads the need of contactless technology, in a way development of remotely sensed management system is getting opportunities in future. Announcement of private sectors involvement in space technology will give the booster to the water related MSME.

13.5.4 New Product Development

Most of the water related measurement techniques are not directly manufactured in India. Prime mister slogan of "vocal for the local" may give booster to MSME to conceptualize & manufacture the product.

13.5.5 Market Competitiveness

Development of new local product is not only sufficient for the MSME sustainability but government policy must be like that it can sustain in long term as it will get sufficient market for the product. Apart from the above-mentioned key challenges; there are certain drivers (Fig. 13.2) which can provide positive path to water related MSME in Post-Covid world.

Fig. 13.2 Sustainable MSME

13.6 Policy Interventions by Governments

1. Government of India announces Collateral Free Automatic Loans worth Rs. 3 lakh crores for MSMEs (Ministry of Micro, Small & Medium Enterprises n.d.)
2. No global tender for procurement up to 200 crores (Ministry of Micro, Small & Medium Enterprises n.d.)
3. Emphasis on reforming and streamlining some critical areas of economy is also very timely which includes land, labour, liquidity and law (Ministry of Micro, Small & Medium Enterprises n.d.)
4. The technology platform CHAMPIONS which stands for Creation and Harmonious Application of Modern Processes for Increasing the Output and National Strength. This ICT based system is set up to help the MSMEs in present difficult situation and also to handhold them to become national and international champions (Ministry of Micro, Small & Medium Enterprises n.d.).
5. MSME SAMADHAAN (MSME Samadhaan n.d.) is the portal where the payments related query can be generated. Till date, 43,008 files were submitted for different purposes.
6. Release of money to MSME from Government related projects within certain time limit (MSME Samadhaan n.d.).

13.7 Future Prospects

13.7.1 Circular Water Economy

In the linear economy, the water sector typically employs the Take-Use-Discharge strategy. In this strategy, water is 'withdrawn' from streams, rivers, lakes, reservoirs, oceans, and groundwater reservoirs as well as harvested directly as rainwater. In the context of water resources management, water utilities are beginning to promote the circular water economy that reduces water consumption, reuses and recycles water and wastewater, and recovers materials, including heat and minerals, from water and wastewater to not only mitigate greenhouse gas emissions but also enhance resilience to climate change from efficiency gained in reducing water consumption and reusing water for various activities (Brears 2020). Services considering the circular water economy will become core as well challenge for the industry. To provide circular water economy for any services, management aspects will become very prompt for the sector.

13.7.2 Core Asset Management Via Digital Platform

Pandemics has given the challenge to MSME and tested the capability to accomplish critical tasks under a varied range of circumstances, and to convert assets or modify structures to introduce new ways of doing so. Water related MSME will go for the asset management via digital platform. For example, numerical modelling can manage from the computer with respect to physical modelling. This sort of activities may be essentially promoted among water policy maker group for their acceptability. In the same time, expert in these areas have to perform with similar reliability to win the confidence of the system.

13.7.3 Procurement Policy

Agriculture is the sector where water consumption is highest in India. Different discussions with policy makers have made the impression that crop procurement policy can lead to change water sector. Procurement policy can lead farmer to better market & income for less water consumable crops. In India, some the states the like Punjab have started the movement "Pani Bachao Paisa Kamao" and get the better results in the case of water conservation PSPCL (n.d.). These types of intervention can lead physiological reforms and lead better security for MSME sector.

13.7.4 Demand Side Management

The water sector is now reconnoitring the value of applying demand management strategies to reduce peak water use through behavioural and technical solutions. Behavioural interventions may be a useful approach in changing the daily peak demand patterns to reduce the pressure on network pumping energy costs during peak use times.

13.7.5 Technological Adaptability in Agriculture

Since the water sector moving towards the circular economy so technological interventions in the agriculture and irrigation sector will having higher opportunities for MSME sectors (Verner et al. 2017). Measure-Tack-Act (MTA) can be the form of intervention in the Agricultural water sector.

13.8 Concluding Remarks

The remainder of 2020 could be one of the most challenging years but at the same time it can open new opportunities as well. The water industry, however, faces an enormous opportunity to rise to the occasion by realizing step changes in terms of investment levels and operational improvements that will prepare cities for ongoing operations in the face of future disruptive events. MSME's in the water sector have an opportunity to disrupt the market and create their own space in the water industry by offering new technologies and state-of-the-art approaches.

References

Beal CD, Gurung TR, Stewart RA (2016) Demand-side management for supply-side efficiency: modeling tailored strategies for reducing peak residential water demand. Sustain Prod Consump 6:1–11

Brears RC (2020) The circular water economy. Developing the circular water economy. Palgrave Pivot, Cham, pp 31–44

KPMG (n.d.). https://home.kpmg/in/en/home/insights/2020/03/the-business-implications-of-coronavirus.html

Ministry of Micro, Small & Medium Enterprises (n.d.). https://msme.gov.in/

PSPCL (n.d.). https://www.pspcl.in/2019/08/pspcl-launches-second-phase-of-paani-bachao-paisa-kamao-scheme/

MSME Samadhaan (n.d.). https://samadhaan.msme.gov.in/

Statista (n.d.). https://www.statista.com/statistics/1107798/india-estimated-economic-impact-of-coronavirus-by-sector/

Verner D et al (2017) Frontier agriculture for improving refugee livelihoods: unleashing climate-smart and water-saving agriculture technologies in MENA. World Bank, Washington, DC

Chapter 14
Transportation and a Pandemic: A Case Study of COVID-19 Pandemic

Yasir Ali, Anshuman Sharma, and Md. Mazharul Haque

Abstract People use various modes for travelling and interact with each other during their trips for business, shopping, education and tourism. The travel increases the risk of spreading communicable diseases and may escalate the spreading rate at a level that can make an endemic to become a pandemic. On the other hand, the measures, for example, lockdown, taken by governments to control the pandemic, negatively impact travels and, in the worst case, bring the transportation system to a standstill. Thus, it is important to understand this relationship between a pandemic and transportation to control the spread of a disease through transportation and to suppress the negative impacts of that disease on transportation. This book chapter explores this relationship using the COVID-19 pandemic as a case study. Firstly, the chapter highlights the role of transportation in spreading any disease, including COVID-19. Secondly, it summarizes the immediate impacts of COVID-19 on transportation and puts forward ways to recover transport systems after a pandemic. Thirdly, the chapter provides insights into how policymakers can use the scenarios like lockdown to reshape transport systems. Finally, this chapter suggests some proactive approaches that transport agencies can adopt during and after a pandemic.

Keywords Transportation · COVID-19 · Pandemic · Public transport · Air travel

---

Y. Ali
Institute of Transport and Logistics Studies, The University of Sydney Business School, University of Sydney, Sydney, NSW, Australia
e-mail: yasir.ali@sydney.edu.au

A. Sharma (✉)
School of Civil Engineering, Faculty of Engineering, Architecture and Information Technology, The University of Queensland, Brisbane, QLD, Australia
e-mail: anshuman.sharma@uq.edu.au

M. M. Haque
School of Civil and Environmental Engineering, Science and Engineering Faculty, Queensland University of Technology, Brisbane, QLD, Australia
e-mail: m1.haque@qut.edu.au

M. K. Goyal, A. K. Gupta (eds.), *Integrated Risk of Pandemic: Covid-19 Impacts, Resilience and Recommendations*, Disaster Resilience and Green Growth,
https://doi.org/10.1007/978-981-15-7679-9_14

14.1 Introduction

Transportation plays an important role in our daily life as well as in the country's economic growth and development. In particular, transportation contributes about 5.2% of a country's gross domestic product (Rodrigue and Notteboom 2020). Of many advantages, transportation networks provide social opportunities and benefits in terms of better accessibility to markets, employment, and other services.

Apart from assisting people in movement from one place to another, modern transportation systems are also a threat and can act as an opportunity for spreading familiar or completely new diseases (Guimera et al. 2005). An increase in the human movement due to tourism, education, employment, economic activities has increased the risk of spreading diseases. In the last 500 years, various examples have been found that suggest how do transportation networks magnify and carry global (communicable) pandemics. For was transported via port cities and killed about more than 200 million people during fourteenth to seventeenth centuries (Duplaix 1988; Eckert 2000). Similarly, in the last 185 years, Vibrio cholera travelled via trade route and resulted in pandemic (Sack et al. 2004), and about 30,000 cases emerged in just a few months in Dhaka, Bangladesh (Faruque et al. 2003). Unlike the aforementioned diseases, the Influenza pandemic has a low fatality rate (Palese 2004) and was considered as a principal infectious disease travelled by the global transport network. Each year, about 20% of children and 5% of adults are reported to develop symptoms of Influenza (Nicholson 2003). More recently, Coronavirus disease (COVID-19) is caused by a newly discovered coronavirus (Haghani et al. 2020). As per WHO, unfortunately, COVID-19 affects different people differently. The most common symptoms include fever, dry cough, and tiredness. Fortunately, most of the infected individuals recover without hospitalisation. How much transportation has pushed the spread of COVID-19 can be guessed by two observations (1) WHO declared COVID-19 as a global emergency on 30th January 2020, i.e., in only 1 month of China's intimation of a pneumonia like disease to WHO (31st December 2019); and (2) in just 5 months, globally, there have been 5,939,234 confirmed cases of COVID-19, including 367,255 deaths, reported to WHO as of 31st May 2020 (almost 216 countries and territories are affected by COVID-19). Top ten most affected countries are the United States, Brazil, Russia, the United Kingdom, Spain, Italy, India, France, Germany, and Peru (as of 31st May 2020).

An intertwine relation of disease and transportation (as humans carry and spread the disease by human-to-human interactions during travelling) can be observed. As mentioned, transportation plays a huge role in spreading a disease (like COVID-19). Likewise, pandemics have also seen to hamper transportation systems. For instance, with the emergence of COVID-19, the metro of Wuhan, China was shut down; India imposed a complete shutdown of all its transport modes for weeks; in Luzon, Philippine, public transport was suspended; free transportation service to elderly people in some parts of Turkey was suspended, and many others. These examples reveal the severity of the impact of a pandemic on the transportation system.

This book chapter, therefore, aims to study how novel coronavirus (COVID-19) impacts the transportation system and what should be the necessary steps in the current time and post COVID-19 era. To this end, this book chapter consists of four sections. Section 14.2 highlights the role of transportation in spreading any disease. Section 14.3 summarises the immediate impacts of a pandemic on transportation. Section 14.4 provides insights into the post-COVID-19 era of transportation and some steps to move forward, and finally, Sect. 14.4 concludes this study.

14.2 Role of Transportation in Spreading the Infectious Diseases

The Spanish Flu in 1918, the deadliest out of the ten major influenza pandemics that have occurred in the last 300 years, infected around one-third of the world's population. The modern transportation system was behind the quick and extensive spread of the Spanish Flu. The modern transportation system also remains a primary mode of spatial diffusion of communicable diseases. The large volume of people travel for various purposes such as business, tourism, and education, among others, and interact with the people during their trips thereby, increasing the risk of spreading communicable diseases.

Air travel has become one of the most critical factors in the global spread of diseases because of its ubiquity. Previous studies have demonstrated the potential role of air travel in spreading diseases like Influenza (Grais et al. 2004; Fadel et al. 2008) and severe acute respiratory syndrome (Olsen et al. 2003; Ali and Keil 2006). Figure 14.1 depicts the increasing global air passenger volume (international and domestic) in the last 18 years. Notably, in the year 2018, the global air passenger

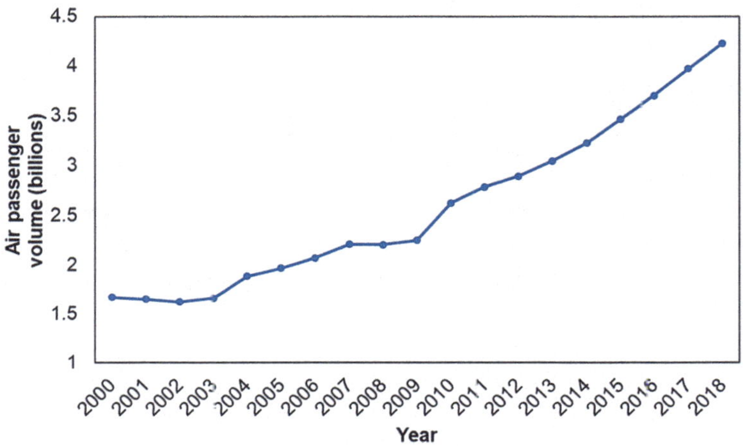

Fig. 14.1 Air passenger travel across different years (WHO 2020)

volume amounted to over 4.2 billion. Following the simple mathematical model by Bradley (1989), one can conclude that the risk of being infected quadruples if the size of the aircraft (seating capacity) is doubled. This risk is even higher for respiratory diseases such as severe acute respiratory syndrome.

Public transport (e.g., buses, tram, metro, railways, ferry) is not far behind air travel. China, India, Japan, Russia, and France are the top five countries in the list of passenger-kilometres travelled by rail with 1346 (year 2017), 1161 (year 2017), 432 (year 2016), 130 (year 2018), and 96 (year 2017) billion passenger-kilometres per year (excluding metros), respectively (Wikipedia 2020). In 2017, it was reported that 178 metros provided ridership to 53,768 million passengers; in the same year, 1.53 billion trips were made on Paris metro (UITP 2018). Considering such a huge gathering of people, public transport certainly becomes another hotspot for the spread of diseases. High density, daily commuters, poor ventilation, high respiratory contact rates, overcrowding, and closed windows are the contributing factors towards the transmission of diseases in public transport (Horna-Campos et al. 2007; Andrews et al. 2013).

14.2.1 How Does Transportation Influence the Rate and the Extent of Pandemic

Before the twentieth century, it would take months for the pandemic to spread primarily due to limited national/international transport connectivity and slow transport systems (Curve A in Fig. 14.2). The emergence of modern transportation systems had two significant impacts on pandemics. Firstly, it influences the rate with which a disease spreads. As shown in Fig. 14.2, transportation reduces the time to reach half of the pandemic's full extent (T (mid)). Moreover, the transmission rate also increases, i.e., more people can get infected in less time. Secondly, the extent to

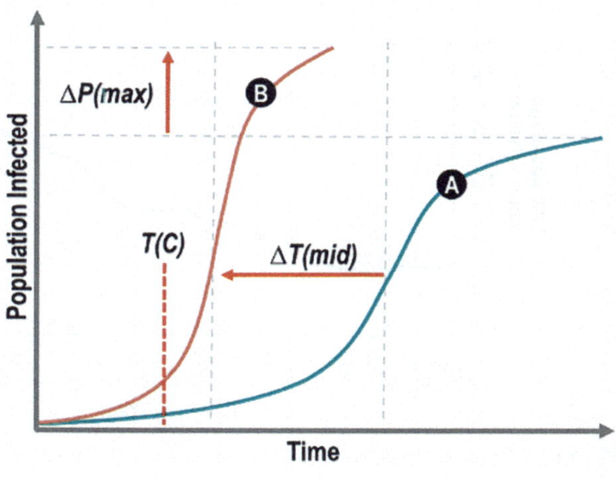

Fig. 14.2 Impact of transportation on the rate and extent of a pandemic (Luke and Rodrigue 2008)

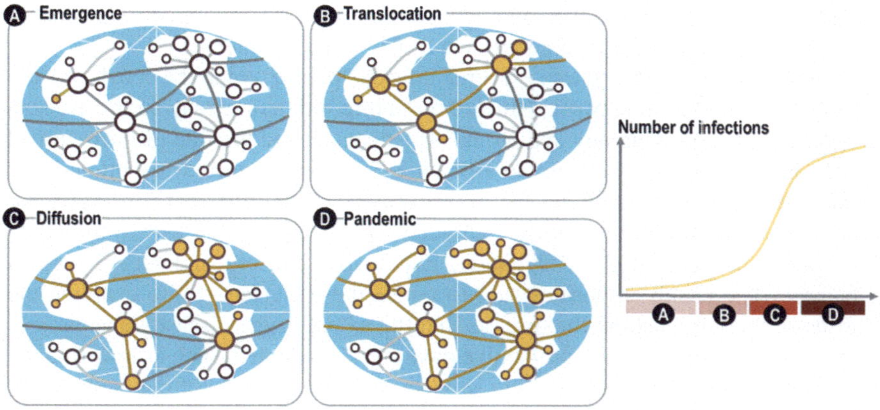

Fig. 14.3 A typical example of spreading a pandemic via global transportation network (Luke and Rodrigue 2008)

which the pandemic may spread. In today's world, most countries have a well-connected transportation network, both internal and external, to a large population that may get infected (P(max)). T(C), a critical parameter, is defined as the time when the translocation phase is in the initial stage. This parameter implies, as nicely put by (Luke and Rodrigue 2008), that "*after a certain amount of time the potential lethality of disease is realised and governments and individuals start to react with a variety of mitigating measures (quarantines, travel restrictions, institutional closures, absenteeism, etc.), which aim at lessening the growth rate of new infections*". Since transportation has pushed the rate and extent of the pandemic, it is challenging to realise the true potential early, and when it is recognised it may be too late.

Figure 14.3 describes a hypothetical scenario describing how a global transportation network could spread a pandemic. The assumptions in this scenario as described by Luke and Rodrigue (2008) is "*a virulent strain of influenza in the line of the Spanish Flu (H1N1) with an incubation phase of about 3 to 4 days and that can easily be transmitted from humans to humans*." The spread of virus through global transportation can be divided into four phases, as described below:

1. **Emergence**: This concerns with an area where the disease has emerged, and it is still not noticed by government officials. The geographic location of this area, the economic activity in the area, and most importantly, transportation connectivity to national and international cities become critical factors that influence the spread of the emerged disease. In the COVID-19 context, these factors are quite evident. China, at the time of COVID-19 outbreak, is one of the world's most dominant trading power. Beijing Capital was ranked second and Shanghai Pudong was ranked ninth in the world's ten busiest airports by passenger traffic in the year 2018. Approximately 600 million (year 2018) is the air passenger volume in China (world bank data), implying trips in enormous numbers for

trade and commerce, tourism, and education purposes. Undoubtedly, the spread of COVID-19 from the emergence area as China was immense.

2. **Translocation**: In this phase, several infected people enter the local and the global transportation network. Most of these infected people may be in a passive state, i.e., not showing any symptoms. The distance and the rate of virus spread will depend on how well the area of emergence is connected to national and international places, implying that translocation will be different for a different area of emergence. The infected individuals will transmit the disease to people during transit and at the destination. Eventually, clusters of infected regions emerge. The health authorities start to notice these clusters, try to examine the disease, the virus behind the disease, and the extent of the disease transmission. Once initial results are out, preventive measures are adopted, such as lowering the overall frequency of transportation services or permanently shutting them down.

   In the case of COVID-19, the capital city of Hubei province, Wuhan, has a population of above 11 million and serves as an important traffic hub in China. Another critical point to note is that the Chinese New Year festival, when the largest movement of people happens in China, coincided with the COVID-19 outbreak. In this festival season, around five million residents left Wuhan before the authorities ordered the lockdown (23rd January 2020) (Al-Jazeera 2020). It is still unclear how many of those were in the incubation phase. On 31st December 2019, China alerted WHO of cases of pneumonia in Wuhan; on 13th January, the first case outside China was reported (in Thailand), and within a few days cases were reported in the U.S., Nepal, France, Australia, Malaysia, Singapore, South Korea, Vietnam, and Taiwan.

3. **Diffusion**: At this stage, the disease has diffused to almost all the major cities of the world, and most of the world population is on the verge of getting infected. From these cities, the disease spreads to close communities mostly through public transport. This spreading is also known as 'community spread'. In this phase, the rate and the extent of the spread are the largest. The extent of the pandemic is global now, and it has become difficult to control the disease. The general public is now aware of this pandemic. Authorities have ordered an emergency. Preventive measures are adopted, and restrictions are being imposed to suppress the rapid spread of the disease and to support the medical system in catching up with the extent of the disease spread.

   Clear diffusion state of COVID-19 can be witnessed. WHO declared the COVID-19 as a global emergency on 30th January 2020. In China alone, the disease was spread in 31 provinces, and the death toll amounted to 170. A Chinese man from Wuhan dies in the Philippines on 2nd February—the first death outside China. The number of infections in China passed 82,000 on 27th February 2020. The countries infected by the COVID-19 up till 1st March 2020 included India, Pakistan, Italy, Philippines, Russia, Spain, Sweden, and the United Kingdom, Australia, Canada, Germany, Japan, Singapore, the U.S., the UAE, Vietnam, Kuwait, Bahrain, Iraq, Oman, Qatar, Norway, Romania, Greece, Georgia, Afghanistan, North Macedonia, Brazil, Estonia, Denmark, Northern

Ireland, the Netherlands, Lithuania, and Wales in addition to those mentioned earlier.

4. **Pandemic**: At this stage, an epidemic has turned into a pandemic. Very few places are not affected either by chance, quarantine, isolation, or containment. The transportation for the general public almost comes to a halt. The transportation operates only to provide essential services such as food and medical supplies, and for emergencies. How countries would react to a pandemic will depend on the virulence and lethality of the pandemic, level of preparedness of policymakers, availability of essentials such as food, clinical supplies, and energy, and understanding of the pandemic. The impact of a pandemic could range from less severe to extremely severe at a social, economic, and political level. It can increase morbidity and mortality and induce fear towards workplaces and social gatherings. A pandemic can seriously damage the economy of a country with short-term to long-term fiscal shocks. Moreover, the political and social fabric could collapse due to lockdown or due to a shortage of essential supplies.

   The geographical spread of the COVID-19 from Wuhan has turned it into a pandemic from an epidemic. The COVID-19 has been reported in 216 countries till now (31st May 2020). There are only 12 unaffected countries namely, Kiribati, Marshall Islands, Micronesia, Nauru, North Korea, Palau, Samoa, Solomon Islands, Tonga, Turkmenistan, Tuvalu, and Vanuatu (Al-Jazeera 2020). The air passenger travel in these unaffected countries is less than 0.1% of the global air passenger travel, implying that low transport connectivity has shielded these countries from COVID-19 (see Fig. 14.4 for a reference).

14.3 Immediate Impacts of the Pandemic on Transportation: Learnings from COVID-19

Societal and psychological factors are two driving forces of transportation choices and patterns during a pandemic like COVID-19. The former represents an external stimulus that affects an individual's life, whereas the latter indicates how internal factors such as motivations, habits, emotions affect an individual's travel behaviour. Although the full effects of COVID-19 cannot be anticipated on travel pattern at this moment, transportation is likely to be impacted, and short-term shifts in transportation patterns are observed. During the initial days of COVID-19 spread, travel patterns remained fairly the same before COVID-19. However, just after a few days, people are instructed to work from home, leading to commuting volumes decline significantly. For instance, on 24th March 2020, ridership on New York subways declined about 87% compared to the same time in 2019 (Eno 2020).

While similar trends were observed for air travel, opposite trends are observed for cycling. People are seen to be frequently cycling for a short trips, reflecting their attitudes and perceived safety. For example, New York's cycling is increased by

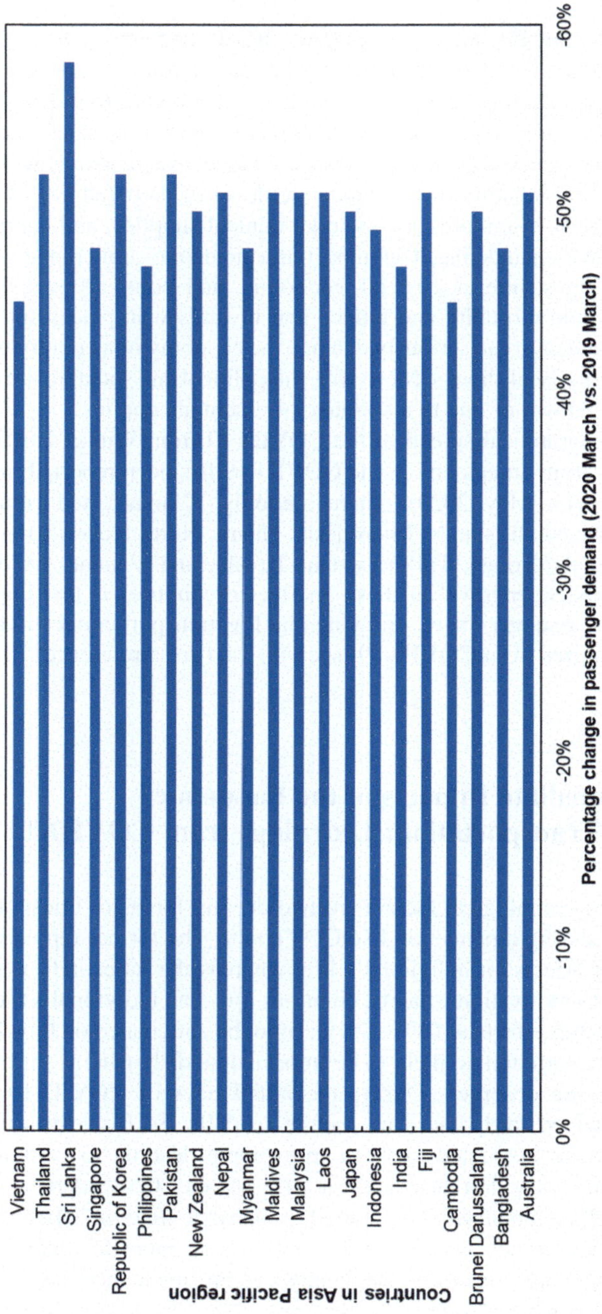

Fig. 14.4 Impact of COVID-19 on air travel (IATA 2020)

two-third of the trips between March 1 and March 11 compared to the same time in 2019 (Eno 2020). Similar trends are also observed in other parts of the world.

14.3.1 Impact on Air Travel

Air travel serves as a basic mode of transportation for international travel, with approximately four billion trips worldwide in 2018 (IATA 2020). Air travel moves people across borders with the strictest immigration policies and security clearances. With countries closing its borders as an immediate step to contain the spread of COVID-19, air travel is declining and in a surviving situation. More specifically, global airlines are reported to face approximately USD314 billion loss during 2020, with a 55% decline in revenue compared to 2019. Similar figures are also reported for airlines in the Asia Pacific region (IATA 2020). These revenue drops are due to several travel restrictions imposed by various countries. Furthermore, 11.2 million jobs associated with the aviation industry are at risk in the Asia Pacific region. Figure 14.4 demonstrates the change in passenger demand for the Asia Pacific region. The impacts of COVID-19 on different aspects of aviation is as follows: passenger demand (both international and domestic) reduced from 35% to 65% in 2020 compared to 2019 (IATA 2020); airport revenues are expected to lose over USD97 billion compared to usual annual business; compared to 2019, revenue from passenger travels (both international and domestic) are estimated to decline by 48% in 2020; with travel restrictions imposed on worldwide tourism destinations, tourism agencies are estimated to face a loss of USD 910 to 1170 billion in 2020. Of note, tourism earned about USD 1.5 trillion in 2019; global merchandise trade volume is suffered by 13–32% in 2020 compared to 2019; it is projected that world GDP would fall by −3% in 2020, which is worse than the financial crisis during 2008–2009 (IATA 2020).

14.3.2 Impact on Public Transport

Public transport connects people to their jobs, homes, schools, entertainment, and other important services, essentially for those people who cannot travel by other modes. With the surge in COVID-19 cases, there is an increasing need for maintaining social distance in order to ensure the safety of both driver and passenger. Due to the fear of catching the virus while using public transport, it has become less attractive to the general public, resulting in decreased patronage. For instance, as COVID-19 cases rose during March 2020 in Australia, about 80% decrease in public transport usage is observed compared to January 2020 (see Fig. 14.5a for a reference (WSP 2020)). A similar drop in public transport usage is seen across different

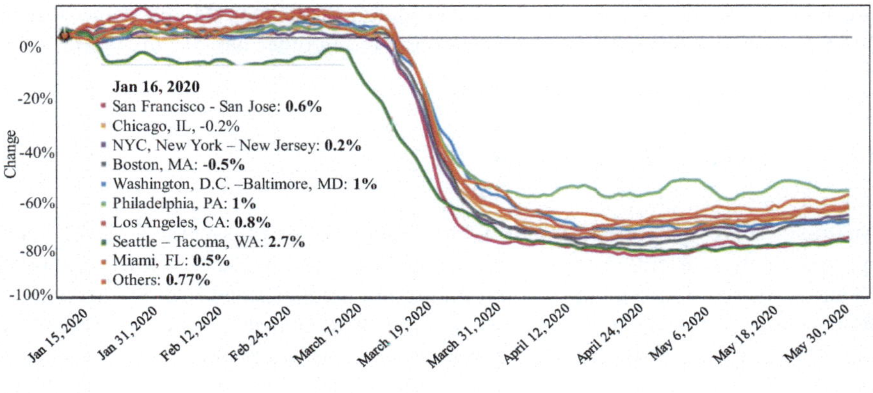

(a) Australia [24]

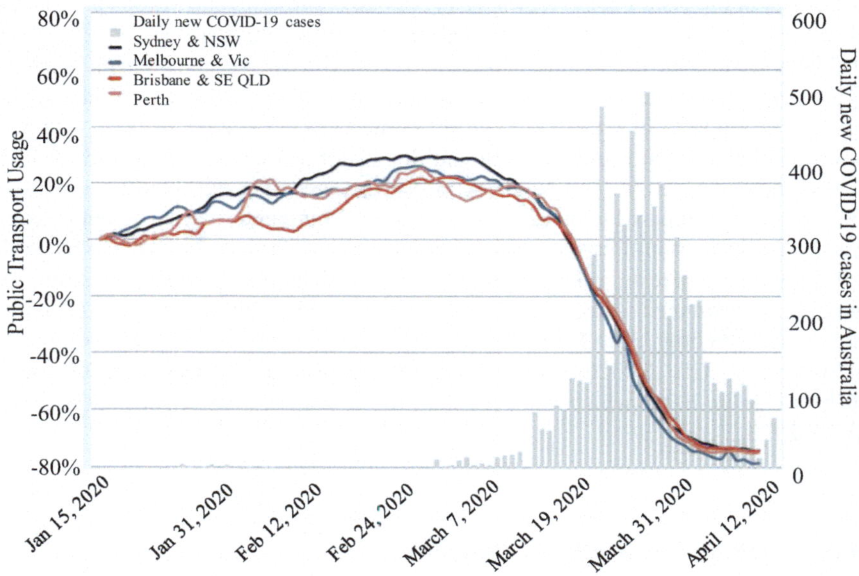

(b) U.S [25]

Fig. 14.5 Public transport usage in Australia and the U.S. during COVID-19. (**a**) Australia (WSP 2020). (**b**) U.S. (Moovit 2020)

countries (see the U.S for an example in Fig. 14.5b). With travel restrictions applied, public transport agencies are expected to bear higher operating and capital costs with decreased passenger demand.

14.3.3 Impact on Private Transport

As the COVID-19 pandemic continues and with physical distancing measures are in practice, the burden of travel is likely to fall on other transport modes that minimise the risk of infection. As a result, people who own a private car are going to use it extensively, more specifically within suburbs for grocery shopping and other essential services. While those people who relied on public transport and did not own private cars, they are likely to shift to other modes such as bicycle or walking. For instance, some Chinese cities reported increased usage of private cars, walking, and biking after this pandemic (WSP 2020).

14.3.4 Impact on Non-motorised Transport

The impact of COVID-19 on non-motorist transport (e.g., cycling) is no less than other modes of transport. Avoiding public transport and maintaining social distancing have exploded in cycling patterns. People are considering cycling as a resilient and reliable alternate mode of transport, which is evident in different countries of the world such as China, Germany, Ireland, the United Kingdom, and the United States (WSP 2020). With increased cycling volume and infrastructure demand, transport agencies are providing emergency bike lanes.

To promote cycling, various strategies, apart from providing additional lanes, are adopted. For instance, some states of the U.S. have closed a few streets for vehicular traffic and allowed cyclists to use these streets. Similarly, New Zealand started pop-up bike lanes and sidewalk widening projects to encourage cycling.

14.3.5 Effective Policy Decisions for the Transport Sector to Survive the COVID-19 Crises

To deal with the COVID-19 crisis, various countries are framing effective policies for the transport sector. For instance, Philippine national railways have placed markers on areas inside the trains, as well as on station platforms where passengers are allowed to either sit or stand in line. Similarly, Auckland road transport recently installed 12 km of pop-up spaces to ensure enough room in busy locations. Temporary spaces were adjusted to make sure people have enough room to walk along the footpath, give people space to queue outside restaurants, and wait at intersections while maintaining physical distancing. Furthermore, some of the cities have turned roads into bike lanes. For instance, Bogota has turned 100 km of road to the bike lane. Other cities have also implemented similar strategies such as Berlin, Mexico, Budapest, and Vancouver, etc. In Brisbane, Australia, some effective policy decisions are implemented. Firstly, passengers have to board and alight from rear doors only (earlier passenger were allowed to board from the front door only). Secondly, a strict social distancing rule has to be exercised when using public transport, i.e., passenger are expected to sit with at least one seat gap. Thirdly, buses and trains are disinfected regularly. Fourthly, and interestingly, no major reduction in frequency of buses or trains is observed. Finally, maximum of two passengers are allowed to use Uber, Didi, or other personal mobility hire services. Similar policy decisions are implemented in other cities of Australia.

A discussion with experts suggests some potential policies that may assist the transport sector to survive in this COVID-19 crisis. For instance, smart lockdown, limited or no public transport (intercity and intracity), limited/controlled freight transport with all precautionary measures such as disinfectants/masks, social distances of drivers, and accompanying staff. Similarly, promoting non-motorised transportation such as cycling and walking, and opening vehicular traffic streets for non-motorised modes may be a strong alternative.

14.4 Transportation Post-COVID-19

By implementing proper strategies and containment plans until the vaccine for COVID-19 develops, transportation may continue to struggle. There are a few facets of transportation post-COVID-19, which are described below. It is worth mentioning here that this section is written with the help of available online resources (cited wherever required) and discussion with the experts via a short expert survey containing essay type questions. Experts views and insights on the future of transport post COVID-19 are combined and discussed herein.

Expected medium-term shifts in transportation patterns: In the foreseeable future, when there is no strict restriction on social distancing, and personal travel may resume immediately post-lockdown, travel patterns may gradually resume to pre-lockdown levels. However, the immediate effects of COVID-19 have seen a massive decline in ridership and revenue loss. As a result, transport agencies are likely to take longer to get services running or may result in unreliable service, which may affect riders' trust towards public transport. A similar perception is observed for air travel. For instance, in a recent survey, about 50% of the respondents revealed that using public transport may be a high risk to their health due to COVID-19. While about 67% of the respondents indicate that airlines and airports are taking appropriate precautions, about one-fourth of respondents reported that they do not believe the travel industry (including airline and public transport) are adopting necessary precautions to minimise COVID-19 risks (Eno 2020).

Potential long-term shifts in transportation patterns: Speculating the precise long term effects of COVID-19 on overall transportation systems is difficult at this moment. Some systems may recover in some time by maintaining a pre-lockdown level of service, while the other systems may never rebound. For instance, ride-hailing, micro-mobility, and carsharing services may find a post-outbreak economy inhospitable for their business models.

While increased fears of sharing transportation modes with other people may force commuters to use other modes, cycling and other personal non-motorised modes are likely to see massive growth in the future. Bike purchase is going to be a more affordable alternative than a new automobile, whereas car sales have already declined compared to last year and likely to decline further. Thus, cycling is expected to be a sustainable mode of travel and may be a longer-term behavioural change in people travel patterns (Eno 2020).

Although governments have immediately asked for maintaining social distancing (e.g., maintaining 1.5 m distance), it is highly challenging and difficult to maintain such a gap with hugely populated areas. In Delhi, for instance, more than eight million trips are made by public transport (Bus and Metro) in a day. It has always been a challenge to manage the gap between demand and supply of commuting. This gap may be further widened due to the COVID pandemic requiring social distancing practices. In view of this, a systematic and strategic approach needs to be adopted to move ahead during the COVID-19 pandemic. A gradual change in demand/supply, and adoption of circumstances by commuters is expected. The great challenge on both sides needs to be dealt with carefully. On one side, it is expected that more commuters may prefer to travel by personalised vehicles (car and two-wheelers). On the other side, public transport capacity (fleet and frequency) has to be augmented, which requires massive investment from the government side. The current guidelines being developed by metro authorities in India address some issues of social distancing, but a long-term planning may be required.

14.4.1 Is "Work from Home" a Blessing in Disguise?

To minimise exposure to and control the spread of COVID-19, a new normal could be work from home. Technology giants are embracing such a policy measure. For instance, Google has asked its employees to work from home for the foreseeable future. The option of working from home has shown some promising results in controlling major externalities such as crashes. For instance, the state of Kerala, India, which accounts for a large share of road accidents due to the high density of population and overcrowding of roads, has witnessed a significant drop in the number of road crashes and fatalities due to the ongoing lockdown. According to the police data, only 105 road crashes were reported in the 21 days of the lockdown as opposed to 1787 incidents during the same period in 2019 (Vishnu Varma et al. 2020). Furthermore, a total of 13 people has lost their lives this year as compared to 185 persons last year during the same period. The above scenario of crash reduction and road deaths during this lockdown period of 2 months is also similar for many other states in India. However, in the long run, work from home may sustain only when it is taken as an integral policy instrument by governments to reduce travel demand and transportation externalities such as emissions, fuel consumption, and congestion.

14.4.2 Steps for Improving Public Perception About Public Transport and Recover Its Patronage

Following the frequent perception of 'new normal', the use of public transport in the near future (or foreseeable future) may decline with a greater number of people staying and working from home and preferring to use their private cars for travel to ensure their safety.

In general and especially in the COVID-19 period, public transport systems may be looked at as a revenue-generating means, though some public transport operators are running in public-private partnership mode. The public transport operators could be under a lot of burden and face social, economic, and financial consequences as ridership is likely to decrease, and other costs of operation may increase in the post-lockdown phase. In order to limit the social, economic, and financial consequences impacting the public transport system as a whole, the government should adopt exceptional measures, prioritise political and financial support for maintaining a sustainable urban transport system.

Recovering public transport patronage may require a variety of interventions, including direct subsidy to offset losses and need for maintaining social distancing inside public transport facilities and vehicles, measures to sanitise the physical assets, and priority for movement of public buses on the streets to make them faster and reliable options. Importantly, by doing so, the transport agencies must develop confidence in commuters that it is a safe option to use. Similarly, several different measures can also be adopted, such as stronger marketing, intensifying 'adopt public transport campaigns' and increase in fares for the higher classes of transport. Other measures, like adopting transit efficiencies through technological interventions, can also be useful to regain lost revenues. To summarise, in a post-COVID-19 world, public transport may need to be aligned and accessible for people to prefer transit over private cars.

Recently, a few transport agencies have framed policies for public transport for the post-COVID-19 era. Table 14.1 summarises the policies framed by different agencies for their respective countries. We report these guidelines from the bus transit perspective, however, similar guidelines are provided for other public transport services as well. The Council of Scientific and Industrial Research-Central Road Research Institute (CSIR-CRRI), India, has recommended spreading its fleet during peak times and increasing the supply of buses to cater for the demand. For ensuring the safety of passengers on board, the number of passengers are reduced proportionately to the length of the bus (e.g., for 18 m bus, the maximum allowable passengers are 18). Similarly, advanced booking and online payment of fares are introduced to minimise interaction with fare collection machines and operators.

Table 14.1 Summary of different policies by different agencies

Department	Fleet management	Maximum allowable passenger	Passenger boarding/alighting	Fare payment	Others
CSIR-CRRI, India (2020)	Peak spreading, optimum usage rather than increasing fleet, increasing the supply of buses by hiring school buses, tourist buses, other mini tourist buses	For 18 m bus = 18 passengers; For 12 m bus = 12 passengers	Boarding = Rear door Alighting = Rear door	Advanced booking, online tickets include 1-day pass, e-ticket, smart card and other e-payment methods (Paytm/ BHIM/UPI)	Increased bus dwell time
SMART-SUT (Parashar 2020)	No reduction in supply; if possible, increase supply of buses	For 12 m bus = 18–20 passengers; For 9 m bus = 12–14 passengers	Boarding = Rear door Alighting = Front door	Monthly passes; smart card based	Better communication of adjustments of schedules, timetables, and intervals
WSP [Scenario 1: Strict distancing; Scenario 2: Moderate distancing; Scenario 3: Relaxed distancing], Australia and New Zealand (2020)	Largely same frequency, peak spreading, and targeted increase in public transport service	In terms of percentage of seating capacity: Scenario 1: 16%; Scenario 2: 32%; Scenario 3: 53%	Boarding = Rear door Alighting = Rear door	Smart card based	NA
ITDP, India (2020)	Increase supply of buses by hiring school buses, tourist buses, other mini tourist buses	N.A.	Separate doors for boarding and alighting	Bus passes, incentives for buying passes, e-ticketing	Better communication of adjustments of schedules, timetable and intervals

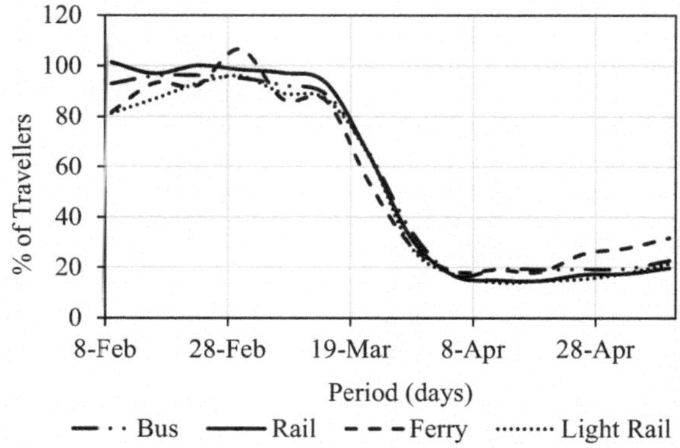

Fig. 14.6 Percentage of travelers in Australia in 2020 compared to the equivalent week in 2019 (ABC 2020)

WSP has recommended three scenarios post COVID-19, namely, strict, moderate, and relaxed distancing. During these three scenarios, public transport may run at a reduced capacity level. More specifically, seating capacities during scenarios 1, 2, and 3 could be 16%, 32%, and 53%, respectively. To minimise interaction with driver (particularly in buses), rear doors will be used for boarding and alighting. For fare collection, a smart card system will be used.

Institute for Transportation and Development Policy (ITDP), India, has recommended an increased supply of buses (by hiring school buses, tourist buses, other mini tourist buses) and asked for using different doors for alighting and boarding. In addition, an incentive-based program may be introduced for online purchase of passes to encourage riders and minimise their interactions. Lastly, to recover their pre-lockdown patronage, passengers should be better informed about the timetable of services and intervals.

The above policies have been mainly framed for the South Asian context (Indian in particular), while developed countries like Australia have started to recover slowly because of effective policy decisions during lockdown (mentioned before). State governments are encouraging their residents to work from home wherever possible to cope with ordinary peak hour demand. Even before lockdown restrictions are eased, commuters were seen to use public transport a little more during mid-May (see Fig. 14.6 for a reference). These arguments are by no means directed to compare India and Australia.

14.4.3 COVID-19 Scenario: An Opportunity to Transform Transport Systems

Almost all cities of COVID-19 affected countries have come to a standstill due to lockdown. This section highlights how do policy makers can utilise these lockdown scenarios in future to transform the transport system.

The present COVID-19 situation is a great opportunity for dealing with many transport problems in reducing trips and reviving sustainable modes (walk, bicycle and public transport), which otherwise is a very daunting and challenging task. Substantial reduction in transportation due to COVID-19 has brought environmental benefits such as cleaner air and lower carbon emissions.

The policymakers can systematically implement some of the travel demand management measures to reshape the transport system. These measures can encourage short length trips by non-motorised transport (walk, bicycle etc.) and intermediate public transport modes (rickshaws, autos, etc.) by providing proper quality of infrastructure in terms of continuous footpaths, walkways, and dedicated path/lane. Other demand management measures may include priority to sustainable modes at intersections, dedicated lanes for public transport modes, and staggered days and/or working hours for offices, markets, shopping, schools to spread peak demand.

When streets are empty, in such situations, this opportunity can be easily utilised by authorities to re-imagine and re-assign transport network space to encourage sustainable modes like walking, cycling, and public buses. These physical interventions may be easily implemented during the lockdown situation when roads are empty, and there is no disruption to the movement of traffic. Also, there could be better acceptability to hard policy decisions of government during such situations that may eventually improve the sustainability of the transport system.

The population of a city/region/state or even country, can now be tested and more importantly be expected to be accepted by the people. For now, people are seen to be realising the severity of the situation, and with lockdown continuing and impacting them severely in most countries, they are ready to bargain significantly if offered their freedom back with such restrictions. Some of the examples are as follows:

(a) *People can be encouraged to cycle*: with most governments around the globe returning to 'new normalcy' in the phased wise system, this is a time when a lot of otherwise pedestrian dominated areas (shopping streets/market areas/CBDs) may be declared to be vehicle free/open to cycles only as a part of 'precautionary measures' by the government. With businesses shut for a long time now, this significantly impacted shopkeepers too, for being allowed to open again, can be expected to be more cooperative than ever before, or perhaps ever again.
(b) *This is a good time to establish new acceptable social order*: in developing countries like India, social status is one factor that repulses people away from cycling/walking to short distances, the other being comfort, of course. Nevertheless, if it is in the guidelines issued by the local government, it can be expected to be followed, if not forever, but for a significant time in the near future. This strategy would give enough time to researchers to conduct all the related studies for the scenarios that were only possible to be simulated hitherto.
(c) *Glorify the positive impacts of COVID on the environment*: as people are already witnessing and appreciating lower air pollution, reduced greenhouse gas emission caused by transportation, it may desirable to not let these impacts fade away so quickly from the minds of people.

Fig. 14.7 A typical example of the Milan model

(d) *Promote minimised person-to-person contacts*: maintaining social distancing at all time, online tickets and reservations, fast TAGs, and online payments can be promoted more than ever before.
(e) *Hygiene in public transport vehicles/public places*: strict penalties for littering inside public transport systems and places should be placed. People have always wanted clean public spaces to be their 'Right'. Now is the time they realise it to be their 'responsibility'.
(f) *Zero tolerance for overloading private buses*: it is believed to be a lethal practice and a challenge for road safety always, private bus operators shall be made clear about the straight cancelling of licenses in case overcrowding of buses is observed any time.

To summarise, the picture would remain grim for the public transport sector for months to come. Operators can utilise this time to maintain and upgrade with some post-COVID 19 arrangements/sanitising and screening arrangements.

Whether we can capitalise this opportunity for the future development is the burning question? Policymakers of Milan, the capital of Lombardy, Italy has already announced a plan named "Strade Aperte (Open Streets)" to capitalise on the nation-wide lockdown due to COVID-19. As per the article in The Guardian (Laker 2020) "*The Strade Aperte plan, announced on 21st April 2020, includes low-cost temporary cycle lanes, new and widened pavements, 30 kph (20 mph) speed limits, and pedestrian and cyclist priority streets*". A total of 35 km of streets will be transformed as per the plan. Figure 14.7 displays a comparison of Corso Buenos Aires before and after the Strade Aperte plan. Janette Sadik-Khan, a former transportation commissioner for New York City, perfectly puts (Laker 2020) "*The Milan plan is so important because it lays out a good playbook for how you can reset your cities now. It's a once-in-a-lifetime opportunity to take a fresh look at your streets and make sure that they are set to achieve the outcomes that we want to achieve: not just moving cars as fast as possible from point A to point B, but making it possible for everyone to get around safely.*"

14.4.4 *Contagious Virus Response Planning for Transport Agencies*

In response to COVID-19, most countries reacted to the situation as it panned out. Once an epidemic becomes a pandemic, it becomes excessively difficult to contain the spread with the reactive response approach. A combination of continuity of operations plan and a contagious virus research plan can provide the transportation policymakers a comprehensive response planning. As per American Public Transportation Association (APTA) standards "*the purpose of a Continuity of Operations Plan is to prepare a transit agency so that it can provide essential agency functions following a significant emergency event that limits or restricts the availability of personnel, facilities or technical systems. Once an outbreak reaches pandemic proportions, the contagious virus research plan will likely need to be used in conjunction with the continuity of operations plan to provide comprehensive response planning.*" Figure 14.8 describes the basic elements of a continuity of operations plan, which are detailed in (APTA SS-SIS-RP 2013). On the other hand, the elements of a contagious virus research plan are identification of alert phases that trigger specific action, information and education program, disinfection program, sanitary aids to limit the spread, vaccine/antiviral medications, service reduction, shutdown, and restoration. Figure 14.9 depicts the last element of a contagious virus research plan in detail, which is summarised form of a detailed discussion presented in APTA SS-S-SEM (2013).

14.5 Conclusions

This book chapter sheds light on how COVID-19 impacts the transportation system. A review of the role of transportation in the spreading of infectious diseases is presented. More specifically, four stages are identified, namely, emergence, translocation, diffusion, and pandemic. Moreover, the current impacts of COVID-19 on air travel, public transport, private transport, and non-motorised transport are discussed, and effective policies framed by the transport sector to deal with COVID-19 are highlighted. COVID-19 is found to result in a modal shift from public transport to private transport and non-motorised transport modes like cycling and walking. This travel pattern may remain for the foreseeable future to avoid the spreading COVID-19.

This book chapter has also discussed transportation in the post-COVID-19 era, where work from home would be a new normal for a class of workers who did not require physical presence. Such an alternative would minimise motorist trips, leading to reduced gas emissions and improved air quality. Moreover, a number steps for improving public transport perception and patronage post-COVID-19 are identified. It has also been discussed how COVID-19 scenario could be used as a tool for future planning and can assist in preparing ourselves for any future pandemic. A nice

Fig. 14.8 Basic continuity of operations planning guidelines for transportation agencies

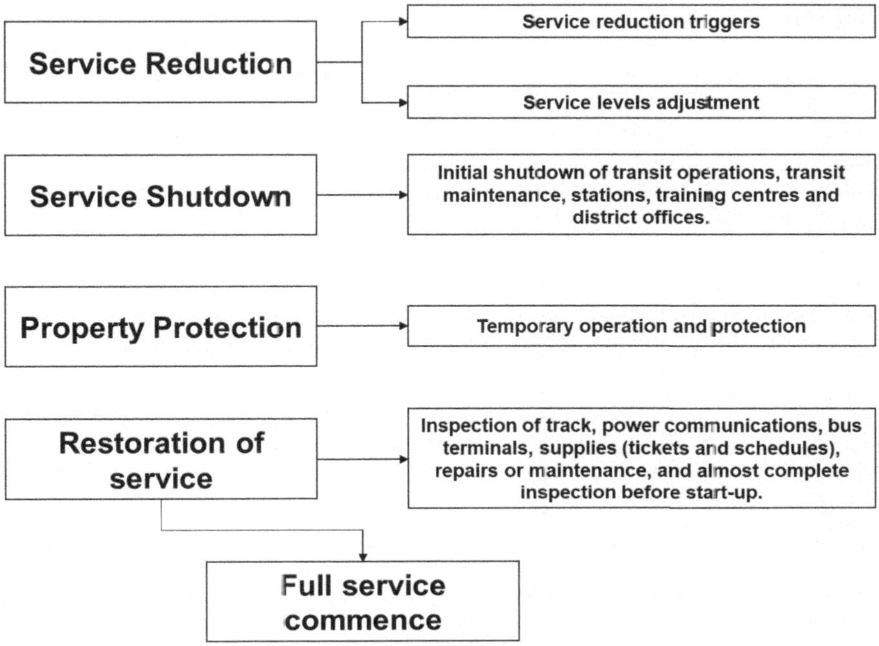

Fig. 14.9 Basic contagious virus response plan for transportation agencies

un-published white paper can also be referred for sustainable transportation intervention in the post COVID-19 period (Verma et al. 2020).

This book chapter can be particularly useful for transport policymakers, operators, agencies, and users who are in transitioning stages to develop future line-of-action in combating this global pandemic (COVID-19) and how to attract the lost patronage. The learnings from COVID-19 will assist us in becoming more prepared to tackle scenarios like this in the future.

Acknowledgement The authors would like to acknowledge the precious time given by the following experts (in alphabetical order) in responding to a short survey on COVID-19 and transportation: Abdul Rawoof Pinjari (Associate Professor and Chairman, Centre for infrastructure, Sustainable Transportation and Urban Planning (CiSTUP), Indian Institute of Science (IISc), Bangalore, India), Ashish Bhaskar (Associate Professor, Queensland University of Technology, Australia), Ashish Verma (President, Transportation Research Group of India (TRG) and Associate Professor, Indian Institute of Science (IISc), Bangalore, India), Carlo Prato (Professor and Acting Head of School of Civil Engineering, The University of Queensland, Australia), Manorajan Parida (Professor and Deputy Director, Indian Institute of Technology Roorkee, India), Muhammad Irfan (Professor and Dean, National University of Sciences and Technology, Pakistan), Satish Chandra (Professor and Director, CSIR-Central Road Research Institute, India). We also acknowledge various online resources that have helped us to write this chapter.

References

ABC (2020) How public transport usage is creeping back after coronavirus lockdown

Ali SH, Keil R (2006) Global cities and the spread of infectious disease: the case of severe acute respiratory syndrome (SARS) in Toronto, Canada. Urban Stud 43(3):491–509

Al-Jazeera (2020) Timeline: how the new coronavirus spread

Andrews JR, Morrow C, Wood R (2013) Modeling the role of public transportation in sustaining tuberculosis transmission in South Africa. Am J Epidemiol 177(6):556–561

APTA SS-SIS-RP (2013) Standard for a continuity of operations plan for transit agencies. In: APTA standards development program. American Public Transportation Association, Washington, DC

APTA SS-S-SEM (2013) Developing a contagious virus response plan. In: APTA standards development program. American Public Transportation Association, Washington, DC

Bradley D (1989) The scope of travel medicine: an introduction to the conference on international travel medicine. In: Travel medicine. Springer, Berlin, pp 1–9

CSIR (2020) Guidelines for public transport and feeder modes considering social distancing norms. In: Commuting in urban area during COVID-19 pandemic. CSIR, New Delhi

Duplaix N (1988) Fleas: the lethal leapers. Natl Geogr 173(5):672–694

Eckert EA (2000) The retreat of plague from Central Europe, 1640–1720: a geomedical approach. Bull Hist Med 74(1):1–28

Eno (2020) How might personal transportation behaviors change as a result of COVID-19, and what does that mean for policy?

Fadel F, Khalil El Karoui M, Knebelmann B (2008) Spread of a novel influenza A (H1N1) virus via global airline transportation. Nat Rev Immunol 8:153–160

Faruque SM, Chowdhury N, Kamruzzaman M, Ahmad QS, Faruque A, Salam MA, Ramamurthy T, Nair GB, Weintraub A, Sack DA (2003) Reemergence of epidemic Vibrio cholerae O139, Bangladesh. Emerg Infect Dis 9(9):1116

Grais R, Ellis JH, Kress A, Glass G (2004) Modeling the spread of annual influenza epidemics in the US: the potential role of air travel. Health Care Manag Sci 7(2):127–134

Guimera R, Mossa S, Turtschi A, Amaral LN (2005) The worldwide air transportation network: anomalous centrality, community structure, and cities' global roles. Proc Natl Acad Sci U S A 102(22):7794–7799

Haghani M, Bliemer MC, Goerlandt F, Li J (2020) The scientific literature on Coronaviruses, COVID-19 and its associated safety-related research dimensions: a scientometric analysis and scoping review. Saf Sci 129:104806

Horna-Campos OJ, Sánchez-Pérez HJ, Sánchez I, Bedoya A, Martín M (2007) Public transportation and pulmonary tuberculosis, Lima, Peru. Emerg Infect Dis 13(10):1491

IATA (2020) COVID-19 impact on Asia-Pacific aviation worsens

ITDP (2020) COVID response: post-lockdown guidelines for bus operations

Laker L (2020) Milan announces ambitious scheme to reduce car use after lockdown

Luke TC, Rodrigue J-P (2008) Protecting public health and global freight transportation systems during an influenza pandemic. Am J Disaster Med 3(2):99–107

Moovit (2020) Impact of Coronavirus (COVID-19) on public transit usage

Nicholson KG (2003) Origin of influenza pandemic, Influenza. Lancet 362:1733–1745

Olsen SJ, Chang H-L, Cheung TY-Y, Tang AF-Y, Fisk TL, Ooi SP-L, Kuo H-W, Jiang DD-S, Chen K-T, Lando J (2003) Transmission of the severe acute respiratory syndrome on aircraft. N Engl J Med 349(25):2416–2422

Palese P (2004) Influenza: old and new threats. Nat Med 10(12):S82–S87

Parashar L (2020) Standard operating procedures (SOPs) for bus transport for post COVID-19 lockdown

Rodrigue J-P, Notteboom T (2020) Transportation and economic development. In: The geography of transport systems. Routledge, New York

Sack DA, Sack RB, Nair GB, Siddique AK (2004) Cholera. Lancet 363:223–233

UITP (2018) World metro figures. In: Statistics brief. UTIP Advancing Public Transport, Brussels

Verma A et al (2020) Post COVID-19 world: sustainable transportation interventions and solutions required. Un-published white paper, Transportation engineering Lab, Department of Civil Engineering, IISc Bangalore, India

Vishnu Varma RAA, Agarwala T, Pisharody RV, Deb D (2020) Covid-19 lockdown has some unintended gains: fall in crime rate, fatalities on roads, suicides. The Indian Express

WHO (2020) WHO Coronavirus disease (COVID-19) dashboard. WHO, Geneva

Wikipedia (2020) List of countries by rail usage

WSP (2020) Public transport and COVID-19: how to transition from response to recovery. White paper

Chapter 15
Integrated Risk of Virus Outbreaks

Gilbert Hinge, Alacrity Muksor, Manish Kumar Goyal, Ravi Kumar Goyal, and Mredu Goyal

Abstract The initial outbreak of Corona Virus Disease in 2019 (COVID-19) in Wuhan, China, has led to its spread in almost all countries of the World. In an attempt to slow down the virus' quick spread, the government of major nations has enforced shutdown of its various borders, travel restriction, lockdowns, and recommend quarantine and social distancing. The actions taken have, therefore, impacted the multiple sectors of the economy. To understand these impacts, the present chapter attempt to sum up the sectors that are being hit by the outbreak and sectors that have benefitted from it. The chapter also highlights briefly the overall effect of the risk outbreaks on global economies and actions taken to mitigate the impact of COVID19 on the economy. A review of media reports and literature was carried out to meet the objective of the present chapter. Findings showed that, though some sectors, such as companies that offer stay-at-home products and the pharmaceutical sector benefitted from the outbreak, others such as chemical industries and medical supply industries are both positively and negatively affected. To build back the resilience in the global economy, aggressive fiscal responses were made. However, considering the magnitude of the crisis, the likelihood that the economy will get back to normal would be slower.

Keywords COVID-19 · Global economies · Risks

G. Hinge
Department of Civil Engineering, Assam Downtown University, Guwahati, Assam, India

A. Muksor
Department of Humanities and Social Sciences, Indian Institute of Technology, Guwahati, Assam, India

M. K. Goyal
Civil Engineering, Indian Institute of Technology, Indore, Madhya Pradesh, India

R. K. Goyal (✉) · M. Goyal
School of Entrepreneurship Skills, Bhartiya Skill Development University, Jaipur, Rajasthan, India

M. K. Goyal, A. K. Gupta (eds.) *Integrated Risk of Pandemic: Covid-19 Impacts, Resilience and Recommendations*, Disaster Resilience and Green Growth,
https://doi.org/10.1007/978-981-15-7679-9_15

15.1 Integrated Risk of COVID-19 Pandemic

The initial outbreak of Corona Virus Disease in 2019 (COVID-19) in Wuhan, China, has led to its spread in almost all countries of the World. As a result of growing case notification, the outbreak was declared a Public Health Emergency of International Concern on January 30, 2020. On March 11, 2020, it was declared a pandemic by the World Health Organization (WHO) (Zhang 2020; Felipe 2020). In an attempt to slow down the quick spread of the virus (Burkert and Loeb 2020), the government of significant countries which constitute the World's largest economy have enforced shutdown of its various borders, travel restriction, lockdowns and recommend quarantine and social distancing (Al Jazeera 2020; Gov.UK 2020), thus putting fear on the probable economic recession (Buck et al. 2020). In India, a nationwide lockdown began on March 25, 2020 (Dun and Bradstreet 2020).

In a highly globalized world, the disease outbreak's economic risks are becoming more and more prominent. The slowing down of the Chinese economy is interrupting production as most of the World's companies depend on China for inputs; the functioning of global supply chains has also been disrupted (McKibbin and Fernando 2020). Globally consumption is severely hit, which adversely affected investment; exports have also declined, thus driving the global economy to a state of collapse or disaster (Dhar 2020). Hazard, exposure, vulnerability, and resilience constitute the four concepts whose interactions lead to economic consequences (Noy et al. 2020). In this case, the hazard is the novel Coronavirus, known as SARS-CoV-2 (Severe Acute Respiratory Syndrome Corona Virus-2). The economic risks are determined by the exposure and vulnerability to the virus and the resilience of the economy.

In an attempt to understand the impact of this outbreak on the economy, this chapter will summarize the sectors that are being hit by this outbreak and sectors that benefit from it. The chapter also highlights briefly the overall effect of the risk outbreaks on global economies and actions taken to mitigate the impact of COVID19 on the economy-globally and India, in particular. A review of media reports and literature was carried to meet the objective of the present chapter.

15.2 How Did the Outbreaks Affect Different Sectors?

Coronavirus has affected many people's lives globally, it has also caused deep economic distress. Most of the companies struggle to keep afloat and, in turn, affecting their revenues and profits amid the Covid-19 crisis (Fitz Gerald et al. 2020). However, the impact differs across sectors as the country eases down the lockdown in some sectors (Whiteside 2020). In terms of industry, the aviation sector is hardest hit, while pharmaceutical and streaming industries are in a better position (Find out how coronavirus is impacting all industries; aviation, hospitality worst hit 2020).

15.2.1 Industries Negatively Affected by the Virus Outbreak

Mentioned below are sectors that are negatively affected surrounding the COVID-19 outbreak.

15.2.1.1 Aviation

After the government enforced strict instruction to maintain social distance and restriction on unnecessary travel (Al Jazeera 2020; Gov.UK 2020), the travel industry sector was walloped with lots of cancelation and decline in demands (Upadhyay 2020). Air travel was suspended across Asia and Africa (Coronavirus travel updates: which countries have restrictions and FCO warnings in place? 2020). The U.S. barred all foreign nationals from entering its states (Ellis 2020). A similar suspension has been implemented across Europe (Britons urged to avoid non-essential travel abroad 2020).

As a result of these events, the share prices of airlines have fallen by almost 25% as per the press release, no 12 by IATA [The International Air Transport Association] dated March 5, 2020. This drop-in in its share price is more than a 21% percentage point compared to the drop that happened during the Severe Acute Respiratory Syndrome crisis occur in the year 2003 (IATA 2020). India's top airline, Indigo, report a loss of around Rs. 1750 crores, while Spicejet's was losing by around Rs. 1050 crores (Mudgill 2020). Airlines from the U.K. have called for emergency bailed out, encompassing £7.5 billion to prevent shutdown (Jones 2020). Travel markets have declined across the World, including South Korea (−46%), Taiwan (−38%), Hong Kong (−75%) (Jones 2020).

The sector worldwide incurs higher fixed costs and due to high competition from low-cost carriers, the margin of safety is very low (Shared and Orelowitz 2020). As reported by global times (Reuters 2020) the aviation industry predicts decrease in revenues for $314 billion which is more than half of total revenue expected this year besides hitting approximately 25 million jobs.

Further, the directions by the ministry of civil aviation to refund cash for cancelled tickets (booked from March 25 for travel up to May 17) has pushed online travel aggregators and travel agents into a deep liquidity crisis. As the low-cost airlines are giving refunds only in the "virtual wallets" and credit shells for future travel (Kulkarni 2020).

However, the recent severe decline in prices of crude is an affirmative news for the sector. But, the advantage of lower prices of aviation fuel cannot be derived until the airlines operate at optimal capacity and utilization (The Economist 2020). Since a sudden change in consumer behavior related to travelling is observed, therefore, the sector will take a long time to revive even in the post lockdown period as people are expected to still avoid travelling for some time.

15.2.1.2 Tourism

The tourism industry represents 10% of the world GDP as per the statistic from 2000 to 2019 provided by the world travel and tourism industry (Moodie Davitt Pax Index: the latest traffic figures from airports across the world 2020). However, as a part of the COVID 19 lockdown, tourists were not allowed to visit any tourist destinations, thus driving this industry to a complete halt. The World Travel and Tourism Council [WTTC] has cautioned that the job of more than 50 million people working in the industry might be at risk (Lock 2020) and Asia will be the most affected continent as projected by the WTTC.

The recent statistic data provided by the government of India indicated that the Foreign Tourist Arrivals [FTA] has dropped by 9.3% monthly as a result of this outbreak. This statistic has further worsened as the Indian government announced the suspension of all Tourist Visa till April 15 to stop the spread of the virus (Faus 2020). Globally, a country like Vietnam which welcomed a record number of 1.5 million Chinese Visitors just in the first quarter of 2019 is projected to suffer a $5 billion loss (Bureau 2020) while the country like the Philippines is expected to suffer from a loss of approximate 0.3–0.7% of its Country year's GDP (Vietnam's tourism sector could lose US$ 5bln due to COVID 19 2020).

The tourism industry unswervingly supports generation of one in ten of the global jobs (Haywood 2020). In the wake of the unfathomable 'domino effect' of the pandemic, a vicious decimation in tourism-allied employment, cascading throughout supply chains is projected (World Travel and Tourism Council 2020). The most evident and immediate impressions are on all its geographical fragments including inbound, outbound, domestic as well as nearly all verticals comprising adventure, luxury and leisure, MICE (Meetings, Incentives, Conferences and Exhibitions), heritage, cruise and business (Dash 2020).

As reported by the International Monetary Fund, 2020 is one of the worst years for the world economy since the Great Depression of the 1930s (Gopinath 2020). Economies reliant on tourism particularly Southeast Asia and European countries are among those being hardest hit. The countries like Thailand, Malaysia, Singapore are estimated to lose US$3–6bn in tourism-related revenues (Euronews.com 2020). However, the combined GDP for the 19 Eurozone countries is likely to contract by 7.5% (GlobalData 2020).

15.2.1.3 Sports Industry

The virus has affected all areas of life, and the sports industry is no exception. The spread of the disease has led many sports events around the globe and professional league of various sports to be suspended or deferred hoping for the condition to be better (Suspension of Premier League extended until April 30 amid coronavirus pandemic 2020). Deferment and suspension of games affect not only the passionate fans of these games but also the people working in the various sports industry, the athletes themselves, and the media industry. Most of these significant sports events depend on media coverage to generate income. As shown by the sports revenue data, the global value of sports media is around $50bn accounted for only the ten significant sports leagues (Hall 2020). Thus a temporary lockdown means no games, which implies no T.V. rights, thus affecting the industry.

The Board of Control for Cricket in India has suspended cricket's most popular league known as the Indian Premier League (IPL 2020 suspended until further notice due to Covid-19, season will commence only when it is safe: BCCI 2020). Euro 2020 football's most awaited event has been postponed for 12 months (Sport-by-sport look at the impact of corona virus around the world 2020). The Tokyo Olympics has also been delayed, which could badly hit the Japanese economy, and it is the third-largest economy in the World (Zaharia 2020). The postponement affected the preparation and lives of around 11,000 Olympic and 4400 Paralympic athletes apart from the huge monetary cost of organizing the Games (pre postponement). The total official cost is 1.35 trillion yen ($12.6 billion) (Dasgupta 2020). Sponsorship cost invested by Japanese business houses alone is at record to be 348 billion yen ($3.3 billion). Moreover, the postponement of the games is projected to add a further cost of about $2.7 billion (Dasgupta 2020). These are a few examples; however, all other sports events from golf, tennis, boxing, and others have canceled their games to curb the spread of this disease.

Furthermore, new regulations of social distancing and risk averse techniques may completely transform the fundamentals of sports (Evans et al. 2020), such as use of saliva to shine the ball in cricket that helps it to swing and tackle, blocking, group hugs, etc. in football, hockey and other sports. Further, the thrill and stimulating sportsman spirit injected by the yelling spectators would not any more instigate the sports person. Besides, implications of accessing restrictions to training venues, cancellation of events, and loss of earnings may commensurate psychological consequences. Emotions, psychological burdens, anxiety, traumatic experiences and resilience in the athletic community are consequential corollaries of the pandemic (Parnell et al. 2020).

As per the World Economic Forum, the estimated global value of the sports industry was $471 billion in 2018, which was a 45% upsurge since 2011. The year 2020 was expected to be a vital year for world sports, with the Olympics, the FIFA U-17 Women's World Cup, the European Championship of football 2020, T20 Cricket World Cups, all programmed to be held (Dasgupta 2020).

15.2.1.4 Food Sector

Even before the virus outbreak, there were indications that the world food supply struggles to cope with climate change, which threatened many people's lives and livelihood (Reinhart and Subbaraman 2020). The COVID 19 outbreak further intensifies this impact leading to a surge in food supply prices (Bloomberg 2020). Food distribution and food retailing have been put under tremendous pressure because of people's panic buying (Bachelor 2020). Panic buying and food hoarding mean low-income people who can't afford to follow this are suffering (Stockpiling 'will hit vulnerable', food bank warns 2020).

In response to this, the Indian government has made an effort to provide food to millions of its migrant workers (India to provide free food grains to millions of migrant workers 2020). Likewise, the U.K. government offers free meals and food parcels to its elderly citizens and students of low-income families (Food parcels coming for vulnerable individuals 2020), leading to global food demand. These global food demands are healthy, helping the supermarkets; however, this could soon be affected because of the disruption in supplies and distribution. Moreover, the labor shortage resulting from travel restriction will affect the migrant farmworkers' normal seasonal cycle (Reinhart and Subbaraman 2020).

Furthermore, stores such as restaurants, pubs, cafes, caretakers have been forced to close, putting many of their workers to lose their jobs (UK pubs and restaurants told to shut in virus fight 2020). Challenger, Gray, and Christmas firm confirm in its report that more than 600 food service jobs were straight cut after the outbreak, and over 7.4 million could lose their jobs (Restaurant closures may cost 7.4 million jobs reports says 2020).

Furthermore, the constrained mobility of delivery officials creates auxiliary challenges. Moreover, if present circumstances extend, it will be very challenging, specifically for small self-regulating restaurants because of high fixed costs coupled with low financial liquidity involved.

The lockdown period has reshaped the consumer behavior, especially their eating habits. Today, they are searching for healthier food options and are preferring to cook at home to evade any risk of infections of the virus. All these abrupt alterations have directly affected the food aggregators and forced them to undergo unfortunate downsizing exercise. A work force cut of 13% by Zomato and 1100 employees laid off by Swiggy are imminent decisions to the pandemic (Rao 2020).

15.2.1.5 Agriculture Sector and Small Businesses

Agriculture constitutes the primary occupation of people living in rural areas. Accounting for 60.4% of employment in developing countries is a source of livelihood for many and contributes to up to two-thirds of these countries' gross domestic product (GDP) (International Labour Organization 2020). As per the United Nations report, 2019, 69% of India's population lives in rural areas (Khan 2020). As they depending on agriculture, the movement restrictions could badly hurt this sector due to difficulty in availing agricultural inputs (Khan 2020; Kumar 2020). After relaxations were made in a phase-wise manner, the sector could now continue; however, with low incomes, the product's demand would decrease. Due to low demand, prices would fall, or the supply would eventually reduce at the source. The fall in price is not right in the case of urban areas, as shortages could lead to price rise (Kumar 2020). The decrease in demand at source due to the closure of restaurants, institutions, and restrictions of movement has led to the reduction of farmers' income—about 75% fall in wholesale prices of food items was reported (International Labour Organization 2020; Sally and Bhosale 2020).

The complete shutdown of exports has impacted India's agricultural sector to a large extent as it is a significant exporter of crops. According to the Agricultural and Processed Food Products Export Development Authority (APEDA) in 2018–2019, India's agri-exports worth Rs. 685 billion (Khan 2020). The Micro, Small and Medium Enterprise MSME and SME's which constitutes small factories, small businesses, traders and shops which provides jobs to low-skilled laborers are also being affected by the lockdowns (Kumar 2020). The extension of lockdowns could be a risk for these small and unorganized sectors of the economy, thereby creating a loss to countries' GDP.

Micro, small and medium enterprises (MSMEs) entails a leading share of enterprises globally, extending from micro enterprises to medium-sized businesses. They encompass a projected 70% of worldwide employment and are accountable for nearly 50% of GDP in developing countries (Raney 2020). The damage caused by the pandemic is expected to keep the MSMEs economically sick for a long period of time. Most of the MSMEs are globally featured to be operating on thin margins (even in normal times) and are regularly suffering from liquidity problems due to limited cash reserves (Sipahi 2020). Thus these are smarting under severe pain in difficult pandemic times. Presently, it is impossible for them to tap credit lines, as opposed to large businesses that have established relationships with banks for lending and efficient working capital management systems. These enterprises, mass employers of the economy, have little remedy principally when demand for their products and services vanishes (Raney 2020; Sipahi 2020). In developing countries, MSMEs are more inclined towards the manufacturing service and tourism sectors that are highly vulnerable to the COVID-19 outbreak. Moreover, informal and unregistered enterprises, which represent the majority of MSMEs in developing countries, are the worst hit by the lock-down. Without policy measures to recover their resilience, MSME survival may be endangered (Goyal 2020).

15.2.1.6 Hospitality

The hospitality industry has been struck by the outbreak, with hourly workers being the most affected as the industries get less business. Hotel industry revenue per room has dropped by 13–29% in India's top cities during January–March (Covid-19 hits hotel industry; revenue per room drops 13–29% in Jan–March in top cities 2020). While in the United States, the revenue industry fell by 11% in March, in China, the rates drop by 89% in January 2020 (Durbin 2020). Hotel occupancy in Germany has fallen by more than 36% as per the survey conducted in March 2020. Italy's cities, such as Rome, are the most affected with the hotel occupancy rate of just about 6% (Coronavirus hits German hotel industry hard: more than every-3-guests stay away 2020). Overall the outbreak has led to the overall slump in the hotel industry.

With no dine-in services and big drop in online orders the restaurant industry has been severely impacted due to the pandemic (Jain 2020). As reported by the National

Restaurant Association of India (NRAI), the estimated loss for the Indian restaurant industry in the year of pandemic is approximately $9 billion (Deb 2020). Besides, even after the lockdown is over the restaurants are supposed to operate at half of its capacity to ensure social distancing directly hitting their bottom line in a gigantic way.

Furthermore, the Hospitality industry is the locus of all other industries and is directly or indirectly affected by as well as affects all of them. Abrupt global lockdown has halted corporate movements, MICE tourism, big fat destination weddings, etc. bringing the hospitality sector to a standstill. Once the global economy begins to recuperate, new travel configurations are expected (Deb 2020) to emerge. A psychological influence of the pandemic on consumer travel behaviors is predicted. Drive-to-resort destinations, avoiding large group gatherings, lesser dense markets and change in lodging preferences may develop. The recovery rate of the hospitality sector will depend upon the pace the sector adopts to modify itself according to the changed preferences (Jain 2020; Deb 2020).

15.2.1.7 Real Estate

Ambiguity clouds are hovering over the real estate industry worldwide. The coronavirus outbreak has caused widespread concern and economic hardship for the real estate sector across the globe (Nicola et al. 2020). Both the real sectors i.e. residential and commercial sectors are projected to be hit in terms of launches, prices as well as sales. Stringency of cash liquidity in the market has stopped the new constructions (Babar 2020). Besides, industry veterans believe that the extent of the impact is difficult to assess completely and thus the sales may take additional hits in the future. The recovery time and path is also expected to be slow and painful (Infographic: how commercial real estate owners can respond to Covid-19 2020).

Since, a fewer footfall is expected in malls and large commercial spaces in near future, the rental value and income both are likely to either remain stagnant or slide. The leasing activity is also predicted to remain below the par during this period (Babar 2020).

As per a research study, due to lockdown stretching for more than 2 months, US may lose $4 trillion amount of all business establishments which is a quarter of it (Infographic: how commercial real estate owners can respond to Covid-19 2020). Even Canada is heading towards a recession. Moreover, the housing market of Australia is also predicted to be affected. Singapore is amongst the most hit countries by the COVID19 and this is also destructive from the housing business outlook as Singapore is counted among the prominent investors in Asia-pacific real estate markets. Further, mounting unemployment and deferred building approvals direct a decline in the construction segment (Ozili and Arun 2020).

However, the commercial real estate is predicted to get a fresh start specifically in India, during the post covid period (Chauhan and Kapoor 2020). In the long run India will comprehend interest both for equity as well as for global players to come and set their industrial units and offices in India. When this happens, a lot of infrastructure will be required and if the real state sector manages to leverage the opportunity, the losses incurred may be recovered soon (Chauhan and Kapoor 2020). Although investors worldwide have postponed new acquisitions decisions, some institutions with high net worth may decide for investments in real estate with a long term perspective in view (Babar 2020). Nevertheless, credit arrangements can prove to be a challenge for investors. According to a report, construction lending is encountering additional challenges than the complete financial markets (Chauhan and Kapoor 2020). The reason is obviously the doubts across construction timelines and social distancing norms that have developed lenders and underwriter's to be conservative and extremely selective towards borrowers worldwide (Chauhan and Kapoor 2020).

15.2.1.8 Automotive Industry

The sun shine sector, automotive industry is expected to unleash a host of challenges post pandemic worldwide. The fast-evolving consumer behavior and dwindling sales are the major concerns of the industry. The industry experts view that the future road is filled with trials amid endless uncertainties (Collie et al. 2020).

Plant operations were suspended among automakers in US, China and Europe. Besides, interrupted production and faded demand all the showrooms globally were shut and society was confined to their homes (Deloitte 2020). The statistics reveal that in US declines of 45% and 38% respectively were recorded in April and March. However, May sales were 30% declined (ETAuto 2020). In May 2020, the sales for western Europe decreased down for 556,000 units which accounted for 57% of the YOY sales. Similarly, new car registrations in UK were reduced by 89%. In Europe, a projected production loss of approximately two million vehicles was recorded resulting in the loss of one million jobs (ETAuto 2020). Even worse scenario was recorded in India, after nil sales in the month of April, only 37,000 units shipment was sent to dealers in May, and a total 85% deterioration YoY was recorded after even after opening up of the country in May month. Also, a major loss was suffered by the Indian passenger vehicle industry which weakened to 18% with only 27.75 lakh units sold till the end of March.

The life of a consumer in the automotive sector revolves around economics and that's the reason that the discounts and deals especially during festive season have been a major marketing strategy of the sector (COVID-19 and the automotive industry 2020). But, in present times maintaining liquidity through conserving cash is the primary goal. Moreover, the Thus flash and features will no more entice the consumers else value for money and low operating cost will supersede. Another main reverberation of the current crisis would be the eminence of Work From Home (WFH) culture which will directly affect the vehicle demand of both companies as well as individuals.

The impact of the pandemic to the worldwide automotive industry will stay for a longtime. While the present effect of OEM production shutdown and supply chain disruptions will disappear post lockdown but lower consumer spending power will definitely affect the sales in the automotive sector for some time till the global economy revives. Whereas, factors like uneven regional recoveries, financial crunch, delayed product development, reducing dependency on china and regulatory changes will have a long-term impact on the sector.

15.2.2 *Industries Positively Affected by the Virus Outbreak*

Though many sectors are negatively affected by COVID 19, there are, however, few companies that are benefiting from the outbreak or at least not getting aspect of a hit. Mentioned below are sectors that are positively affected surrounding the COVID-19 outbreak:

15.2.2.1 Companies that Offer Stay at Home Products

Services that offer stay at home products are the bright spot for people isolating at home during the lockdown. Most of the streaming giant—Netflix, Amazon Prime, Voot, Hotstar, etc. seen record growth during this lockdown. The Internet Exchange operator DE-CIX reported a 265% surge in Over the Top Media Service since February 2020 (Streaming platforms who are on a new high-are hosting a web premiers and might even expand to gaming 2020). Disney+ Hotstar, which was launched just after the lockdown was announced, gained more than 50 million users within a week (Streaming platforms who are on a new high-are hosting a web premiers and might even expand to gaming 2020). Netflix, the biggest among the streaming giant, which was earlier going down, saw itself growing by more than 8% during the lockdown (Singh 2020). Likewise, other Indian streaming networks like Zee5 saw a 35% growth during the lockdown and its foreign version zee5 global gains more than a 60% increase in its subscriber across the World (Streaming in the times of Covid-19 2020).

15.2.2.2 Pharmaceutical Sector

As economies across the world are experiencing the effect of Covid-19, with big companies dropping their shares, workers losing their jobs; pharmaceutical companies are taking center stage in a fight against this outbreak (The impact of big pharma on Covid-19 2020). Doors for companies involving drugs and vaccine development have opened up. U.S. companies such as Vir Biotechnology, Johnson, Novavax have reported cooperative intends to build up a viral immunization against this disease (NIH clinical trial of investigational vaccine for COVID-19 begins 2020).

Although Big Pharma companies such as AstraZeneca (AZ) is expected to drop in revenue due to China's economic position, global data suggest that AZ will be able to overcome this loss. This could be due to the prospect of success in developing calquence (acalabrutinib), a drug agent against COVID 19 (The impact of big pharma on Covid-19 2020).

In India the stocks of pharma companies have outperformed in this COVID times. The S&P BSE Healthcare index is ready to mark its biggest monthly gain in 21 years. A massive spike in export orders is recorded for most Indian pharma companies. This has triggered a favorable currency movement (appreciation of 5.9% in US$/rupee and 3.6% in Euro/rupee during the quarter) directly growing the topline advancement (TPL 2020).

15.2.2.3 Chemical Industries

The fall in oil prices, coupled with the coronavirus outbreak, has hit the chemical industry hard (Global chemical industry and the impact of COVID-19: regional analysis and key player profiles 2020; COVID-19's impact on the chemical industry 2020). The restrictions imposed led to lowered demand for chemicals used in production facilities of several end-user industries, a decline in productivity, and operational and supply chain disruptions (Global chemical industry and the impact of COVID-19: regional analysis and key player profiles 2020; Coronavirus update (Live)-Worldometer 2020). The crisis has also led to the reduction in capital and operational expenditure of manufacturing operations of leading chemical manufacturers. Labor shortages and unavailability of raw materials due to disruptions in transport have scaled-down these manufacturing operations to 40–60% capacity. Consequently, in February 2020, the global production of chemicals fell by 2.4% (COVID-19's impact on the chemical industry 2020).

Nevertheless, the growing demand for protective packaging, disinfectants, antiseptics, and personal protective equipment could keep operations in the chemical industry running, thus responding positively (Global chemical industry and the impact of COVID-19: regional analysis and key player profiles 2020; COVID-19's impact on the chemical industry 2020) to the crisis. For example, the production of medical shields by plastic manufacturers who conventionally made sports gears, increase in production of hand sanitizers by Dow, Huntsman, and INEOS (UK chemicals firm plans to build two hand sanitizer factories in 10 days 2020), and so on.

The promotion of personal hygiene as preventive measure in controlling the spread of Coronavirus has led to an increase in demand for hand sanitizers, among other personal hygiene products. The increasing demand has led to panic buying, and as per the analysis made by Fortune Business Insights, the hand sanitizer market is expected to grow at a faster rate. This short-term impact of COVID19 could shoot up the global hand sanitizer market's annual growth rate from 5.06% to 45.71% in 2020, thereby will generate USD 1.87 billion in 2020 (Impact of Covid-19 on the global hand sanitizer market size, share, industry analysis and regional forecast, 2019–2026 2020). The reason for the growth may be attributed to the fact that liquor manufacturing companies are also now producing hand sanitizers (COVID-19 impact on medical supplies market by type [intubations, personal protective equipment, infusion, radiology, wound care supplies], end user [hospitals, clinics] - global forecast to 2021 2020) due to its growing demand at present.

15.2.2.4 Medical Supplies Industries

As per the data shown by Worldometer, the number people suffering from COVID19 is increasing globally every day, with 5,397,342 cases reported as on May 24, 2020. India is no exception, with 131,423 cases on the same date; it is the country with the third-largest number of cases in Asia after Turkey and Iran (Coronavirus update (Live)-Worldometer 2020). However, as India is the second-most populous country of the World with a total population of 1.21 billion (as per Census 2011) (Cenus Info India 2011), such an increased in number of COVID19 case is anticipated.

The increase in the number of cases globally and in India means more people need treatment; hence the market for COVID-19 related medical supplies increases. However, there are also restrain on the growth of the medical supplies market because of the disruption in the supply chain of logistics which are not COVID19 related and postponement of non-urgent treatment meant to create spaces for critically ill COVID19 patients (COVID-19 impact on medical supplies market by type [intubations, personal protective equipment, infusion, radiology, wound care supplies], end user [hospitals, clinics] - global forecast to 2021 2020; Freeman 2020). Thereby, Original Equipment Manufacturers [OEMs] of non-critical devices used for aesthetic surgery and ophthalmology could have economic risks shortly (Freeman 2020). Nevertheless, the global medical supplies market is expected to rise from USD ~78 billion estimated in 2019 to USD ~100 billion by 2021 (COVID-19 impact on medical supplies market by type [intubations, personal protective equipment, infusion, radiology, wound care supplies], end user [hospitals, clinics] - global forecast to 2021 2020; Williamson et al. 2020).

The positive impact on global medical supplies is due to the growing demand for disinfectants, increased requirement of personal protective equipment including different types of masks, rising demand for ventilators, sterilization supplies, diagnostic supplies for rapid diagnosis of COVID19 and other medical supplies required for increasing the healthcare capacity (COVID-19 impact on medical supplies market by type [intubations, personal protective equipment, infusion, radiology, wound care supplies], end user [hospitals, clinics] - global forecast to 2021 2020; Williamson et al. 2020).

Among others, the important players in the global medical supplies market include Getinge Group and Mölnlycke Health Care AB in Sweden; Medtronic plc in Ireland, Johnson & Johnson, Baxter International, Inc., STERIS Corporation and Advanced Sterilization Products in the U.S.; B. Braun Melsungen AG and Fresenius Medical Care AG & Co. KGaA in Germany; and Smith & Nephew in U.K. (COVID-19 impact on medical supplies market by type [intubations, personal protective equipment, infusion, radiology, wound care supplies], end user [hospitals, clinics] - global forecast to 2021 2020).

15.2.2.5 Ed-Tech Industry

Mass adoption of online learning among students has created new impetus for the edtech sector. A paradigm shifts in the education system due to the pandemic-led lockdown has resulted in a spike in the growth of edtech startups (Coronavirus update (Live)-Worldometer 2020). Learning online is the smart option in this phase of social distancing, leading to sharp rise in new users for ed-tech players. Taking figures of March 2020, BYJU's noted 6 million new students accessing free lessons on its app, while Unacademy logged 1.4 billion watch mins. Moreover, Toppr saw 100% progress in free user engagement (Cenus Info India 2011).

To minimize the risk of hindrances due to prolonged lockdown on learner's academic progress, these digital learning platforms have emerged as the most viable tool for educators. Numerous ed-tech platforms are proposing high rebates on several courses and even free classes are being offered (Cenus Info India 2011). Consequently, this online learning and virtual classes are being viewed as essential elements of the future educational system. Both educators and learners are of the view that the disruptions and novelties imminent during this COVID period will play a crucial role in driving progression of the ed-tech industry (Coronavirus update (Live)-Worldometer 2020; Cenus Info India 2011).

15.2.2.6 E-Retail

The new social distancing norms are shaping the contemporary consumer behavior. An abrupt shift of buyers towards building and supporting novel online buying behaviors as well as habits is being observed. It is anticipated that online grocery, apparels shopping will substitute store and mall visits enduringly until a cure for the disease is discovered (Freeman 2020). Even many loyal supporters of the mom-and-pop model are gradually switching to online retail platforms like Grofers, Milkbasket and BigBasket. Despite the pandemic effect most e-retailers are witnessing a 20–80% increase in their order volumes (Freeman 2020).

A recent study in US (Williamson et al. 2020) revealed that about 24% of consumers studied believed that they shopping in a mall would not be comfortable for them for at least 6 months and 16% said that in the next 3 months, they will feel comfortable. Another report by Digital Commerce 360 analysis found that 16.2% of total retail sales in US was represented by online spending for the first quarter, which is 15% higher as compared with for the same quarter in 2019 and grades the second-highest online share for any quarter in past, after 17.8% in Q4 2019(Verma 2020).

Based on market research, the predictions for progress in the e-commerce market in India before the pandemic was expected to reach 200 billion US dollars by the year 2026. But the recent developments post covid indicates that the figures may be attained more earlier than predictions due to sudden change in consumer behavior amid dreads of possible infection (Columbus 2000).

15.2.2.7 Telecom

The COVID-19 outbreak has transformed the world's ways of living, working and connecting. As the world economy is learning new ways to cope up with the long lasting impact of the pandemic, "work from home" and "social distancing" have emerged as the buzzwords in the present business environment, with the telecom sector being one of the major drivers helping the economy to overcome the shock. Working from home, video conferencing and telecom technology have swiftly arisen as a crucial facilitator for every single task during this period of restrictions. As per insights by an empirical study globally, 20% upsurge in international voice traffic in March 2020 was observed as compared to March 2019, whereas decline in roaming traffic by 30% was recorded (Meyers 2020). Additionally, due to worldwide lockdown with people forced to stay at home resulted in increasing the average length of calls by more than 30% in March and over 60% in April 2020 compared to 2019. Similarly, according to a report on Indian telecom sectors, it has witnessed an overall traffic jump by 10% in March 2020 (Ali 2020).

Telecom sector is also finding new opportunities for growth in the novel scenario. Skyrocketing network usage has pushed telecom companies globally to take a variety of measures to improve the customer experiences (Basu 2020). In the UK, for better consumer experience operators have increased capacity and free minutes. Network reliability and infrastructure is the new focus. All this is leading to collaborations which might be unthinkable weeks ago. For instance, operators across the US are increasing their capacity by borrowing spectrum from competitors. The sector is also helping the governments to contain the impact of COVID-19 by using consumer data to track and contain the infection.

15.3 Effect of the Risk Outbreaks on Global Economies

The essential measures to contain the COVID-19 virus has triggered an unparalleled global economic crisis. The clouds of uncertainty are still hovering over the world economy, with no clues of extent to losses incurred including its severity and length. The further amplification of the crisis is expected to distress the global financial stability (Basu 2020). It is a foregone inference that the world economy is going to be slipped into a recession in 2020. Only uncertainty is about the depth, length and width of the contraction in economic activities. Due to unparalleled halt in universal economic activities, 2020 would be the spectator of the greatest "Global Recession" after World War II (Robuck 2020).

The initial outbreak of SARS-CoV-2 in China, which is the second-largest economy of the World, has brought devastation to the entire global economy (Zaharia 2020; Fitch sees global economy shrinking 3.9 per cent in 2020 2020), as discussed above sector-wise. The lockdown imposed has disrupted the global supply chains due to the halt in imports and exports globally, thereby creating the likelihood of debt distress (Global financial stability report, April 2020: markets in the time of COVID-19 2020; Izvorski et al. 2020) for the government of different countries. The severity of the disruptions could be exacerbated if there is prolonged restrictions in economic activities and if there is absence of competent and prompt fiscal response (Global economy could shrink by almost 1% in 2020 due to COVID-19 pandemic: United Nations 2020). According to the United Nations, the extension of economic activities' restrictions could lead to a decline of the global economy by 1% or further, from the previous forecast of 2.5% growth in the current year (Global economy could shrink by almost 1% in 2020 due to COVID-19 pandemic: United Nations 2020). However, Fitch Ratings stated that there would be a contraction in the world GDP by up to 3.9% in the same year due to the massive decline in Asian economies, especially China and India (Fitch sees global economy shrinking 3.9 per cent in 2020 2020). The percentage fall in the global GDP is equivalent to a USD 2.8 trillion falls in global income levels from that of 2019. The Eurozone GDP is expected to decline by 7%, the U.S. by 5.6%, and British by 6.3% in 2020 (Fitch sees global economy shrinking 3.9 per cent in 2020 2020).

After the lockdown was enforced, there is a decrease in the Indian economy's ability to function, with just 25% of its capacity. With the extension of lockdowns, it is anticipated that India's GDP could drop by 35% or more in 2019–2020 (Kumar 2020).

The lockdowns have hit the service sector hard, especially in the developed countries, and thereby prompting substantial job losses. Similarly, according to Japan's Purchasing Managers' Index [PMI] surveys, the services sector and factory activity are hardest hit in March 2020 (Zaharia 2020). The banking sector, no wonder, is affected by the present crisis due to the rise in levels of debts and bad loans (Dun and Bradstreet 2020). Whereas in developing countries, the economic risks are primarily due to the halt in tourism and commodity exports, as countries are closing their borders and imposing restrictions on movement (Global economy could shrink by almost 1% in 2020 due to COVID-19 pandemic: United Nations 2020).

After the outbreak of the pandemic, sharp decline in the prices of risk assets and jump in credit spreads is observed. The World Bank has estimated that GDP in all developing areas (ECA, LAC, MENA and Africa) except Asia will contract. Chief short-term funding markets, comprising the global market for U.S. dollars, are also strained. Significant deterioration in market liquidity, with the spike in volatility, has been subsidizing to unforeseen asset price changes (Global financial stability report, April 2020: markets in the time of COVID-19 2020).

15.4 Actions Taken to Mitigate the Impact of COVID19 on the Economy

The Governments of different countries are developing various fiscal policies to build back resilience in the economy after taking stock of the economic crisis. At the G20 summit, world leaders came forward in the fight against coronavirus and pledged a "united front" promising to pump in USD 5 trillion into the global economy. The measures injected to counter the pandemic amid forecasts of a deep recession are aimed to bring stability in the global economy (Mishra 2020).

Resilience building, which includes aggressive fiscal and monetary support given by governments and central banks around the World (Dhar 2020; Zaharia 2020), aims to reduce economic risks (Noy et al. 2020) posed by the COVID19 pandemic. After arriving at a consensus, prominent policymakers listed several key measures that include a halt on debt payment to assist developing countries and emerging economies of the World (Guardians of global economy come up short in fight against virus 2020).

Central banks throughout the globe have played a crucial role to aid the economy. Eased monetary policy along with injecting additional liquidity in markets and enhancement of the provision of U.S. Dollar Liquidity by swap line arrangements, are the major tools embraced globally (Georgieva 2020). To minimize the risk of deep economic recession, a comprehensive multilateral response or well-designed fiscal stimulus packages (Dhar 2020; Global economy could shrink by almost 1% in 2020 due to COVID-19 pandemic: United Nations 2020) that prioritize public health spending and providing financial assistance to the most affected families, especially the migrant workers, is crucial (Global economy could shrink by almost 1% in 2020 due to COVID-19 pandemic: United Nations 2020).

The International Monetary Fund (IMF), being the guardian of the global economy, has requested for assistance by more than 100 of its 189 member countries to cope with the crisis. However, measures undertaken by the IMF do not meet the demand (Guardians of global economy come up short in fight against virus 2020) to fight against COVID19. But a sum of about $160 billion is committed by the World Bank to assist developing countries in coping with the global emergency (Guardians of global economy come up short in fight against virus 2020). For India, the World Bank's total commitment is USD 2 billion (World Bank approves USD 1 billion loan to support India's fight against COVID-19 2020).

Several countries have announced stimulus packages to fight against COVID19. In the U.S., Coronavirus Aid, Relief, and Economic Security (CARES) Act 2020 amounts to $2.2 trillion (approximately 10% of its GDP); Germany with a stimulus package of €1.1 trillion almost a third of its GDP is the highest among other European countries; recently Japan has also pledged $990 billion, (about 19% of its GDP) (Dhar 2020) for reviving the economy and for other health-related expenditure.

In India, several relaxation measures were undertaken since April 20, 2020, to revive the economy—after almost 1 month since the lockdown was declared, i.e., on

March 24, 2020. A stimulus package valued at approximately 0.8% of GDP ($22.6 billion) was announced on March 26, 2020 (Dhar 2020; International Monetary Fund, policy responses to COVID19 2020; Ahmed 2020). The package is in addition to the commitment made for health spending of about Rs. 150 billion (about 0.1% of GDP) to deal with the management of COVID19 cases (International Monetary Fund, policy responses to COVID19 2020).

Along with the previously announced monetary and fiscal measures, the PM on May 13 announced a stimulus package of Rs. 20 lakh crore (around 10% of GDP) (International Monetary Fund, policy responses to COVID19 2020; India's Rs 20 lakh crore Covid relief package one among the largest in the world 2020; Summing up Modi's Covid stimulus: big takeaways from the big Covid package 2020) which will focus on different sectors such as SMEs, MSMEs, laborers, industries, banking sector among others (India's Rs 20 lakh crore Covid relief package one among the largest in the world 2020; Summing up Modi's Covid stimulus: big takeaways from the big Covid package 2020; Mulye 2020). Along with this, numerous monetary and macro-financial measures were also undertaken by the Reserve Bank of India (RBI) since February 8, 2020, thereby accounting for a liquidity injection of around 4% of the country's GDP (International Monetary Fund, policy responses to COVID19 2020).

Despite the unprecedented macro policy responses, with massive job losses and growing economic pressures on SMEs, MSMEs, and various other sectors, the likelihood of the economy to get back on its feet would be slower (Pandemic throws global economy into deeper decline 2020).

15.5 Summary

The outbreak of the contagion since December 2019 in China has left the World to a standstill till today and is not showing any sign that it will end sooner despite the preventive measures undertaken. The number of cases keeps on increasing, similarly the number of deaths. On May 27, the Worldometer showed that the U.S. has the highest number of cases among all World countries. In India, the number of cases are also increasing despite the measures undertaken to control its spread. The actions undertaken globally and in India to control the spread of the virus, however, affected the already sinking global economy. The fall in oil prices, coupled with the economic inactivity due to lockdowns, has brought a disaster to the economy. The disruption in the functioning of different sectors, as discussed above, has hit the global supply chains and international trade. Though there are few sectors that benefitted from the crisis, the majority are affected. An estimate advocates that the crisis may erase all the achievements made in poverty reduction during the past 3 years, pushing up to 100 million individuals globally into extreme poverty.

To mitigate the pandemic's impact on the economy, aggressive fiscal measures were undertaken globally along with relaxation in some sectors, such as the agriculture and other segments of the primary sector. Stimulus packages were announced by different countries of the World to revive the economy that is moving towards a deep recession. Despite these, due to prolonged restrictions on movement and economic inactivity earlier, the global economy is unlikely to get back to normality. Thus, considering the magnitude of the crisis, the global economy will require a longer duration to recover, which has not left any country of the world untouched.

Conclusively with an optimistic note, "if history is any indication, the world is bound to recover sooner than later". Every domain of the global economy will face constraints such as reduced demand and delayed business decisions, but it will recover ultimately. Although, the calamity has appeared as a challenge but will eventually evolve as an opportunity.

References

Ahmed A (2020) India outlines $23 billion stimulus to help poor hit by lockdown. https://www.weforum.org/agenda/2020/03/india-stimulus-support-lockdown-pandemic-covid19-epidemic-economics/. Accessed 25 May 2020

Al Jazeera (2020) Coronavirus: travel restrictions, border shutdowns by country. https://www.aljazeera.com/news/2020/03/coronavirus-travel-restrictions-border-shutdowns-country-200318091505922.html. Accessed 30 May 2020

Ali F (2020) Ecommerce trends in charts during the coronavirus pandemic. https://www.digitalcommerce360.com/2020/05/20/ecommerce-during-coronavirus-pandemic-in-charts/. Accessed 30 May 2020

Babar K (2020) Covid-19 impact: real estate sentiments hit lowest level. The Economic Times. https://economictimes.indiatimes.com/wealth/real-estate/covid-19-impact-real-estate-sentiments-hit-lowest-level/articleshow/75175857.cms. Accessed 14 Jun 2020

Bachelor L (2020) MPs in plea to government over UK's Covid-19 stockpiling. The Guardian. https://www.theguardian.com/world/2020/mar/21/mps-plea-government-uk-covid-19-stockpiling-coronavirus. Accessed 17 May 2020

Basu P (2020) The impact covid19 pandemic is exerting on e-commerce. https://www.orfonline.org/expert-speak/the-impact-covid19-pandemic-is-exerting-on-e-commerce-65128/. Accessed 30 May 2020

Bloomberg (2020) Countries are starting to hoard food due to coronavirus outbreak, threatening global trade. The Economic Times. https://economictimes.indiatimes.com/news/international/world-news/countries-are-starting-to-hoard-food-amid-coronavirus-outbreak-threatening-global-trade/articleshow/74805865.cms. Accessed 17 May 2020

Britons urged to avoid non-essential travel abroad (2020) BBC News. https://www.bbc.com/news/uk-51924405. Accessed 20 May 2020

Buck T, Arnold M, Chazan G, Cookson C (2020) Coronavirus declared a pandemic as fears of economic crisis mount. Financial Times. https://www.ft.com/content/d72f1e54-6396-11ea-b3f3-fe4680ea68b5. Accessed 20 May 2020

Bureau TT (2020) Key recommendations by ICC to contain COVID-19 impact on Tourism & Hospitality. https://www.todaystraveller.net/single-post/2020/04/06/Key-recommendations-by-ICC-to-contain-COVID-19-impact-on-Tourism-Hospitality. Accessed 21 May 2020

Burkert A, Loeb A (2020) Flattening the COVID-19 curves. Scientific American Blog Network. https://blogs.scientificamerican.com/observations/flattening-the-covid-19-curves/. Accessed 20 May 2020

Cenus Info India (2011). https://censusindia.gov.in/2011census/censusinfodashboard/index.html. Accessed 26 Apr 2020

Chauhan AS, Kapoor S (2020) Impact of coronavirus on Indian real estate. https://www.99acres.com/articles/impact-of-coronavirus-outbreak-on-indian-real-estate.html#comment-21021. Accessed 14 Jun 2020

Collie B et al (2020) COVID-19's impact on the automotive industry. https://www.bcg.com/en-in/publications/2020/covid-automotive-industry-forecasting-scenarios.aspx. Accessed 14 Jun 2020

Columbus L (2000) How COVID-19 is transforming E-commerce. https://www.forbes.com/sites/louiscolumbus/2020/04/28/how-covid-19-is-transforming-e-commerce/#6603f8bc3544. Accessed 20 May 2020

Coronavirus hits German hotel industry hard: more than every-3-guests stay away (2020). https://www.hospitalitynet.org/performance/4097569.html. Accessed 10 May 2020

Coronavirus travel updates: which countries have restrictions and FCO warnings in place? (2020) The Guardian. https://www.theguardian.com/travel/2020/mar/19/coronavirus-travel-updates-which-countries-have-restrictions-and-fco-warnings-in-place. Accessed 20 May 2020

Coronavirus update (Live)-Worldometer (2020). https://www.worldometers.info/coronavirus/. Accessed 24 May 2020

COVID-19 and the automotive industry (2020) Ilo.org. https://www.ilo.org/sector/Resources/publications/WCMS_741343/lang%2D%2Den/index.htm. Accessed 10 Jun 2020

Covid-19 hits hotel industry; revenue per room drops 13–29% in Jan–March in top cities (2020) The Economic Times. https://economictimes.indiatimes.com/industry/services/hotels-/-restaurants/covid-19-hits-hotel-industry-revenue-per-room-drops-13-29-in-jan-mar-in-top-cities-report/articleshow/75709902.cms?from=mdr. Accessed 20 May 2020

COVID-19 impact on medical supplies market by type [intubations, personal protective equipment, infusion, radiology, wound care supplies], end user [hospitals, clinics] - global forecast to 2021 (2020). https://www.marketsandmarkets.com/Market-Reports/covid-19-impact-on-medical-supplies-market-158189339.html. Accessed 20 May 2020

COVID-19's impact on the chemical industry (2020). https://www.gep.com/mind/blog/covid-19s-impact-on-the-chemical-industry. Accessed 20 May 2020

Dasgupta A (2020) Sports has the Coronavirus and a cure is needed urgently. Times of India Blog. https://timesofindia.indiatimes.com/blogs/a-good-sport/sports-has-the-coronavirus-and-a-cure-is-needed-urgently/. Accessed 14 Jun 2020

Dash J (2020) Covid-19 impact: tourism industry to incur Rs 1.25 trn revenue loss in 2020. https://www.business-standard.com/article/economy-policy/covid-19-impact-tourism-industry-to-incur-rs-1-25-trn-revenue-loss-in-2020-120042801287_1.html. Accessed 8 Jun 2020

Deb A (2020) What will the post-pandemic future of Indian F&B industry look like? https://www.timesnownews.com/business-economy/industry/article/what-will-the-post-pandemic-future-of-indian-fb-industry-look-like/585112. Accessed 9 June 2020

Deloitte (2020) Understanding the sector impact of COVID 19. https://www2.deloitte.com/global/en/pages/about-deloitte/articles/covid-19/understanding-covid-19-s-impact-on-the-automotive-sector.html. Accessed 14 Jun 2020

Dhar B (2020) Covid-19: Govt's fiscal package is underwhelming. https://www.thehindubusinessline.com/opinion/covid-19-india-must-ramp-up-economic-support-measures/article31297536.ece#. Accessed 20 May 2020

Dun and Bradstreet (2020) India may register recession in third quarter of this fiscal, shows report. The Economic Times. https://economictimes.indiatimes.com/news/economy/indicators/india-may-register-recession-in-third-quarter-of-this-fiscal-shows-report/articleshow/75937359.cms. Accessed 20 May 2020

Durbin AD (2020) Airbnb, hotels seek US government aid as demand flattens. https://apnews.com/8c9ace544023a0f0a58688ab74a2d61a. Accessed 20 May 2020

Ellis TN (2020) Coronavirus travel restrictions around the world: a country-by-country list. USA Today. https://www.usatoday.com/story/travel/news/2020/03/17/coronavirus-travel-bans-countries-impose-travel-restrictions/5058513002/. Accessed 20 May 2020

ETAuto (2020) Opinion: COVID-19 impact on Indian automotive industry - taking action in troubled times - ET Auto. ETAuto.com. https://auto.economictimes.indiatimes.com/news/industry/opinion-covid-19-impact-on-indian-automotive-industry-taking-action-in-troubled-times/74944847. Accessed 10 Jun 2020

Euronews.com (2020) 'Wish you were here': tourism sector 'hardest hit' by COVID-19. https://www.euronews.com/2020/04/16/coronavirus-in-europe-tourism-sector-hardest-hit-by-covid-19. Accessed 8 Jun 2020

Evans AB et al (2020) Sport in the face of the COVID-19 pandemic: towards an agenda for research in the sociology of sport. Eur J Sport Soc 17(2):85–95. https://doi.org/10.1080/16138171.2020.1765100

Faus J (2020) This is how coronavirus could affect the travel and tourism industry. World Economic Forum. https://www.weforum.org/agenda/2020/03/world-travel-coronavirus-covid19-jobs-pandemic-tourism-aviation/. Accessed 21 May 2020

Felipe L (2020) Five things you should know now about the COVID-19 pandemic. UN News. https://news.un.org/en/story/2020/03/1059261. Accessed 15 May 2020

Find out how coronavirus is impacting all industries; aviation, hospitality worst hit (2020). https://www.cnbctv18.com/finance/find-out-how-coronavirus-is-impacting-all-industries-aviation-hospitality-worst-hit-5512211.htm. Accessed 20 May 2020

Fitch sees global economy shrinking 3.9 per cent in 2020 (2020) The Economic Times. https://economictimes.indiatimes.com/news/economy/indicators/fitch-sees-global-economy-shrinking-3-9-pc-in-2020/articleshow/75320075.cms. Accessed 21 May 2020

Fitz Gerald A, Kwiatkowski K, Singer V, Smit S (2020) An instant economic crisis: how deep and how long?. https://www.mckinsey.com/business-functions/strategy-and-corporate-finance/our-insights/an-instant-economic-crisis-how-deep-and-how-long. Accessed 20 May 2020

Food parcels coming for vulnerable individuals (2020) BBC News. https://www.bbc.com/news/av/uk-politics-51995298/coronavirus-minister-urges-food-shoppers-to-be-considerate. Accessed 19 May 2020

Freeman T (2020) COVID-19's impact on the medical device supply chain. https://www.mpo-mag.com/issues/2020-04-01/view_columns/covid-19s-impact-on-the-medical-device-supply-chain/. Accessed 28 May 2020

Georgieva K (2020) The global economic reset—promoting a more inclusive recovery. https://blogs.imf.org/2020/06/11/the-global-economic-reset-promoting-a-more-inclusive-recovery/. Accessed 12 Jun 2020

Global chemical industry and the impact of COVID-19: regional analysis and key player profiles (2020). https://www.prnewswire.com/news-releases/global-chemical-industry-and-the-impact-of-covid-19-regional-analysis-and-key-player-profiles-301045312.html. Accessed 20 May 2020

Global economy could shrink by almost 1% in 2020 due to COVID-19 pandemic: United Nations (2020) The Economic Times. https://economictimes.indiatimes.com/news/international/business/global-economy-could-shrink-by-almost-1-in-2020-due-to-covid-19-pandemic-united-nations/articleshow/74943235.cms. Accessed 21 May 2020

Global financial stability report, April 2020: markets in the time of COVID-19 (2020) IMF. https://www.imf.org/en/Publications/GFSR/Issues/2020/04/14/Global-Financial-Stability-Report-April-2020-49020#:~:text=Summary%3A,(COVID%2D19)%20pandemic.&text=The%20pandemic%20led%20to%20an,in%20emerging%20and%20frontier%20markets. Accessed 14 Jun 2020

GlobalData (2020) Southeast Asian tourism likely to be worst hit due to coronavirus, says GlobalData. GlobalData. https://www.globaldata.com/southeast-asian-tourism-likely-to-be-worst-hit-due-to-coronavirus-says-globaldata/. Accessed 10 Jun 2020

Gopinath G (2020) The great lockdown: worst economic downturn since the great depression. IMFBlog. https://blogs.imf.org/2020/04/14/the-great-lockdown-worst-economic-downturn-since-the-great-depression/ Accessed 14 Jun 2020

Gov.UK (2020) Guidance on social distancing for everyone in the UK. https://www.gov.uk/government/publications/covid-19-guidance-on-social-distancing-and-for-vulnerable-people/guidance-on-social-distancing-for-everyone-in-the-uk-and-protecting-older-people-and-vulnera ble-adults. Accessed 30 May 2020

Goyal R (2020) Impact of corona on MSMEs. EH News Bureau. https://www.expresshealthcare.in/blogs/impact-of-corona-on-msmes/421713/. Assessed 13 Jun 2020

Guardians of global economy come up short in fight against virus (2020) The Economic Times. https://economictimes.indiatimes.com/news/international/business/guardians-of-global-econ omy-come-up-short-in-fight-against-virus/articleshow/75226938.cms. Accessed 22 May 2020

Hall S (2020) This is how COVID-19 is affecting the world of sports. https://www.weforum.org/agenda/2020/04/sports-covid19-coronavirus-excersise-specators-media-coverage/. Accessed 17 May 2020

Haywood KM (2020) A post-COVID future: tourism community re-imagined and enabled. Tour Geogr:1–11. https://doi.org/10.1080/14616688.2020.1762120

IATA (2020) IATA updates COVID-19 financial impacts - relief measures needed. https://www.iata.org/en/pressroom/pr/2020-03-05-01/. Accessed 21 May 2020

Impact of Covid-19 on the global hand sanitizer market size, share, industry analysis and regional forecast, 2019–2026 (2020). https://www.fortunebusinessinsights.com/impact-of-covid-19-on-hand-sanitizer-market-102719. Accessed 20 May 2020

India to provide free food grains to millions of migrant workers (2020). https://www.aljazeera.com/news/2020/05/india-provide-free-food-grains-millions-migrant-workers-200514141450778.html. Accessed 19 May 2020

India's Rs 20 lakh crore Covid relief package one among the largest in the world (2020) The Economic Times. https://economictimes.indiatimes.com/news/economy/finance/latest-stimu lus-package-among-largest-in-the-world/articleshow/75701976.cms?from=mdr. Accessed 22 May 2020

Infographic: how commercial real estate owners can respond to Covid-19 (2020) Bain. https://www.bain.com/insights/how-commercial-real-estate-owners-can-respond-to-covid-19-infographic/. Accessed 14 Jun 2020

International Labour Organization (2020) International Labour Organization [ILO] sectoral brief. COVID-19 and the impact on agriculture and food security. https://www.ilo.org/sector/Resources/publications/WCMS_742023/lang%2D%2Den/index.htm. Accessed 19 May 2020

International Monetary Fund, policy responses to COVID19 (2020). https://www.imf.org/en/Topics/imf-and-covid19/Policy-Responses-to-COVID-19. Accessed 22 May 2020

IPL 2020 suspended until further notice due to Covid-19, season will commence only when it is safe: BCCI (2020) India Today. https://www.indiatoday.in/sports/ipl-2020/story/ipl-2020-suspended-until-further-notice-confirms-bcci-1667634-2020-04-16. Accessed 17 May 2020

Izvorski I et al (2020). https://www.brookings.edu/blog/future-development/2020/04/20/a-policy-framework-for-mitigating-the-economic-impact-of-covid-19/. Accessed 14 Jun 2020

Jain D (2020) Effect of COVID-19 on restaurant industry–how to cope with changing demand. https://doi.org/10.2139/ssrn.3577764

Jones R (2020) UK airlines call for multibillion bailout to survive Covid-19 crisis. The Guardian. https://www.theguardian.com/world/2020/mar/15/uk-airlines-call-for-multibillion-bailout-to-survive-covid-19-crisis. Accessed 21 May 2020

Khan T (2020) How covid-19 is impacting the rural market. https://www.financialexpress.com/brandwagon/how-covid-19-is-impacting-the-rural-market/1940102/. Accessed 19 May 2020

Kulkarni M (2020) Coronavirus lockdown: airline travel agents face severe cash crunch. https://www.deccanherald.com/business/coronavirus-lockdown-airline-travel-agents-face-severe-cash-crunch-834871.html. Accessed 10 June 2020

Kumar A (2020) Covid-19 crisis: understanding the state of economy during and after the lockdown. Economic and Political Weekly 55(19)

Lock S (2020) Share of GDP generated by the travel and tourism industry worldwide from 2000 to 2019. https://www.statista.com/statistics/1099933/travel-and-tourism-share-of-gdp/. Accessed 22 May 2020

McKibbin W, Fernando R (2020) The global macroeconomic impacts of COVID-19: seven scenarios. https://www.brookings.edu/wp-content/uploads/2020/03/20200302_COVID19.pdf. Accessed 25 May 2020

Meyers A (2020) When consumers say they'll feel ok about dining out and other activities. https://morningconsult.com/2020/04/10/consumer-expectations-normal-activities-comfortable/. Accessed 20 May 2020

Mishra A (2020) G20 nations vow to inject $5 tn to fight Covid-19, global slump. Livemint. https://www.livemint.com/news/india/g-20-members-express-readiness-to-do-whatever-it-takes-to-fight-covid-19-11585236885510.html. Accessed 14 Jun 2020

Moodie Davitt Pax Index: the latest traffic figures from airports across the world (2020). https://www.moodiedavittreport.com/moodie-davitt-pax-index-the-latest-traffic-figures-from-airports-across-the-world/. Accessed 21 May 2020

Mudgill A (2020) IndiGo, SpiceJet face major air turbulence after Corona hit, may soon run out of fuel. https://economictimes.indiatimes.com/markets/stocks/news/indigo-spicejet-face-major-air-turbulence-after-corona-hit-may-soon-run-out-of-fuel/articleshow/75025420.cms. Accessed 21 May 2020

Mulye P (2020) Who benefits, and by how much, from India's $266 billion economic stimulus. https://qz.com/india/1856488/details-of-modis-coronavirus-package-atmanirbhar-bharat/. Accessed 22 May 2020

Nicola M et al (2020) The socio-economic implications of the coronavirus and COVID-19 pandemic: a review. Int J Surg 78:185–193

NIH clinical trial of investigational vaccine for COVID-19 begins (2020). https://www.nih.gov/news-events/news-releases/nih-clinical-trial-investigational-vaccine-covid-19-begins. Accessed 15 May 2020

Noy I, Doan N, Ferrarini B, Park D (2020) The economic risk from COVID-19 is not where COVID-19 is. https://voxeu.org/article/economic-risk-covid-19-not-where-covid-19. Accessed 20 May 2020

Ozili PK, Arun T (2020) Spillover of COVID-19: impact on the Global Economy. Available at SSRN 3562570

Pandemic throws global economy into deeper decline (2020) The Economic Times. https://economictimes.indiatimes.com/news/international/business/pandemic-throws-global-economy-into-deeper-decline/articleshow/75319594.cms. Accessed 21 May 2020

Parnell D, Widdop P, Bond A, Wilson R (2020) COVID-19, networks and sport. Manag Sport Leisure:1–7. https://doi.org/10.1080/23750472.2020.1750100

Raney HM (2020) COVID 19 pandemic: impact on MSMEs. Stud Indian Place Names 40 (68):396–401

Rao R (2020) Coronavirus impact: Swiggy lays off 1,100 employees, future of cloud kitchens uncertain. https://www.businesstoday.in/sectors/jobs/coronavirus-impact-swiggy-lays-off-1100-employees-future-of-cloud-kitchens-uncertain/story/404195.html. Accessed 10 Jun 2020

Reinhart CM, Subbaraman R (2020) How can we Prevent a COVID-19 food crisis? https://www.weforum.org/agenda/2020/05/preventing-a-covid-19-food-crisis/. Accessed 27 May 2020

Restaurant closures may cost 7.4 million jobs reports says (2020). https://www.bloomberg.com/news/articles/2020-03-17/restaurant-closures-may-cost-7-4-million-jobs-report-says. Accessed 29 Apr 2020

Reuters (2020) Battered consumer confidence to slow airline recoveries: IATA. Global Times. http://www.globaltimes.cn/content/1186553.shtml. Accessed 10 June 2020

Robuck M (2020) COVID-19's impact will evolve the telecommunications industry. https://www.fiercetelecom.com/telecom/covid-19-s-impact-will-evolve-telecommunications-industry. Accessed 28 May 2020

Sally M, Bhosale J (2020) Prices of vegetable, pulses, grains, sugar fall on poor demand. The Economic Times. https://economictimes.indiatimes.com/news/economy/indicators/prices-of-vegetable-pulses-grains-sugar-fall-on-poor-demand/articleshow/75899116.cms. Accessed 29 May 2020

Shared I, Orelowitz B (2020) The airline industry and COVID-19: saving for a rainy day. Am Bankruptcy Inst J 39(5):36–58

Singh M (2020) Disney + Hotstar has about 8 million subscribers. https://techcrunch.com/2020/04/08/disney-plus-hotstar-subscribers-india/. Accessed 20 May 2020

Sipahi E (2020) COVID-19 and MSMEs: a revival framework. Res J Adv Humanities 1(2):7–21

Sport-by-sport look at the impact of corona virus around the world (2020) Independent. :https://www.independent.co.uk/sport/sport-football-basketball-rugby-olympics-cancelled-coronavirus-impact-around-the-world-a9398186.html. Accessed 17 May 2020

Stockpiling 'will hit vulnerable', food bank warns (2020) BBC News. https://www.bbc.com/news/uk-england-beds-bucks-herts-51856592

Streaming in the times of Covid-19 (2020). https://economictimes.indiatimes.com/magazines/panache/streaming-in-the-time-of-covid-19-youtube-follows-netflix-limits-quality-to-sd-for-beating-bandwidth-stress/articleshow/74807651.cms. Accessed 20 May 2020

Streaming platforms who are on a new high-are hosting a web premiers and might even expand to gaming (2020). https://www.businessinsider.in/entertainment/news/disney-hotstar-amazon-prime-videos-zee-5-gained-a-lot-this-lockdown-and-it-might-not-be-temporary-either/articleshow/75443378.cms. Accessed 20 May 2020

Summing up Modi's Covid stimulus: big takeaways from the big Covid package (2020) The Economic Times. https://economictimes.indiatimes.com/news/economy/policy/summing-up-modis-covid-stimulus-takeaways-so-far-from-the-mother-of-all-incentives/articleshow/75758840.cms. Accessed 22 May 2020

Suspension of Premier League extended until April 30 amid coronavirus pandemic (2020) Goals. https://www.goal.com/en/news/suspension-of-premier-league-extended-until-april-30/1sno8nlw1xtdk1idv45pigu54s. Accessed 17 May 2020

The Economist (2020) Coronavirus is grounding the world's airlines. https://www.economist.com/business/2020/03/15/coronavirus-is-grounding-the-worlds-airlines. Accessed 14 June 2020

The impact of big pharma on Covid-19 (2020). https://www.pharmaceutical-technology.com/comment/covid-19-pharmaceutical-companies-impact/. Accessed 20 May 2020

TPL (2020) While global stock markets take a COVID-19 beating, Indian drugmakers doing well. https://www.thepharmaletter.com/article/while-global-stock-markets-take-a-covid-19-beating-indian-drugmakers-doing-well. Accessed 20 May 2020

UK chemicals firm plans to build two hand sanitizer factories in 10 days (2020) The Guardian. https://www.theguardian.com/world/2020/mar/24/uk-chemicals-firm-plans-to-build-two-hand-sanitiser-factories-in-10-days. Accessed 7 May 2020

UK pubs and restaurants told to shut in virus fight (2020) BBC News. https://www.bbc.com/news/uk-51981653. Accessed 29 Apr 2020

Upadhyay P (2020) Coronavirus: travel industry hit hard with cancellations, decline in demand. India Today. https://www.indiatoday.in/business/story/coronavirus-travel-industry-hit-hard-with-cancellations-decline-demand-1652050-2020-03-03. Accessed 20 May 2020

Verma P (2020) Covid-19 impact: Ed-tech companies reap dividend as schools, universities go online. https://economictimes.indiatimes.com/industry/services/education/ed-tech-cos-reap-dividend-as-schools-univs-go-online/articleshow/74782879. Accessed 20 May 2020

Vietnam's tourism sector could lose US$ 5bln due to COVID 19 (2020) Vietnam Times. https://vietnamtimes.org.vn/vietnams-tourism-sector-could-lose-us-5bln-due-to-covid-19-17829.html. Accessed 21 May 2020

Whiteside P (2020) Coronavirus lockdowns unlocked: which countries are easing restrictions. Sky News. https://news.sky.com/story/coronavirus-lockdowns-unlocked-which-countries-are-easing-restrictions-11974027. Accessed 20 May 2020

Williamson B, Eynon R, Potter J (2020) Pandemic politics, pedagogies and practices: digital technologies and distance education during the coronavirus emergency. J Learn Media Technol 45(2):107–114

World Bank approves USD 1 billion loan to support India's fight against COVID-19 (2020) The Hindu. https://www.thehindu.com/business/businesslive-15-may-2020/article31588355.ece. Accessed 23 May 2020

World Travel and Tourism Council (2020) The domino effect of COVID-19. https://wttc.org/NewsArticle/Latest-research-from-WTTC-shows-a-50-percentage-increase-in-jobs-at-risk-in-Traveland-Tourism. Accessed 12 Jun 2020

Zaharia M (2020) Surveys show coronavirus pandemic savaging global economy. The Economic Times. https://economictimes.indiatimes.com/news/international/business/surveys-show-coronavirus-pandemic-savaging-global-economy/articleshow/74786359.cms. Accessed 27 Apr 2020

Zhang J (2020) Coronavirus spread now a global emergency declares World Health Organization. UN News. https://news.un.org/en/story/2020/01/1056372. Accessed 25 May 2020

Part III
Role of Various Measures and Resilience

Chapter 16
Understanding Public Health Interventions: Isolation, Quarantine, Social Distancing

Aditi Madan and Anil Kumar Gupta

Abstract The purpose behind this chapter is to provide terminology of terms frequently used inter-changeably in the context of recent COVID-19 pandemic such as "Isolation, Quarantine and Social Distancing". It discusses in detail what these terms mean, how they are different from each other and the overall goals and benefits of adopting these measures. It also discusses right time for initiating social distancing measures during an outbreak. Further, it throws light on approaches adopted by the countries while responding to the pandemic. The chapter goes further to discuss effectiveness of social distancing as an effective strategy for containing the spread of disease as well as impacts and challenges posed by such strategies adopted to limit the transmission of disease among people.

Keywords Isolation · Quarantine · Social distancing · Pandemic · Covid-19 · Strategies

16.1 Terminology and Differences: Isolation, Quarantine, Social Distancing

COVID-19 outbreak caused by novel coronavirus has spread across the world with most of the infected people suffering from mild, flu-like symptoms with few becoming seriously ill and dying (Nussbaumer-Streit et al. 2020). The virus spreads easily through cough and sneeze among the people in close contact. There have been several words which have been used to contain this spread of coronavirus such as

A. Madan (✉)
Institute for Human Development, New Delhi, Delhi, India

A. K. Gupta
Environment and Disaster Risk Management and Centre for Excellence on Climate Resilience, National Institute of Disaster Management, New Delhi, Delhi, India

M. K. Goyal, A. K. Gupta (eds.), *Integrated Risk of Pandemic: Covid-19 Impacts, Resilience and Recommendations*, Disaster Resilience and Green Growth,
https://doi.org/10.1007/978-981-15-7679-9_16

community spreading, social distancing, self-isolation, lockdown (Shaw et al. 2020), containment, quarantine etc. All these terms are just different ways of limiting spread of disease to attain containment of disease through public health interventions. Supporting this, Miao and Huang (2012) point out that adoption of health-protective behaviours can limit the spread of disease transmission during a pandemic. Such public health interventions aim at preventing spread of disease by maintaining physical distance between people to contain disease transmission from person to person through droplets.

Though terms "quarantine and isolation" have been often used interchangeably, both terms refer to unique situations. A major difference between isolation and quarantine is that isolation is a preventive control measure applied to sick people, whereas quarantine is applicable to people who might get sick usually due to exposure to an infected person. Until there is no effective treatment or vaccine found for COVID-19, these non-medical public health measures need to be implemented to restrict the transmission of disease among people (shown in Table 16.1).

16.1.1 Isolation

'Isolation' refers to separation of the infected persons with contagious diseases from the non-infected persons general population and usually occurs in hospital settings (Wilder and Freedman 2020). Simply put, it implies separating people with confirmed infections from people who are not sick, so that they can get better without infecting anyone else. Isolation curbs the freedom of movement of people who have been infected or suspected of being affected with a communicable disease or condition.

According to Briscese et al. (2020), complying by isolation measures in response to the pandemic lowers the probability of contracting virus and infecting others, helping in curbing the spread of the disease. Thus, convincing the public to isolate is critical in the wake of such pandemics. Supporting this, one of the study reviewed by Fong et al. (2020) reports that one of ban on public gatherings, in combination with other interventions, for a median of 4 weeks, could reduce the weekly death rate. Further, isolation of patients is particularly effective in limiting the transmission of disease if early detection is possible before overt viral shedding (Wilder and Freedman 2020). Fong et al. (2020) states that while assuming that a high level of compliance with home isolation is possible for symptomatic persons, voluntary home isolation could be a preferred strategy to limit transmission compared to other preventive measures, which have not shown effectiveness in several randomized controlled trials.

Wilder and Freedman (2020) points out that a longer incubation period helps in providing more time for identification of cases and putting them into isolation. The incubation time of COVID-19 has a median of 5 days (Li et al. 2020), however, at

Table 16.1 Public Health Measures for limit spread of Outbreaks, adapted from Cetron and Simone

	Definition	Objective	Setting	Challenges	Remarks
Isolation	Separation of ill persons with contagious diseases from non-infected persons	To interrupt transmission to non-infected persons	Effective for infectious diseases with high person-to-person transmission where peak transmission occurs when patients have symptoms	Early case detection is paramount	Largely ineffective for infectious diseases where asymptomatic or pre-symptomatic infections contribute to transmission
Quarantine	Restriction of persons who are presumed to have been exposed to a contagious disease but are not ill, either because they did not become infected or because they are still in the incubation period	To reduce potential transmission from exposed persons before symptoms occur	Quarantining is most successful in settings where detection of cases is prompt, contacts can be traced within a short time frame with prompt issuance of quarantine	Quarantined persons will need psychological support, food and water, and household and medical supplies	Financial compensation for work days lost should be considered Voluntary is preferred over mandatory quarantine, but law enforcement may need to be considered if quarantine violations occur frequently
Community containment	Intervention applied to an entire community, city or region, designed to reduce personal interactions and movements. Such interventions range from social distancing among (such as cancellation of public gatherings, school closures; working from home) to community-use of face masks to locking down entire cities or areas (cordon sanitaire)	To reduce intermixing of unidentified infected persons with non-infected community members	Social distancing is particularly useful in settings where community transmission is substantial	Ethical principles and codes are needed to guide community containment practice and policy Community containment to protect the population's health potentially conflicts with individual rights of liberty and self-determination	Law enforcement is needed in most settings. Therefore such restrictive interventions should be limited to the actual level of risk to the community

this stage, it remains unknown how frequently pre-symptomatic cases result in secondary cases.

Given that influenza patients can transmit while remaining asymptomatic or before clinical symptoms emerge, isolation is usually too late to be sufficiently effective in limiting disease transmission during an influenza pandemic (Wilder and Freedman 2020).

16.1.2 Quarantine

World Health Organization's (WHO) recommends 'Quarantine' as one of the preventive interventions for controlling the disease. Quarantine limits the freedom of movement of people who have been, or are suspected of being exposed to a contagious disease. It denotes the separation of people who are suspected of being exposed to the infection/contagious disease with chances of getting infected and then spreading the disease (ECDC 2004). It implies restricting movement of people exposed to infection but not sick yet. Self-quarantining includes staying home, not welcoming others to home, avoiding going out of house, washing hands, staying home if sick etc. According to Parmet and Sinha (2020), in public health practices, quarantine refers to separating persons/communities exposed to a contagious disease while Cetron and Landwirth (2005), mentions that quarantine helps in restricting movement of those suspected to have been exposed to an infectious disease but are not ill as they did not get infected or are still asymptomatic but in the incubation period.

Quarantine remains one of the oldest and most effective tools for curbing spread of contagious disease outbreaks. This public health intervention was used when ships arriving from plague-infected ports to the Venice port were forced to anchor and wait for 40 days before disembarking the passengers in Italy in the fourteenth century (Cetron and Simone 2004). The term 'Quarantine' in Italian refers to 'quaranta' meaning 40. Forty days provided adequate time to complete the incubation period and to identify symptomatic cases (Wilder and Freedman 2020).

It usually can be applied at the individual or group level and requires being restricted to the home or a designated institutional facility through voluntary or mandatory quarantine, needing all individuals to be monitored for the occurrence of any symptoms and immediate isolation in case of any symptoms. Although, Lessler et al. (2009) suggests 4 days of quarantine as sufficient, including two incubation periods of influenza, however there is lack of evidence-based research addressing the optimal duration of quarantine. On the other hand, Gemert et al. (2011) reports an increased risk of infection with a longer duration of quarantine. Quarantining has proved most successful in settings with prompt detection of cases and contact tracing within a short period of time with immediate issuance of quarantine and voluntary compliance to it (Wilder and Freedman 2020). Voluntary compliance to home quarantine helps reduce stress on the emergency healthcare system. Although family clusters of infections may occur, however, the numbers of affected people are likely to be far lower than in institutional settings.

16.1.3 Social Distancing

Nussbaumer-Streit et al. (2020) defines social distancing as a measure where people without symptoms maintain a distance from each other physically by adopting social distancing practices including changes in behaviour that prevent disease transmission by reducing contact rates between susceptible and infected individuals (Reluga 2010). Social distancing is an important public health intervention that refers to actions taken to stop or minimise physical contact between individuals and thereby reduce or slow down the spread of a highly contagious disease (ECDC 2020). Social distancing places no locational constraints, rather it is a behavioural practice to lower the risk in most circumstances. Social distancing behaviours are more likely to occur on individual scale than on large scale. Use of social distancing measures to reduce direct and close contact between people in the community includes closure of schools or office buildings and suspension of public markets, and cancellation of gatherings (Wilder and Freedman 2020) closure of childcare facilities, religious services, entertainment venues, and places of large gatherings. These measures have played a critical role in mitigating previous pandemics, including Spanish Flu of 1918–1919 pandemic (Markel et al. 2007; Caley et al. 2008), and are a key part of current pandemic preparedness plans (Qualls et al. 2017; European Centre for Disease Prevention and Control 2009). Individual level measures involve practice of social distancing like teleconferencing instead of in-person work meetings and more extreme form of social distancing steps.

16.2 When to Initiate Social Distancing Measures

There is uncertainty over the effectiveness of most social distancing measures as a means of limiting the transmission of COVID-19 virus among people. Quarantines and travel bans are usually the first response to limit the spread of new contagious diseases. However, these old preventive measures are usually of limited utility for highly transmissible diseases, and can be counterproductive if imposed in a haphazard manner. However, due to the high reproduction and transmissibility rate of the virus, the impact of such measures on peak magnitude of epidemic and delay in peak likely depends on how early the preventive measures are implemented in the local context, with reference to the epidemiological situation.

Some of the available literature on past pandemics and experiences of China with COVID-19 indicates that the early, decisive, rapid, coordinated and comprehensive implementation of social distancing measures are likely to be more effective in reducing the spread of the virus than delayed actions (WHO 2020; Hatchett et al.

2007). Lai et al. (2020) estimates that if several non-pharmaceutical interventions had been conducted 1, 2 to 3 weeks earlier in China, the number of corona positive cases could have been reduced by 66%, 86%, and 95% respectively, while also considerably reducing the number of areas affected. Thus, social distancing is useful in where community transmission is believed to have occurred with unclear linkages between cases, and where restrictions placed only on persons known to have been exposed is considered insufficient to limit further spread of contagious disease. On the other hand, quarantining has proved most successful in settings with prompt detection of cases and contact tracing within a short period of time with immediate issuance of quarantine and voluntary compliance to it (Wilder and Freedman 2020).

There is neither a one size fits all approach of selecting the best time to implement social distancing measures involving isolation, quarantine, closures, cancellations of events with large gatherings etc. The detection of corona positive cases and/or deaths outside of known chains of transmission provides a signal to implement social distancing measures in different scenarios such as localised outbreaks and widespread transmission of COVID-19. For examples, closure of schools and opening of workplaces can result in parents seeking support of caregivers or grandparents to take care of (possibly infected) children while at work, resulting in higher chances of transmission of disease to individuals already at high risk of getting infected (Adlhoch et al. 2020).

Available literature supports the layering of multiple preventive measures under social distancing at once, rather than one by one (Hatchett et al. 2007) to increase the overall effectiveness of individual measures (Adlhoch et al. 2020) like teleconferencing, working from home instead of in-person work meetings and others. Citing example of China, Roosa et al. (2020) states that implementation of several social distancing strategies, including city-wide lockdowns, screening measures at railway stations and airports, active case finding, and isolation of suspected cases, appear to have slowed down the transmission of COVID-19 outside of Hubei province (Roosa et al. 2020).

16.3 Approach to Covid-19: Countries Setting Examples

In the recent pandemic, measures of isolation, quarantine, social distancing and community containment have been used at a massive scale across China (Wilder and Freedman 2020), where the coronavirus first emerged. Retrospective analyses of the 1918–1919 Spanish Flu pandemic reflects that delay in introducing social distancing measures are linked with excess deaths (Bootsma and Ferguson 2007; Richard et al. 2007). In support, Goh et al. (2006) states that quarantine was an effective measure during the SARS epidemic in 2003. Further, as part of observational studies, Li et al. (2013) estimated that quarantine policy in Beijing during the pandemic of influenza A (H1N1) reduced the number of cases at the peak of the epidemic and also delayed the epidemic peak, though at high economic and social costs (Li et al. 2013). On the other hand, Gemert et al. (2011) reported an increased risk of infection among household members concurrently quarantined with an isolated person and projected increased risk of infection with a longer duration of quarantine. Although, strong evidence is still not available for the effectiveness of such practices (Viner et al. 2020), however, individual adoption of such measures has collective benefits with much less disruption, financial costs or harms (Viner et al. 2020) besides helping in augmenting government credibility for defence against a future pandemic of such scale.

Some of the more common approaches have seen governments issue advisories or guidelines on social distancing as a preventive strategy for the countries as a whole while other have imposed complete or partial restrictions on all non-essential internal movement calling for a lockdown. Many countries have opted for strict measures to limit the spread of the disease. According to BBC (2020), over 100 countries worldwide imposed either a full or partial lockdown by the end of March 2020. In Columbia, the last number on the national ID determined whether one was allowed to leave the house or not while separate dog-walking hour was introduced in Serbia with Panama restricting the movement of people by gender on different days. Unlike most of Europe, Belarus refrained from placing any restrictions on sports events while Sweden banned gatherings of more than 50 people but schools for children under 16 years of age remained open and pubs and restaurants continued to offer table service. Austria enforced use of masks in public following the footsteps of countries such as Czech Republic, Slovakia and Bosnia and Herzegovina.

16.4 Social Distancing: An Effective Strategy for Containment

It is also a well-established fact that social distancing is one of the most effective strategies to contain the spread of any communicable disease to prevent people's exposure to the infected patient specially in case of a fast-spreading epidemic as it helps to buy time for healthcare system preparedness by delaying the timing of the peak of infections, to reduce the size of the epidemic peak, and to spread infections over a longer time period, enabling better management and lesser impact. The goal of social distancing of "flattening the curve," i.e. reducing the peak number of cases at any one time during the outbreak of disease (shown in Fig. 16.1). Several social distancing appear to have slowed down the transmission of COVID-19 outside of Hubei province (Roosa et al. 2020).

This helps to ensure that the healthcare system is not overburdened as a large number of people getting sick over a course of few days could overwhelm a hospital due to shortage of resources. On the other hand, if the same large number of patients arrived at the hospital at a slower rate, the graph would present a longer, flatter curve as fewer patients would arrive at the hospital each day prolonging the outbreak somewhat such that the hospitals have to time be more prepared and have better

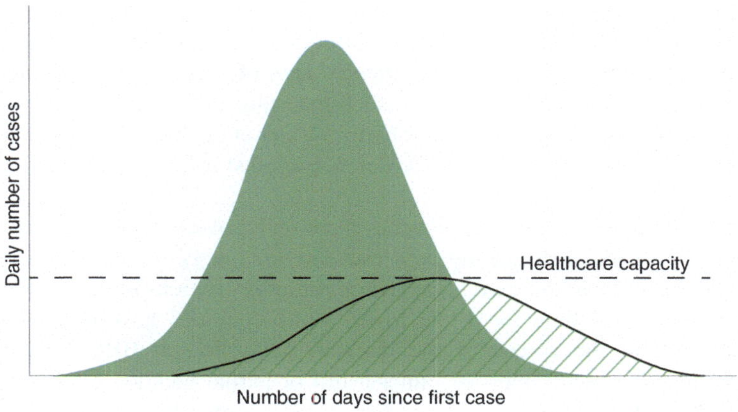

Fig. 16.1 Flattening the curve through social distancing measures. (Adapted from European Centre for Disease Prevention and Control (ECDC) 2020)

chance of being able to keep up with adequate supplies of rooms, equipment, medical supplies and doctors for all patients needing care.

Social distancing measures act as a means of reducing transmission and delaying spread and it can dramatically slow the rate of spread of the disease, easing the burden on the healthcare system and by providing community critical time needed to be prepared and to enhance the capacity of health infrastructure. Given its novelty, information on the effects of social distancing during the current COVID-19 pandemic is limited but emerging and although limited, the best available evidence supports social distancing measures as a means of reducing transmission and delaying spread.

16.5 Impact and Concerns

Quarantine and restricting movement of some uninfected and non-infected persons raises major ethical concerns regarding freedom of movement with the evidence on the effectiveness of quarantine being very limited with no solid rationale for the intervention. The ethical concerns are further exacerbated with increased risks of infection among quarantined persons (Miyaki et al. 2011; Gemert et al. 2011; Ferguson et al. 2006). Thus, voluntary/self-quarantine is likely to be a preferred strategy for containment of disease over mandatory quarantine in most of the scenarios (WHO 2007). Prolonged quarantine can result in significant burden on social services and working persons (Rothstein and Talbott 2007).

Three studies were identified in the Fong review (Fong et al. 2020) regarding avoiding crowd and restricting movement. Limited evidence suggests that avoidance of mass gatherings may be beneficial, with considerable uncertainty on the size of the crowd that would constitute a gathering. According to Rashid et al. (2015)

reviewers reported that there was "not a universal consensus from modelling studies on the impact of restrictions on movement" as one of their study showed that strict internal mobility restrictions (more than 50% reduction in mean travel frequency) applied in the initial stages of the pandemic for a period of 2–4 weeks could result in moderate delays of 1–1.5 weeks. However, another study cited by them showed that "weak travel restrictions" (10% travel restrictions) could possibly increase attack rates due to preventing travellers transmitting more infections within their local area.

In the absence of a vaccine or a medication, one of the most vital strategies adopted for slowing and controlling the spread of the pandemic is social distancing (Ferguson et al. 2020). Social distancing practices can reduce the severity of an outbreak, but the benefits of social distancing depend on the extent to which it is accepted and complied by the individuals. The effectiveness of social distancing as a control measure can be limited by the reluctance of the individuals to pay the costs inherent in social distancing (Reluga 2010).

Social distancing is likely a huge challenge as we, humans, are innately social beings living in a community environment with the deep-seated human instinct to connect with others (Baumeister and Leary 1995). Social connection amongst the human beings helps people regulate emotions, cope with stress and remain resilient during difficult times (Jetten et al. 2012, 2017; Rimé 2009). On the contrary, engaging in social distancing can exacerbate feelings of isolation and loneliness resulting in stress, anxiety and harmful effects on mental, cardiovascular and immune health (Haslam et al. 2018; Hawkley and Cacioppo 2010) while people subject to quarantine or self-isolation remain at risk of suffering from confusion and anger (Brooks et al. 2020) with explosive emotional tendencies when several household members endure them for a long period of time. Even for households free from the virus, the pandemic is likely to be the cause for chronic anxiety and economic difficulties. Bavel et al. (2020) emphasizes that self-isolation can exacerbate the impacts and can further increase social isolation and relationship difficulties. He further adds that such measures can aggravate feelings of loneliness and could result in negative long-term health consequences. Some of the studies also point out that forced proximity is a risk factor for aggression (Ellemers and Jetten 2013; Greenaway et al. 2015) and domestic violence (Owen 2020).

Thus, the success of social distancing interventions that are implemented over an extended period may depend on ensuring that people avoid physical contact but maintain emotional and social contact—from a distance—with friends, family and colleagues (ECDC 2020) to address challenges of stress, anxiety, isolation etc. Telephonic and virtual healthcare care facilities could be expanded to for people quarantined at home. Online-based communication is a key tool for ensuring a successful social distancing strategy. It is suggested that people take such steps to reduce stress and anxiety arising out of social distancing measures since humans are after all social creatures. Thus, the term 'social distancing' should be replaced with an alternative term 'physical distancing' to highlight and promote that social connection and meaningful interactions are possible even when people are physically separated (Bavel et al. 2020).

Confinement of movement with no specific end date could increase the perceived severity of the situation, potentially increasing compliance. However, lockdown and restrictions imposed by the government could also surge anxiety and have other psychological costs as it may signal uncertainty and people might start falling prey to behavioural fatigue (Ferguson et al. 2020) in the case of longer quarantines. Further, for the government, it is critical to announce an anticipated end date as a means of facilitating public acceptance and compliance of social distancing measures, as soon as possible. In addition, plans should be drawn up and communicated clearly to the public.

References

Adlhoch C, Baka A, Ciotti M, Gomes J, Kinsman J, Leitmeyer K et al (2020) Considerations relating to social distancing measures in response to the COVID-19 epidemic. European Centre for Disease Prevention and Control, Stockholm

Baumeister RF, Leary MR (1995) The need to belong: desire for interpersonal attachments as a fundamental human motivation. Psychol Bull 117:497–529

Bavel VJJ, Baicker K, Boggio PS, Capraro V, Cichocka A, Cikara M, Crockett MJ, Crum AJ, Douglas KM, Druckman JN, Drury J (2020) Using social and behavioural science to support COVID-19 pandemic response. Nat Hum Behav, 1–12

BBC News (2020) What are social distancing and self-isolation? BBC. https://www.bbc.com/news/uk-51506729. Accessed 16 Mar 2020

Bootsma MC, Ferguson NM (2007) The effect of public health measures on the 1918 influenza pandemic in US cities. Proc Natl Acad Sci 104(18):7588–7593

Briscese G, Lacetera N, Macis M, Tonin M (2020) Compliance with COVID-19 social-distancing measures in Italy: the role of expectations and duration (No. w26916). National Bureau of Economic Research, Cambridge, MA

Brooks SK et al (2020) The psychological impact of quarantine and how to reduce it: rapid review of the evidence. Lancet 395:912–920

Caley P, Philp DJ, McCracken K (2008) Quantifying social distancing arising from pandemic influenza. J R Soc Interface 5:631–639. https://doi.org/10.1098/rsif.2007.1197

ECDC (2004) Factsheet on isolation and quarantine. https://www.cdc.gov/sars/quarantine/fs-isolation.pdf

Cetron M, Landwirth J (2005) Public health and ethical considerations in planning for quarantine. Yale J Biol Med 78:329–334

Cetron M, Simone P (2004) Battling 21st-century scourges with a 14th century toolbox. Emerg Infect Dis 10:2053–2054

ECDC (2020) Considerations relating to social distancing measures in response to COVID-19 – second update. https://www.ecdc.europa.eu/sites/default/files/documents/covid-19-social-distancing-measuresg-guide-second-update.pdf

Ellemers N, Jetten J (2013) The many ways to be marginal in a group. Pers Soc Psychol Rev 17:3–21

European Centre for Disease Prevention and Control (2009) Guide to public health measures to reduce the impact of influenza pandemics in Europe: the ECDC Menu. European Centre for Disease Prevention and Control, Stockholm

Ferguson NM, Cummings DA, Fraser C, Cajka JC, Cooley PC, Burke DS (2006) Strategies for mitigating an influenza pandemic. Nature 442:448–452. https://doi.org/10.1038/nature04795

Ferguson NM et al (2020) Impact of non-pharmaceutical interventions (NPIs) to reduce covid-19 mortality and healthcare demand. Imperial College COVID-19 Response Team, London. https://doi.org/10.25561/77482

Fong MW, Gao H, Wong JY, Xiao J, Shiu EY, Ryu S, Cowling BJ (2020) Nonpharmaceutical measures for pandemic influenza in nonhealthcare settings—social distancing measures. Emerg Infect Dis 26(5):976

Gemert VC, Hellard M, McBryde ES, Fielding J, Spelman T, Higgins N, Lester R, Vally H, Bergeri I (2011) Intrahousehold transmission of pandemic (H1N1) 2009 virus, Victoria, Australia. Emerg Infect Dis 17(9):1599

Goh KT, Cutter J, Heng BH et al (2006) Epidemiology and control of SARS in Singapore. Ann Acad Med Singap 35:301–316

Greenaway KH, Jetten J, Ellemers N, van Bunderen L (2015) The dark side of inclusion: undesired acceptance increases aggression. Group Process Intergroup Relat 18:173–189

Haslam C et al (2018) The new psychology of health: unlocking the social cure. Routledge, New York

Hatchett RJ, Mecher CE, Lipsitch M (2007) Public health interventions and epidemic intensity during the 1918 influenza pandemic. Proc Natl Acad Sci 104:7582–7587

Hawkley LC, Cacioppo JT (2010) Loneliness matters: a theoretical and empirical review of consequences and mechanisms. Ann Behav Med 40:218–227

Jetten J, Haslam C, Haslam SA (eds) (2012) The social cure: identity, health and well-being. Psychology Press, New York

Jetten J et al (2017) Advancing the social identity approach to health and well-being: progressing the social cure research agenda. Eur J Soc Psychol 47:789–802

Lai S, Ruktanonchai N, Zhou L, Prosper O, Luo W, Floyd J. (2020) Effect of non-pharmaceutical interventions for containing the COVID-19 outbreak: an observational and modelling study. medRxiv

Lessler J, Reich NG, Brookmeyer R, Perl TM, Nelson KE, Cummings DA (2009) Incubation periods of acute respiratory viral infections: a systematic review. Lancet Infect Dis 9:291–300. https://doi.org/10.1016/S1473-3099(09)70069-6

Li X, Geng W, Tian H, Lai D (2013) Was mandatory quarantine necessary in China for controlling the 2009 H1N1 pandemic? Int J Environ Res Public Health 10:4690–4700. https://doi.org/10.3390/ijerph10104690

Li Q, Guan X, Wu P et al (2020) Early transmission dynamics in Wuhan, China, of novel coronavirus-infected pneumonia. N Engl J Med 382:1199–1207

Markel H, Lipman HB, Navarro JA, Sloan A, Michalsen JR, Stern AM et al (2007) Nonpharmaceutical interventions implemented by US cities during the 1918–1919 influenza pandemic. JAMA 298:644–654. https://doi.org/10.1001/jama.298.6.644

Miao YY, Huang JH (2012) Prevalence and associated psychosocial factors of increased hand hygiene practice during the influenza A/H1N1 pandemic: findings and prevention implications from a national survey in Taiwan. Tropical Med Int Health 17:604–612. https://doi.org/10.1111/j.1365-3156.2012.02966.x

Miyaki K, Sakurazawa H, Mikurube H, Nishizaka M, Ando H, Song Y et al (2011) An effective quarantine measure reduced the total incidence of influenza A H1N1 in the workplace: another way to control the H1N1 flu pandemic. J Occup Health 53:287–292. https://doi.org/10.1539/joh.10-0024-FS

Nussbaumer-Streit B, Mayr V, Dobrescu AI, Chapman A, Persad E, Klerings I et al (2020) Quarantine alone or in combination with other public health measures to control COVID-19: a rapid review. Cochrane Database Syst Rev 4:CD013574

Owen L (2020) Five ways the coronavirus is hitting women in Asia. BBC News. https://www.bbc.com/news/world-asia-51705199

Parmet WE, Sinha MS (2020) Covid-19—the law and limits of quarantine. N Engl J Med 382(15):e28

Qualls N, Levitt A, Kanade N, Wright-Jegede N, Dopson S, Biggerstaff M et al (2017) CDC Community Mitigation Guidelines Work Group. Community mitigation guidelines to prevent pandemic influenza—United States. MMWR Recomm Rep 66:1–34. https://doi.org/10.15585/mmwr.rr6601a1

Rashid H, Ridda I, King C et al (2015) Evidence compendium and advice on social distancing and other related measures for response to an influenza pandemic. Paediatr Respir Rev 16:119–126

Reluga TC (2010) Game theory of social distancing in response to an epidemic. PLoS Comput Biol 6(5):e1000793

Hatchett RJ, Mecher CE, Lipsitch M (2007) Public health interventions and epidemic intensity during the 1918 influenza pandemic. Proc Natl Acad Sci 104:7582–7587

Rimé B (2009) Emotion elicits the social sharing of emotion: theory and empirical review. Emot Rev 1:60–85

Roosa K, Lee Y, Luo R et al (2020) Short-term forecasts of the COVID-19 epidemic in Guangdong and Zhejiang, China: February 13–23, 2020. J Clin Med Res 9:596. https://doi.org/10.3390/jcm9020596

Rothstein MA, Talbott MK (2007) Encouraging compliance with quarantine: a proposal to provide job security and income replacement. Am J Public Health 97(Suppl 1):S49–S56. https://doi.org/10.2105/AJPH.2006.097303

Shaw R, Kim YK, Hua J (2020) Governance, technology and citizen behavior in pandemic: lessons from COVID-19 in East Asia. Progress Disaster Sci, 100090

Viner RM, Russell SJ, Croker H, Packer J, Ward J, Stansfield C, Mytton O, Bonell C, Booy R (2020) School closure and management practices during coronavirus outbreaks including COVID-19: a rapid systematic review. Lancet Child Adolesc Health

WHO (2020) Report of the WHO-China Joint Mission on coronavirus disease 2019 (COVID-19). World Health Organization, Geneva

Wilder-Smith A, Freedman DO (2020) Isolation, quarantine, social distancing and community containment: pivotal role for old-style public health measures in the novel coronavirus (2019-nCoV) outbreak. J Travel Med 27(2):taaa020. (IMP Paper)

World Health Organization (2007) Ethical considerations in developing a public health response to pandemic influenza. Report no. WHO/CDS/EPR/GIP/2007.2. WHO, Geneva

Chapter 17
Technological Advancement and Pandemic

Nikhil Kumar, Saket Dubey, Manish Kumar Goyal, Carlos Jimenez-Bescos, and Amin Talei

Abstract On 31 December 2019, the World Health Organization (WHO, Pneumonia of unknown cause in China, 2020) was notified of a cluster of cases with unknown reasons of the people being affected by pneumonia in the Wuhan City, China. On January 7, 2020, the Chinese authorities and experts declared the novel coronavirus as the root cause. It was in that very first instance labeled as Novel Coronavirus (WHO 2019) on February 11, 2020. A new type that has not been found before in human beings, Coronaviruses (CoV) are a diverse family of viruses that cause infections ranging from common colds to greater serious and grave illnesses. Augmenting their surveillance, countries around the world are tracking and finding new 2019-nCoV (novel coronavirus) cases. Use of modern technologies along with healthcare/medical systems is pervasive around the globe for an effective response and reduction of disease outbreak. Along with the evolution of biomedical response, an advanced technological response is also evolving to combat the pandemic. Technologies such as Big data, artificial intelligence, blockchain, remote sensing and others are linked to medical technologies for a prompt, rapid and effective response (Shaw et al., Governance, technology and citizen behavior in pandemic: lessons from COVID-19 in East Asia, Prog Disaster Sci 6:10009, 2020). This chapter covers a brief history of epidemics and technology, technology in healthcare, manufacturing production and surveillance under COVID-19 control measures.

Keywords Healthcare · Artificial intelligence · Manufacturing · Geographical Information System (GIS) · Remote sensing

N. Kumar (✉) · S. Dubey · M. K. Goyal
Discipline of Civil Engineering, Indian Institute of Technology, Indore, Madhya Pradesh, India

C. Jimenez-Bescos
Architecture and Built Environment, University of Nottingham, Nottingham, UK

A. Talei
Discipline of Civil Engineering, School of Engineering, Monash University Malaysia, Selangor, Malaysia

M. K. Goyal, A. K. Gupta (eds.), *Integrated Risk of Pandemic: Covid-19 Impacts, Resilience and Recommendations*, Disaster Resilience and Green Growth,
https://doi.org/10.1007/978-981-15-7679-9_17

17.1 A Brief History of Epidemics and Technology

Evaluation of interlinkages between epidemics, society and technology over the past can provide valuable background information in order to confront the recent challenges of COVID-19. The episodes of epidemics (Cholera, SARs, Ebola and others) around the globe has been sporadic but had a huge impact on our society. Long-term developments in public health strategies and scientific knowledge are evident along with the interaction of our society with these episodes. In terms of public health strategies, a major development has taken place with introduction of strategies such as quarantine, urban sanitation, vaccination and upcoming of modern medicines such as quinine and penicillin (Snowden 2019). The development of vaccine has plummeted the number of cases of smallpox, tetanus, mumps, polio and others. Moreover, some countries also take a stance to deny the presence of epidemic such as China did during SARS outbreak, and similar strategy has been implemented in the past (such as Spanish flu during World War 1) by many countries as well. On the other hand, the interactions with epidemic diseases have enriched scientific knowledge and further yielded the development of modern biomedical paradigm of epidemics, and foundation of new disciplines such as immunology, parasitology, tropical medicine and microbiology (Snowden 2019). To summarize, epidemics expose the intrinsic vulnerabilities of the global system and lessons from previous episodes of epidemics are conducive for formulation of more efficient, cost-effective public health strategies in the future. Here, we discuss briefly lessons learnt in combating SARS and Ebola.

17.1.1 Lessons from SARS (Severe Acute Respiratory Syndrome)

SARS is a respiratory disease with the ability of human transmission, it can spread rapidly through air; and has an asymptomatic incubation period (>7 day) (WHO n. d.) and hence, it has taken a heavy toll on healthcare staff and caregivers. The first case of SARS was reported in Guangdong, China in November 2002. It erupted as an international health emergency in March 2003 and till July 2003, SARS had already affected 8098 people and caused 774 deaths and led to gross expenditure and business losses of $60 billion to Asian countries alone (Snowden 2019). The major learnings from SARS were importance of early detection, effective public communication, strengthening containment strategy, international collaboration, and advancement in research and development (Mahmoud and Lemaon 2004). Technology can play leading role in early detection of epidemics and enhancing research and development. Firstly, a prompt, rapid, and effective surveillance and containment strategy was the key response to SARS, and it is applicable to any microbial threat. Found that initial delay in detection of SARS and delay in alerting national governments and its health-officials about the disease outbreak, substantially increased the spread of SARS and its impact in many affected countries. Moreover, global outbreak alerts and surveillance networks (such as GOARN and GPHIN) were found to be effective to curb the spread of SARS and helped in dismantling the transmission chain. In case of future epidemics, an integrated global database from various treatment centers would initiate the assessment of effectiveness of different treatment methods and further pave a way towards for more reliable treatment methods through comprehensive research. Secondly, advancement in research and development proved to be a critical lesson from SARS. The identification of SARS coronavirus within months after its emergence was made possible due to a long-term development of research on coronaviruses (Mahmoud and Lemaon 2004). It is to be noted that along with the basic research on viruses and microbes, there is a dire requirement to invest in biomedical research incorporating the zoonotic infectious disease research.

17.1.2 Lessons from Ebola

The Ebola epidemic originated in West Africa in 2014–2016 (WHO n.d.). The devastating Ebola virus causing a rare and deadly disease in humans and non-human primates was originally associated with severe bleeding (hemorrhages) and organ failure leading to fatalities. The hemorrhagic fever caused by Ebola virus disease has been classified to be highly contagious and is now known to have originated from fruit eating bats. Ebola has been devastating in terms of the economic and the social losses in West Africa with the cost of the epidemic in terms of both the social and economic losses being $53 billion. A rapid diagnosis with faster test results of higher standards and quality was the primary concern after the epidemic broke out. After the successful trials of the vaccine in Guinea in 2016, the vaccine was made available after getting an approval from the concerned management teams so that the public health would not be affected globally. After being licensed in Europe and the United States in late 2019 and prequalified by the WHO, it was approved and licensed in The Democratic Republic of Congo and five other African countries in early 2020 (Key lessons from Ebola that can help us defeat coronavirus 2020). In order to contain the virus and slowdown its spread, the major learnings from the Ebola spread have been valuable lessons in taking actions and keeping people apart. The lessons to contain the disease have mainly been banning mass gatherings, closure of institutions of learning and imposing social restrictions on movement by the concerned authorities. In addition to these measures the rapid testing has always been the firsthand response in reducing the spread. The health care workers are always at chances of higher infection rates while providing care and attending to the sick. The need of the healthcare workers in times of an outbreaks should always be building up stocks of protective gear by the governments in action. The long-term plan of recovering and revamping the economy should be the next thing to take up and ponder over after the curve of the outbreak flattens.

To conclude, the overall learnings from the previous infectious diseases outbreak are to strengthen the global surveillance and response system using modern communication technology and international collaboration. And, to strengthen the research and development for substantial improvement in understanding of infectious diseases for effective diagnosis, treatment and prevention. And, also to ensure the availability of the drugs, vaccines, and diagnostic tests needed to combat infectious diseases and infectious disease emergencies through public and private sector cooperation. Additionally, to curb the economic losses linked with episodes of epidemics, global supply chains incorporate sustainability in its systems as comprehensively as possible.

17.2 Technology in Manufacturing Sector

The blowout of COVID-19 has impacted and brought to the knees all the sectors of technology and human endeavor rendering most of them non-operational/inadequate with a reduction in production in many (Smedley 2020). The potential role of science and technology studied in the light of using and improving technology in the pandemonium/hubbub created by the current pandemic is creating ripples in the world. Technology can end jobs, but it does not end work (David 2015). The analysts are finding a platform to base their studies on how the world would be affected and served later on by the technologies both in pre and the post pandemic times. The question of how the social and economic fields will be affected stares us in the face. Now that the companies are coping in one way or the other, many have a high potential of reviving their sales. With the technological sector being disciplined/controlled to control the cash flows and operations in most of the industries, the pharmaceutical/health industry producing sanification products, protective gears and medicines is seen to have had a boost in sales with enhanced manufacture operations and increased market growth. This is where the concept of adopting smart technology in terms of being relevant in the times of the current pandemic steps in. The human labour with the use of smart technology is expected to be swapped by sophisticated software systems (having smart security provisions), automation and greater advances and massive development in production processes taking place in various industries making them efficient and flexible to be operated and remotely controlled at the click of a mouse (Ford 2015). The agricultural and the food sector with the use of sophisticated technology will see a way forward. With a greater twist in competition in the consumer markets, new serious dynamics of thinking, contact free choices of delivering products can bring a drastic change in heightening the sales and bringing the latest technologies to the forefront in the pre pandemic and the post pandemic businesses. Now that the inevitable changes are taking place at a greater scale, people can foresee how the technologies will be affected and embraced to avoid any sorts of crisis in any of the important sectors.

The impact and the role of technologies observing a paradigm shift in the times of crisis is hence central and analyzable so that the relation between the technologies and COVID-19 will be brought to light and lead to an aimed increase in mass production and newer and safer technology adoption (Schulte and Howard 2020). In terms of globalization, the policy makers and the governments will have an upper hand in observing the competitiveness in a greatly changing global market in terms of adopting new policies and strategies to improve and ramp up the economy. Sustainable packages keeping the changes and damages to the climate minimum would be appreciable if digital technologies are embraced and made to prosper under government guidelines. Communication networks and production technologies need to be overhauled and given a makeover so that the technological aspects see an improvement for economic growth to seize the opportunity. Swarming information sources are involved currently researching the impacts of this dreaded virus and the associated effects on technology, economy and psychology (Leso et al. 2018). The latest new technologies which were otherwise uncommon will be seen to be adopted at a common and a broader level so that all the sources of the entry of the virus will be traced and checked as much as possible.

COVID-19 in terms of on-site industrial manufacturing businesses is posing to be a great task for the manufacturers where the workers employed find it challenging to carry out the tasks remotely (Haseeb et al. 2019). Causing extensive concern and economic suffering for consumers, communities and businesses across the world, major industrial units have shut their facilities. The social distancing-oriented meetings will involve teleconferencing and an eruption in other means of communication to reach out to people for best practices so that robotic solutions in many industries and services are given a forefront to reduce the social interactions. Small and Medium-sized Enterprises (SMEs) will see the need to engage themselves in growth and competition by incorporating newer and cheaper technologies to be a part of the technological revolution. Digital skills and new potentials in all the sectors will be the key to meet the consumer demands keeping in mind the social impacts of technology and to flourish in the times when the pandemic will become a way of life until a vaccine is developed (Ting et al. 2020). Now that the pandemic that caused greater fear and anxiety turned into an economic crisis gradually, the flow of commodities got hit and the global supply chains got affected critically. The worldwide, nationwide and local public healthcare facilities and infrastructure developments in addition to save the lives of people geared up to deal with the pandemic and hence saw a further upgradation under pressure in their technologies and services.

17.2.1 Health Upscaling Production

Taking into account the human health factors greater technological advances have been made and will be made until there comes a stability of a certain level in the economic crisis balancing the supply and demand blows in the face of the economy that is the major challenge in the current crucial times. Research efforts in biological research will swing into action to deliver new innovations and subsequent solutions to foreseeable challenges. Biological research will be focused on pathogens giving a great lead to the pharmacies and pharmaceutical sectors to peak their sales in the post pandemic times.

17.2.2 Global Supply Chain Distortion

The worldwide spillover experienced is currently affecting each and every sector/ industry with suspended services due to the lockdown giving a jolt to the global market businesses (Elliott 2020). This hints at accepting the supply chain distortions and preparing for it. Ideally, pandemic control measures should be taken based upon the marginal social costs-benefits of the control measure. Guan et al. (2020) shows that the losses induced by COVID-19 in the global supply chain are mostly dependent upon no. of countries using lockdown as a measure, and lockdown duration has been found to be more sensitive to losses than strictness of the lockdown. It also suggests that earlier, stricter and shorter lockdowns can minimize overall losses and a gradual lifting of lockdown restrictions might decrease the overall losses. However, the losses in complex global supply chain might increase due to indirect causes of COVID-19, as the positive externalities of control measures offset losses and might lead to market failures and underinvestment. On the other hand, local and traditional trade due to the confinements at a worldwide scale has led to the integration of the financial activities of the confined countries because of the importance given to the use of local and national products resulting in financial integration. And, many countries that are import-dependent had a reduction in the flow of commodities because of the safety considerations and hiked prices due to the pandemic. The reliability of many industries on China hit the global supply chain because of many countries being dependent on China as the supply chains are markedly affected by the disruption faced in businesses across the world. Guan et al. (2020) estimated that a strict lockdown in China might decrease the 21% of its China's GDP and further decrease global GDP by 3.5%. Pharmaceutical components and automotive industries in India are the worst hit because of these materials and commodities being imported into India and this has hampered the processes where large productions could have been made.

17.3 Technology in Health Infrastructure

In the face of recent COVID-19 outbreak (WHO 2019), world woke up to realization of limitations of the health care system. It seemed that there was an urgent need of health care revolution to face such crisis. With recent eased commute and digital revolution sudden outburst of contagious disease was inescapable. Congregation of patients in the medical facilities and their waiting areas during crisis led vulnerable population with multiple chronic conditions to make difficult choice between risking

COVID-19 exposure during a visit to medical facility or postponing needed care. Upsurge of virus spread in the twenty-first century call for an urgent transform of health care facilities by unleashing digit technologies, contactless treatments, telemedicine, disinfecting robots, AI based diagnosis and tracing. Though some of these technologies have existed for decades, they have had poor traction into the market and health facilities due to heavy governmental regulations and sparse supportive payment structure (Flannery and Jarrin 2018). Additionally, application of IoT, big data analytics, AI and blockchain have been proven to be rewarding in monitoring, surveillance, detection and prevention of diseases (Ting et al. 2020). Though, these tools do not have significant application all around the world due to lack of technical expertise, still thanks to technological advancements, presently we are well equipped than any era in the past to respond to a global outbreak. In 2002, during the outbreak of SARS, it took almost a year to decode the genome of the virus. Whereas during COVID-19 it was decoded within a month. The diagnosis period during earlier outbreaks used to be few days whereas during recent times, results were available within a few hours. Contact tracing was almost impossible and protection kits for the medical staff and healthcare workers were less equipped. Therefore, this section of the book has been dedicated to application of technological developments in the health care infrastructure.

17.3.1 Advancements in Medical Facilities

(a) **Protection kits**: As epidemic spreads, hospitals become overwhelmed with high number of infectious patients seeking healthcare, in these situations prevention of healthcare workers to and from spread of disease needs protective equipment which includes face shields, masks, gowns, goggles and respirators (Livingston et al. 2020). Recent outbreaks have seen significant advancements in these protections kits including use of face shields instead of goggles, highly impermeable gowns, particulate respirators, head covers and rubber boots. These equipment's are now even designed in a way to allow safe reuse (disinfecting fabrics) as well as immediate change to reduce delays (Hersi et al. 2015). During epidemics, rapid supply of these equipment's in response to the enormous demands create a supply demand gap which may be fatal in some cases. The best way to meet these demand supply gap resides with effective local production and immediate organizational interventions.

(b) **Sanitization robots**: Preliminary way of preventing contagious diseases is by avoiding the places of most frequent visits such as airports, bus stops, hospitals, etc. But it is not possible for everybody such as health workers and cleaning staffs to avoid such visits and therefore it becomes vital to keep these places disinfected, recent technological development including the use of IoT and AI have led to development of many robots that disinfects entire airports or health care infrastructure without human interventions (Yang et al. 2020). These robots usually make use of short wavelength ultraviolet lights and are capable of disinfecting almost everything you point at them (Shayak and Rand 2020).

(c) **Modified ventilators**: Severe patients with viral infections such as COVID-19 develop pneumonia which reduces the lungs capabilities to inhale oxygen and in some cases require additional aids using ventilators. Due to excessive number of patients during an outbreak scarcity of ventilators leads to fatal consequences and therefore, numerous groups have proposed modifications in existing ventilators to support multiple patients (Herrmann et al. 2020). Several Modifications have also been advised to avoid aerosolizing the spread of viral droplets (Marchese et al. 2020). These modifications have been found to be feasible in multiple cases and may lead to Significant changes in ventilation technologies in future.

17.3.2 Advancements in Testing and Treating Procedures

In the context of testing during an outbreak, we have come a long way, but there is still a long way to go. As of now we are only capable of testing high priority, these are the people who are in hospital, those who are sick, healthcare workers and first responders who are symptomatic. We are not in a condition to test everybody who wants a test all around the world. Though china has tested ten million people within 10 days during May of 2020, applying this level of testing globally may even not be possible in near future. Current testing procedure involves taking nasopharyngeal swab from patients, extract the genetic material and convert the RNA to detectable DNA and then detect the presence of SARS-CoV-2 genes. Rapid testing procedures have been available for a while now, which are capable of diagnosing the viral infection within several minutes but these are still a limited resource for countries like Cambodia, Angola and Bangladesh. Most of the countries are still residing with swab tests and with global outbreak self-swab tests will become a necessity to assess the number of people infected in the country. Immunological testing also referred to as analyzer are capable of determining if a person was infected in the past and has developed required immune system to fight the virus, this would be very useful as it might act as a digital immunity passport that will contain protected health information one can carry and act as certified immune who cannot contract the virus and spread it. Specialist dogs are even being trained to detect COVID-19, by their sniffing abilities. These dogs have been found to be successful in detecting malaria, cancer and Parkinson's disease (Fischer 2017). Current believe still resides with antibody tests and scientist believe that these will help in understanding the full impact of the disease and enable practical decisions in the community such as when to social distancing protocols.

17.3.3 Efficient Disposal of Bio-Medical Waste

Bio-medical waste usually refers to waste generated by biomedical research facilities, medical laboratories and health care facilities. Various reports have suggested about the mis-management of bio-medical wastes especially in Africa, Asia and Middle East (Caniato et al. 2015). As a consequence, these wastes are discarded with municipal waste or openly burned leading to bio-medical hazards. Recently many advanced technologies have been developed for minimization and efficient disposal of bio-medical wastes, usually incineration of these waste materials leads to production of combustible air pollutants such as carbon monoxide and hydrochloric acid, to deal with this, technologies such as steam autoclave, microwave irradiation, solar disinfection are being developed, these alternatives are both eco-friendly and cost effective and forms the base for revolutionizing medical waste management (Thakur and Katocha 2012). A new surge is also seen in recent times which focuses on energy conversion when dealing with incineration of waste materials. New bio-medical waste companies are arising and government keeps frequently changing the regulations for bio-medical wastes and health care facilities are expected to keep themselves updated with current regulations.

17.4 Technology in Monitoring

Technologies such as big data analysis, geographical information system, artificial intelligence have played an important role during the recent COVID-19 pandemic, tasks such as rapid aggregation of multisource data, effective visualization of pandemic information, spatial tracking of suspected patients, transmission predictions, management of supply and demand of material resources, proper communication of information and social-emotional guidance were effectively handled by use of these technological aids (Zhou et al. 2020). High amount of data generated while performing these tasks provided solid spatial information needed for policy formulation, rapid decision making and applying preventive and control measures. Additionally, newer technologies adaptation such as extensive use of drone and facial recognition, remote scanners have enhanced our traceability and cellular applications are helping us to fight misinformation during these crucial times. Though technologies cannot prevent the onset of pandemics, but can significantly reduce the spread and educate and empower those on the frontlines, consequently, lessening the impact. Therefore, this section of the book has been dedicated to demonstrate the role of the state of the art advanced technologies in the tracing, monitoring and prediction of virus outbreaks.

17.4.1 Use of Mobile Applications During Pandemic

During the recent COVID-19 pandemic various mobile applications played very vital role. A few types of applications that played major part are discussed in this section.

(a) **Web based dashboard monitor**: These applications used data available from multisource to track new corona virus cases including direct communication with the hospitals, social media feeds and various news services (Boulos and Geraghty 2020). Most of these dashboards confirmed with the centers of disease control to authenticate the information that is being portrayed on their web page. The applications were a real time hub of huge database on the spatial and temporal spread of virus and were accompanied by intellectually designed infographics. These dashboards were first shared in the month of January, 2020 and initially contained information on total number of corona virus case, active number of corona virus cases and total number of deaths.
(b) **Health care applications**: With the increase in number of infectious patients, health care facilities started to crumble and was unable to support the number of patients. During these times patients were advised for home quarantine, and therefore these apps enabled the option to the patients to connect with the medical professionals and doctors within a very short interval of time, these applications also helped in keeping track of available medical facilities in the nearby hospitals such as available number of beds, ventilators, medical staffs etc.
(c) **Tracking applications**: Recent advantage of mobile technology was most profoundly seen with GIS based tracking, as these applications used available location information of the individuals in tracking down people who came in contact with infectious patients (Jhunjhunwala n.d.). Most of the countries developed their own tracking applications and were extensively used to determine probable chances of infection in various patients. These application were also used to allot e-pass to the people during lockdown period and restricted unnecessary crowds from the places of frequents visits. Most of these applications also contained information of preliminary precautions such as reminders to wash hands, keep clean hygiene, sanitize the utilities and also shared frequent warnings such as “avoid touching your face”.
(d) **Conferencing applications**: As schools were shut down, company employees were instructed to work from home and hospitals were overwhelmed with patients, video conferencing apps such as Google Meets, Zoom played a key role by enabling easy and rapid communication. School lectures were being carried out via video conferencing, board meetings and even medical consultation were being carried out online.

17.4.2 Artificial Intelligence and Remote Sensing

COVID-19 has become a global pandemic all over the world affecting human health, industries, economy, pollution, water quality, and other many variables. Artificial intelligence and remote sensing has broad applications; it can be a useful tool to deal with the current scenario.

Viruses are not visible to naked eyes, sensors using remote sensing techniques can be used to track the movement of virus, so these geospatial data can be useful for monitoring and modeling of disease and its impact on human life and environment, these geospatial data enables region-specific characterization of disease transmission, risk factors and forecasts of pathogen prevalence (Anastassopoulou et al. 2020; Biswas et al. 2020). After modeling dynamics of disease, remote sensing data can be used to monitor the spread of disease and spatiotemporal pattern of impact can be analyzed.

Artificial Intelligence is a technological aid which can help us fight COVID-19, it can enhance our tracing, screening and prediction capabilities. It can even be deployed for early detection and diagnosis of the infection (Vaishya et al. 2020). Recent studies have also shown that AI can also be deployed for drug and vaccine development which will significantly reduce the workload on medical infrastructure. AI-based models have been applied to predict mortality rates, recovery rates and economic loss by analyzing the data acquired from the past. The tools developed using AI can provide evidence-based medical solutions to cope up with rapidly mutating virus genomes.

17.4.3 Unmanned Aerial Vehicle

Technologies have emerged secondary after frontline workers such as doctors, nurses, medical staffs, local police etc., to assist nations in this global pandemic, the COVID 19. Drones have played major roles in helping local authorities and the country as a whole to combat the spread of corona virus outbreaks in urban and even rural areas. In many countries, drones are employed for surveillance and monitoring infected areas, and readily provide information to governments for formulating new action plans in advance (Kumar et al. 2020). They can help locate buildings that can be used as temporary medical care during this time, to provide more places for treatment and ease the burden of already packed hospital. In areas outreach by local authorities, they best serve their purpose to observe and collect information about the conditions of the people in those areas. They are effective specially to protect local policemen from coming in direct contact with others for obtaining necessary information. In some areas, like European countries and China, drones have ensured that people should follow the rules of restricting social gathering and maintaining social distancing. The drones are also mounted with speakers or microphones to broadcast the latest information and news in cities, rural areas and regions that lack proper communication means. This is extremely important to make everyone aware of the virus, how it spreads, infect, symptoms of the disease and the necessary steps to follow in such a situation. It can be used to catch people not following protocols and confront them not to move around without masks and social distancing. As they fly at a certain height above, they avoid close contact with people thereby avoid spreading the virus from one person to another and to the operator. Another benefit of this technology is that drones can be of service as to disinfect the infected and non-infected areas to curb the spread of the virus. Especially in red zone areas where the risk of getting infected by health workers is very high. They can be used to disinfect large scale area at comparatively faster rate than traditional methods. In this way, many places can be disinfected at a time especially if multiple drones are used and can be done more regularly. In order to help authorities to keep in check with their citizens the drones can be used to as thermal scanners to take temperature measurements of citizens as preliminary check-ups. This will help to alert local health care's and neighbouring residents of possible infections and to take the necessary precautions. The most important thing here is that it helps to avoid direct contact risk further spread of the infection. In a more subtle way, drones are also used as means of transporting goods and necessary items, such as medical supplies, test kids, essential goods, food items, groceries, etc. Using drones as deliveries make it the safest and faster way for transport than by vehicles travelling on roads during this pandemic. For remote areas this has provided an easy alternative for delivering medical kits and essential food items. These are some of the benefits of using drones during the COVID 19 pandemic. This technology has helped made a difference to combat and slow down the spread of the infection.

17.4.4 Blockchain Technology

The Information and Communication Technology (ICT) has undergone great advances in the recent times and this has encouraged the progression of conventional computer-aided manufacturing industry to smart industry featuring data driven decision making (Lade et al. 2017). Blockchain also referred to as decentralized and data driven decision making has emerged as a revolutionary technology with a wide range of applications and uses and is evolving into a secure and efficient network for secure data distribution in applications such as smart manufacturing, finance, operations management in supply chain, food industry for food safety and data tracking, energy market and healthcare services. Blockchain technology is a way to facilitate and to share factual data, track relevant data records and help speed up the treatment process in the healthcare system (Azaria et al. 2016; Petre 2017; Radanović and Likić 2018; Randall et al. 2017; Vora et al. 2018). Extended and spread over the whole distributed system, a blockchain is principally a distributed ledger (Zheng et al. 2018). For the management of healthcare, Griggs et al. (2018) developed a structure based on the blockchain system which is commonly known as blockchain based system. It has been considered as a new and efficient way of data sharing in healthcare as it helps in the management of the supply chain of medicines. It helps the healthcare researchers to get the details about the genetic code as it retains the identity. This activity of using blockchain technology to control the transmission of the diseases facilitates the treatments of many diseases which may prove fatal and removes the role of intermediaries. In addition to this, blockchain technology also allows the management authorities to trace, track and reach out to the patients disease-ridden by transmittable viruses such as Severe Acute Respiratory Syndrome (SARS), Middle East Respiratory Syndrome (MERS), Novel Coronavirus (Wu et al. 2020) The records in the healthcare sector are digitized using the blockchain technologies and in this way many services are made available to the patients and the healthcare sector which includes the physicians and the medical fraternity. The confidential records of importance include the patient records, payments and claims, availability of a database for audit purposes and all sorts of information related to healthcare. A blockchain supports the Contract system and a smart contract system can also be built to control the transmission of COVID-19. This can help in clubbing information and forming it into a network of data for large scale data processing and healthcare management for simplifying the complex and cumbersome processes and procedures in medical environment. This technology has gained widespread support from the researchers around the globe as their need of detailed data sets and data from diverse backgrounds will be covered by the availability of a lifetime record of individuals in the database. It has also gained the confidence of the patients in terms of addressing their privacy concerns of the medical information that they share from the start to the end of the disease that affected them and in this way checks privacy leakage and security breaches (Esposito et al. 2018). By remaining connected to the local and international data groups and databases, we can greatly monitor the spread of the disease and hence contain it.

17.4.5 Strategical Decision Making

Big data analysis driven on geospatial platforms are being applied for smooth and reliable decision making during the pandemic (Ienca and Vayena 2020). The use of various standard based services and analytical tools are being used to assess the current situation of pandemic on both micro and macro scales and the output from these tools are being employed in devising area specific strategies to handle the socio-economic consequences (Weible et al. 2020). Community engagement is possible with the help virtual interactions using social media applications and the issues from the lowest levels are being communicated through proper channel to the advisory committees. This integration of demographic information and geospatial data is essential for decision making, governance and development.

References

Anastassopoulou C, Russo L, Tsakris A, Siettos C (2020) Data-based analysis, modelling and forecasting of the COVID-19 outbreak. PLoS One 15:e0230405

Azaria A, Ekblaw A, Vieira T, Lippman A (2016) Medrec: using blockchain for medical data access and permission management. In: 2016 2nd international conference on open and big data (OBD). IEEE, New York, pp 25–30

Biswas SK, Ghosh JK, Sarkar S, Ghosh U (2020) COVID-19 pandemic in India: a mathematical model study. Preprints 2020:202005050

Boulos MNK, Geraghty EM (2020) Geographical tracking and mapping of coronavirus disease COVID-19/severe acute respiratory syndrome coronavirus 2 (SARS-CoV-2) epidemic and associated events around the world: how 21st century GIS technologies are supporting the global fight against outbreaks and epidemics. Int J Health Geogr 19:8

Caniato M, Tudor T, Vaccari M (2015) International governance structures for health-care waste management: a systematic review of scientific literature. J Environ Manag 153:93–107

David H (2015) Why are there still so many jobs? The history and future of workplace automation. J Econ Perspect 29:3–30

Elliott L (2020) Prepare for the coronavirus global recession | Business. The Guardian. https://www.theguardian.com/business/2020/mar/15/prepare-for-the-coronavirus-global-recession. Accessed 7 Jun 2020

Esposito C, De Santis A, Tortora G, Chang H, Choo K-KR (2018) Blockchain: a panacea for healthcare cloud-based data security and privacy? IEEE Cloud Comput 5:31–37

Fischer S (2017) Sniffing for cancer: nano noses hold promise for detecting lung cancer and other diseases. IEEE Pulse 8:20–22

Flannery D, Jarrin R (2018) Building a regulatory and payment framework flexible enough to withstand technological progress. Health Aff 37:2052–2059

Ford M (2015) The rise of the robots: technology and the threat of mass unemployment. Simon and Schuster, New York

Griggs KN, Ossipova O, Kohlios CP, Baccarini AN, Howson EA, Hayajneh T (2018) Healthcare blockchain system using smart contracts for secure automated remote patient monitoring. J Med Syst 42:130

Guan D, Wang D, Hallegatte S, Davis SJ, Huo J, Li S, Bai Y, Lei T, Xue Q, Coffman D (2020) Global supply-chain effects of COVID-19 control measures. Nat Hum Behav 4:1–11

Haseeb M, Hussain HI, Ślusarczyk B, Jermsittiparsert K (2019) Industry 4.0: a solution towards technology challenges of sustainable business performance. Soc Sci 8:154

Herrmann J, da Cruz AF, Hawley ML, Branson RD, Kaczka DW (2020) Shared ventilation in the era of covid-19: a theoretical consideration of the dangers and potential solutions. Respir Care. https://doi.org/10.4187/respcare.07919

Hersi M, Stevens A, Quach P, Hamel C, Thavorn K, Garritty C, Skidmore B, Vallenas C, Norris SL, Egger M (2015) Effectiveness of personal protective equipment for healthcare workers caring for patients with filovirus disease: a rapid review. PLoS One 10:e0140290

Ienca M, Vayena E (2020) On the responsible use of digital data to tackle the COVID-19 pandemic. Nat Med 26:463–464

Jhunjhunwala A (n.d.) Role of telecom network to manage COVID-19 in India: Aarogya Setu. Trans Indian Natl Acad Eng. https://doi.org/10.1007/s41403-020-00109-7

Key lessons from Ebola that can help us defeat coronavirus (2020) Time. https://time.com/5806459/five-key-lessons-from-ebola-that-can-help-us-win-against-coronavirus-everywhere/. Accessed 7 Jun 2020

Kumar A, Gupta PK, Srivastava A (2020) A review of modern technologies for tackling COVID-19 pandemic. Diabetes Metab Syndr Clin Res Rev 14:569–573

Lade P, Ghosh R, Srinivasan S (2017) Manufacturing analytics and industrial internet of things. IEEE Intell Syst 32:74–79

Leso V, Fontana L, Iavicoli I (2018) The occupational health and safety dimension of Industry 4.0. Med Lav 110:327–338

Livingston E, Desai A, Berkwits M (2020) Sourcing personal protective equipment during the COVID-19 pandemic. JAMA 323:1912–1914

Mahmoud A, Lemaon S (2004) Lessons from SARS for future outbreaks - learning from SARS - NCBI Bookshelf (n.d.). https://www.ncbi.nim.nih.gov.books/NBK92465/. Accessed 7 Jun 2020

Marchese M, Capannolo A, Lombardi L, Di Carlo M, Marinangeli F, Fusco P (2020) Use of a modified ventilation mask to avoid aerosolizing spread of droplets for short endoscopic procedures during coronavirus COVID-19 outbreak. Gastrointest Endosc. https://doi.org/10.1016/j.gie.2020.03.3853

Petre A (2017) Blockchain use cases in healthcare. Accessed 10 Mar 2018

Radanović I, Likić R (2018) Opportunities for use of blockchain technology in medicine. Appl Health Econ Health Policy 16:583–590

Randall D, Goel P, Abujamra R (2017) Blockchain applications and use cases in health information technology. J Health Med Informatics 8:8–11

Schulte P, Howard J (2020) The impact of technology on work and the workforce. https://www.ilo.org/safework/events/safeday/33thinkpieces/WCMS_681603/lang%2D%2Den/index.htm. Accessed 7 Jun 2020

Shaw R, Kim Y, Hua J (2020) Governance, technology and citizen behavior in pandemic: lessons from COVID-19 in East Asia. Prog Disaster Sci 6:100090

Shayak B, Rand RH (2020) Self-burnout-a new path to the end of COVID-19. medRxiv

Smedley (2020) Manufacturing and the COVID-19 pandemic -news analysis. https://connectedworld.com/manufacturing-and-the-covid-19-pandemic/. Accessed 7 Jun 2020

Snowden FM (2019) Epidemics and society: from the black death to the present. Yale University Press, London

Thakur Y, Katocha SS (2012) Emerging technologies in biomedical waste treatment and disposal. Chem Eng 29:787–792

Ting DSW, Carin L, Dzau V, Wong TY (2020) Digital technology and COVID-19. Nat Med 26:459–461

Vaishya R, Javaid M, Khan IH, Haleem A (2020) Artificial Intelligence (AI) applications for COVID-19 pandemic. Diabetes Metab Syndr Clin Res Rev 14:337–339

Vora J, Nayyar A, Tanwar S, Tyagi S, Kumar N, Obaidat MS, Rodrigues JJPC (2018) Bheem: a blockchain-based framework for securing electronic health records. In: 2018 IEEE Globecom workshops (GC Wkshps). IEEE, New York, pp 1–6

Weible CM, Nohrstedt D, Cairney P, Carter DP, Crow DA, Durnová AP, Heikkila T, Ingold K, McConnell A, Stone D (2020) COVID-19 and the policy sciences: initial reactions and perspectives. Policy Sci:1–17. https://doi.org/10.1007/s11077-020-09381-4

WHO (2019) Ebola then and now: eight lessions from West Africa that were applied in the Democratic Republic of the Congo (n.d.). https://www.who.int/news-room/feature-stories/detail/ebola-thenand-now. Accessed 7 Jun 2020

WHO (2020) Pneumonia of unknown cause in China. WHO, Geneva

WHO (n.d.) Severe acute respiratory syndrome (SARS). https://www.who.int/csr/sars/en/. Accessed 7 Jun 2020

Wu F, Zhao S, Yu B, Chen Y-M, Wang W, Song Z-G, Hu Y, Tao Z-W, Tian J-H, Pei Y-Y (2020) A new coronavirus associated with human respiratory disease in China. Nature 579:265–269

Yang G-Z, Nelson BJ, Murphy RR, Choset H, Christensen H, Collins SH, Dario P, Goldberg K, Ikuta K, Jacobstein N (2020) Combating COVID-19—the role of robotics in managing public health and infectious diseases. Sci Robot 5(40):eabb5589

Zheng Z, Xie S, Dai H-N, Chen X, Wang H (2018) Blockchain challenges and opportunities: a survey. Int J Web Grid Serv 14:352–375

Zhou C, Su F, Pei T, Zhang A, Du Y, Luo B, Cao Z, Wang J, Yuan W, Zhu Y (2020) COVID-19: challenges to GIS with big data. Geogr Sustain 1:77–87

Chapter 18
Application of Nanotechnology in Detection and Prevention of COVID-19

Ajay K. Kushwaha, Hemen Kalita, Aditya Bhardwaj, Siddhartha Suman, Biswajit Dehingia, and Rajneesh Kumar Mishra

Abstract Over the years, due to climate change, developing cities, industrial transformation, excessive farming, food habits, and ease of traveling have increased the chances of spreading of a pathogen in case of any pandemic. Scientists are trying to devise strategies to stop or mitigate any potential outbreak of any unknown virus: SARS-CoV-2, an invisible virus, a profound way to distort human lifestyle globally, causing COVID-19. Herein, the chapter discusses the advancements in nanotechnological solutions for the detection and prevention of COVID-19. The structure, characteristics, detection methods, possible mechanisms, sensors, and portable test kits of the SARS-CoV-2 virus are reviewed. Various nanomaterials-based formulations and promising approaches for the detection and clinical diagnosis of SARS-CoV-2 infections are carefully summarized. It discusses how nanotechnology can be used as a tool to alleviate the SARS-CoV-2 virus and explore the applications of nanomaterials for antiviral, antifungal coatings on personal protective equipment for protection. Studies revealed that there is a broad scope to develop various nanotechnology-based solutions to combat the invisible enemy of humankind.

Keywords SARS-CoV-2 · Coronavirus · PPEs · Facemask · Antivirus · Coatings · Nanomaterials

---

A. K. Kushwaha (✉) · A. Bhardwaj · S. Suman
Department of Metallurgy Engineering and Materials Science, IIT Indore, Indore, Madhya Pradesh, India
e-mail: akk@iiti.ac.in

H. Kalita (✉) · B. Dehingia
Department of Physics, Gauhati University, Guwahati, Assam, India
e-mail: hemenkalita@gauhati.ac.in

R. K. Mishra
Department of Physics, Yeungnam University, Gyeongbuk, South Korea

M. K. Goyal, A. K. Gupta (eds.), *Integrated Risk of Pandemic: Covid-19 Impacts, Resilience and Recommendations*, Disaster Resilience and Green Growth,
https://doi.org/10.1007/978-981-15-7679-9_18

18.1 Introduction

Respiratory infections in humans have a vital impact on morbidity and mortality all over the world. The health conditions in the present scenario have a significant setback due to the outbreak of the novel coronavirus. There are plenty of respiratory diseases caused by different viruses such as a human respiratory syncytial virus (HRSV) cause the URI, bronchiolitis, pneumonia, and can be detected by culture, Ag detection, RT-PCR. While human parainfluenza viruses (HPIV), mostly found in infants and children, causing bronchiolitis, URI, bronchitis, croup, and pneumonia (Pillai et al. 2020). Further, human rhinovirus (HRVs) are the immensely generic respiratory pathogens of humans; it is non-enveloped, positive-stranded RNA virus. HRVs cause asthma, URI, and COPD exacerbation diseases and can be detected from culture, RT-PCR (Carter et al. 2020a). Similarly, SARS Cov and SARS-CoV-2 are viruses that cause zoonotic disease and get transferred to humans from animals (Pimentel et al. 2009). There are seven different kinds of coronaviruses, where SARS and SARS-CoV-2 are most integral out of all viruses. These viruses are unpredictable and quite dangerous because it keeps on changing the structure when it encounters species. The effects of these viruses are hypoxia, pneumonia, and even damage to the liver, kidney, and especially lungs and can be fatal (Schmitt et al. 2020). Identifying these viruses in humans is still in progress; however, few early symptoms of the virus infection in humans can be observed by body temperature ($\geq$100.3 °F), cough, difficulty in breathing, shortness of breath, etc. (Zhu et al. 2020a). Some laboratory diagnoses are also completed for the confirmation for decisive virus attack, i.e., blood test, stool test, nasal secretion test, and pneumonia test (Burrell et al. 2017). The magnitude of transmission and the total number of cases of SARS-CoV-2 clearly shows that virus is affecting human life at a grander scale and widespread pandemic (Peeri et al. 2020). The outbreak of SARS-CoV-2 has made a significant impact on social life; thus, the ways to prevent diseases need new inventions, and nanotechnology can be the central front of such devices.

The size of SARS-CoV-2 is in the range of 60–140 nm belong to the category of nanomaterials; thus, nanotechnology can provide effective weapons to combat COVID-19 (Sportelli et al. 2020). Nanotechnology can provide various solutions to develop better testing kits, antiviral coating, PPEs (masks, gowns, goggles, shoes), disinfecting spray solutions, clothing, packing materials, multifunctional surface, AC filters, drug discovery, vaccine developments (Fig. 18.1) (Yu et al. 2020). After considering the widespread of the subject area, this chapter focus on detection and prevention methods only. The early detection of the COVID-19 and the use of personnel protective equipment is the primary tool to reduce the community transmission of this deadly disease. The COVID-19 virus can also be detected in a similar manner as other coronaviruses, responsible for SARS (Severe acute respiratory syndrome). However, the detection mechanism depends on the biological structure of the SARS-CoV-2, which is unique. Hence it becomes a more widely spread virus and requires better selectivity and sensitivity to detect. At present nucleic acid testing, protein testing, chest chromatography, antibody test, etc. are

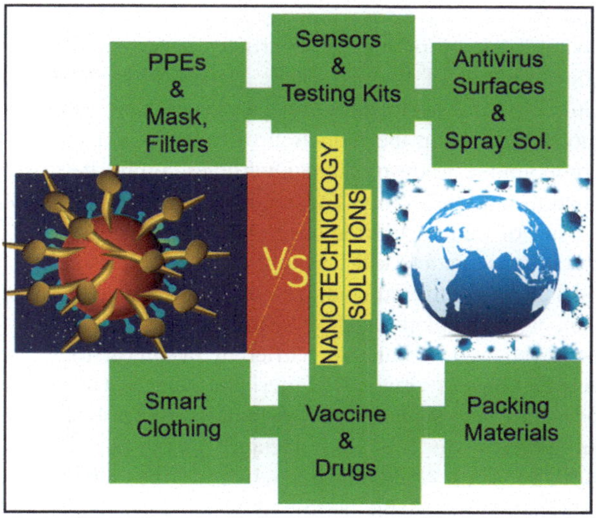

Fig. 18.1 Nanotechnology solutions to combat with COVID-19

the main methods for the detection of SARS-CoV-2 (Ai et al. 2020; Udugama et al. 2020; Zhou et al. 2020; Lu et al. 2020). Hamming et al. develop a naked eye detection method with the help of specific oligonucleotide capped plasmonic nanoparticles (Hamming et al. 2004). Luminescent immunoassay is also a serology-based technique that has been reported for SARS-CoV-2 detection. Wrapp et al. designed a peptide-based immunoassay for COVID-19 detection (Wrapp et al. 2020). Seo et al. reported graphene-based field-effect transistor for sensing of SARS-CoV-2 by detecting antigen proteins in which the SARS-CoV-2 spike protein antibody was on graphene using a coupling agent 1-pyrene butyric acid N-hydroxysuccinimide ester (Seo et al. 2020). In summary, across the globe, more than 100 COVID-19 test kits are being, and scientists are also putting their effort into utilizing the benefits of nanomaterials to achieve improved sensitivity, rapid and point of care detection.

The transmission of COVID-19 occurs mainly in three ways, direct contact of an infected person with a healthy person (via respiratory droplets), a healthy person comes into contact with aerosol formed from a respiratory droplet of an infected person, touching of the contaminated surface by hand then face. These three modes of transmission need to be passivated to reduce the spread of COVID-19, and it can be achieved by using appropriate PPEs, spraying antiviral solutions, and developing antiviral surfaces. Facemask is the primary requirement to combat coronavirus. However, the effectiveness of the surgical mask is doubtful, and N-95 mask not very convenient and economical for the entire population. Hence new solutions are required to develop appropriate facemasks. One solution could be the use of woven and nonwoven fabrics with light active antiviral dyes/materials that offer excellent antimicrobial and antiviral properties. Nanoparticles of noble metals have been extensively investigated as antiviral material and coatings in which silver nanoparticles have shown exceptional ability. It is observed that an increase in the coating density of Ag-SiO_2 particles exponentially increases the reduction of the concentration of the virus (Joe et al. 2014; Mori et al. 2013). Nanocrystals of $CuFeO_2$ have useful antiviral properties, and investigation shows Cu^+ species is more effective in the reduction of bacteria and viruses than copper metal or Cu^{2+} compound. Carbon-metal nanoparticles composite was patented in 2019 as antiviral coating materials (Jose-Yacaman et al. 2009). Recently, researchers from the University of Central Florida have proposed nanomaterials based protecting suites that catch the virus and kill it within a few seconds under UV light excitation (Sudipta Seal 2020). Antiviral coatings and hydrophobic surfaces have been designed using nanocarbon fullerene C-60 (Siddiquie et al. 2020). Graphene oxide has been deployed as a breathable barrier in facemask to protect from SARS-CoV-2 and similar kinds of viruses (Palmieri and Papi 2020). The heat comfort in facemask can be achieved using nanofibers on nanoporous polyethylene-face mask that can reduce an approximate 87% IR reflectance (Yang et al. 2017). Recently, scientists from KAUST Saudi Arabia have newly developed a flexible nanoporous template for the fabrication of reusable N-95 hydrophobic masks (El-Atab et al. 2020). In summary, the advanced nanotechnology solutions can support the fight against the invisible enemy by putting further intense efforts. The early detection and protection measures are a promising solution to reduce the spread of SARS-CoV-2 because compulsory social confinement is not feasible for a longer time due to the essential works and social nature of human beings.

18.2 Detection of Coronaviruses

18.2.1 Coronaviruses

In the past few months, coronavirus has created havoc in the world. The coronavirus was reported for the first time in the year 1960 (Su et al. 2016). The projection of the virus under electron microscopy looks like a corona or crown, hence it was named as coronavirus in the year 1968 (Almeida et al. 1968). The coronaviruses belong to coronaviridae family, which are basically a type of enclosed single-strand RNA viruses (size 26–32 kb) (Lu et al. 2020; Weiss and Navas-Martin 2005). These viruses mostly affect the birds, humans as well as other mammals, and cause severe acute respiratory diseases such as sore throat, sneezing, fever, coughing, running nose. In extreme cases, respiratory illnesses can also lead to the death of an individual (Su et al. 2016). The coronaviruses are mainly divided into four types of viruses called as α, β, γ, -coronavirus. HCov-NL63, HCov-229E are types of α-coronavirus, while HCov-OC43 and HCov-HKU1 are types of β-coronavirus. The SARS Cov (severe acute respiratory syndrome) and MERS Cov (middle east respiratory syndrome) and the most recent SARS-CoV-2 (COVID-19) are γ-coronaviruses. The typical structures of the prominent coronaviruses are represented in Fig. 18.2. The typical structures of coronaviruses consist of four

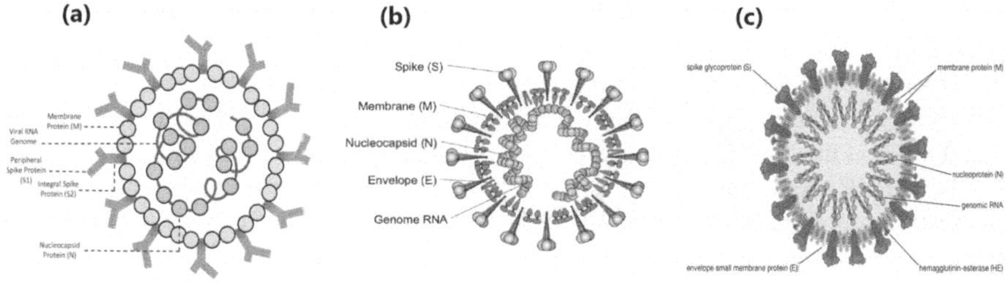

Fig. 18.2 Typical structures of coronaviruses (**a**) SARS (Reproduced with permission Yonesi and Rezazadeh 2020. Copyright MDPI). (**b**) MERS (Reproduced with permission Xu et al. 2019a. Copyright Taylor & Francis). (**c**) SARS-CoV-2 (Reproduced from Britannica encyclopedia)

types of proteins, namely spike (S), nucleocapsid (N), membrane (M) and envelope (E). In the past last two decades, several coronaviruses are noticed, the most prominent coronaviruses being the SARS in China in 2002 (Peiris and Poon 2008) and the MERS in Saudi Arabia in 2012 (Assiri et al. 2013). Recently, in December 2019, another deadly virus belonging to this family called SARS-CoV-2 resurfaced the first time in Wuhan, China, and spread like wildfire and has affected millions of human lives globally. The infection of these viruses may have initial symptoms of fever, myalgia, chills and malaise; later stages causes were shortness of breath, diarrhea, and tachypnea.

The major symptoms for the MERS-Cov were identified to be pneumonia and gastrointestinal damage, causing kidney failure in extreme cases. The other symptoms are similar to SARS Cov (Assiri et al. 2013). SARS-CoV-2 is the most recently known coronavirus that also has similar symptoms of SARS Cov. In SARS-CoV-2, initial symptoms are respiratory diseases such as common cold, shortness of breath, sore throat, fever, dizziness, headache. In later stages, SARS-CoV-2 causes acute respiratory infection, damaging the several parts of the body, especially lungs and in extreme cases can even cause the death of an individual. The knowledge of coronavirus and its biological structures is very important to develop detection and protection methods.

18.2.2 Methods for Detection of Coronaviruses

Understanding the structure of coronaviruses and its effects on the human being are imperative to find out the best detection methods for these viruses. Typically, after the collection of the swab sample from the infected people, the positive-strand RNA is extracted from the virus. After that, the amplification of the nucleic acid is done by adopting different types of methods. Now the main task is to detect whether amplification takes place or not. The detection of the amplification of nucleic acid is done by gel electrophoresis, various spectrophotometric techniques such as fluorescent measurement, optical density measurement by turbidimeter etc. (Balboni et al. 2012; Thi et al. 2004; Poon et al. 2004). Mostly, PCR (polymerase chain reaction) based detection is employed due to excellent sensitivity, production of sufficient amounts of species, and reproducibility. PCR method is an enzymatic based method used for DNA analysis in research laboratories and clinics (Shen et al. 2020). In PCR, the amplification of DNA to be analyzed is carried out with the help of the oligonucleotide primer. One cycle of the PCR process can be divided into three stages: denaturation, primer hybridization, and DNA polymerization (Erlich and Davm Gelfand 1991). In the denaturation stage, double-stranded DNA is converted into single-stranded DNA with the help of temperature treatment. Then oligonucleotide primer is allowed to hybridize with separated strands in annealing followed by the extensions of the primer to get the DNA polymerase. Since Coronavirus consists of positive-strand RNA, therefore, real-time RT-PCR (Reverse transcription-PCR) is more favored for the detection (Poon et al. 2004; Stadler et al. 2003). In real-time RT-PCR, targeted RNA has reversed transcript to complementary DNA (cDNA), and then amplification of the cDNA is carried out using specific oligonucleotide primer.

Another commonly used method for coronavirus detection is LAMP (Loop-mediated isothermal amplification). Herein amplification of the nucleic acid is carried out in constant temperature. This method provides efficient amplification of nucleic acid due to the obviation of the thermal cycle, which takes a lot of time (Notomi et al. 2000). However, this method does not require expensive chemicals and equipment, which reduces the cost of virus detection. Poon et al. suggested a rapid and less costly LAMP method for SARS Cov detection. Here gel electrophoresis is adopted for the detection of the amplified viral RNA (Poon et al. 2004). RCA (Rolling circle amplification) is another isothermal based method selected for the amplification of viral nucleic acid for detection (Xu et al. 2019b). Chapin and Doyle suggested an ultrasensitive RCA method for the rapid quantification of the microRNA, which can be effectively used for the detection of viral RNA (Chapin and Doyle 2011). Microarray-based detectors are also one of the methods used for the rapid detection of the SARS virus. Rong et al. reported a 60mer oligonucleotide microarray for the diagnosis of SARS Cov. Recently, concerning point-of-care (POC) detection Zhang et al. designed Cas13a for the rapid detection of DNA or RNA associated with the infected virus (Gootenberg et al. 2017). The serological detection method is also adopted for the detection of coronavirus, which is an indirect method of detection (Chan et al. 2004). Basically, the immune response of a body is probed to detect the presence of the infectious virus in this method since our body gives immune responses to any viral infection by producing antigens and antibodies (Long et al. 2020). Since the viruses have a size similar to nanomaterials, hence the role of nanotechnology is significant to improve the detection efficiency and selectivity.

18.2.3 Sensors for Detection of Coronaviruses

The ease of detection in the future is based on the development of high-tech sensors that can correspond to the analytical requirement by applying simple and highly sensitive devices. To counter the requirement in the detection of SARS virus, various kinds of sensors have been studied. The scientists and doctors came together to develop sensors for fast and easy detection techniques; studies show that different sensors are available for SARS virus detection. Nowadays, revisiting hybrid technology improves the old school process of detection and gives a well-defined approach for virus detection as well. The presence of nanomaterials has been introduced to the sensor technology for a long time. In a recent situation, due to COVID-19 outbreak, the detection of viruses is an essential point of concern.

At present, various sensors are available for the detection and monitoring of the virus. Sensors based on different platforms like electrochemical, plasmonic-based biosensors, Field effect transistor-based sensors, colorimetric sensors, screen printed based sensors, paper-based sensors have been developed, as shown in Fig. 18.3, the use of nanomaterial, bio probe, hybrid technology has also been explored. A simple electrochemical detection was employed with a Geno sensor based on the gold substrate. The surface of the material served as a static and transduction for the analytes. Study on 30-mer short lysine-rich region sequence, which was like the SARS virus was taken. The relation between alkaline phosphate permits indirect signals of electrochemical detection. The square wave voltammetry signals on an electrochemical setup show a detection limit of 6 pM for the corresponding DNA sequence (Abad-Valle et al. 2005). A biochip system based detection sensor was developed using the idea of two physical implications, which is ellipsometry and surface plasmon resonance detection. The ellipsometer detected the surface concentration of the biomedical reactions with the help of various features. These factors are thickness and concentration of the multilayer samples, the interaction of the biomolecules, and the changes occurring due to the kinetic behavior of the molecule were also detected with the surface plasmon resonance-based method (Lee et al. 2004). Dual functional systems have been used to identify SARS virus with more efficient parameters; the reason behind this response is due to the PPT effect and LSPR sensing. This is carried at two different angles of incidence; the plasmons generate different wavelengths that improve the stability, sensitivity, and reliability towards the detection of the virus (Qiu et al. 2020).

New kinds of genosensors were designed using disposable screen-printed electrodes made up of carbon, and gold nanostructures were incorporated for better performance. The reaction between the thiol and gold takes place very quickly, and the response of the sensors shows linear behavior to the biotinylated target, the detection limit of 2.5 pmol/L is seen while operating these sensors (Martínez-Paredes et al. 2009). Electronic devices such as Field effect transistors were also studied for building sensors for detections; a device was fabricated using indium oxide (In_2O_3) nanowires which helps in detecting the analytes. The study was done on fibronectin-based protein, which was behaving as a capturing agent to specific

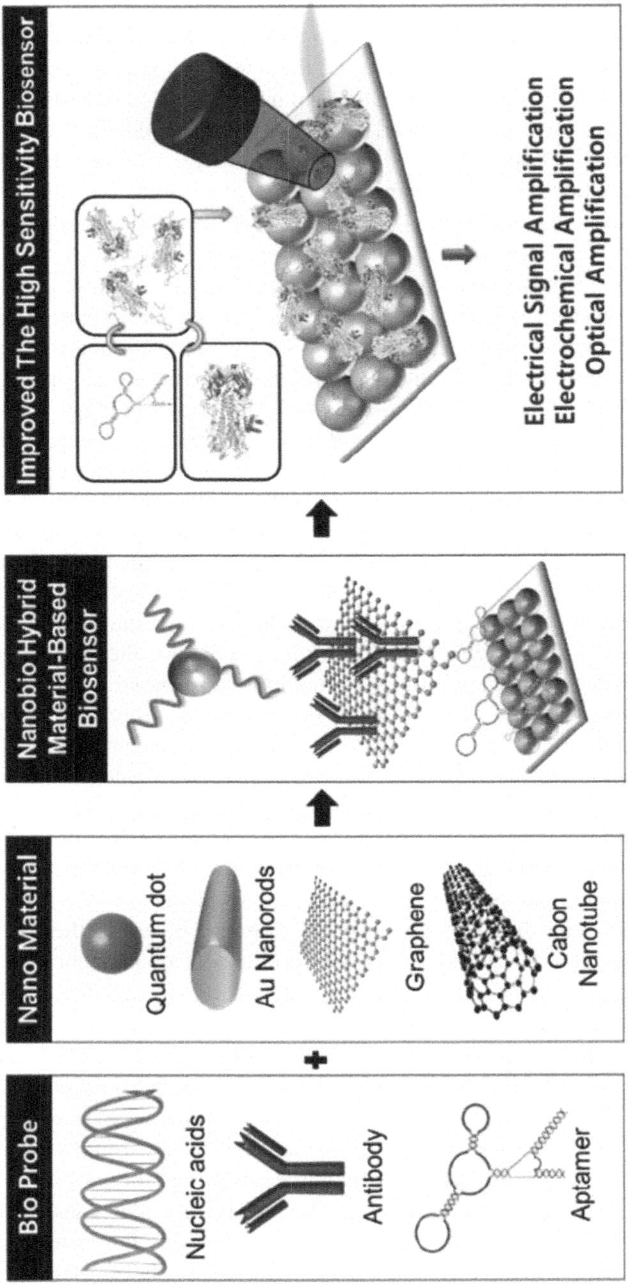

Fig. 18.3 Different scenarios of SARS virus detection with the physical presence of nanomaterial and bio measures for developing sensors. (Reproduced with permission Lee et al. 2018. Copyright MDPI)

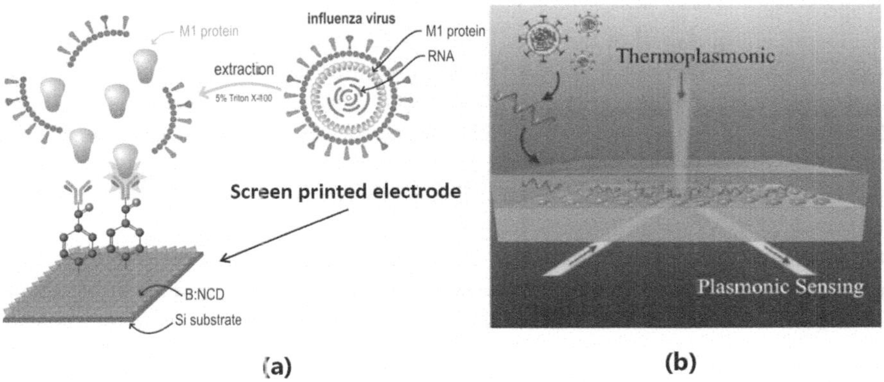

Fig. 18.4 Different sensors available for virus detection. (**a**) Electrochemical based screen-printed electrode sensor for detection of the virus (Reproduced with permission Siuzdak et al. 2019. Copyright Sensors and Actuators B). (**b**) Plasmonic photothermal biosensors (Reproduced with permission Qiu et al. 2020. Copyright ACS Publications)

nucleocapsid protein; the captured protein biomarker is associated as a SARS coronavirus. The system was able to enhance the capability of sensing the virus with high selectivity and sensitivity (Ishikawa et al. 2009). Some sensors developed using the underlying phenomenon is shown in Fig. 18.4 (Mahari et al. 2020), the screen printed electrode is also high in demand due to its nature shown in Fig. 18.4a (Layqah and Eissa 2019), a different technique using plasmonic sensing, is depicted in Fig. 18.4b, where it shows the detection of viruses using photonics phenomenon (Qiu et al. 2020). Quantum dots-based technology has been into the picture for fast and efficient device performance; using QDs RNA aptamer was used to detect SARS Cov N protein, and the detection limit of the device was obtained was 0.1 pg/mL (Roh and Jo 2011). Colorimetric sensors are an integral part of biosensors where for detecting a particular analyte, the color change depicts the information of the particular presences. The result is visible in a quick time and is even visible with the naked eye. This is an important way of determining the virus present in the system and can be admitted in a timely manner (Zhao et al. 2020).

18.3 Detection of SARS-CoV-2 Virus

18.3.1 Structure and Characteristic of SARS-CoV-2 Virus

Novel severe acute respiratory syndrome (SARS)-like coronavirus (SARS-CoV-2) has occurred recently and quickly transmitting in humans. Thus, understanding of the fundamental behavior of the SARS-CoV-2 virus is needed to come with practical solutions to tackle the pandemic. The SARS-CoV-2 particles consist of a spherical shape system where the system contains single-stranded RNA associated with a nucleoprotein in a shell of the virus, which encloses a genetic material of different matrix proteins (Shang et al. 2020). The RNA relates to the intimate of coronaviridae and order Nidovirales. SARS-CoV-2 (COVID-19) virus belonged to the betacoronavirus family that was initially observed in Wuhan city, China (Andersen et al. 2020). The structure of the virus can be defined with the presence of structural proteins, namely envelope (E), spike (S), membrane (M), and nucleocapsid (N). The influence of S, M and E proteins are generally for the development of envelope for the virus; however, protein M is responsible for the shape of the virus. The efforts of protein N are related to the RNA formation inside the envelope, and the influence of protein N is majorly in the process that is related to the viral genome and is also responsible for the replication of Cov to host viral infection (Lan et al. 2020).

The study on structural properties shows the effect of current virus conditions and the information that can be gathered to work in understanding the behavior of viruses. Another critical factor for understanding the virus would be the biological characteristics. Various studies in the biological parameter can influence the investigation on a big note; the coronavirus incubation period is of 2–14 days, which shows the effect on the growth of the virus. The main biological sequence of the virus is positive-sense single-stranded RNA that affects the cell structures of different species. Finally, the spread speed of COVID-19 is exceptionally high as compared to the other family of the coronavirus, and the seasonal occurrence of these viruses is high in winter seasons (Meo et al. 2020).

18.3.2 *Novelty of SARS-CoV-2 Virus as Compare to Other SARS Viruses*

There are certain vital similarities between the SARS-CoV-2 virus and SARS Cov of 2002, such as (1) both belong to β-coronavirus group (2) the primary host carrier for both viruses is considered to have been initiated from bats (3) almost 76% amino acid sequence similarity of the S glycoproteins (4) the RNA sequencing and the process of binding of S-protein to the target cells is also identical (5) the symptoms such as acute respiratory infections, fever, cough, cold, sore throat are similar for both virus (Gorbalenya et al. 2020; Schröder 2020). However, based on the research findings of scientists on various patients infected with SARS-CoV-2, it was found that both the viruses are not as identical as they seem, it was found that the virus is much firmly related to bat like coronaviruses mainly—bat-SL-CoVZC45 and bat-SL-CoVZXC21 (~89% similarity). There are specific vital parameters that make the new virus novel as compared to its predecessor viruses. The changes in receptor-binding motif and three short insertions in the N-terminal domain separates both the virus characteristics (Zhou et al. 2020). The S protein of the SARS-CoV-2 is more abundant as compared to the SARS virus. The R_0 value of the SARS-Cov-2 is higher in magnitude (2–3 times) as compared to SARS Cov (<1)[10] (R_0 defines how contagious the virus is and how much it can affect other people in its vicinity) (Ceccarelli et al. 2020). Still, the complete description of the COVID-19 is inadequate to describe its novel characteristics precisely, only based on certain experimental studies by researchers can we quantify the differences between the two viruses. Walls and co-workers (2020) studied about the SARS-CoV-2, they found that the entry of the virus is based upon the interaction of S proteins with the angiotensin-converting enzyme (ACE2), which helps in maintaining the blood pressure by controlling the volume of fluids that are present inside the body, and enters the human body. There are different cleavage site analogous to enzyme twin, where it is observed that some proteins are inactive when they are synthesized. So to activate the function of the protein, enzyme furin, breaks the inactivated proteins into different sections to activate the protein (Coutard et al. 2020) at the S_1/S_2 boundary that differentiates the virus with previous coronaviruses. The S proteins of SARS-CoV-2 use the function of the furin enzyme to activate its proteins, and further spread the virus in the whole body. This key feature of the SARS-CoV-2 makes it a novel coronavirus. Still, the research is going on to identify the exact features of SARS-CoV-2 which makes it a novel virus. However, the virus is still very fatal and has already seen to be very powerful as it has affected millions of lives, and the magnitude of spreading of the virus is also rapid. The virus is more prone to affect the elderly, children, and people with past illnesses, and weak immune systems are severely affected.

18.3.3 Detection Mechanism of SARS-CoV-2 Virus

The general mechanism for the detection of SARS-CoV-2, which is the causative virus of COVID-19, is quite similar to the other coronaviruses responsible for SARS (Severe acute respiratory syndrome). The image of various regions of the COVID-19 infected lungs in comparison to a normal healthy lung shows the variation in opaqueness, which leads to the detection of pneumonia initially (Ai et al. 2020; Udugama et al. 2020). However, later on, polymerase chain reaction (PCR) could analyze the additional nucleic acid, which suggested that the cause of the pneumonia was not from known pathogens. By analyzing the bronchoalveolar lavage (BAL) fluid from various patients, it was confirmed that the genetic structure of this unknown pathogen is very similar to the acute bat coronavirus SARS Cov and MERS Cov (Zhou et al. 2020; Lu et al. 2020). The COVID-19 is caused by the novel coronavirus pathogen SARS-CoV-2. The upper respiratory system is mainly the lung epithelium and is infected by the SARS-CoV-2 (Kostarelos 2020). The cure and the vaccine of the COVID-19 are still in progress, yet there is no approved vaccine available (Zhang et al. 2020; Liu et al. 2020). Hence, detection and investigation play a pivotal role in the confinement of COVID-19. Depending on the detection results, various plans can be assessed and carried out like spotting, quarantine of the patients. Besides, discovering and tracking normal people is also important, which might come in contact with the infected person. The COVID-19 was first detected in a patient sample from Wuhan, China, where the epithelial cells of the airway were cultured with fluid from bronchoalveolar lavage (BAL), which contains the virus. Transmission electron microscopy (TEM) observed that the diameter of the COVID-19 virus is nearly 60–140 nm with spikes of protein in the surroundings (Moore et al. 2020; Zhu et al. 2020b).

The detection mechanism of a virus primarily depends on the biological properties of the virus. The genetic structure of SARS-CoV-2 contains single-stranded RNA with around 30,000 nucleotides lengthwise (Udugama et al. 2020; Zhou et al. 2020). Figure 18.2c shows to enter into the cell; the SARS-CoV-2 interacts with the angiotensin-converting enzyme 2 (ACE2) receptor. The ACE2 mRNA is seen in various human organs such as the skin, lungs, stomach, small intestine, etc. The cell infection and entry into the host cell by SARS-CoV-2 is initiated by the S protein via components of S protein binding to the host cell receptor. The S protein initially in a metastable state undergoes a dramatic structural change to fuse the viral membrane into the host cell (Wrapp et al. 2020). The key target for vaccines and diagnostics is the (S) spike glycoprotein. It is observed that the upper portion of the respiratory tract tissue is not the primary place for the infection of SARS-Cov-2 (Hamming et al. 2004). Computed tomography (CT) scan shows that as compared to the upper region of the lungs, the lower region shows greater opaqueness, which concludes more ACE2 in the lower portion. In order to develop efficient methods for detection of SARS-Cov-2, the scientist and the researchers need to have a better knowledge of its biological properties.

18.3.4 Nanotechnology for Pioneering Sensors for the Detection of SARS-CoV-2

Researchers and scientists have been trying to explore different types of detection methods by taking into consideration various factors like enhancement of sensitivity, less time consuming, and point-of-care diagnosis. Few methods are nucleic acid testing, protein testing, chest chromatography, etc. At the present moment, for the detection of SARS-CoV-2, the nucleic acid testing method is primarily used (Ai et al. 2020). In this method, extracted viral RNA is mixed with a master mixture. This mixture is prepared by mainly adding nuclease-free water, oligonucleotide primer, fluorescence quencher probe, and a reaction mixture together (Udugama et al. 2020). The reaction mixture primarily contains polymerase, reverse transcriptase, nucleotide, and other additives. The extracted viral RNA is amplified to an adequate amount with the help of the master mixture by using various methods such as real-time RT-PCR (Zhou et al. 2020). During the reaction process, the amplification of the viral nucleic acid is detected by analyzing whether fluorescence occurs or not due to cleavage of the fluorescence quencher probe, as shown in Fig. 18.5. The amplification of the viral nucleic acid is detected by analyzing whether fluorescence occurs or not due to the cleavage of fluorescence quencher probes. Enhancement of

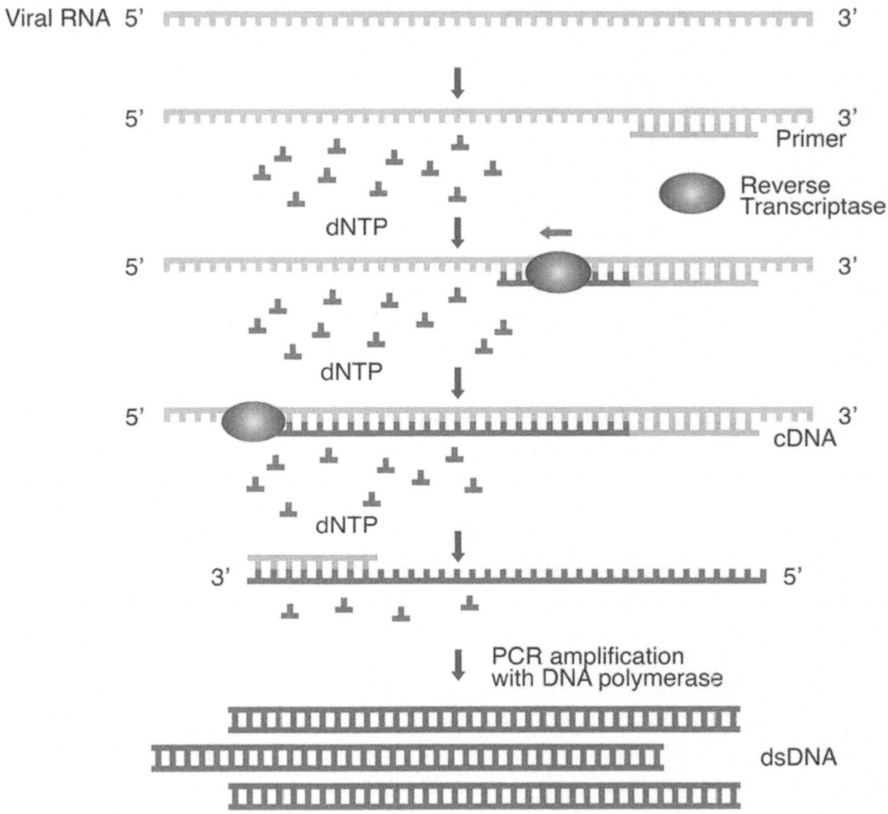

Fig. 18.5 Schematic of amplification of viral RNA using RT-PCR method for SARS-CoV-2 detection. (Reproduced with permission Carter et al. 2020b. Copyright ACS Publications)

the fluorescence during the process signifies the presence of the infected virus in collected samples. In this mechanism, two types of primers are employed: one is a universal prime, which can detect all kinds of coronavirus along with SARS-CoV-2, and another one is used to detect, especially SARS-CoV-2 (Udugama et al. 2020).

Reverse transcription loop-mediated isothermal amplification (RT-LAMP) assay is another method of COVID-19, which is highly sensitive and cost-effective in comparison to the RT-PCR method (Park et al. 2020). This method is able to amplify trace amounts of targeted genetic sequence at the isothermal condition and does not require any equipment for thermocycler, which enhances the efficiency of the amplification process for detection. Apart from this, the method gives visible results without using instruments with the help of dyes, which sense the amplification target nucleic acid (Zhang et al. 2020). Amplicon-based metagenomic sequencing is another method of both amplicon sequencing and metagenomics sequencing-based method used to detect the SARS-CoV-2 virus. Amplicon based sequencing allows us to detect the evolution of the SARS-CoV-2 virus, including its molecular epidemiology. On the other hand, the metagenomic sequencing is adopted for the detection of other microbiomes, i.e., bacteria, and fungi of the patient (Moore et al. 2020; Carter et al. 2020b).

Serological methods have also been explored for COVID-19 detection. When a virus infects an individual, antibodies are generated in response to that by the body's immune system. The generated antibodies, namely immunoglobulin M (IgM) and immunoglobulin G (IgG) are sensed in this method to analyze the presence of the virus. IgM antibody is the first body defense form at the first stage of infection, and if the infection is continued for several days, then the IgM antibody switches over to the IgG antibody. Therefore, the serological method also helps to study the post-infection immunity of the body. ELISA (Enzyme-Linked Immunosorbent Assay) is one of the techniques that can be employed for the qualitative and quantitative detection of antibodies. Initially, the microplate is coated with the protein of the virus to be detected. Then the antibodies of the suspected sample are allowed to bind with that protein if present and form a complex. An additional antibody is used to get the information on the formation of the protein- antibody complex with the help of colorimetric or fluorescence emission. Recently, Adams et al. tested SARS-CoV-2 antibodies using this method (Adams et al. 2020). Luminescent immunoassay is also a serology-based technique that has been reported for SARS-CoV-2 detection. Cai et al. designed a peptide-based immunoassay for COVID-19 detection (Cai et al. 2020). A COVID-19 detection method is proposed using sensors available in our smartphone (Maghdid et al. 2020). The available processor and sensors (camera, sound sensor, and temperature sensor, etc.) in the smartphone is used to detect primary syndrome of COVID-19 patients such as fever, dry cough, nausea, short breath, chest image, fatigue, etc. An artificial intelligent (AI) framework is designed to investigate the signal recorded by the smartphone sensor for the detection of COVID-19. AI treatment reveals many potentials in the COVID-19 detection process since it involves less complexity for used and takes a short time for detection. Recently, the chest computed tomography method is also found widely used in COVID-19 detection due to the shortage of diagnostic kits. X-rays measurements of the chest of infected individuals are taken at various angles to analyze. Abnormal features of the scanned X-ray are analyzed, which are diverse depending on the severity of infection in the lungs (Bernheim et al. 2019; Pan et al. 2020). Depending on the duration of infection, different features such as ground-glass opacities, crazy paving pattern, and consolidation are visible in scan images. Although several methods have been demonstrated to detect the SARS-CoV-2 successfully, however, the selectivity and quick sensing is still a challenging task. The application of nanotechnologies and nanomaterials further can improve the performance of existing and upcoming sensing technologies. There are several notable works have been presented to demonstrate the effective and efficient sensing of SARS-CoV-2 using nanomaterials, as discussed below.

The nanoparticles demonstrated a sharp impact on the study and development of the various types of tests against viruses at a crucial time. Pillai et al. reported the biological viewpoint properties of inorganic nanoparticles, which results in an outstanding effect on antibacterial and antifungal investigations (Pillai et al. 2020). They piloted the synthesis, optical properties, and biocompatible nature of inorganic nanoparticles, which makes them appropriate in medical treatments. Furthermore, these inorganic nanoparticles illustrate valuable compatibility and positive achievements towards respiratory viruses, which further needs quick attention and investigation for practical use. Carter et al. discussed exclusive protein self-assembling nanoparticles for the noteworthy contribution towards the progress and manufacturing of bioengineered self-assembling nanoparticles based vaccines in the recent pandemic (Carter et al. 2020a). Pimentel et al. studied the peptide nanoparticles, which is a new immunogen for the latest study and investigation of the SARS vaccine (Pimentel et al. 2009). Schmitt et al. polymeric nanoparticles for stable and flexible drug delivery via low systemic treatment in vivo in rats (Schmitt et al. 2020).

The nanoparticles-based biosensors are the most essential and critical techniques, which can be useful in the detection of individuals infected with pathogens. Zhu et al. investigated the role of nanotechnology for the detection of SARS-CoV-2 by using a nanoparticles-based biosensor (Zhu et al. 2020a). Currently, the biosensors were adequately examined for other microorganisms in the literature. However, these quick and responsive diagnostic procedures for SARS-CoV-2 are inadequate and, therefore, need immediate validation due to its long development period and very high infectivity. The conventional detection approaches depend on nucleic acid detection. They have numerous shortcomings which are (a) low sensitivity and more prolonged experimental procedures, (b) time lapse between sample collection and report evaluations, (c) false severe negative rates, and (d) the deficiency of characteristics which outcomes in misdiagnosis in patients cause other viral infections/diseases. Zhu et al. reported the fast and accurate diagnostic process for diagnosing SARS-CoV-2 patients, which results in suitable dealing and simplifying the control of infections, as illustrated in Fig. 18.6 shows the schematic representation of a complete detection test of SARS-Cov-2 patients.

In a Study Moitra et al. developed a naked eye detection method, as shown in Fig. 18.6, with the help of specific oligonucleotide capped plasmonic nanoparticles and targeting nucleocapsid phosphoprotein gene (N Gene) of the SARS-CoV-2 (Moitra et al. 2020). Gold nanoparticles (AuNPs) capped with specific well-designed thiol-modified antisense oligonucleotides (ASOs) are used to interact with the targeted N gene of SARS-CoV-2; as a result, agglomeration of capped gold nanoparticles is observed with the change in surface plasmon resonance of gold nanoparticles. Further, RNaseH addition to the agglomerated gold nanoparticle leads to the cleavage of RNA-DNA hybrid structure and results as precipitation due to the additional aggregation of gold nanoparticles, which can be detected by the naked eye. This proposed method is very less time consuming, i.e., after the isolation of RNA from the sample, it takes only 10 min to detect the virus.

Seo et al. reported FET (Field effect transistor) based biosensor using an exciting two-dimensional nanomaterial 'Graphene' (Seo et al. 2020). A graphene-based FET is designed in such a way that they sense SAR Cov-2 antigen proteins. For this purpose, graphene material is coated with the SARS-CoV-2 spike protein antibody

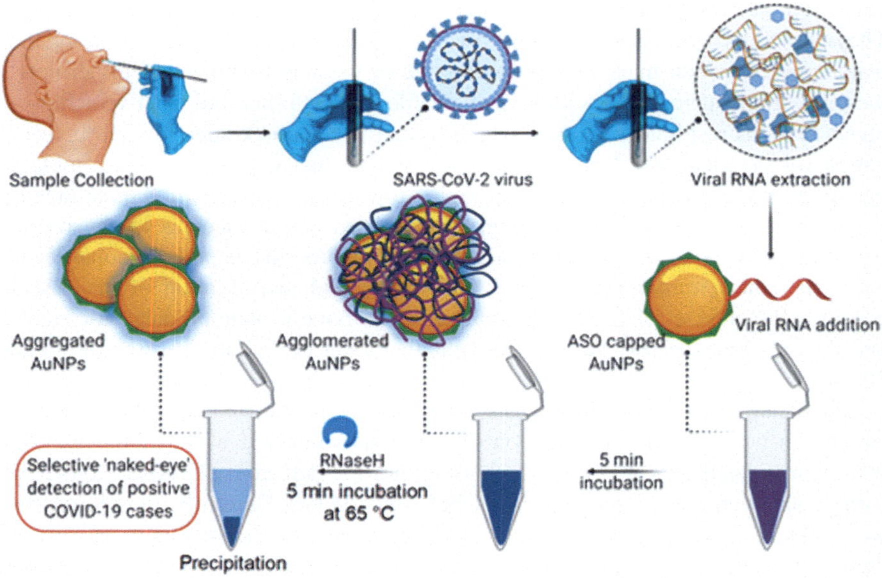

Fig. 18.6 Schematic of SARS-CoV-2 detection with the help of thiol-modified ASOs capped gold nanoparticles. (Reproduced with permission Moitra et al. 2020. Copyright ACS Publications)

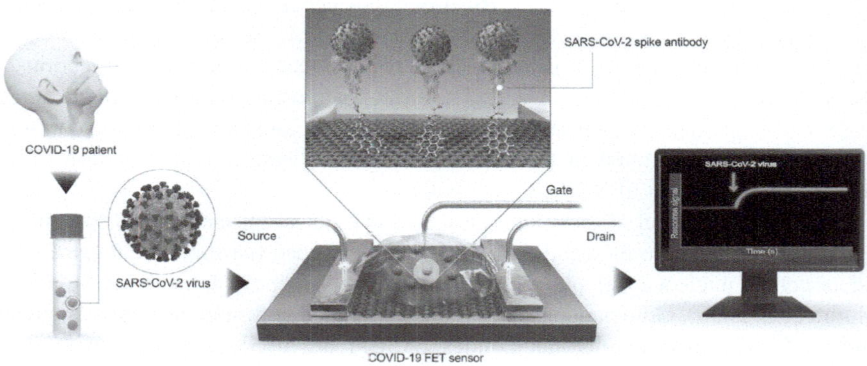

Fig. 18.7 Schematic of FET based biosensor for detection of SARS-CoV-2 virus. (Reproduced with permission Seo et al. 2020. Copyright ACS Publications)

by using a coupling agent 1-pyrenebutyric acid N-hydroxysuccinimide ester (PBASE) as shown in Fig. 18.7. When targeted spike antigen proteins interact with the SARS-CoV-2 spike antibody, then a change in electrical signal is detected, which signifies the presence of the virus in the targeted samples. Apart from this, Qui et al. reported plasmonic photo-thermal based biosensor using two-dimensional gold nanoislands coated with cDNA having a specific genetic sequence extracted from

SARS-CoV-2 (Qiu et al. 2020). The biosensor has high sensitivity and can detect up to a very lower concentration of 0.22 pM of targeted SARS-CoV-2 sequence.

In the present worldwide pandemic situation due to COVID-19, many scientists, doctors contribute intense and continuous effort to impede the rapid spreading of the SARS-CoV-2 virus. Although some improvements and developments are found to tackle this pandemic situation, there are still many challenges to be addressed for accurate controlling of the diseases. One most existing challenge is related to the detection of the causative virus SARS-CoV-2 in the suspected patients. In the last few months, several detection techniques based on RT-PCR, serological, and immunological assay are developed, and many are in the developing stage. But, most of the existing tests can be run in only laboratories having high instrumental facilities, need expertise human resources, and a long time for test results. In many cases, it is also found that the RT-PCR method, which has been used frequently for SARS-CoV-2 detection, gives false-negative results of confirmed positive cases (Xie 2020).

On the other hand, serological tests are found convenient for rapid testing with POC diagnosis, but it is found more after several days of infection. Although a chest CT scan is employed for the detection purpose due to the lack of detection kits, it is unable to detect at the early stage of viral infection. Therefore, at the present moment, it is a great challenge to develop a testing method/device which can be highly sensitive and accurate, affordable, less expensive, robust with POC facility, and user friendly to control the COVID-19 outbreak. In some cases, patients are found to carry the SARS-CoV-2 virus in the human body without showing any syndrome. Therefore, it is very crucial to develop ultrahigh sensitive testing kits for the early detection of the SARS-CoV-2 virus. Few research groups are developing biosensors for highly sensitive detection, but these are in the testing stage now.

18.3.5 Test Kits for SARS-CoV-2 Detection

Intense global spreading of COVID-19 disease leads to the urgent requirement of rapid testing kits to control and combat the causative virus (SARS-CoV-2). Various research laboratories and many companies worldwide have been developing test kits for COVID-19 detection based on the different methods of detection, mainly focusing on the enhancement of sensitivity, rapid and point-of-care (POC) detection. More than one hundred COVID-19 test kits are being developed in different countries of the world within the last few months by giving continuous effort for further improvement.

It is observed that PCR based detection kits have been predominantly used in detection COVID-19 (Carter et al. 2020b). Test kits developed based on this method used different targeted regions of gene of the causative virus such as ORF1ab, M gene, N gene, N2 gene, ORF 8 and so on extracted from different sample sources, e.g., nasal swab, throat swab, oropharyngeal swab, etc. At the early stage of the COVID-19 outbreak, CDC-US (Centers for Disease Control and Prevention) developed "CDC 2019-Novel Coronavirus Real-Time RT-PCR Diagnostic Panel" for

SARS-CoV-2 testing and can probe 264 samples per day (Redfield 2020). Along with this, some other test devices such as the TaqPath COVID-19 Combo kit, COVID-19 RT-PCR, Cobas SARS-CoV-2 test also developed by many other groups (FDA 2020; Barrett et al. 2020). Credo Diagnostics Biomedical Pte. Ltd. Developed at test kit based on RT-PCR method named Vita PCR SARS-CoV-2 assay, which can detect the presence of SARS-CoV-2 virus with 20 min using nasal and oropharyngeal swabs sample of suspected patient and can test 2000 samples per day [credodxbiomed.com]. Most of the testing devices are capable of running only in complex laboratory systems. Looking at the necessity of the point-of-care (POC) diagnosis due to the severe global outbreak of the virus, some groups have been developing several testing devices with the facility of POC diagnosis. Abbott Diagnostics Scarborough, Inc. designed ID NOW COVID-19 testing device with POC diagnosis based on isothermal nucleic acid amplification process (FDA 2020). The RdRP gene of the SARS-CoV-2 is used as a targeted gene in this testing device, and it takes only 5 min to positive results and 13 min for negative results, but one of the limitations of this test is that it can test only one sample per run. Xpert Xpress SARS-CoV-2 test and Accula SARS-CoV-2 test are another two testing devices developed by Cepheid and Mesa Biotech Inc., respectively, having a POC diagnosis facility (Moran et al. 2020; Hogan et al. 2020).

Demonstration of the testing kit based on serology and immunology for the rapid detection of COVID-19 is underway. Testing kits developed based on this is designed in such a way that it can detect IgM or IgG antibodies and viral antigens produced in the patient's body. In general, blood serum and plasma of infected individuals are used in most of the testing; however, a few testing devices used nasal swab for detection SARS-CoV-2 (Carter et al. 2020b; Mertens et al. 2020). qSARSnCov-2 IgG/IgM Rapid Test is the first serology based test approved by EUA (Emergency use authorization) and able to get test results within 15–20 min (FDA 2020; No 1988). Recently, several numbers of serology and immunology-based testing kits developed for rapid detection of SARS-CoV-2, such as the COVID-19 ELISA IgG Antibody Test, SARS-CoV-2 IgG assay, Anti-SARS-CoV-2 ELISA (IgG) and so on (FDA 2020). Although, most of the recently developed testing is improved for rapid, there are only a few numbers of testing having POC diagnosis facilities which need to be concerned.

18.4 Application of Nanomaterials and Coatings for Protection from COVID-19

At present, the world is facing a challenging task to overcome the COVID-19 pandemic situation. The COVID-19 disease is caused due to the novel beta coronavirus (SARS-CoV-2), creating acute respiratory tract infection. Observing the novelty of the virus and severe effect on human health, leading to large percentages of causality and the absence of any competent medical treatment/vaccines. It has

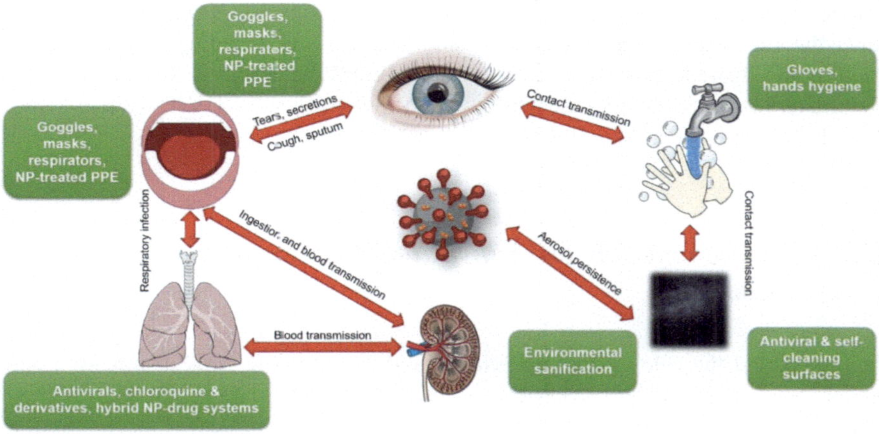

Fig. 18.8 Preventive and protective measures against COVID-19. (Reproduced with permission Sportelli et al. 2020. Copyright MDPI)

become imperative to think about protection from the rapid spreading of SARS-CoV-2 to reduce the impact of COVID-19. Sportelli et al. have summarized the set of prevention measures/guidelines that can help to protect humans from the exposure of SARS-CoV-2, as given in Fig. 18.8 (Fathizadeh et al. 2020). The guidelines suggest the three significant steps, a regular hand was and personnel hygiene, improving immunity and a good lifestyle, and the use of personal protective equipment (PPE) (Fathizadeh et al. 2020). The use of adequate PPE is the critical point of discussion to protect healthcare persons, essential service staff, and other COVID-19 fighters. Transmission of SARS-CoV-2 happened via three significant ways, airborne transmission, surface transmission, and human to human via respiratory droplets (Fathizadeh et al. 2020). When an infected person coughs, sneezes and talks, the respiratory droplets containing the virus can infect others. These droplets can spread through the air and cause airborne severe transmission results in the spreading of disease. Airborne transmission and direct human to human transmission can be effectively reduced by avoiding crowded places, use of face masks, and by early identification of asymptotic patients to execute the self-isolation.

The surface transmission of SARS-CoV-2 is also a significant factor in spreading COVID-19. The virus can sustain from a few hours to several days on the surface, and if a healthy person comes into contact with that infected surface, it can spread to that individual. Frequent cleaning and disinfection are required mainly in commonly used surfaces such as door handles, toilets-fittings, office-desks, switches, handlers, railings, and other commonly used items. However, cleaning is not a very viable solution, it requires the deployment of the regular workforce and dedicated time-interval of cleaning, even after this there could be a chance of infections due to the gap between cleaning time. Furthermore, long term exposure to cleaning chemicals may degrade the surface of the materials and can damage the look, mechanical properties, and lifetime. Therefore, antiviral surface coatings can be an effective solution to prevent the spreading of the SARS-CoV-2 virus via surface transmission. Since the virus is in the nanometer size, nanomaterials-based coatings can effectively capture and degrade the SARS-CoV-2 virus. Worldwide various investigations of nanomaterials-based coating are underway to fight with the COVID-19 pandemic; few of them are discussed in the next section of the chapter.

18.4.1 Nanomaterials for Antiviral and Anti-infective Coatings

To understand the role of nanomaterials-based coatings for virus disinfection and cleaning, let us get a short overview of antiviral surface coatings (Botequim et al. 2012). Figure 18.9 demonstrates the mechanism of virus (influenza) inactivation, it could not be universal, but the analogy can help us to understand the role of antiviral coatings. Herein, when virus particles interact with N,N-dodecyl, methyl-PEI-coated surface, it breaks the protection layer of the virus, then the leakage of RNA into the solution occurs Fig. 18.9. Hsu et al. had demonstrated the damage of the influenza virus when it contacted N,N-dodecyl, methyl-PEI coated surface as visualized the viral particles on the surfaces using scanning electron microscopy (SEM) in Fig. 18.9d–f. There is no damage in the influenza virus observed on a plain silicon wafer Fig. 18.9d. While the interaction of the viruses to the coated surface shows

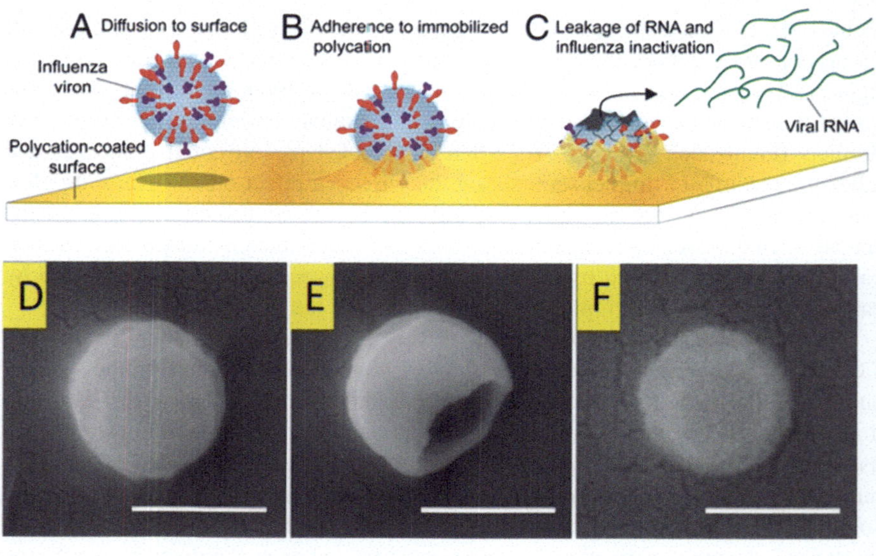

Fig. 18.9 Schematic of inactivation of influenza virus by N,N-dodecyl,methyl-PEI coatings (**a–c**). SEM images of the WSN strain of influenza virus after exposure to plain (**d**) and N,N-dodecyl, methyl-PEI-coated (**e** and **f**) silicon substrate. (Reproduced with permission Hsu et al. 2011. Copyright NCBI)

substantial structural damage leading to a gaping hole with a 54% success rate Fig. 18.9e; otherwise, no noticeable damage occurred, as presented in Fig. 18.9f.

The invention and technology for antibacterial and antiviral coatings have been patented by Donald E. Snyder., in which substrate is coated with a composition of an anti-pathogenic agent consisting essentially of PVP-I and N-9 in a different ratio (Donald and Snyder 2000). To achieve antiviral properties, a dual or multilayered coating was proposed where all layers or at least one layer must have the active ingredients. Woven and nonwoven fabrics with light active antiviral dyes/materials offer good antimicrobial and antiviral properties. Upon exposure to light, the dye generates the singlet oxygen that kills the viruses (John and Wilson 2001). The influenza virus can efficiently be killed within a few minutes using branched or linear N,N-dodecyl methylpolyethylenimines (PEIs), and certain other hydrophobic PEI derivatives (Haldar et al. 2006). The polymeric coating is exceptionally well in the case of two specific strains of influenza virus and other enveloped viruses. Nanoparticles of noble metals have investigated as antiviral material and coatings in which silver nanoparticles have shown excellent ability. The silver nanoparticles with an average size of 10 nm are demonstrated as an inhibitor for the H1N1 virus, electron microscopic images, and other analyses were reported to explain that silver nanoparticles can reduce the H1N1 influenza virus (Xiang et al. 2011). In another study, the effect of silver nanoparticles (AgNPs) on HIV-1 is investigated and observed that the spatial arrangement of nanoparticles, center to center distance of nanoparticles and exposure to sulfur-bearing residues significantly control the virus-killing properties (Elechiguerra et al. 2005). Silver nanoparticles can also reduce the Tacaribe virus effectively, interaction with silver nanoparticles during the pre-infection of the virus reduces the RNA production. While post-infection interaction of AgNPs is only useful if it is at the beginning of the virus replication (Speshock et al. 2010).

Spark discharge synthesized silver nanoparticles were tested as an antiviral coating in air filter against the aerosolized virus, and the results were used to prepare a model for comparing antiviral behavior of different antiviral materials (Joe et al. 2016). Antiviral filter based on silver and silica nanoparticles was also demonstrated under continuous airflow condition, and the antiviral performance of the filter was measured with respect to virus deposition time (Joe et al. 2014). It is observed that an increase in the coating density of Ag-SiO_2 particles exponentially increases the reduction of the concentration of the virus. Yellow or brown floc like a composite of AgNPs/chitosan was reported to reduce the H1N1 virus. The composite is formulated using 3–12 nm-sized silver nanoparticles that are embedded into the chitosan matrix without forming any aggregates. The size of silver nanoparticles significantly changed the antiviral properties of the composite, and more potent properties were observed in smaller nanoparticles (Mori et al. 2013). Silver nanoparticles can also be used as an inhibitor for hepatitis B virus (HBV) replication; monodisperse silver nanoparticles with an average size of 10–50 nm were tested for in vitro HBV activities and result-in reduction of extracellular HBV DNA formation. These results indicated that nanoparticles of silver and other Noble metal and metal-composite could play an essential role in antiviral coatings.

Apart from metallic nanoparticles, other nanostructures based on carbon materials or metal oxide or composites are also reported as antiviral material with variable performance. Hydrothermally grown nanocrystals of $CuFeO_2$ shows active antiviral properties. It is said that Cu^+ species is more effective in the reduction of bacteria and viruses as compared to copper metal or Cu^{2+} compound (Qiu et al. 2012). Alumoxane/ferroxane nanoparticles were used to prepare a bi-functional nanocomposite coating that actively binds or deactivate the virus. Though the iron-oxide does not have active virus immobilization properties, the surface functional group surprisingly increases the antiviral performance (Maguire-Boyle et al. 2012). Similarly, the technology for carbon-metal nanoparticles composite based antiviral coating materials was patented in 2019, which consists of silver nanoparticles with a size range of 8–25 nm and micron size carbon particles (Jose-Yacaman et al. 2009). Cuprous compounds (Cu_2O, Cu_2S, CuI, CuCl) have also been demonstrated as highly effective antiviral materials, the investigation has shown that the infection is reduced up to five times of magnitude within 30 min using Cu_2O loaded glass substrate (Sunada et al. 2012). Figure 18.10a represents the schematic diagram of the testing procedure of antiviral properties of copper compounds, and the results show that Cu_2O effectively reduces the virus within 30 min Fig. 18.10b. Reactive oxygen species (ROS) are produced due to reaction with Cu^{2+}, and this ROC reacts with virus particles and damages the proteins and nucleic acid of the virus.

Carbon nanohorns are also investigated as antiviral materials using propidium iodide (PI), which intercalates dual strand DNA (dsDNA) under NIR light activation targeting the T7 bacteriophage (T7 phage) as the virus which has dsDNA with size 55 nm wide and 100 nm long (Miyako et al. 2008). In a similar direction, Cu(II)-TiO_2 is proposed as an antiviral material that shows high activities under visible light activation as compared to dark conditions. Experimental results show that 1 h visible light activation followed by 1-h incubation in dark lead to 99.9% disinfection from viral particles (Joe et al. 2014). After overlooking these materials, it can be concluded that a variety of nanomaterials is available that can be used to prepare antiviral coating to protect humankind from viruses. However, specific selectivity and required deactivation efficiency need to be regulated. Hence further effort in the development of best-suited nanomaterials for a particular virus is required.

18.4.2 Nanomaterials for Personal Protective Equipment (PPE) Development

The outbreak of COVID-19 has recently attracted great attention towards personal protective equipment (PPE), and the supply-demand gap has become very crucial worldwide. In the last 2 months, most of the countries were suffering from the shortage of appropriate PPE to fight with COVID-19. The high-performance PPE is an important and essential tool to protect the frontline worker and common people

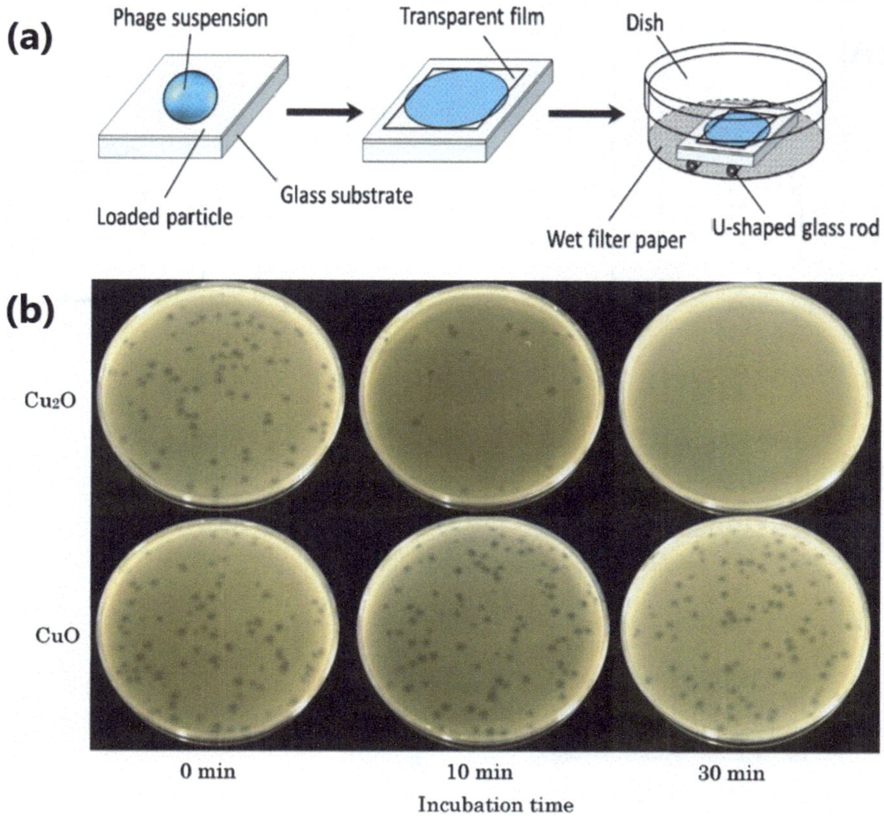

Fig. 18.10 (**a**) Schematic diagram of the evaluation method on antiviral effect and (**b**) Comparative study of Cu_2O and CuO on plaque assay against bacteriophage Qβ. (Reproduced with permission Sunada et al. 2012. Copyright Elsevier)

from COVID-19. The facemask, gloves, gown, boot, helmet/face shield, goggles, etc. are considered as PPE to protect the body parts from the exposure of the virus. Since SARS-CoV-2 infects the respiratory system of the human body, there is a requirement to safeguard the nose, mouth, eyes, and ears along with transmission through our hand to face. Therefore, to fight with SARS-CoV-2, protection of face is primarily required that can be done using facemask and face shield. Moreover, the transmission through touching infected surfaces and aerosols can be restricted using a medical gown and gloves. When we talk about the quality of PPE, apart from adequate protection from viruses, other factors such as comfort, ergonomics, multifunctionality, esthetic look, and adaptable environmental condition also need to be cared (Dolez and Mlynarek 2016). The self-contamination capability is another critical parameter for antiviral PPEs, and it can be developed using various functional materials. The metal, metal-oxide, enzymes, N-halamines, metal ions, light active, and surface-active nanomaterials have been demonstrated to achieve the self-contamination capability.

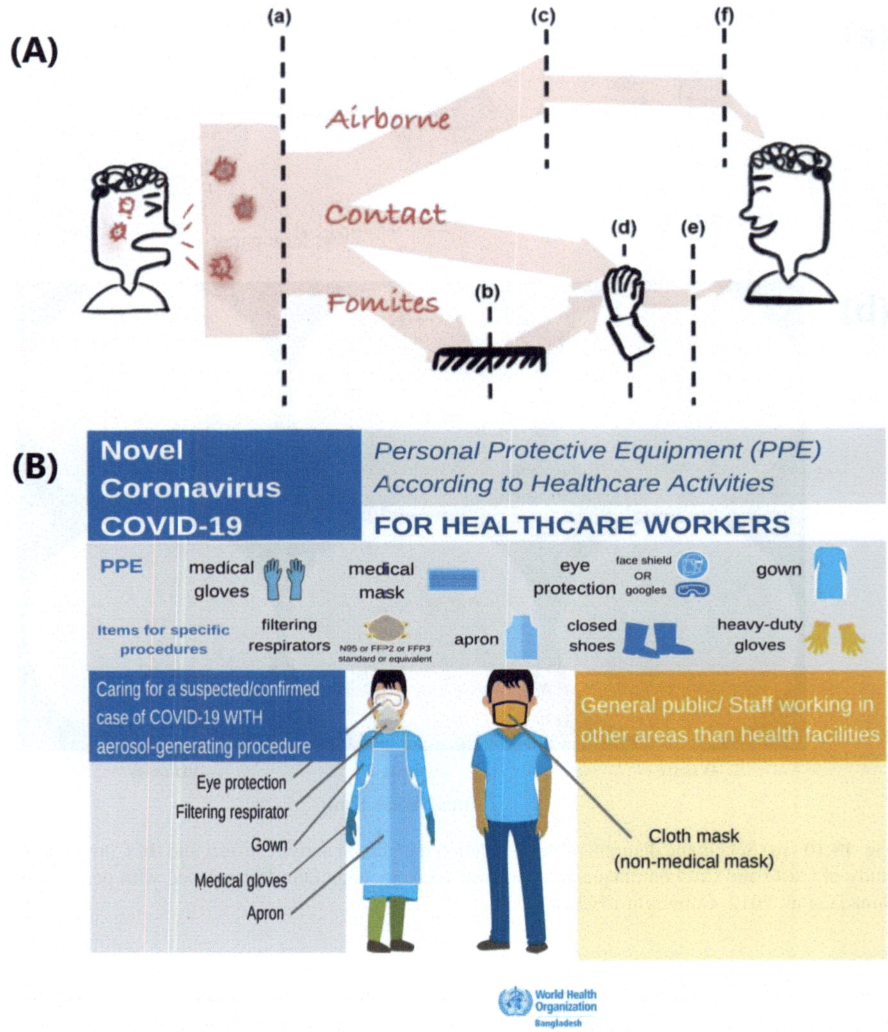

Fig. 18.11 (**a**) Schematic illustration of how a virus spreads (Reproduced with permission Huang et al. 2020. Copyright ACS Publications) and (**b**) Type of PPE recommended by the World Health Organization (WHO) for healthcare workers and the general public (Figure Curtsey: WHO website)

The prevention from the virus is the main priority for frontline medical workers as well as the general public to get protection from droplets, and aerogels containing SARS-CoV-2 virus is explained in Fig. 18.11a, b. Considering these facts, the role of nanomaterials for the fabrication of advanced, effective, and efficient facemask and PPE is very important. The design of antivirus and anti-germ face mask was patented, which includes three protective layers (Su Jen Chou 2005). The outer layer consists of a hairy structure; the intermediate layer is of active carbon materials and the inner layer of anti-moist cloth materials. A large number of face masks are made using petrochemical-derived materials which are non-degradable and meant for singly use only. These kinds of face masks are not environmentally friendly and cause pollution (Das et al. 2020). Polymers were also used to fabricate the facemask, which works on a self-powered electrostatic adsorption process. The incorporation of graphene materials in facemask offers excellent superhydrophobic properties and shows active solar-driven virus sterilizing behavior.

It has been observed that the outbreak of COVID-19 suddenly increased the demand of facemask and other PPEs, due to shortage of supply at the initial stage (month of March 2020), it was recommended that facemask is not necessary for other than healthcare personnel. However, it is the fact that coronavirus is related to SARS, and the facemask is essential for prevention from it. Therefore, a later face mask is recommended for everyone, which again increases the demand for facemasks. Then after the researchers/medical experts came with the idea of using cloth-based home-made masks for the general public. The scientific validation of cloth-based masks has also become equally important to understand the level of protection from cloth-based face masks (Das et al. 2020). Very recently, a team of researchers led by Prof Supratik Guha in the USA has investigated the aerosol filtration efficiency of respiratory cloth masks, as shown in Fig. 18.12a (Konda et al. 2020). In this study, the combination of cotton−silk, cotton−chiffon, cotton−flannel are studied, and the filtration efficiency was found >80% for aerogel particles that are smaller than 300 nm size. The filtration efficiency for larger particles (more than 300 nm) is achieved to >90%. Usually, aerogel filtration Fig. 18.12a, b depends upon gravity sedimentation, inertial impaction, interception, diffusion, and electrostatic attraction. In this study, it was proposed that the filtration efficiency is due to the combination of mechanical and electrostatic filtration. This study also emphasizes the proper fitting of the facemask; the report says that the filtration efficiency can be around 30% if there is leakage. One more study is just reported on May 21, 2020, in ACS nano by Lustig et al., they extensively investigated more than 70 fabrics to test the effectiveness of aerosols blocking viruses like nanoparticles as shown in Fig. 18.12b (Lustig et al. 2020). They consider the fact that the size of SARS-CoV-2 is in the range of 60–140 nm with polydisperse nature. The study suggests that the design of many-layered fabric can reduce the risk of virus inhalation from aerosols, the combination of two outer layers with two inner layers of nonwoven polypropylene or two layers of Kona quilting cotton with four layers of OLY-FUN effectively control the virus transmissions. The effective face mask can be designed by applying at least two layers: one is absorbent (hydrophilic layers), and the second is a barrier (hydrophobic layers).

Recently, researchers from the University of Central Florida have proposed nanomaterials based on protecting suites that catch the virus and kill it within a few seconds. The nanomaterials first capture the virus, and then under UV light excitation, it starts a chemical reaction to kill the virus (Sudipta Seal 2020). This kind of self-disinfecting PPE will be very useful in hospitals in the present scenario of COVID-19, where medical practitioners come into the direct contact of the virus carrier during their treatment. The role of active materials in facemask and PPE is

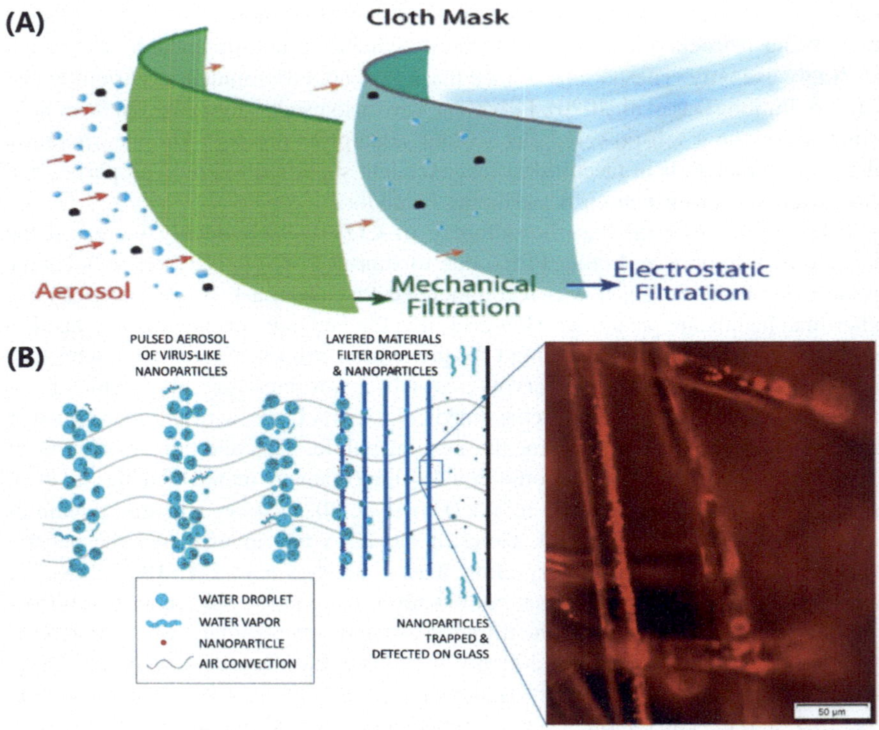

Fig. 18.12 (**a**) Aerosol filtration mechanism in two-layer cloth facemask (Reproduced with permission Konda et al. 2020. Copyright ACS Publications) and (**b**) multilayer facemask (Reproduced with permission Lustig et al. 2020. Copyright ACS Publications)

very crucial for deactivation of the virus. Suitably selected material can destroy the self-assembly kind of virus structure which consist of proteins, genetic and lipid membrane (Jones 2020). The application of polymeric nanofiber was reported to fabricate face masks for haze pollution. It can be further designed applying antiviral materials for virus protection (Li and Gong 2015). Metal-oxide nanoparticles (TiO_2 and MgO) are integrated with textiles to achieve the anti-chemical and anti-biological warfare cloths (Borkow et al. 2010). To improve the protection level of facemask, utilization of nanofiber is demonstrated, the high surface area of fibers improves the capture efficiency. The small pore size, high permeability, and low weight of nanofibers offer a wide range of filtration. Further addition of gold nanoparticles with electrospun fiber can be one of the best alternatives for mask materials (Fathi-Azarbayjani et al. 2010). You et al. have developed carbon nanotube-based filtration materials that have shown similar performance as commercially available face mask materials (Zou and Yao 2015). Single-wall carbon nanotubes (SWNTs) used to prepare a base filter with a pore size of 10 nm and results show 87% filtration efficiency with the loading of 0.2 mg/cm^2. Anti-influenza facemask was designed using copper oxide nanomaterials that filtered above 99.85% of aerosolized viruses of human influenza A virus (H1N1) and avian influenza virus (H9N2) (Zou and Yao 2015). Since the size of SARS-CoV-2 is in the range of 100 nm, it belongs to the nanomaterials. Hence the application of nanomaterials in face mask designing and fabrication are very relevant and necessary for adequate protection.

18.4.3 Proposed Nanotechnology Solutions to Combat with COVID-19

There is a vast scope and, at the same time, a tough challenge for the nano-scientists to offer various solutions to fight with COVID-19 and similar diseases. The nanotechnology and nanomaterials can provide a very practical solution to combat from COVID-19 due to the size of SARS-CoV-2 in the range of 60–140 nm (Sportelli et al. 2020). The application of nanotechnology in the development of antiviral and anti-infection materials for fabrication of personal protective equipment like masks, gowns, goggles, shoes, and disinfecting spray solutions, clothing, packing materials, multifunctional surface, AC filters, etc. to reduce the spread of COVID-19 (Yu et al. 2020). The combination of nanotechnology with 3D printing technology can offer to fabricate the user-specific mask and other PPE, so the leakage and fitting issues can be resolved effectively. Cotton Fabric has been demonstrated as an excellent behavior to block the virus-like particles present in artificial aerosols (Lustig et al. 2020). Through the integration of antiviral nanomaterials with cotton fabrics, self-cleaning, and highly protective facemask can be designed. Nanocarbon fullerene C-60 is another potential material for antiviral coatings and hydrophobic surfaces (Siddiquie et al. 2020). Fullerene is capable of degrading lipids by the generation of singlet oxygen when fullerene is excited with UVA light. Due to a large active surface area, graphene is well known for virus capturing properties, and functionalized graphene further enhances the capacity leading to disinfection after applying heat (Palmieri and Papi 2020). Graphene oxide can be designed as a breathable barrier and can apply in facemask to protect from SARS-CoV-2 and similar kinds of viruses (Palmieri and Papi 2020). The researchers from Ben-Gurion University of the Negev have claimed that metal nanomaterials and polymers based anti-coronavirus surface has been developed by them (Be'er Sheva Anti-Coronavirus Surface Coating Based on Nanomaterials n.d.).

The facemask is the primary weapon to fight with SARS-CoV-2 virus; however, the comfort of a face mask is a point of discussion. The county link India, where temperature can rise to 48 °C, thermal management is an important task, that can be controlled via nanotechnology solutions. The nanofibers on nanoporous polyethylene-based face mask can effectively reduce the radiative heating, approximate 87% IR reflectance is reported in fiber/Ag/nanoPE mask (Yang et al. 2017). A

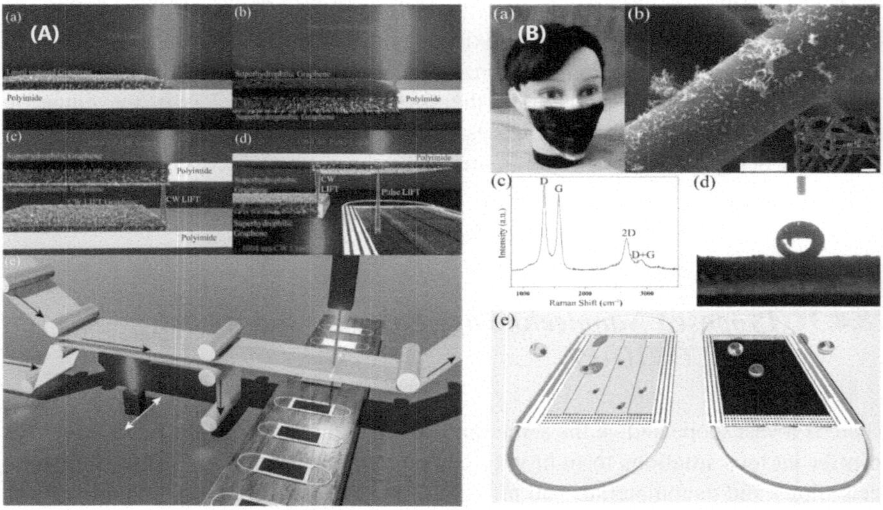

Fig. 18.13 (**a**) Laser-induced graphene synthesis method and roll-to-roll production of a graphene-coated mask, (**b**) Demo of Laser-fabricated graphene mask and its structural properties with hydrophobic nature. (Reproduced with permission Zhong et al. 2020. Copyright ACS Publications)

group researcher from KAUST Saudi Arabia has recently developed a flexible nanoporous template for the fabrication of reusable N-95 hydrophobic masks (El-Atab et al. 2020). Initially, a nanoporous silicon template was fabricated using the KOH etching method; then, it was used as a hard mask for reactive ion etching to transfer the pattern on a flexible polymeric substrate. Fabricated pours are up to 5 nm in size, hydrophobic in nature along with antifouling and self-cleaning properties. Recently, the laser-induced process is reported to fabricate graphene mask with roll to roll process for large scale production. Laser light was directed on polyimide film to get the instant graphene in bottom and top as shown in the Fig. 18.13, and this graphene can be further coated on a surgical mask to achieve antiviral and hydrophobic properties.

18.5 Summary and Future Perspective

A study on the effect of viruses and their symptoms is essential in the current scenario for the best outcome when related to the living environment. Apart from that, the sources of different respiratory diseases caused by various viruses also need to be understandable. The rate of transmission of the SARS-CoV-2 virus is fatal, and data are still inconclusive as to how the virus originated. However, even the widespread affected persons indicate the pandemic nature of SARS-CoV-2. While the typical viruses such as the common cold and cough, seasonal viruses affect the human life form in one way or the other, but the nature of spreading of the SARS-CoV-2 is much more dangerous than other viruses. Proper research is desired to study about all the coronaviruses, to be better prepared and equipped in future origination of some other unknown viruses. SARS-CoV-2 shows a close resemblance to SARS Cov (~79%) of 2002 and bat coronaviruses (~85%). However, the nature of transmission of the virus, the number of affected individuals, and the mechanism of the entry of the virus inside the host cells show unique and novel characteristics of the viruses compared to other coronaviruses. Even though currently many methods such as PCR method, LAMP method, RCA method is commonly used for SARS-CoV-2 detection, but still issues such as reliability, accuracy, uniform testing method requires significant attention to improve the detection rate of viruses and that too at a much faster rate. Gold nanoparticles capped with thiol oligonucleotides and graphene-based FET biosensors, AI-based frameworks in the smartphones, chest computed tomography methods are other recent methods proposed for SARS-CoV-2 detection. The fight with SARS-CoV-2 is weaker due to the lack of appropriate personal protective equipment. There is an urgent need for smart and interactive PPE kits for health workers and corona warriors; nanotechnology can offer several solutions to this problem. The metal nanoparticles, carbon nanostructures, metal-oxide nanoparticles, and polymers-based materials have shown excellent potential as antiviral coatings. However, more accurate scientific solutions need to be discovered with the help of nanotechnology to design face mask and PPEs, considering the user comfort and safety level. Moreover, nanomaterials and composite can be developed to achieve sanitizing surfaces for everyday use, coating solution, and packaging materials.

References

Abad-Valle P, Fernández-Abedul MT, Costa-García A (2005) Genosensor on gold films with enzymatic electrochemical detection of a SARS virus sequence. Biosens Bioelectron 20:2251–2260. https://doi.org/10.1016/j.bios.2004.10.019

Adams E, Ainsworth M, Anand R, Andersson M (2020) Antibody testing for COVID-19: a report from the National COVID Scientific Advisory Panel. medRxiv. https://doi.org/10.1101/2020.04.15.20066407v2

Ai T, Yang Z, Hou H, Zhan C, Chen C, Xia L (2020) Correlation of chest CT and RT-PCR testing in coronavirus disease 2019 (COVID-19) in China: a report of 1014 cases. Radiology. https://doi.org/10.14358/PERS.80.2.000

Almeida JD, Berry DM, Cunningham CH, Hamre D, Hofstad MS, Mallucci L, McIntosh K, Tyrrell DAJ (1968) Virology: coronaviruses. Nature 220(5168):650

Andersen KG, Rambaut A, Lipkin WI et al (2020) The proximal origin of SARS-CoV-2. Nat Med 26:450–452. https://doi.org/10.1038/s41591-020-0820-9

Assiri A, McGeer A, Perl TM et al (2013) Hospital outbreak of middle east respiratory syndrome coronavirus. N Engl J Med 369:407–416. https://doi.org/10.1056/NEJMoa1306742

Balboni A, Gallina L, Palladini A et al (2012) A real-time PCR assay for bat SARS-like coronavirus detection and its application to Italian greater horseshoe bat faecal sample surveys. Sci World J. https://doi.org/10.1100/2012/989514

Barrett ES, Horton DB, Roy J et al (2020) Prevalence of SARS-CoV-2 infection in previously undiagnosed health care workers at the onset of the U.S. COVID-19 epidemic. medRxiv. https://doi.org/10.1101/2020.04.20.20072470

Be'er Sheva Anti-Coronavirus Surface Coating Based on Nanomaterials (n.d.). https://www.hospimedica.com/covid-19/articles/294782200/scientists-develop-anti-coronavirus-surface-coating-based-on-nanomaterials.html

Bernheim A, Mei X, Huang M et al (2019) Chest CT findings in coronavirus disease-19 (COVID-19): relationship to duration of infection. Radiology 2020:200463. https://doi.org/10.1148/radiol.2020200463

Borkow G, Zhou SS, Page T, Gabbay J (2010) A novel anti-influenza copper oxide containing respiratory face mask. PLoS One 5. https://doi.org/10.1371/journal.pone.0011295

Botequim D, Maia J, Lino MMF et al (2012) Nanoparticles and surfaces presenting antifungal, antibacterial and antiviral properties. Langmuir 28:7646–7656. https://doi.org/10.1021/la300948n

Burrell CJ, Howard CR, Murphy FA (2017) Coronaviruses. In: Fenner White's medical virology. Academic, Cambridge, MA, pp 437–446. https://doi.org/10.1016/B978-0-12-375156-0.00031-X

Cai X, Chen J, Hu J et al (2020) A peptide-based magnetic chemiluminescence enzyme immunoassay for serological diagnosis of coronavirus disease 2019 (COVID-19). J Infect Dis. https://doi.org/10.1093/infdis/jiaa243

Carter DC, Wright B, Jerome WG et al (2020a) A unique protein self-assembling nanoparticle with significant advantages in vaccine development and production. J Nanomater. https://doi.org/10.1155/2020/4297937

Carter LJ, Garner LV, Smoot JW et al (2020b) Assay techniques and test development for COVID-19 diagnosis. ACS Cent Sci 6:591–605. https://doi.org/10.1021/acscentsci.0c00501

Ceccarelli M, Berretta M, Venanzi Rullo E, Nunnari G, Cacopardo B (2020) Differences and similarities between Severe Acute Respiratory Syndrome (SARS)-CoronaVirus (CoV) and SARS-CoV-2. Would a rose by another name smell as sweet? Eur Rev Med Pharmacol Sci 23:7–9. https://doi.org/10.26355/eurrev

Chan PKS, Ng KC, Chan RCW et al (2004) Immunofluorescence assay for serologic diagnosis of SARS. Emerg Infect Dis 10:530–532. https://doi.org/10.3201/eid1003.030493

Chapin SC, Doyle PS (2011) Ultrasensitive multiplexed microRNA quantification on encoded gel microparticles using rolling circle amplification. Anal Chem 83:7179–7185. https://doi.org/10.1021/ac201618k

Coutard B, Valle C, de Lamballerie X et al (2020) The spike glycoprotein of the new coronavirus 2019-nCoV contains a furin-like cleavage site absent in CoV of the same clade. Antivir Res 176:104742. https://doi.org/10.1016/j.antiviral.2020.104742

Das O, Neisiany RE, Capezza AJ et al (2020) The need for fully bio-based facemasks to counter coronavirus outbreaks: a perspective. Sci Total Environ 736:139611. https://doi.org/10.1016/j.scitotenv.2020.139611

Dolez PI, Mlynarek J (2016) Smart materials for personal protective equipment: tendencies and recent developments. Elsevier Ltd, Amsterdam

Donald E. Snyder J (2000) Antibacterial antiviral coatings, coating process and parameters thereof. Us006120784a

El-Atab N, Qaiser N, Badghaish HS et al (2020) A flexible nanoporous template for the design and development of reusable anti-COVID-19 hydrophobic face masks. ACS Nano. https://doi.org/10.1021/acsnano.0c03976

Elechiguerra JL, Burt JL, Morones JR et al (2005) Interaction of silver nanoparticles with HIV-1. J Nanobiotechnol 3:1–10. https://doi.org/10.1186/1477-3155-3-6

Erlich HA, Davm Gelfand JJS (1991) Recent advances in the polymerase chain reaction. Science 252(5013):1643–1651

Fathi-Azarbayjani A, Qun L, Chan YW, Chan SY (2010) Novel vitamin and gold-loaded nanofiber facial mask for topical delivery. AAPS PharmSciTech 11:1164–1170. https://doi.org/10.1208/s12249-010-9475-z

Fathizadeh H, Maroufi P, Momen-Heravi M, Dao S, Köse Ş, Ganbarov K, Pagliano P, Esposito S, Kafil HS (2020) Protection and disinfection policies against SARS-CoV-2 (COVID-19). Infez Med 2:185–191

FDA (2020) FDA combating COVID-19 with medical devices. US Food Drug Administration

Gootenberg JS, Abudayyeh OO, Lee JW et al (2017) Nucleic acid detection with CRISPR-Cas13a/C2c2. Science 356:438–442. https://doi.org/10.1126/science.aam9321

Gorbalenya AE, Baker SC, Baric RS et al (2020) The species Severe acute respiratory syndrome-related coronavirus: classifying 2019-nCoV and naming it SARS-CoV-2. Nat Microbiol 5:536–544. https://doi.org/10.1038/s41564-020-0695-z

Haldar J, An D, De Cienfuegos LÁ et al (2006) Polymeric coatings that inactivate both influenza virus and pathogenic bacteria. Proc Natl Acad Sci U S A 103:17667–17671. https://doi.org/10.1073/pnas.0608803103

Hamming I, Timens W, Bulthuis MLC et al (2004) Tissue distribution of ACE2 protein, the functional receptor for SARS coronavirus. A first step in understanding SARS pathogenesis. J Pathol 203:631–637. https://doi.org/10.1002/path.1570

Hogan CA, Garamani N, Lee AS, Tung JK, Sahoo MK, Huang CH, Bryan Stevens M, Zehnder J, Pinsky BA (2020) Comparison of the Accula SARS-CoV-1 2 test with a laboratory-developed assay for detection of SARS-CoV-2 RNA in clinical nasopharyngeal specimens. J Clin Microbiol. https://doi.org/10.1128/JCM.00708-20

Hsu BB, Wong SY, Hammond PT et al (2011) Mechanism of inactivation of influenza viruses by immobilized hydrophobic polycations. Proc Natl Acad Sci U S A 108:61–66. https://doi.org/10.1073/pnas.1017012108

Huang H, Fan C, Li M et al (2020) COVID-19: a call for physical scientists and engineers. ACS Nano. https://doi.org/10.1021/acsnano.0c02618

Ishikawa FN, Chang HK, Curreli M et al (2009) Label-free, electrical detection of the SARS virus n-protein with nanowire biosensors utilizing antibody mimics as capture probes. ACS Nano 3:1219–1224. https://doi.org/10.1021/nn900086c

Joe YH, Woo K, Hwang J (2014) Fabrication of an antiviral air filter with SiO_2-Ag nanoparticles and performance evaluation in a continuous airflow condition. J Hazard Mater 280:356–363. https://doi.org/10.1016/j.jhazmat.2014.08.013

Joe YH, Park DH, Hwang J (2016) Evaluation of Ag nanoparticle coated air filter against aerosolized virus: antiviral efficiency with dust loading. J Hazard Mater 301:547–553. https://doi.org/10.1016/j.jhazmat.2015.09.017

John E Wilson CB (2001) Light-activated antimicrobial and antiviral materials. United States Patent

Jones ST (2020) How materials can beat a virus. J Mater Sci 55:9148–9151. https://doi.org/10.1007/s10853-020-04678-4

Jose-Yacaman M, Schroder KA, Martin KM, Willauer DL (2009) Anti-viral uses of carbon and metal nanomaetrial compositions. US Patent

Konda A, Prakash A, Moss GA et al (2020) Aerosol filtration efficiency of common fabrics used in respiratory cloth masks. ACS Nano. https://doi.org/10.1021/acsnano.0c03252

Kostarelos K (2020) Nanoscale nights of COVID-19. Nat Nanotechnol 15:343–344. https://doi.org/10.1038/s41565-020-0687-4

Lan J, Ge J, Yu J et al (2020) Structure of the SARS-CoV-2 spike receptor-binding domain bound to the ACE2 receptor. Nature 581:215–220. https://doi.org/10.1038/s41586-020-2180-5

Layqah LA, Eissa S (2019) An electrochemical immunosensor for the corona virus associated with the Middle East respiratory syndrome using an array of gold nanoparticle-modified carbon electrodes. Microchim Acta:186. https://doi.org/10.1007/s00604-019-3345-5

Lee C, Lin C, Lin S et al (2004) From an integrated biochip detection system to a defensive weapon against the SARS-CoV virus: OBMorph. MRS Online Proc Lib Arch 820:1–12

Lee T, Ahn JH, Park SY et al (2018) Recent advances in AIV biosensors composed of nanobio hybrid material. Micromachines 9:1–17. https://doi.org/10.3390/mi9120651

Li X, Gong Y (2015) Design of polymeric nanofiber gauze mask to prevent inhaling PM2.5 particles from haze pollution. J Chem 15. https://doi.org/10.1155/2015/460392

Liu C, Zhou Q, Li Y et al (2020) Research and development on therapeutic agents and vaccines for COVID-19 and related human coronavirus diseases. ACS Cent Sci 6:315–331. https://doi.org/10.1021/acscentsci.0c00272

Long QX, Liu BZ, Deng HJ et al (2020) Antibody responses to SARS-CoV-2 in patients with COVID-19. Nat Med. https://doi.org/10.1038/s41591-020-0897-1

Lu R, Zhao X, Li J et al (2020) Genomic characterisation and epidemiology of 2019 novel coronavirus: implications for virus origins and receptor binding. Lancet 395:565–574. https://doi.org/10.1016/S0140-6736(20)30251-8

Lustig SR, Biswakarma JJH, Rana D et al (2020) Effectiveness of common fabrics to block aqueous aerosols of virus-like nanoparticles. ACS Nano. https://doi.org/10.1021/acsnano.0c03972

Maghdid HS, Ghafoor KZ, Sadiq AS et al (2020) A novel AI-enabled framework to diagnose coronavirus COVID 19 using smartphone embedded sensors. Design Study:1–7

Maguire-Boyle SJ, Liga MV, Li Q, Barron AR (2012) Alumoxane/ferroxane nanoparticles for the removal of viral pathogens: the importance of surface functionality to nanoparticle activity. Nanoscale 4:5627–5632. https://doi.org/10.1039/c2nr31117h

Mahari S, Roberts A, Shahdeo D, Gandhi S (2020) eCovSens-ultrasensitive novel in-house built printed circuit board based electrochemical device for rapid detection of nCOVID-19. bioRxiv. https://doi.org/10.1101/2020.04.24.059204

Martínez-Paredes G, González-García MB, Costa-García A (2009) Genosensor for SARS virus detection based on gold nanostructured screen-printed carbon electrodes. Electroanalysis 21:379–385. https://doi.org/10.1002/elan.200804399

Meo SA, Alhowikan AM, Khlaiwi TAL et al (2020) Novel coronavirus 2019-nCoV: prevalence, biological and clinical characteristics comparison with SARS-CoV and MERS-CoV. Eur Rev Med Pharmacol Sci 24:2012–2019. https://doi.org/10.26355/eurrev_202002_20379

Mertens P, De Vos N, Martiny D et al (2020) Development and potential usefulness of the COVID-19 Ag Respi-Strip diagnostic assay in a pandemic context. Front Med 7:225. https://doi.org/10.3389/FMED.2020.00225

Miyako E, Nagata H, Hirano K et al (2008) Photoinduced antiviral carbon nanohorns. Nanotechnology 19. https://doi.org/10.1088/0957-4484/19/7/075106

Moitra P, Alafeef M, Dighe K et al (2020) Selective naked-eye detection of SARS-CoV-2 mediated by N gene targeted antisense oligonucleotide capped plasmonic nanoparticles. ACS Nano. https://doi.org/10.1021/acsnano.0c03822

Moore SC, Penrice-Randal R, Alruwaili M et al (2020) Amplicon based MinION sequencing of SARS-CoV-2 and metagenomic characterisation of nasopharyngeal swabs from patients with COVID-19. medRXiv. https://doi.org/10.1101/2020.03.05.20032011

Moran A, Beavis KG, Matushek SM et al (2020) The detection of SARS-CoV-2 using the Cepheid Xpert Xpress SARS-CoV-2 and Roche cobas SARS-CoV-2 assays. J Clin Microbiol. https://doi.org/10.1128/JCM.00772-20

Mori Y, Ono T, Miyahira Y et al (2013) Antiviral activity of silver nanoparticle/chitosan composites against H1N1 influenza A virus. Nanoscale Res Lett 8:93. https://doi.org/10.1186/1556-276x-8-93

No C (1988) Cellex qSARS-CoV-2 IgG/IgM Rapid Test

Notomi T, Okayama H, Masubuchi H et al (2000) Loop-mediated isothermal amplification of DNA. Nucleic Acids Res 28:e63. https://doi.org/10.1093/nar/28.12.e63

Palmieri V, Papi M (2020) Can graphene take part in the fight against COVID-19? Nano Today 33:100883. https://doi.org/10.1016/j.nantod.2020.100883

Pan F, Ye T, Sun P et al (2020) Time course of lung changes at chest CT during recovery from coronavirus disease 2019 (COVID-19). Radiology 295:715–721. https://doi.org/10.1148/radiol.2020200370

Park GS, Ku K, Baek SH et al (2020) Development of reverse transcription loop-mediated isothermal amplification (RT-LAMP) assays targeting SARS-CoV-2. J Mol Diagn 22:729–735. https://doi.org/10.1016/j.jmoldx.2020.03.006

Peeri NC, Shrestha N, Rahman MS et al (2020) The SARS, MERS and novel coronavirus (COVID-19) epidemics, the newest and biggest global health threats: what lessons have we learned? Int J Epidemiol. https://doi.org/10.1093/ije/dyaa033

Peiris JSM, Poon LLM (2008) Severe acute respiratory syndrome (SARS). Elsevier ltd, Amsterdam

Pillai AM, Sivasankarapillai VS, Rahdar A et al (2020) Green synthesis and characterization of zinc oxide nanoparticles with antibacterial and antifungal activity. J Mol Struct 1211:128107. https://doi.org/10.1016/j.molstruc.2020.128107

Pimentel TAPF, Yan Z, Jeffers SA et al (2009) Peptide nanoparticles as novel immunogens: design and analysis of a prototypic severe acute respiratory syndrome vaccine. Chem Biol Drug Des 73:53–61. https://doi.org/10.1111/j.1747-0285.2008.00746.x

Poon LLM, Leung CSW, Tashiro M, Chan KH, Wong BWY, Yuen KY, Guan Y, Peiris JSM (2004) Rapid detection of the severe acute respiratory syndrome (SARS) coronavirus by a loop-mediated isothermal amplification assay. Clin Chem 50(6):1050–1052. https://doi.org/10.1373/clinchem.2004.032011

Qiu X, Liu M, Sunada K et al (2012) A facile one-step hydrothermal synthesis of rhombohedral $CuFeO_2$ crystals with antivirus property. Chem Commun 48:7365–7367. https://doi.org/10.1039/c2cc33475e

Qiu G, Gai Z, Tao Y et al (2020) Dual-functional plasmonic photothermal biosensors for highly accurate severe acute respiratory syndrome coronavirus 2 detection. ACS Nano. https://doi.org/10.1021/acsnano.0c02439

Redfield RR (2020) Regulation of COVID-19. US Food and Drug Administration

Roh C, Jo SK (2011) Quantitative and sensitive detection of SARS coronavirus nucleocapsid protein using quantum dots-conjugated RNA aptamer on chip. J Chem Technol Biotechnol 86:1475–1479. https://doi.org/10.1002/jctb.2721

Schmitt V, Kesch C, Jackson JK et al (2020) Design and characterization of injectable poly(lactic-co-glycolic acid) pastes for sustained and local drug release. Pharm Res 37. https://doi.org/10.1007/s11095-019-2730-4

Schröder I (2020) COVID-19: a risk assessment perspective. ACS Chem Health Saf. https://doi.org/10.1021/acs.chas.0c00035

Seo G, Lee G, Kim MJ et al (2020) Rapid detection of COVID-19 causative virus (SARS-CoV-2) in human nasopharyngeal swab specimens using field-effect transistor-based biosensor. ACS Nano 14:5135–5142. https://doi.org/10.1021/acsnano.0c02823

Shang J, Ye G, Shi K et al (2020) Structural basis of receptor recognition by SARS-CoV-2. Nature 581:221–224. https://doi.org/10.1038/s41586-020-2179-y

Shen M, Zhou Y, Ye J et al (2020) Recent advances and perspectives of nucleic acid detection for coronavirus. J Pharm Anal 10:97–101. https://doi.org/10.1016/j.jpha.2020.02.010

Siddiquie RY, Agrawal A, Joshi SS (2020) Surface alterations to impart antiviral properties to combat COVID-19 transmission. Trans Indian Natl Acad Eng. https://doi.org/10.1007/s41403-020-00096-9

Siuzdak K, Niedziałkowski P, Sobaszek M et al (2019) Biomolecular influenza virus detection based on the electrochemical impedance spectroscopy using the nanocrystalline boron-doped diamond electrodes with covalently bound antibodies. Sensors Actuators B Chem 280:263–271. https://doi.org/10.1016/j.snb.2018.10.005

Speshock JL, Murdock RC, Braydich-Stolle LK et al (2010) Interaction of silver nanoparticles with Tacaribe virus. J Nanobiotechnol 8:1–9. https://doi.org/10.1186/1477-3155-8-19

Sportelli MC, Izzi M, Kukushkina EA et al (2020) Can nanotechnology and materials science help the fight against sars-cov-2? Nanomaterials 10:1–12. https://doi.org/10.3390/nano10040802

Stadler K, Masignani V, Eickmann M et al (2003) SARS —beginning to understand a new virus. Nat Rev Microbiol 1:209–218. https://doi.org/10.1038/nrmicro775

Su Jen Chou HMT (2005) Mask having antivirus and anti-germ effect. United States Pat Appl Publ 1:6–9

Su S, Wong G, Shi W et al (2016) Epidemiology, genetic recombination, and pathogenesis of coronaviruses. Trends Microbiol 24:490–502. https://doi.org/10.1016/j.tim.2016.03.003

Sudipta Seal GP (2020) Anti-viral surface coating to prevent the spread of COVID-19 through touch. PCI Magazine

Sunada K, Minoshima M, Hashimoto K (2012) Highly efficient antiviral and antibacterial activities of solid-state cuprous compounds. J Hazard Mater 235–236:265–270. https://doi.org/10.1016/j.jhazmat.2012.07.052

Thi H, Thai C, Le MQ et al (2004) Development and evaluation of a novel loop-mediated isothermal amplification method for rapid detection of severe acute respiratory syndrome coronavirus development and evaluation of a novel loop-mediated isothermal amplification method for rapid detection. J Clin Microbiol 42:1956–1961. https://doi.org/10.1128/JCM.42.5.1956

Udugama B, Kadhiresan P, Kozlowski HN et al (2020) Diagnosing COVID-19: the disease and tools for detection. ACS Nano 14:3822–3835. https://doi.org/10.1021/acsnano.0c02624

Walls AC, Park YJ, Tortorici MA et al (2020) Structure, function, and antigenicity of the SARS-CoV-2 spike glycoprotein. Cell 181:281–292.e6. https://doi.org/10.1016/j.cell.2020.02.058

Weiss SR, Navas-Martin S (2005) Coronavirus pathogenesis and the emerging pathogen severe acute respiratory syndrome coronavirus. Microbiol Mol Biol Rev 69:635–664. https://doi.org/10.1128/mmbr.69.4.635-664.2005

Wrapp D, Wang N, Corbett KS et al (2020) Cryo-EM structure of the 2019-nCoV spike in the prefusion conformation. Science 367:1260–1263. https://doi.org/10.1126/science.aax0902

Xiang D-x, Chen Q, Pang L, Zheng C-l (2011) Inhibitory effects of silver nanoparticles on H1N1 influenza A virus in vitro. J Virol Methods 178:137–142. https://doi.org/10.1016/j.jviromet.2011.09.003

Xie X (2020) Chest CT for typical COVID-19 pneumonia. Radiology. https://doi.org/10.14358/PERS.81.12.21

Xu J, Jia W, Wang P et al (2019a) Antibodies and vaccines against Middle East respiratory syndrome coronavirus. Emerg Microbes Infect 8:841–856. https://doi.org/10.1080/22221751.2019.1624482

Xu M, Ye J, Yang D et al (2019b) Ultrasensitive detection of miRNA via one-step rolling circle-quantitative PCR (RC-qPCR). Anal Chim Acta 1077:208–215. https://doi.org/10.1016/j.aca.2019.05.028

Yang A, Cai L, Zhang R et al (2017) Thermal management in nanofiber-based face mask. Nano Lett 17:3506–3510. https://doi.org/10.1021/acs.nanolett.7b00579

Yonesi M, Rezazadeh A (2020) Plantas as a prospective source of natural antiviral compounds and oral vaccines against COVID-19 coronavirus. Preprints. https://doi.org/10.20944/preprints202004.0321.v1

Yu Y, Bu F, Zhou H et al (2020) Biosafety materials: an emerging new research direction of materials science from COVID-19 outbreak. Mater Chem Front. https://doi.org/10.1039/d0qm00255k

Zhang Y, Odiwuor N, Xiong J et al (2020) Rapid molecular detection of SARS-CoV-2 (COVID-19) virus RNA using colorimetric LAMP. medRxiv 2:2020.02.26.20028373. https://doi.org/10.1101/2020.02.26.20028373

Zhao VXT, Wong TI, Zheng XT et al (2020) Colorimetric biosensors for point-of-care virus detections. Mater Sci Energy Technol 3:237–249. https://doi.org/10.1016/j.mset.2019.10.002

Zhong H, Zhu Z, Lin J et al (2020) Reusable and recyclable graphene masks with outstanding superhydrophobic and photothermal performances. ACS Nano. https://doi.org/10.1021/acsnano.0c02250

Zhou P, Yang XL, Wang XG et al (2020) Discovery of a novel coronavirus associated with the recent pneumonia outbreak in humans and its potential bat origin. https://doi.org/10.1101/2020.01.22.914952. Now published in Nature. https://doi.org/10.1038/s41586-020-2012-7

Zhu X,Wang X, Han L, Chen T, Wang L (2020a) Reverse transcription loop-mediated isothermal amplification combined with nanoparticles-based biosensor for diagnosis of COVID-19. medRXiv

Zhu N, Zhang D, Wang W et al (2020b) A novel coronavirus from patients with pneumonia in China, 2019. N Engl J Med 382:727–733. https://doi.org/10.1056/NEJMoa2001017

Zou Z, Yao M (2015) Airflow resistance and bio-filtering performance of carbon nanotube filters and current facepiece respirators. J Aerosol Sci 79:61–71. https://doi.org/10.1016/j.jaerosci.2014.10.003

Chapter 19
Resilience of Healthcare System to Outbreaks

Manish Chaturvedi and Rajneesh Mohan Siwan

Abstract The infrastructure for public health system and emergency management is still in a nascent stage for many of the third world countries. Plethora of challenges, such as multiple novel and evolving microorganisms with potential to cause public health emergencies, antimicrobial resistance, economic crisis and extreme weather events have increased the stress on the public health system. Resilience is an upcoming concept in the health systems which highlights the role of public health emergency management as health plays the central role. The chapter highlights the complexities of COVID-19, model on healthcare system and model on resilient health system. While discussing the models adopted by various countries to tackle this pandemic, it has been brought out that NPI and TTT are effective strategies to reduce the transmission. Novel approaches of all hazard preparedness and policy for one health approach are vital to control future pandemics.

19.1 Introduction

The twenty-first Century began with infectious diseases outbreaks which threatened the health of the entire globe. These events increased the stress on various public health systems and institutions with varying capacities spread over the world. The infrastructure for public health system and emergency management is still in a nascent stage for many of the third world countries. The burden of disease scenario in many of the countries reaffirms the 'Triple burden' of disease as persisting communicable diseases, emerging new diseases like Zika, Ebola, Nipah and

M. Chaturvedi (✉)
Planning and Evaluation, National Institute of Health and Family Welfare, Munirka, New Delhi, Delhi, India
e-mail: manish@nihfw.org

R. M. Siwan
South Delhi Municipal Corporation, Minto Road, New Delhi, Delhi, India

M. K. Goyal, A. K. Gupta (eds.), *Integrated Risk of Pandemic: Covid-19 Impacts, Resilience and Recommendations*, Disaster Resilience and Green Growth,
https://doi.org/10.1007/978-981-15-7679-9_19

Coronavirus disease and the growing chronic diseases pound humanity across the globe. During the course of these emerging diseases millions of lives have been lost and implementation of Non- pharmacological interventions in the form of enforced lockdowns has led to economic shocks. Plethora of challenges, such as multiple novel and evolving microorganisms with potential to cause public health emergencies, antimicrobial resistance, economic crisis and extreme weather events have increased the stress on the public health system.

Resilience is an upcoming concept in the health systems. The concept has been conceptualized as the capacity of a community or system to absorb a shock, while still retaining the basic functions, to which a system should return after being shocked. The other school of thought narrates it as adaptive and transformative capabilities that allow a system to transform its characteristics to soften future shocks while still retaining its basic structure, or may even modify its structure to eliminate risks altogether (Fridell et al. 2020).

There are sundry terminologies used like disaster risk reduction, disaster management, emergency management etc. but as the heath plays the central role in managing the risks and reducing the consequences of both routine and emergency situations due to all-types of hazards, a new terminology has come into wider use, namely the Public Health Emergency Management (PHEM), combining emergency management and public health system. The health sector may play a paramount role in communicable disease management but their crucial role has only been visualized in curbing consequences of all other disasters. However, health sector resilience should focus on research and evidence based myriad experiences of previously experienced natural and man-made disasters for quick response in periods of disturbances. This can be understood by the quote of Director-General WHO: "Strong and resilient health systems are the best defense not only against outbreaks and pandemics, but also against the multiple health threats that people around the world face every day."

19.2 Quandary of COVID-19 Emergency

The countries started preparedness, in the backdrop of New Year 2020, on the report of the novel coronavirus, with epidemic potential, from Wuhan City of China. In the next few weeks, countries observed widespread closure that had brought the world to a virtual standstill. As of 7th April, COVID-19 had affected more than 200 countries, territories and areas; more than four out of five people in the global workforce of 3.3 billion were affected by workplace and school closures (Singh 2020).

The incidence of emerging communicable diseases in humans has increased in the recent past and will continue to do so in the near future. After emergence of COVID-19 in Wuhan city of China in December 2019, the World Health Organization (WHO) declared the COVID-19 epidemic, on March 11, 2020, as a pandemic (Dikid et al. 2020). Globally more than 200 countries are affected with more than

23 lac confirmed cases and 157,847 deaths as of 20th April 2020 which shows the quick spread in a span of around 90 days from the first case reported (WHO 2020a).

This recent pandemic of COVID-19 has led to a huge loss to mankind irrespective of developed or developing country. Most of the countries are struggling to provide a rational response to manage psycho-social impact on the community simultaneously ensuring routine life with least chances of spread in the community. On the other hand, the general public is abysmal and afraid of the present COVID-19 menace, as it led to collapse of basic health-care services. If COVID-19 affected half the world's current population over the course of a year with around 1% case fatality rate (CFR), the mortality would be around 35 million. (Burrows and Engelke 2020).

Migrant labourers and non-skilled self-employed workers lost their wages, amidst COVID-19, the closure of cities and lack of clear guidelines exacerbated these workers' anxieties, and compelled them to make the long walk home, from the cities to the hinterland due to the absence of transport. Furthermore, this hindered food-related logistic services, disrupted the entire food supply chains which adversely affected the availability of food especially in remote areas. These situations are worsened due to restriction in movement of agricultural labor and may lead to critical challenges to food production, thus jeopardizing food security.

The challenged health system of developing countries was further compromised due to closures and bifurcated healthcare services (COVID and non-COVID). Poor availability of infection prevention equipment and ventilators with their operators ruined large-scale public health measures, and efforts to replenish supplies have been thwarted by trans-border closure. This catastrophe has also revealed that rapid response medical teams at national level are not adequately resourced for mobilization. In the areas most affected to date, the number of deaths occurring per day has overwhelmed the capacities of mortuary and funeral services. These will be further complicated by an increasing number of individuals seeking care for chronic, pre-existing medical problems such as hypertension, cancer, cardiac disease, and diabetes, especially in more developed countries with advanced health systems. Routine immunization services are disrupted due to interplay of multiple factors like human resources are diverted for other priorities, prolonged lockdown and sealed borders which led to disruption of logistics supply chain and lastly because of social distancing parents are not bringing their children to the health centers. This may lead to compromising immunity of children to communicable diseases and may end in increased child mortality.

19.3 Factors Affecting Disease Transmission

Historically, communicable diseases contributed the major section of human morbidity and mortality till the twentieth century, when non-communicable diseases began to rival. The communicable diseases usually affect all ages but have more affinity for unimmunized young populations. The spread of any pathogen depends upon the interaction of epidemiological triad i.e. agent, host and environment. To initiate any cycle of infection, an agent or pathogen should interact with the host or human through an external environment, and after reaching inside the host, the agent multiplies to make the host diseased depending upon the internal environment of the host. The external environment through a variety of transmission systems in operation determines the spread of the disease. With the advent of international trade, local and international boundaries have lost their significance and made the whole world a single unit which is providing a platform to spread the disease through a chosen transmission system. Globalization causing excessive, erratic, changes in the ecological, biological and social conditions that shape the burden of infectious diseases in specified population (Kieny et al. 2014).The international health regulations guidelines given by WHO in 2005, enforced to control the international spread of diseases for periods of novel infectious disease, suggested health offices at each point of entries with a national focal point for regulating and monitoring the movement at trans borders.

The COVID-19 is transmitted through respiratory droplets of >5–10 μm in diameter. The droplet transmission occurs when a person who is in close contact (within 1 m) with a person has respiratory symptoms, not following cough etiquettes, becomes at risk of having infection through mouth, nose and eye. COVID-19 may transmit through air borne mode in specific settings in which some procedures or support treatments may generate aerosols like endotracheal intubation, open suctioning and administration of nebulized treatment etc. (WHO 2020a). The growing public transport system in developing nations provides an opportunity for airborne disease transmission due to overcrowding and in transit microenvironments. These further enhance chances of disease transmission through increased contact period and reduced circulations in transport infrastructures with increased mobility.

19.4 Elements of Resilient Health System

ADMASS is adopted from the suggested elements by Kruk et al. (2015), which names seven elements of resilient health structure as Awareness, Diversity, Mutualism with Malleability, Auto regulation, accorded in unison with shield and scrutiny in the following Fig. 19.1. This should be promoted among hospital managers and PHEM experts.

For a resilient health system, risk assessment in terms of human, physical, and information assets for assessing areas of strength and vulnerability through Hazard, Vulnerability, Capacity and Risk Assessment (HVCRA) Survey involving the local community for prioritizing potential risk. This includes assessment of person or population data like disease or beneficiary data; place or geographic locations like proximity to flood zones or industries; time to reach the nearest health facility and prompt public health care delivery system. *Risk awareness* needs robust health information systems that can report on the status of the system and impending health threats in real time for sound planning to simulate the logistics of a response to crisis.

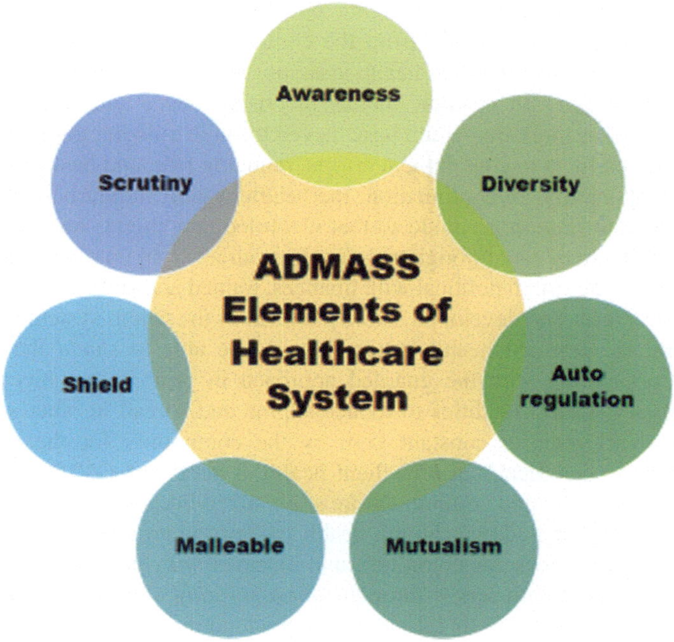

Fig. 19.1 Elements of Healthcare System

Health care delivery systems should be planned and equipped to address *diverse* needs of patients and locations. The systems that have the capacity to address a broad range of health challenges through tiered approach rather than a targeted few are more stable and capable of detecting diverse disturbances. For example, in a primary care clinic, a patient with an unfamiliar pattern of symptoms triggers a systematic investigation for new pathogens rather than dismissal because the patient fails to fit into known algorithms. On the other hand, if a group of familiar constellation patients arrive at the emergency of a tertiary care hospital, their functionality may be compromised. The existing three tiered health care delivery system should be strengthened with material and human resources so that the full coverage, with diverse health needs of the community, of the defined boundaries may be achieved. This will enhance the confidence of the community in calm days and enable sound surveillance for faster recognition of new threats.

Resilient health systems may be achieved by bringing together various players including transportation, media, education, law enforcement agencies, water and sanitation with civil society organizations and private players to formulate solutions with *mutualism* to initiate action. In this way the health systems with recognized partners will allow governments to trigger rapid deployment of a wider set of resources.

The health system should be *malleable* to provide services with limited resources in crisis times as well as modernized facilities in normal times for the changing needs of the community in terms of changing epidemiological determinants. Regular evaluations and crucial feedback from the community is required for innovative planning and developing resilient health systems. This will develop a user friendly health system which can cushion shocks in crisis, reduce social and economic disruptions and less mortality would be achieved through utilizing trained workforce of community partners in containment efforts, reducing fear, and hastening resumption of normal activity. In continuation, the health system should be designed and developed in auto-*regulating* mode so that identified new threats may not compromise the continuum of care through transfer of resources from unaffected areas. This could be achieved through political will, finances, trained and skilled man power and good infrastructure. The developed stable platform of the health system will absorb crisis situations. Resilient health systems should be able to *shield* the identified vulnerable population with the guarded approach in prior community identified well-furnished isolation facilities to reduce stigma and spread in other sections of the society. This requires constant vigil by the community for the community through imbibed confidence in a resilient health system. In COVID-19 pandemic the infected workforce and community are stigmatized because of fear of transmission. This may be handled by enabling health systems, in normal days, for isolation facilities in remote areas for communicable diseases with potential of pandemic. Finally any health system serves through robust *scrutiny* of all the community for quick identification of new threats and new epidemiological trends for action in silence. This requires scientific research, health budgeting and user friendly testing approaches. In the current crisis, gene amplification tests, gold standards, not available everywhere but screening tests (Rapid Antigen Test), not validated, are used for surveillance.

19.5 Conceptual Framework in Impending Crisis for Resilient Health System

The Sendai Framework for Disaster Risk Reduction (SFDRR) and sustainable development goals (SDGs), are landmark agreements by the United Nations adopted in 2015, suggesting implementation of Disaster Risk Reduction (DRR) strategies to improve resilience to disasters globally. The SFDRR in contrast to its predecessor, the Hyogo Framework for Action, puts a lot of emphasis on health. It proposes resilient health systems as an opportunity (Olu 2017).

In the Public Health Emergency Management (PHEM) concept, specific standards of knowledge, techniques, and organizing principles of emergency management and public health are pooled together for the effective management of complex health events with serious impacts (Rose et al. 2017). The response to any crisis event is done at twin levels, one, at the administrative structure at national, state and district level while other is in the public health care delivery system. The PHEM standards suggest action in unison in both the structures following set standards and

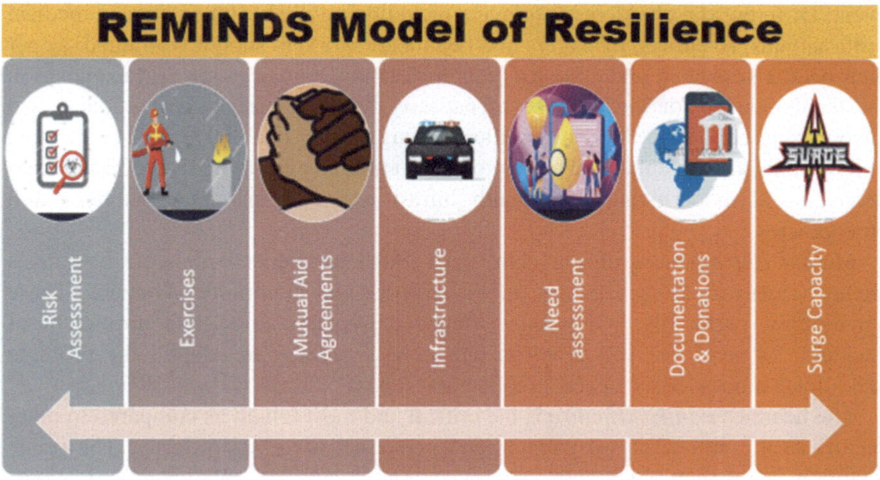

Fig. 19.2 Model of Resilience

guidelines. The response and preparedness of two streams elaborated, in terms of district preparedness (DP) and healthcare preparedness (HP), as REMINDS model, is shown in Fig. 19.2, for resilience in healthcare.

Risk Assessment (HVCRA) involving the local community for prioritizing potential risk at district and healthcare level for preparedness should be done on a regular basis. This assessment guides the planners about upcoming threats, preparedness and recognition of it at various levels concurrently identifying various players for action. Although countries are fundamentally responsible for their health systems, they need the coping capacity to mobilize the full range of available players to rapidly assess and if required draw on higher or external resources as the pathogen does not respect borders.

The pivotal issue in resilience is preparedness through capacity building by *exercises* for enhancing thoughtfulness for prompt action. These include awareness about set standards, mock drills and table top exercises which in a way encourages professionals to acclimatize with the local environment and set standards as well as augmenting teamwork in the available workforce. The customized, consistent mock exercises and drills build the capacity of the individual as per the need and prepared plan of organization. The Global Health Security Agenda (2014) expanded on International Health Regulations (2005) by strengthening Emergency Operation Centers (EOC) is envisaged to be set up at National and State headquarters as part of PHEM standards. In consonance with this initiative, the Center for Disease Control, Atlanta initiated a Public Health Emergency Management Fellowship program to train recommended leaders and practitioners around the globe in PHEM concepts and principles so that emergency management programs may be implemented with orientation of public health. (Rose et al. 2017). The workforce availability, capacities and their deployment at respective locations is the crucial event. Different stakeholders and players including public health engineers, medical professionals, scientists, managers and healthcare workers should be trained, as per their need assessment, and asked to participate in the regular mock drill exercises in the local vicinity for the purpose of capacity building among professionals and encouraging confidence in communities. Strong district healthcare delivery system will be attained with enhanced morale of the community by uninterrupted healthcare services and modernization of health infrastructure in managing public health emergencies.

Mutual aid agencies during calm days like agreements with schools, canvas shelter (tent) providing agencies, medical stores and community kitchens. These agreements, preferably, are without financial transfers through commitment of exchange of services. This leads to enhanced capability to address the problem and allows a broader set of resources to be used.

The healthcare *infrastructure* is the critical and most reliant as compared to other sectors as life and safety of patients and personnel is the top priority. The prime factors in infrastructure resilience are power supply, water supply, communications, medical information and digital infrastructure. The system should sustain without grid power especially in emergency and utility areas by local power generation through non-conventional energy sources. Similarly, water conservation, ground water harvesting and local storage may address the issue of water depletion. Lock-down measures in COVID-19 forced employees to work from home which spiked the demand for telecommunications and internet, leading to outages and limiting the ability of teams to communicate with each other remotely. Additionally surge in utilization of e-commerce portals increased the chances of cyber-attacks, even on hospital management information system. This is compounded by telemedicine portals which flourished in COVID crisis to address the fury of patients. The IT resilience may be attained through forming committees of experts, using fire walls, and registering the clients prior to providing services.

> *This significant demographic shift is accompanied by changing demands on urban infrastructure, social structure, and ability of municipalities to meet the needs of their residents. As people increasingly aggregate, the consequences of failures in our systems affect more people and thus are more catastrophic… Traditional resilience-building initiatives have focused on infrastructure and environmental sectors…too often a commensurate focus on the people served by this infrastructure is lacking in preparedness plans and frameworks.. … This link between community design and healthy people is well understood by public health practitioners and is part of a "health in all policies approach". Such an approach is evident in areas such as community walkability and access to healthy food yet has been exploited less in investments in critical infrastructure or disaster mitigation.* (Wulff et al. 2015)

Need assessment of healthcare providers should be done consistently to identify specific problems and customized solutions by communicating with the health workforce, neighbouring healthcare systems, and state and/or national authorities for real-time tracking of a pandemic. The trained professional, serving as incident commander, shows coordinated response to maximize capacity in the early phase of demand simultaneously as they are trained to alert higher authorities in overwhelming situations. The healthcare workforce trained through national disaster management authorities would provide for early warning signs in relation to every disaster including flood, fire, famine, cyclones etc. Recently systems are also strengthened for complex disasters, like chemical, biological, radiological, nuclear and explosives, through CBRN training centers in selected capitals and few disaster prone areas.

Documentation for collating information to improve future coordination, information sharing and identification of lessons learned. This responsibility lies with the incident commander. The materials include a list of staff and experts, agencies, mapped stakeholders, rapid response plans, copies of memorandum of understanding with agencies, reports of past mitigation exercises and other resources. *Donations* of various forms should be encouraged like fund, blood, equipment, materials, food and services. The issue raised in relation to mismanagement of this noble practice of societies can be handled by prior mapping of donors and their registration. For regularizing donations and creating a global pool for resilience WHO launched WHO Foundation, in May 2020, which is an independent grant-making foundation focused on addressing the most pressing global health challenges. It will fund to promote high-impact initiatives, innovation, and rapid response actions, and health system strengthening (WHO 2020b).

During impending crisis events, healthcare facilities ask for increasing their resources to the maximum capacity to handle unexpected flow of patients, the surge capacity, with the available or reduced resources. The surge planning elaborated by public health practitioners as space, staff, stuff and system (Anesi et al. 2020). Space refers to patient care—bed capacity. This bed capacity includes general ward, specialized care intensive and cardiac units and isolation wards for adults as well as children. In emergency periods of COVID-19 the bed capacity enhanced through postponing or cancelling elective surgeries and admission, diverting patients to lower-acuity hospitals, repurposing licensed beds (e.g., using wards beds for ICU care); laying beds in new areas near to operational areas like corridors and later on in non-clinical areas like educational spaces and reverse triage. Staff refers to the healthcare professionals. These are distributed in various healthcare areas on the basis of their training and acquired skills including critical care and hospital operation's staff. The staff may suffer with scarcity due to co-morbid conditions (Hypertension, Diabetes etc), infection, injury, quarantine, pregnancy, immune-compromised state and family care duties. The staff availability is compromised because of non-availability of public transport and residential facilities in public health infrastructure specifically trans-border interstate staff. The number and morale of the workforce may be maintained through following infection control guidelines, respecting personal commitments, extending shifts and bringing in off-duty usual staff, calling in previously trained but inactive staff (e.g., researchers, retirees); deploying unrelatedly trained or nonclinical staff with just-in-time task-specific training and tiered supervision; and accepting external staff of varying degrees of related training without clinical privileges. Stuff refers to the equipment required for care and support in the health system. In COVID-19 there is unprecedented demand for ventilators which are procured in bulk but lying idle as trained professionals are not available in the system. Sudden surge in demand for specific medicines, increase in turnover time of machines and special care requirements (e.g., PPE). The increased demand further shot up because of lockdown which compromised logistics supply chain management. This situation may be handled through prior agreements with mutual aid agencies like demanding stockpiles from private operators as private hospitals were not operational in lockdown situations. The last S of preparedness refers to a system for the planning and leadership to operationalize and optimize a response effort. This is already elaborated above under elements of resilient healthcare systems. During emergency conditions, the central command hands over to the incident command system, which is constituted by seven professional's viz. incident commander (medical superintendent or any deputed professional), security officer, public relation officer or media manager, officers for finances, logistics, operations and planning. These officials immediately take command, on alert, to meet dynamic demand. The ICS has its mandate for same voice or terminology, modular organization, consolidated action plans, pre designated command office and comprehensive resource management.

19.6 Challenges in Healthcare Preparedness and Approach

The WHO floated *guidelines* for level of care, case and screening definition, treatment modalities for COVID-19 pandemic. The governments customized them as per their capacity to respond but the cascading of this information is a challenge. For COVID-19, three types of *tests* are available, one is specifically to identify infected individuals rapidly by Rapid Antigen Test secondly by (RT-PCR) while the third is for testing antibodies from sera. An effective strategy that tests, through molecular diagnostic testing (RT-PCR), tracks people infected and traces their contacts (TTT), helps to reduce the spread of the virus. This is the only available strategy, to be effective at suppressing the spread of the virus, requiring high level contact tracing. There is an urgent need for testing at inordinate capacity, use of artificial intelligence like distant screening of voice on mobile phones for COVID-19, restricting people in confinement as well as tracing contacts through new approaches like digital tracing. The testing faced many constraints including non-availability of trained personnel for sample collection, time for analysis and availability of reagents. *Personal Protective Equipment* (PPE) demands surged amid COVID-19 pandemic but shortage of supply and inappropriate utilization, because of lack of training, led to panic among health care providers. *Digital platforms* (like IGOT-DIKSHA in India) utilize training which typically begins with infection control guidelines to concerned officials. The hospitals are asked to appoint trained observers for monitoring safety mechanisms like doffing.

Improving the resilience of healthcare services requires planning for *continuity of services* through the optimal management of resources, supporting services, and the workforce. Planning involves early stockpiling of non-perishable resources, power backup, upkeep of the IT system, and rotation of staff. Clear, continuous and compact communication play crucial roles in crisis and reducing confusions. The centre point for *communication* in a hospital is the office of the incident commander which connects with workers and in-charge of local units. The commander may broadly communicate policies and announcements to bring behavior change and reassurance. Considering the dynamic decision making during COVID-19, a searchable centralized document repository may improve compliance with all the updated policies and guidelines. As mentioned above, the public relation officer needs to communicate with patients, families, and the community for reducing visits inside the premises.

The National Governments may be required to run few specialized Centres for Excellence for the high quality of services they provide, and for serving as the benchmarks for secondary and specialty care. Some of these hospitals may turn into *dedicated COVID care centers* to reduce the chance of transmission of infection. Thus apart from addressing overcrowding and efficiency losses, this would also improve the quality of services. Few Nations also have health infrastructure of integrated medicine and indigenous systems of medicine (*AYUSH system*) which may also cater at least as a good quarantine center during crisis situations. The popularity of these medicines and their consumption during illness may enhance recovery rate which may bring breakthrough in overall management of COVID-19. Considering the present situation of lockdown where all the private clinics were closed which led to paucity of qualified healthcare personnel, the most optimal solution is *distance healthcare* (also including tele-diagnosis, telemedicine, tele-consulting and tele-counselling) where expert advice can be made available at some central point and accessed as and when required by telephone or internet. Tele-Preventive Medicine can also be used to collect information from a large number of people (both healthy and sick) to prevent outbreak of disease. Develop a strong facility-based *disease surveillance system* with an integrated electronic backbone. The system will capture surveillance data from communities, dispensaries, mobile health units, school health clinics and hospitals on a real-time platform by user-friendly devices. There is a need to develop brief, standardized data entry protocols suitable for different platforms (mobiles, smartphones, tablets, PCs, etc.) and data flow that can be coordinated. A group of data analysts may be identified, involving interested sections of the health administration, academia and civil society for aiding evidence-based decision-making. Since the vaccine will not be available in near future therefore the current *prevention strategies* of non-pharmacological interventions (NPI) shall continue to work like social distancing, face mask, frequent hand sensitization and lockdown measures. The lockdown measures should be eased out in phased manner so that the system may utilize the only available hope of herd immunity at present and in future expected outbreaks.

19.7 Lessons Learned

WHO guidelines for COVID-19, were discussed at health ministries of various countries with myriad experiences of their past disasters and local environmental conditions, finally operationalized with modifications. To prevent the transmission of disease, lockdown imposed which ranged from complete lockdown as in India to least closures as in Sweden. These models were praised and advocated on the basis of herd immunity. Largely the community of public health experts were of opinion to move from case or cluster containment to district closure and successively proceed to country lockdown, which would have prevented the quantum of deaths due to other complexities like problem of supply chain, non COVID emergencies and economic crisis.

19.7.1 Swedish Model

Sweden has stopped the classroom teaching at universities and high school level, while daycare and junior schools were open, postponed the sports and banned the gatherings without promotion of non-pharmacological interventions aimed to flatten the curve as early as possible, probably envisaging gains through herd immunity. The NPI implementation was easier in Sweden as most of the population lives alone and utilizing the virtual world with hi-speed internet services. People over 70 years of age were advised to stay at home. Patients with influenza like illness (ILI) were advised to stay at home without medical certificates, asked to contact health facilities if condition worsens for medical advice and testing. The citizens brought infection from Italy which culminated in massive spread of infection and deaths especially in Stockholm. The model was appreciated in initial days but with evolving epidemiology of COVID disease elaborated about asymptomatic transmission and old age people with comorbidities.

> *"Though it is mostly the middle-aged who have contacted the virus in the country, most of the deaths have taken place among the elderly. A major portion . . .taken place in place in elderly care homes. . . .many European countries, and also the US, are recording deaths only from hospitals and they do not include the deaths taking place in nursing homes or other long-term care facilities.* (Swan 2020)*".*

19.7.2 India Model

India started preparedness for COVID-19 from early phase of this year in form of procurement of PPEs, ventilators, testing kits, restrictions of flight movement with travel advisories and self-quarantine of 14 days for all international travelers. Indian government imposed 21 days lockdown to contain the spread of the virus. The closure involved everything except essential and emergency services. In India, as of 19 April 2020, a total of 16,116 COVID-19 cases have been reported from 32 states/ union territories (WHO 2020b). This led to failure of logistics supply chain, compromised service delivery to vulnerable populations, wage loss and food distribution but succeeded in containment of virus. But India witnessed a surge in confirmed coronavirus cases after a religious congregation held at Delhi in March which was attended by more than 5000 members including foreigners but the system was able to trace approximately 95% of the members and contact tracing is going on (The Economic Times 2020). Government of India promoted telemedicine for virtual management of patients and launched Arogya Setu app for awareness, self-assessment and to inform users about COVID positive cases in near vicinity and relevant advisories.

India has a pluralistic healthcare system with 476 medical colleges and tertiary care facilities. These generally constitute 5–8% critical care beds of the total beds. ICU care is poor or non-existent at district hospitals in rural India, which cater to 80% of the population. (Yeolekar and Mehta 2008) There are around 7 lac beds in government sector while medical care facilities under AYUSH have around 4000 hospitals (CBHI 2017).

India has gained from the concept of lockdown especially in Bhilwara district of Rajasthan where after reporting of clusters of cases strict lockdown was imposed even on essential services with no cross-border movement. The strict lockdown and containment strategy with home quarantine of contacts given success to this model which was widely appreciated.

But considering the evolving scientific knowledge on options available to control novel coronavirus it is being accepted that humankind will have to "live with the virus" and strategies are being shifted from containment to mitigation. The phased manner unlocks of four phase lockdowns was observed, probably thinking of herd immunity concept in backdrop. Most of the infected cases have been reported from good performing states with comparatively better health care infrastructure like Maharashtra, Tamil Nadu, Delhi and Gujarat.

19.7.3 South Korean Model

South Korea with a population of 50 million could have experienced the worst outcomes but it turned the present crisis into opportunities. South Korea has already learned lessons from MERS which affected South Korea in 2015 when 186 people were infected and 36 killed. South Korea did not impose lockdowns nor sealed its borders but adopted a preventive and rigorous testing strategy using the 3T method, test, trace and treat. Mass scale testing of 18,000 cases a day from high risk areas and clusters were performed. South Korea was successful in identifying the positive cases at an early stage and isolating these cases. With the support of legal policy, the quarantine violators were fined heavily.

19.7.4 Japan Model

Japanese people as a part of their routine life wear masks, use hand sanitisers and maintain "utmost" cleanliness as a part of their culture, so COVID-19 pandemic did not bring many changes in the country. They prefer to follow seclusion and isolation in their normal lives and are busy in their work. They followed a cluster-based approach whereby they track each cluster to the original infection source and persons with high transmissibility were isolated to prevent the spread of infection. Therefore, pinpoint testing was carried out and broad testing of the population was not undertaken in Japan, in contrast to the approaches taken in other countries.

Another important part of the Japan model was the social distancing method whereby they follow "the three Cs," referring to closed spaces with poor ventilation, crowded places with many people nearby and close-contact settings such as close-range conversations. All the three Cs were to be avoided as a measure to prevent spread of novel coronavirus in Japan.

19.8 Way Forward

NPI, only effective strategy, will reduce community transmission and flatten the curve for all the present and upcoming infections with epidemic potential. Simultaneously, TTT strategy i.e. testing, tracking and tracing will also remain effective to reduce the spread of the virus. The evidence required for other routes of transmission of SARS-CoV-2 and possible ways of spreading like air conditioning systems, inanimate objects, asymptomatic transmission, environmental conditions etc.

> *"How long is immunity to COVID-19 likely to last? The best estimate comes from the closely related coronaviruses. . .who had an antibody response, immunity might wane, but is detectable beyond 1 year after hospitalization. . .the possibility that there could be another wave of COVID-19 cases in 3 or 4 years as. . . .MERS CoV immunity . . .detectable for 4 years. . . basic reproduction number (R0) is about 2·2 for SARS-CoV-2. . . .the herd immunity calculation suggests that at least 60% of the population would need to have protective immunity, either from natural infection or vaccination. . . Delivery of efficacious vaccines is not a competitive race to finish,..reliance on comprehensive seroprevalence data and a solid, research-based grasp of correlates of protection will allow policy to be guided by secure, evidence-based assumptions on herd immunity, rather than optimistic guesses.* (Daniel et al. 2020)*"*

Multiple clinical trials are underway to develop treatment modalities and vaccines for COVID-19, but results are still awaited. Smart working and staggered shifts may have to be adopted to mitigate transmission. Webinars and telemedicine need to be implemented for teaching and training. All hazard preparedness is gaining momentum for PHEM and disaster risk reduction including chemical, biological, radiological and nuclear hazards. The resilience system should also show solidarity for antimicrobial resistance and one health. The NPI measures with precautions related with mass gathering may prevent the other projected waves of COVID-19. The implementation of IHR (2005) suggested standards for point of entries is desirable to watch the movement of passengers in relation to health and to quickly contain the emerging and novel infections.

19.9 Conclusion

The foundation of the resilience health system is capacity and risk assessment which guides the planners. The resilient health systems, a system which is able to protect human life with positive outcomes for all in calm and crisis duration would lead to to enhanced confidence in the community. Frequent pandemics and public health emergencies may be mitigated by prompt emergency care only through prior preparedness to prevent overload on the already burdened health system in developing countries. This can be handled by creating a panacea for biological disasters in collaboration with health services, medical research, eminent health care institutes and capacity building institutes with disaster management authorities.

With increasing inequalities within and across countries, while many are reaping benefits, the poor on the whole, appear to be getting poorer (Kieny et al. 2014) and the communicable diseases affect the poor unduly in terms of access to basic amenities including safe water, safe food, sanitation, access to health care and public transport because of lack of information and illiteracy. Therefore, they should be prepared with the wider analysis of the vulnerable population and need assessment.

In summary, REMINDS model is suggested to remind the policy makers and district managers to think about achieving the standards suggested and prepare a performance indicator for readiness in response to disturbances.

All these novel outbreaks remind us of the concept of "one health" which suggests to respect the relationship of wildlife, social life, and the environment. Rapid urbanization, deforestation and industrialization have changed the epidemiological determinants of disease. However, in view of the present policies, an optimistic public health approach inculcating NPIs in our lifestyles with strengthening medical development and research need to be evolved.

References

Anesi GL, Lynch Y, Evans L (2020) A conceptual and adaptable approach to hospital preparedness for acute surge events due to emerging infectious diseases. Crit Care Explor 2(4):e0110. Published 2020 Apr 29. https://doi.org/10.1097/CCE.0000000000000110

Burrows JM, Engelke P (2020) What world post COVID-19. https://www.atlanticcouncil.org/in-depth-research-reports/report/what-world-post-covid-19-three-scenarios/

CBHI, DGHS (2017) Health infrastructure. http://www.cbhidghs.nic.in/WriteReadData/l892s/Chapter%206.pdf. Accessed 25 May 2020

Daniel MA, Daniel CD, Boyton RJ (2020) What policy makers need to know about COVID-19 protective immunity. Published: 2020-05. https://doi.org/10.1016/S0140-6736(20)30985-5

Dikid T, Chaudhary S, Goel K, Padda P, Sahu R, Kumar T, Jain SK, Singh SK, Narain JP (2020) Perspective: responding to COVID-19 pandemic: why a strong health system is required. Indian J Med Res 151:140–145. https://www.ncbi.nlm.nih.gov/pubmed/32317411. https://doi.org/10.4103/ijmr.IJMR_761_20

Fridell M, Edwin S, von Schreeb J, Saulnier DD (2020) Health system resilience: what are we talking about? a scoping review mapping characteristics and keywords. Int J Health Policy Manag 9(1):6–16. Published 2020 Jan 1. https://doi.org/10.15171/ijhpm.2019.71

Kieny MP, Evans DB, Schmets G, Kadandale S (2014) Health-system resilience: reflections on the Ebola crisis in western Africa. Bull World Health Organ 92:850

Kruk ME, Myers M, Varpilah ST, Dahn BT (2015) What is a resilient health system? Lessons from Ebola. Lancet 385:1910–1912

Olu O (2017) Resilient health system as conceptual framework for strengthening public health disaster risk management: an African viewpoint. Front Public Health 5:263. Published 2017 Sep 28. Accessed 25 May 2020. https://doi.org/10.3389/fpubh.2017.00263

Rose DA, Murthy S, Brooks J, Bryant J (2017) The evolution of public health emergency management as a field of practice. Am J Public Health 107(S2):S126–S133. https://doi.org/10.2105/AJPH.2017.303947

Singh K (2020) Editorial: Quo vadis after COVID-19: a new path for global emergency preparedness? WHO South East Asia J Publ Health 9(1):1–4. http://www.who-seajph.org/temp/WHOSouth-EastAsiaJPublicHealth911-4177001_113610. Accessed 27 Apr 2020

Swain A (2020) COVID-19 Strategy—The Swedish Model and Lessons for India. https://practiceconnect.azimpremjiuniversity.edu.in/covid-19-strategy-the-swedish-model-and-lessons-for-india/

The Economic Times (2020) Fluctuations in fresh coronavirus cases in Delhi due to varied contact tracing, other factors: Experts. https://economictimes.indiatimes.com/industry/healthcare/biotech/healthcare/fluctuations-in-fresh-coronavirus-cases-in-delhi-due-to-varied-contact-tracing-other-factorsexperts/articleshow/75180113. Accessed 29 Apr 2020

WHO (2020a) Modes of transmission of virus causing COVID-19: implications for IPC precaution recommendations: scientific brief. https://apps.who.int/iris/handle/10665/331601. License: CC BY-NC-SA 3.0 IGO. Accessed 27 Mar 2020

WHO (2020b) WHO Foundation established to support critical global health needs 27 May 2020 News release Geneva. https://www.who.int/news-room/detail/27-05-2020-who-foundation-established-to-support-critical-global-health-needs. Accessed 28 May 2020

Wulff K, Donato D, Lurie N (2015) What is health resilience and how can we build it? Annu Rev Public Health 36:361–374

Yeolekar ME, Mehta S (2008) Editorial: ICU care in India—status and challenges, vol 56. https://japi.org/april2008/E-221.pdf. Accessed 25 May 2020

Chapter 20
Resilience of Water Bodies to Outbreaks and Disasters

Vikas Chandra Goyal

Abstract Globally the countries are grappling with Covid-19 pandemic and a large scale churning is presently underway. Apart from health aspects, the society is dealing with a variety of issues to mitigate and manage this emergency. Further complicating this outbreak situation is the vagaries of the weather. Water and food security play vital role in the times of disasters and pandemic outbreaks. Our future water planning need to incorporate the critical uncertainties that pandemics and disasters bring and we should manage water in a way that increases resilience and reduces vulnerabilities for both humans and ecosystems. This chapter presents an overview of the role and the importance of the resilience of water bodies during times of outbreaks and disasters. Vulnerabilities of water bodies during outbreaks and disasters, and conservation measures needed thereof are discussed. Issues needing urgent attention of the decision makers are also highlighted.

Keywords Water bodies · Water and wastewater · Water quality · Water conservation · Resilience · Disasters · Outbreaks · Climate change · Vulnerability · Covid-19

20.1 Introduction

It is imperative to have safe water in the aftermath of a health emergency to meet domestic requirement, other WASH activities, and to resume normal farming and commercial activities. During Covid-19 crisis, World Health Organization (WHO) has been advocating water as a critical resource that can help limit the spread of virus if individuals practice personal hygiene. Large scale hospitalization and confinement in widely scattered quarantine facilities has also generated additional water demand.

V. C. Goyal (✉)
National Institute of Hydrology, Roorkee, Uttarakhand, India
e-mail: vcg.nihr@gov.in

M. K. Goyal, A. K. Gupta (eds.), *Integrated Risk of Pandemic: Covid-19 Impacts, Resilience and Recommendations*, Disaster Resilience and Green Growth,
https://doi.org/10.1007/978-981-15-7679-9_20

It is estimated that the daily water demand may increase by about 20–40 lpcd just for hand washing. For a city with population of about ten million (Chennai or Bengaluru, for example), this additional water demand can range between an extra 200–400 million litres of water per day. Moreover, mass movement of migrants from one State to another has shifted the water demand clusters in an unpredicted manner. The water supply systems will have to deal with this critical additional water demand during the summer months. In such scenarios, resilience of water bodies becomes all the more important to ensure access to safe and assured water supply.

20.2 Water Bodies as Source of Water Supply Systems

Urban water supply generally comprises surface water sources (e.g. streams, lakes, tanks), and groundwater sources (e.g. tube wells). Rural water supply systems are largely dependent on the groundwater sources (e.g. dug wells, tube wells and hand pumps). By addressing vulnerability through appropriate measures, sustainability of drinking water sources is ensured to provide safe drinking water in adequate quantity to communities in urban and rural areas. Water security can be achieved through adoption of appropriate source sustainability measures such as revival of traditional systems, enhancing storage capacity of water bodies through desilting etc., conjunctive use of surface and ground water, rainwater harvesting and recharging of drinking water sources (GoI 2019a).

Rivers and streams are considered dynamic actors on landscape, also regarded as the lifeline of human civilization (cGanga and NMCG 2019). These are an important source of drinking water supply as well as irrigation. Rivers are usually connected hydraulically to floodplain wetlands, groundwater aquifers and other water bodies. Unfortunately, most of the rivers in India are presently in a pathetic condition as their flows have dwindled and quality deteriorated over the years.

Restoration and conservation of rivers and streams is extremely important for sustenance of humans and ecology. Rejuvenated rivers and streams also help in flood management. We should start from smaller streams and drains passing through urban and semi-urban areas. The wastewater flowing through these drains can be easily treated using Nature Based Solutions (NBS) and the treated wastewater will then make the streams perennial when these drains join the stream in a lower stretch (ADB 2016; WWAP 2018). While doing so, we have to be vigilant about the quality of wastewater flowing in the drains and take appropriate measures if hazardous and/or toxic contaminants are present (Schellenberg et al. 2020). Presence of terrestrial and aquatic life forms (flora and fauna) is essential for making rivers and streams resilient. Conjunctive use approach in river basins makes rivers resilient.

Groundwater is the major supplier of drinking water (also for irrigation) in most of the areas in India. Although dug wells are prevalent in rural areas, drinking water supply in urban and peri-urban areas is done through abundance of tubewells. Lowering of groundwater levels at alarming rate is seen in most parts of India (about 1500 Blocks are categorized as drought affected, water stressed or over

Fig. 20.1 A river bank filtration site for drinking water supply

Fig. 20.2 Pond as a source of village water supply

exploited). River Bank Filtration (RBF) offers a nice NBS mechanism to make drinking water wells resilient (Fig. 20.1). Managed Aquifer Recharge (MAR) and Aquifer Storage and Recovery (ASR) are also useful technologies in making aquifers resilient.

Ponds are lifeline of the village ecosystem in India, and are of paramount importance in tackling the key issues of water security and disruptions due to disasters and climate change. Although ponds are small wetland structures, strategically located pond networks have the potential to hold water at source, recharge aquifers, and moderate flooding events (Fig. 20.2). Through denitrification and sedimentation processes, pond acts as natural filter by significantly reducing the pollution load of incoming water (Cereghino et al. 2014). Globally, 30% of the surface area of freshwater is occupied by millions of small water bodies less than

10 ha (e.g. lakes, ponds) on the earth's surface represent (EPCN 2008). According to National Wetland Atlas (2011), India has 555,557 small water bodies (including village ponds) with area <2.25 ha. As such, rejuvenation of these water bodies has large potential of achieving water security and climate resilience.

Ponds are also good source of carbon sequestration. A single 500 m^2 pond could sequester 1000 kg of carbon per annum. Ponds, lakes, tanks and other wetland forms also act as cushion/sponge during times of flooding. For such reasons, it is considered imperative to protect and conserve these small water bodies which are our shield against adverse impacts of climate change.

Another important source of water for millions of people and their livestock in the Indian Himalaya Region (IHR) is springs (NITI 2018). Both rural and urban communities in IHR depend on springs for their drinking, domestic, livestock and agricultural water needs. Springs also contribute to the base flow of major Himalayan rivers, and their contribution is possibly more than snow and glaciers. With increasing evidence that springs are drying up or their discharge is reducing throughout the IHR, springs need rejuvenation and conservation so that the flows and quality of spring water is sustained.

Climate change is threatening drinking water supplies in two major ways. First, extreme rainfall (higher intensity but in less duration), or lack of it, leads to either high runoff leading to heavier floods (and less infiltration/storage underground) or frequent droughts. Second, rising sea levels will destroy a significant proportion of the freshwater supplies in coastal regions by inundating aquifers with brackish water that is not safe for human consumption. In case of extreme events and emergencies such as cyclones, earthquakes, floods, disease outbreaks, availability of the drinking water supply poses additional challenges, which require advanced scientific planning.

20.3 Outbreaks and Disasters

A disease epidemic or outbreak is the occurrence of cases of a particular disease in excess of the expected requiring emergency control measures. The threat of communicable disease outbreaks is greater after occurrence of a disaster, especially when large populations have been displaced. Disasters (whether natural or anthropogenic) are highly complex events resulting in mass destructions of humans, livestock, environment (biodiversity), and leading to disruption of the economy and quality of life of the people. Global warming, rising populations, rapid urbanization, development activities without adequate concern for environment, enhanced human-wildlife conflicts, etc. are mainly responsible for the increasing frequency and severity of disasters. The implication of a disaster depends on its severity, vulnerability of the population, and capacity of the local community and organizations to cope up with the disaster (Singh 2020). Vulnerable sections of the population include elderly, children, pregnant women, patients with co-morbidities, migrants (especially low-income groups), and those living in high-density habitations (such as slums).

Water related disasters such as floods, cyclones, hurricanes, landslides, have a far reaching impact on sustainability of the water supply systems and affecting the quality of water. Epidemics are commonly caused by polluted water source (e.g. due to fecal contamination, toxic chemicals), lack of hygiene conditions, contaminated food, etc. (Amar 2010). Some of the diarrheal epidemics associated with disasters include cholera, leptospirosis, viral hepatitis A and E, Acute Respiratory Infections (ARI), meningitis, influenza. People staying within close proximity in crowded shelters with poor ventilation, poor hygiene, poor nutrition, limited access to medical cares, are major risk carriers responsible for spread of the disease.

Consuming contaminated water poses health hazards such as diarrhea, hepatitis, cholera, rotavirus infection, typhoid, fluorosis, trachoma, etc. As the consequence of an outbreak or a disaster, water sources often get contaminated, and unsafe water from these sources must be treated before use. In case of situations like Covid-19, water sources may be contaminated due to passage of coronavirus through wastewater. In the aftermath of a disaster or disease epidemic, the water utilities should aim to ensure supply of safe water to the affected population so that the occurrence and spread of water borne disease is contained.

A high level of preparedness is needed to maintain uninterrupted supply of safe water for survival and sustenance of humans and livestock. In such emergencies, preparedness measures to ensure adequate and safe water supply must cover all aspects of protection of water sources, monitoring of water quality, and treatment of water at treatment plants. In times of emergencies, resilient water systems help communities to cope better, but it is always desirable to conserve and maintain water resources during normal periods. Should a major virus pandemic occur, water and wastewater treatment utilities would need to respond rapidly to minimize occupational and public health risks.

Kouadio et al. (2012) have reviewed potential infectious diseases resulting from major natural disasters (e.g. floods, tsunamis, earthquakes, cyclones) and the associated risk factors. They have concluded that disasters do not carry diseases/epidemics but the after effects such as mass displacement of populations, overcrowded shelters, poor water and sanitation facilities, high exposure to and proliferation of disease vectors, and limited access to required health facilities, cause occurrence of infectious diseases.

Specific risk reduction measures are needed for different disasters. Some key intervention areas for disaster risk reduction (DRR) include proper landuse planning, emergency preparedness and early warning systems, coordinated response strategy with all relevant stakeholders, monitoring of key performance indicators, raising awareness and capacity building of local communities.

20.4 Contamination of Water Sources and Vulnerability Assessment

According to an estimate, about 61,754 MLD sewage is generated in India, of which only 37% is adequately treated (CPCB 2016). Most of this wastewater enters rivers and other water bodies. Enhanced consumption of antibiotics and antimicrobials leads to detection of Antibiotic Resistant Bacteria (ARB) and Antibiotic Resistance Genes (ARG) in municipal wastes receiving effluents from hospitals (Kumar et al. 2019). The occurrence of multi-drug resistant microbes, pathogenic viruses, metals, and Pharmaceuticals and Personal Care Products (PPCPs) in waters is impacting the sustainability of urban water supply systems (Fig. 20.3).

Nature of disaster or outbreak situation might affect the water quality by inclusion of natural pollutants such as dissolved gases and minerals, suspended impurities, PPCPs, fertilizers, pesticides and herbicides. In certain cases, water quality may be jeopardized due to various toxic chemical substances such as cyanide, mercury, lead, copper, cadmium, chromium, hydrocarbons, etc. Biological, nuclear and radiological disasters pose serious threats of contaminating water bodies by dangerous microbes or toxic materials.

Contamination of water bodies may be a two-way process. A contaminated water body may lead to disease in humans and livestock, and infected humans or livestock (through fecal route) may cause contamination of water bodies. In both cases, our effort should be to prevent contamination of water bodies. A coronavirus can impact the water supply system if it survives in human waste and infects other persons, most likely via nasal droplets. Infective human viruses often remain in wastewater treatment plant effluent even after wastewater treatment, although with reduced virus levels. If such effluent reaches water bodies, it is likely to enter the human body through drinking water supply systems.

Wastewater treatment plants receiving sewage from hospitals treating coronavirus patients, and sewage from areas of known high contamination (e.g. isolation

Fig. 20.3 Factory releasing effluent in a drain

centres), may have elevated concentrations of viruses. Surface water treatment plants are most susceptible to the contamination of coronavirus in raw water supply during, and after, an outbreak. Although many wastewater treatment plants may claim to be well equipped to remove and disinfect viruses through disinfection processes, it is important to continuously monitor the disinfection performance.

The already ongoing process of climate change is likely to make the impacts of disasters all the more complex and difficult to cope up. Many times this combination of disasters and climate change will prove a deadly combination. Both surface water and groundwater sources are deteriorating fast in India. Climate variability in terms of extreme weather events and addition of a variety of contaminants make the water sources vulnerable.

Vulnerability analysis is a planning tool to study the ability of a system to withstand, absorb and cope up the impacts of such hazardous occurrences, and for empowering and mobilizing vulnerable communities during emergent situations. Vulnerability mapping of water sources is, therefore, considered a desirable step in the water security assessment of an area. For vulnerability assessment of water sources, the traditional physical, chemical and biological water quality parameters are not adequate and the monitoring must include PPCPs indicators such as Pepper Mild Mottle Virus (PMMcV), caffeine, carbamapzine and acetaminophen. Resilience of water supply systems requires understanding of the vulnerability of water sources from various biological and chemical pollutants (including PPCPs, viruses, metals), and the associated health risks. Understanding and predicting climate change impacts on water sources add another dimension to the problem. Our effort should be to transform the vulnerability into resilience, and move from risk to resilience for a water secure future. Indigenous Traditional Knowledge (ITK) systems of local communities provide coping mechanisms for the resilience of water bodies.

20.5 Monitoring of Water Bodies

Sustainable development of water resources requires reliable data on desired spatial and temporal scales, meaning that comprehensive and long-term monitoring of water resources is a prerequisite to sound planning, designing and better management of this vital resource (water). Monitoring of water bodies is essentially required to provide the data required for evaluating the status and safeguarding the water resources from depletion, and chemical contamination arising from point sources or through diffuse emissions. Bio-chemical oxygen demand (BOD) and Coliform bacterial count (Total and Fecal Coliform) are important indicators of the extent of water quality degradation and presence of pathogens in water bodies. For exhaustive water and wastewater management planning, Schellenberg et al. (2020) have suggested monitoring of four key parameters (pH, BOD, TSS, Fecal Coliform), and information on seasonal flows and existing sewage treatment capacity at Sewage Treatment Plants (STPs).

Fig. 20.4 Real-time water quality monitoring stations of CPCB

Central Pollution Control Board (CPCB), Central Water Commission (CWC), and Central Groundwater Board (CGWB) are the three national agencies responsible for monitoring the water bodies in India. Present water quality monitoring network of CPCB under National Water Quality Monitoring Network (NWMP) comprises 4111 stations on rivers (and their tributaries), lakes, ponds, tanks, canals, creeks and groundwater (wells) in 28 States and 8 Union Territories. Water samples are analyzed for various parameters as per the Guidelines on Water Quality Monitoring (2017) issued by Ministry of Environment, Forest and Climate Change (MoEF & CC) (Fig. 20.4).

CWC maintains hydrological observation network covering all the major river basins of India, and water quality monitoring stations are established at 552 key locations. Three-tier CWC water quality laboratories are located at 318 stations on major rivers of India where all major physico-chemical and bacteriological parameters, including heavy metals/toxic parameters, are being analyzed. CGWB monitors groundwater level from a network of about 22,000 stations (mostly dug wells). Groundwater level at these stations is measured either manually or automatically. Groundwater quality is monitored using analogue/digital instruments to measure pH, electrical conductivity (EC) and temperature at well sites.

Water and wastewater operators should be skilled enough to detect and disinfect coronaviruses (or any others) in water and wastewater treatment systems. In case a major health pandemic occurs (such as Covid-19), water utilities would need to respond quickly to minimize occupational and public health risks. Monitoring of water bodies and effluent from STPs is, therefore, required to know the presence of bacteria and viruses in aquatic environment and in sewage and wastewater treatment plants. Appropriate detection methods are required to detect the presence of both enveloped enteric viruses (e.g. coronavirus) and non-enveloped enteric viruses.

Treatment of wastewater is important to contain the prospect of contamination of nearby water bodies receiving wastewater from potentially polluted sources. Monitoring is also required to check the efficacy of water treatment and disinfection processes for virus removal and inactivation. Altenburger et al. (2019) have reported strategies developed to deal with the identification of chemical contaminants, assess the impact of contamination in water bodies, and quantify cause-effect relationships amongst contaminants. To deal with water contamination monitoring issues in outbreaks and disaster situations, advanced instrumentation and sampling techniques along with availability of analytical methods can be used to obtain results with high precision and in near real-time mode.

20.6 Water Resilience

A common analogy of a resilient system could be a rubber band, which when subjected to stress and strain, ultimately has the ability to return to its original state. The shorter the 'return time', the more resilient the system (Folke et al. 2010). Ecosystems are fundamentally dynamic and, in response to disturbance (such as outbreaks and disasters) systems evolve and adapt to the changes, and may return to a new state of equilibrium. In an alternative perception of resilience, system has ability to absorb disturbances and reorganize itself into a better configuration, while still retaining its fundamental characteristics (Walker et al. 2004).

According to many social media reports, Covid-19 lockdown has a reason to celebrate. About 2 months into the nationwide lockdown (as on 24 May 2020), air and water pollution levels have shrunk and the water bodies have sort of rejuvenated. *Blessings in disguise!* According to the real-time water monitoring data of Central Pollution Control Board (CPCB), the water quality of Ganga river at most of the monitoring units located at various stretches from Uttarakhand to West Bengal was found suitable for bathing and maintaining fisheries and wildlife. But, an important point to ponder is "do we want such changes (resilience of water bodies) to happen at the cost of a pandemic?" A better way is to relook and reconsider our development paradigms, lifestyles and economic pursuits so that there is a symbiotic existence between nature and humanity.

In terms of water resilience, the resilience is expected in terms of quantity of water as well as quality of water. In times of emergencies, resilient water systems help communities to recover faster and cope better. The factors affecting water resilience include hydro-meteorological extreme events (e.g. drought, flood, cyclone, tsunami, landslide), water scarcity, water contamination (Smith 2014). In terms of disasters, communities must evolve and become resilient to shocks from floods, droughts, cyclones, etc. The first impact of climate and disaster induced changes, and extreme events is often on water cycles, impacting access to safe water for population at large. In order to deal with such situations, a dynamic and flexible system approach is preferable in operating and maintaining the water-supply systems. For example, incorporating a broad range of source options within a water supply system provides a shield against failures. For these reasons, decentralized water supply systems are generally considered more sustainable and adaptive to functional risks. Next we shall discuss what makes the water bodies resilient to disasters, extreme events, and disease outbreaks.

20.7 Protection and Conservation of Water Bodies

India has 17% of the world's population and 4% of the world's freshwater resources, making it among the top ten water rich countries. However, with India's significantly low storage capacity of annual rainfall (~8%), the country is currently designated as water stressed by Intergovernmental Panel on Climate Change (IPCC). India's dependence on monsoon amounts to about 80% of its annual rainfall in a short span of 2–3 months (July–September). The monsoon performance decides a 'boom' or 'doom' situation for the farmers as well as for the drinking water supply providers. The resilience of water bodies is also in a way dependent on the monsoon performance.

In the last few decades, waterbodies have been under unrelenting stress caused primarily by rapid urbanization and unplanned growth. Also, waterbodies are being polluted by dumping of untreated sewage effluents into them. Encroachment of waterbodies has been identified as a major cause of flash floods and water logging situations in major Indian cities such as Mumbai, Chennai, Delhi, Gurugram. Years of neglect have led to silt accumulation leading to reduction in the water storage capacity of small dams, tanks, ponds, lakes, reservoirs, etc. We have to plan strategies to establish/highlight how the ecological value of the water bodies be enhanced by including science in the existing approaches (Fig. 20.5). An integrated approach to the rejuvenation of water bodies, which includes desilting, results in healthy habitats for flora and fauna. Before taking up any restorative work of water bodies, we should study the water balance of the system providing an assessment of the supply and demand of water for various uses. A clear understanding of the seasonal variations in demand as well as supply is crucial for sustainable development of water resources.

A pilot project launched in Maharashtra's Marathwada region, and funded by a local philanthropist, is '*Gaal Yukt Shivar*', which aims to desilt over 30,000 small dams and reservoirs over a 4-year period. The Nature Conservancy (a US-based

Fig. 20.5 A protected water harvesting structure in Rajasthan (India)

NGO) is building science-based strategies and creating partnership across stakeholders such as communities, businesses and governments to achieve the objective of developing a holistic framework for desilting of dams, improving biodiversity, accounting for environmental flows, and addressing the catchment treatment measures to control excessive siltation. Such initiatives have the potential to improve water security in water scarce regions and need to be taken up at a scale so that its impact is felt at the highest levels of decision makers, and whatever gaps and barriers exist in the policies should be tackled appropriately.

Another promising development is that the city of Chennai has taken up the challenge of drought-proofing and making Chennai the 'water capital of India'. The city's Chief Resilience Officer had developed an ambitious plan of reviving 210 dead water bodies with the help of volunteers (e.g. local bodies, corporates, NGOs, RWAs). These water bodies, rehabilitated with a budget of INR 512 crore (US$ 68 million), hold a month's water supply and recharge groundwater. Harvesting of runoff water from roads into water collection structures in temple complexes is also planned.

National Institute of Hydrology, Roorkee (India) has recently rejuvenated some village ponds in Uttarakhand and Uttar Pradesh using NBS-technology namely constructed/floating wetland. The storage capacity of ponds is now enhanced three to fourfold, and their functional status (measured in terms of Trophic Status Index) is restored so that these water bodies have become effective as the instruments of water security at the village level. Once restored, these ponds contribute to the local groundwater recharge (with good quality water) and the treated pond water can also be used for limited irrigation purposes (such as agriculture, horticulture, floriculture), leading to improved livelihood for the local community (Fig. 20.6).

The Government of India (GoI) is making efforts for resilience of water bodies in India through *Jal Shakti Abhiyan* (JSA), which is a government-people's joint movement to facilitate water conservation, watershed development, afforestation and renovation of water bodies, especially in water-stressed districts (GoI 2019a). By propagating this endeavor beyond water-stressed districts to cover the entire country, our water bodies will be made climate and disaster proof. Another flagship program of GoI is its *Jal Jeevan Mission (Har Ghar Jal)* through which it proposes to provide piped water supply to all households by 2024 (GoI 2019b). This program will need strategic planning to make sure the water sources can be made resilient to disasters and climate change impacts.

Focus of JSA in 2020 should be on creating water storage capacity and conserving water quality of small water bodies in villages/*Gram Panchayats* (GPs). Desilting of these water bodies will also help in restoring soil health in nearby agricultural fields. GPs need to be trained on sustainable water conservation and development practices, which are linked to livelihood opportunities. Goyal et al. (2020) have demonstrated the utility of developing IWRM plan at district level having components of water management, land management, and livelihood management (Fig. 20.7). The local district government finds such plan as useful input for their District Irrigation Plan (DIP) or District Water Conservation Plan (DWCP) under JSA. CSR funding should be allowed to be used to support response efforts

Fig. 20.6 A rejuvenated village pond in Uttarakhand (India) using constructed wetland

(including innovations, skill development of field staff, etc.) in Covid-19 like situations.

It has been observed that wherever water bodies (e.g. ponds) are used for livelihood purposes (e.g. fishery), these are maintained by the user community. This indicates an important link between resilient water bodies and livelihood of people. Water bodies in villages are especially useful for livelihood purposes as limited source of irrigation for horticultural crops, medicinal plants, etc. Resilience of water bodies also leads to food security. National Green Tribunal (NGT) has directed that government agencies should on priority start mapping of small water bodies using GPS, monitoring the water quality and eutrophication of water bodies. In response to NGT's directive to deploy dedicated staff to take care of the water bodies, Delhi Jal Board (DJB) has recently designated an officer as SE (Water Bodies). Similar action by all ULB/Water Utilities will prove to be effective way to protect and conserve small water bodies in the country.

In yet another recent directive, NGT has asked CPCB to submit guidelines on restoration of water bodies across the country. The green panel emphasized that the protection of water bodies not only enhances the water availability for different uses, it also contributes to groundwater recharge and maintaining e-flow in the rivers. Highlighting the role of *Gram Panchayats* (GPs) (i.e. village government) and District Magistrates in taking up such activities using *MGNREGA* funds, the NGT desired that at least one pond/water body must be restored in every village apart from creation of any new pond/water body. Making water bodies resilient to outbreaks and disasters is also a vital contribution towards achieving the goals of SDG 6.

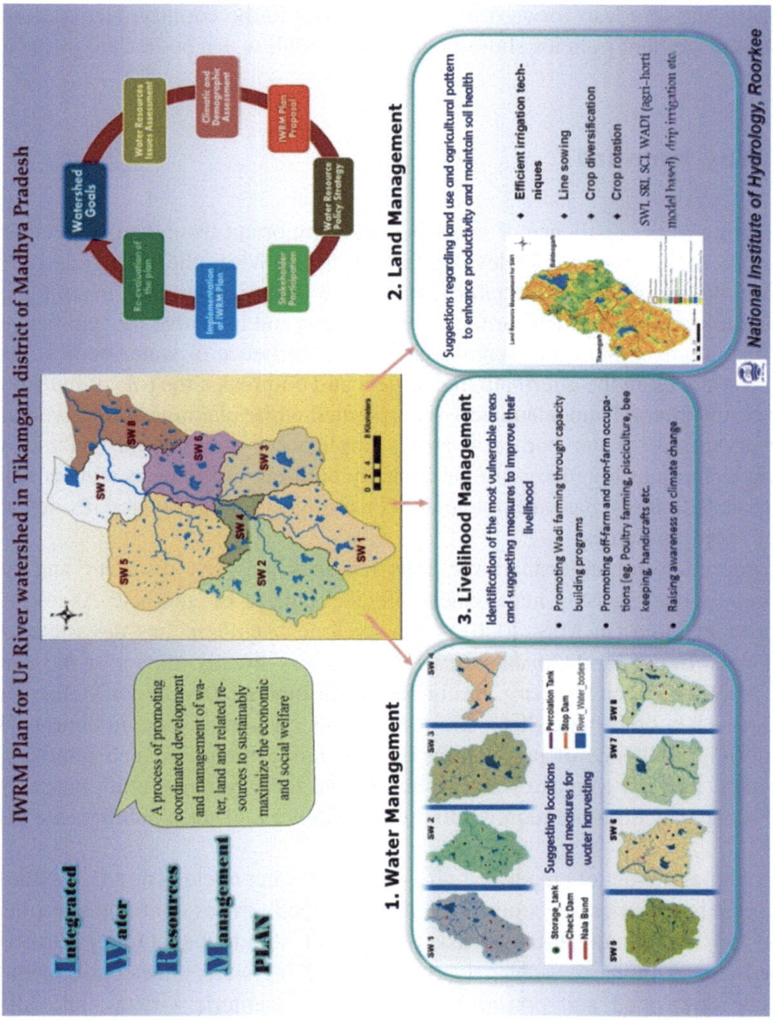

Fig. 20.7 IWRM Plan developed for Ur river watershed in Tikamgarh district (India)

As large scale movement of migrants takes place, e.g. during Covid-19, increased pressure will be felt on the water resources in villages. Rejuvenation of local water bodies assumes highest priority for water security in such situations. India needs comprehensive legislation for protection and conservation of small water bodies. Efforts are also being made in other cities for rejuvenation of lakes, tanks, ponds, streams, urban drains, etc. With corporate funding under CSR, such efforts can be upscaled leading to a leap forward for water security in the country. National/State Wetland Management policies should ensure that small water bodies are protected.

20.8 Way Forward

Understanding of the resilience of water bodies is important from the perspective of disaster and climate compatible development in India. While the ecological role of small water bodies is now well established, science-based information in simple and understandable form is needed for their conservation and to bridge the gap between science and practice. With the increasing need for resilience of water bodies, made even more complex in the aftermath of disasters and outbreaks, the role of hydrology needs to be understood and adequately incorporated while planning for rejuvenation and conservation of water bodies in India. Application of advanced analysis and modelling techniques together with the capacity building of the users and sensitization of the stakeholders, has vast potential of ensuring sustainability and resilience in the water supply schemes.

It is important to understand the inter-linkages between disasters and how responding to one disaster may exacerbate the impact of another. As recently experienced in India (in May 2020), super cyclone '*Amphan*' has devastated the States of West Bengal and Odisha when Covid-19 was already impacting the whole country. Another cyclone '*Nisarg*' has impacted (in June 2020) areas of Maharashtra and Gujarat when both these States are reeling under severe disease infection. This needs building resilience of populations and preparing them for such intersecting disaster risks. In order to make water resources and food systems future proof from multiple hazards, actions are required to build resilience and sustainability into these systems.

The potential impact of enveloped viruses on the water cycle is not much studied and this must be an emerging study area. As enveloped viruses continue to replicate (mutate) and evolve, quantitative risk assessment should be a top priority for enveloped viruses in wastewater, irrigation and drinking waters. Detection methods need to be developed and optimized for enveloped enteric viruses, and utility operators need to be trained on these technologies. Surveillance is essentially required in areas affected by a disaster such as hospitals and clinics, isolation centers, shelters and other locations where victims and patients are treated. In order to mitigate the discharge of PPCPs in hospital effluent, it may be a good idea to install Natural Treatment Systems, such as constructed wetlands, for secondary/tertiary treatment at STPs receiving wastewater from hospitals.

The Covid-19 pandemic has reinforced the critical need for access to safe and sustainable water supply for all, and the water utilities have a great challenge to provide simple and effective mechanisms for the requirement. Operation and maintenance of water supply systems is the responsibility of State government agencies in India. As per standard procedure, the water utilities are required to repair and disinfect immediately after a disaster or disease outbreak situation. Also, the agencies should make efforts for training and capacity building of the line department officials. The training programs should be organized on periodical basis for the officials and staff of the line departments covering the topics of planning and working for management of disaster situations. Training in meticulous handling of disinfection procedure is crucial as negligence or lapses could be fatal.

As climate change impacts combine with outbreaks (Covid-19), long-term disaster-proof and climate smart solutions should be planned. Health pandemics (such as Covid-19) must be mainstreamed in disaster preparedness planning to ensure resilience of humans and ecosystem. UN Office for Disaster Risk Reduction (UNDRR) is emphasizing on planning systemic risk management. Risk assessments are useful in apprising decision makers to plan, prepare for and respond to hazards and outbreaks. Sendai Framework (2015–2030) includes biological hazards, which are quite frequent now and are posing significant risks to humans and livestock. It is highly recommended that the National Disaster Risk Reduction (NDRR) strategies should include biological hazards, such as Covid-19.

Scenario planning exercise is needed to study the potential impacts of biological and chemical pollutants on water bodies in case of disasters and pandemic outbreaks. Also important is to explore the scenario of water vulnerability and resilience based on projected population growth, population migration, land-use changes, and impacts due to climate change. It would be pertinent to study scenarios where communities, after the learnings of Covid-19, start using less antibiotics and pharmaceuticals resulting in lesser loadings of PPCPs. Increasing use of NBS for wastewater treatment is also worth exploring.

Also important in such situations is to create awareness among the community on the risk of contamination of drinking water sources and the resultant health risk of the human and livestock population. In order to avoid the catastrophic impacts of rumors in such situations, people need be made aware in easily understandable forms of the benefits of consuming safe water from protected sources or after proper treatment. Functions of water bodies, impacts of climate change, their role during disasters and outbreaks, and how to make them resilient through conservation measures should be included in the higher education curriculum and research agendas.

With life-changing experience of Covid-19, we may affirm that 'back to normal' is not good enough. Rather, we need to address fragilities, vulnerabilities and recover to a resilient system that should be our new normal. India will now have to drastically improve not only its health infrastructure but also its WASH infrastructure, and make its water bodies resilient to deal with outbreaks and disasters.

References

ADB (2016) Nature-based solutions for building resilience in towns and cities: case studies from the Greater Mekong Subregion. Asian Development Bank, Mandaluyong City, Philippines, p 2016

Altenburger R et al (2019) Future water quality monitoring: improving the balance between exposure and toxicity assessments of real-world pollutant mixtures. Environ Sci Eur 31:12

Amar PK (2010) Ensuring safe water in post-chemical, biological, radiological and nuclear emergencies. J Pharm Bioallied Sci 2(3):253–266

Cereghino R, Boix D, Cauchie H-M, Martens K, Oertli B (2014) The ecological role of ponds in a changing world. Hydrobiologia 723:1–6

cGanga and NMCG (2019) River restoration and conservation (a concise manual and guide). Centre for Ganga River Basin Management and Studies, IIT Kanpur, 35 pp

CPCB (2016) CPCB bulletin, vol 1, July 2016, New Delhi

EPCN (2008) The pond manifesto. European Pond Conservation Network, 20 pp

Folke C, Carpenter SR, Walker B, Scheffer M, Chapin T, Rockström J (2010) Resilience thinking: integrating resilience, adaptability and transformability. Ecol Soc 15(4):20

GoI (2019a) Jal Shakti Abhiyan—a water conservation campaign, July 2019

GoI (2019b) Operational guidelines for the implementation of Jal Jeevan Mission (Har Ghar Jal), December 2019

Goyal VC, Garg A, Patil JP, Thomas T (2020) Formulation of integrated water resources management (IWRM) plan at district level: a case study from Bundelkhand region of India. Water Policy 22:52–69

Kouadio IK, Aljunid S, Kamigaki T, Hammad K, Oshitani H (2012) Infectious diseases following natural disasters: prevention and control measures. Expert Rev Anti Infect Ther 10(1):95–104

Kumar M, Chaminda T, Honda R, Furumai H (2019) Vulnerability of urban waters to emerging contaminants in India and Sri Lanka: resilience framework and strategy. APN Sci Bull 9 (1):57–66

National Wetland Atlas (2011) SAC, ISRO; sponsored by Ministry of Environment and Forests, Government of India

NITI (2018) Report of working group I inventory and revival of springs in the Himalayas for water security. NITI Aayog, New Delhi, 70 pp

Schellenberg T, Subramanian V, Ganeshan G, Tompkins D, Pradeep R (2020) Wastewater discharge standards in the evolving context of urban sustainability—the case of India. Front Environ Sci 8:30. https://doi.org/10.3389/fenvs.2020.00030

Singh Z (2020) Disasters: implications, mitigation, and preparedness. Indian J Public Health 64:1–3

Smith HM (2014) Understanding resilience: implications for the water sector. In: Grafton RQ, Wyrwoll P, White C, Allendes D (eds) Global water: issues and insights. ANU Press, The Australian National University, Canberra, Australia

Walker B, Holling CS, Carpenter SR, Kinzig A (2004) Resilience, adaptability and transformability in social-ecological systems. Ecol Soc 9(2):5

WWAP (United Nations World Water Assessment Programme)/UN-Water (2018) The United Nations world water development report 2018: nature-based solutions for water. UNESCO, Paris

Chapter 21
Strategy in Water Resources Sector in India Post COVID 19

Avinash Mishra

Water is an essential resource for the survival of mankind. Managing and conserving water resources has always been seen a key to any successful civilizations as is evident from the prominent famous ancient towns/cities which have grown on the banks of Rivers i.e. London, Paris, Budapest, Lahore, New Delhi, Cairo, Dresden etc. Various civilizations flourished around the world in the river valleys such as Mesopotamia, Nile, Tigris, Euphrates and Yellow River in China. The Indus Valley Civilization covering parts of north India had flourished under canal irrigation system. The Vedas and other ancient Indian scriptures make references to wells, canals, tanks and dams. In south India, the perennial irrigation commenced with construction of the Grand Anicut by the Cholas as early as second century. The entire landscape in the central and southern India is covered with numerous irrigation tanks constructed many centuries before the beginning of the Christian era. Excellent water supply and proper drainage system were some of the prominent features of these cultures.

Prominently, water management has always been seen as a tradition in India. India boasts a rich and precious knowledge of various water harvesting structures specific to geography and topography. The country has over 45 major traditional harvesting systems, where different kinds of tanks and ponds account for almost 30% of structures (DTE 2020). One of the oldest water harvesting systems is located near Pune in the Western Ghats. Besides water harvesting structures in arid and semiarid region, India also boasts an ingenious system of Bamboo drip irrigation system in rainfall rich region of Meghalaya where about 18–20 L of water collected by tapping stream and spring water by using bamboo pipes (CSE 1997). The traditional knowledge and water systems played an important role in maintaining

A. Mishra (✉)
NITI Aayog, Government of India, New Delhi, Delhi, India
e-mail: amishra-pc@gov.in

M. K. Goyal, A. K. Gupta (eds.), *Integrated Risk of Pandemic: Covid-19 Impacts, Resilience and Recommendations*, Disaster Resilience and Green Growth,
https://doi.org/10.1007/978-981-15-7679-9_21

and restoring ecological balance, acting as a source of drinking water, groundwater recharge, flood control, and also provide livelihood opportunities to people.

Administratively, the Provision of clean drinking water is a mandate of the Government of India as per the Constitution of India, with Article 47 conferring the duty of providing clean drinking water and improving public health standards. Sustainable Development Goal-6, ensuring access to water and sanitation for all, also pushing countries to provide reliable, potable and affordable service of water to all. However, for millions of people water is still a cause of constant worry in a context as water tables are constantly falling and water quality rapidly diminishing. Rising population, incessant urbanization and changing climatic trends have also created a pressure on existing water resources. It had been estimated that municipal or domestic water demand in India would see a higher growth trend in near future, where, by 2035, the country needs to be prepared for such grave situation of rising population and reducing resource availability in urban India.

Along with this, the present crisis of worldwide outbreak of Corona Virus and collapsing of the World Economy has brought us to the point where we need to bring some serious changes in the patterns of our pre-existing notions of growth. Growth and development are not the words which we can afford to use in describing the economy of the Nation but they have multi-faceted aspects including social, political, institutional wellbeing. While talking about growth we cannot ignore the existence of the dreadful phrases like water scarcity and climate change. In order to achieve the sustainable development in the country, we have to not only look into the problems of these concepts but also have to find out the solutions for them which may result in overhauling of the present mode of operations in water sector.

21.1 COVID 19: Pandemic

The understanding of the CoVID-19 virus largely comes from the last outbreaks of virus in the similar fashion. The coronavirus family is not new to the world. The outbreak of SARS and MERS in 2003 and 2012 respectively were part of the coronavirus family. However, the extent of infection, mortality rate, geographical spread and the affected age group majorly differs in SARS and MERS epidemics. The SARS-COV-2 is an advanced and unknown variation of the coronavirus.

The presence of coronavirus in water has always been a matter of concern. It is believed that the CoVID-19 virus cannot be transmitted through water and the disinfectants have the potential to destroy the virus in water (WHO 2020). However, recent traces of virus found in non-potable water supply in some European nations and the United States has aroused the curiosity if virus can sustain in water.

Certain studies decoding the characteristics of coronaviruses concur that coronaviruses are less stable in water where inactivation of virus in water highly depends on temperature, level of organic matter, and presence of antagonistic bacteria. However, the presence of organic matter and suspended solids in sewage wastewater while collecting faecal matter from large number of households makes wastewater more susceptible to virus. More specifically, in the 2003 SARS outbreak, the virus was detected in untreated sewage for up to 14 days. In all, studies suggest that higher solids do provide protection for coronaviruses in water but simple methods of water purification practices such as chlorination may be sufficient to inactivate coronaviruses. Furthermore, the conditions that could facilitate faecal oral transmission are still not clear (Mishra and Sharma 2020).

As India is battling this crisis, water scarcity can pose a serious threat to policymakers. Due to increased awareness, people are washing their hands about 5–7 times a day and disinfecting their houses at least once in 2–3 days. Due to this, average water consumption has increased nearly 1.5 times. For example, in Delhi, the average water consumption varies from 150 lpcd to 175 lpcd; now, it will accelerate. With our already strained resources, it will be difficult for service providers to provide water in adequate quantum to everyone. Moreover, a great volume of disinfectants is being used, which creates risks of leaks. The composition of grey and wastewater from normal households has changed due to frequent use of disinfectants. Safe disposal of wastewater will be another issue for India, which has a deplorable sewer network (Mishra 2020).

The presence of coronavirus in wastewater appears to be a threat urging facilities to take precautionary steps while dealing with wastewater. World Health Organization has also given certain recommendations for water, sanitation and hygiene measures for patients and caregivers to avoid infection risks. WHO has suggested using standard and well maintained plumbing system with sealed bathroom drains to prevent aerosolized faecal matter to enter into ventilation system. The faulty plumbing system and poorly designed air ventilation system contributed to the spread of SARS epidemic in a high-rise building in Hong Kong in 2003 affecting over 300 people. Certain hygiene practices are strictly need to be maintained to avoid any further spread of virus including hand washing hygiene, use of PPE and safe and treated disposal of affected sewage and wastewater. In this context and present scenario, India needs to take certain structural and reformative measures to be pandemic and crisis ready in future (Mishra and Sharma 2020).

21.2 Structural Reforms in Current Scenario

The outbreak of such Pandemic calls for adequate water supply for life, health, hygiene and continuance of economic activity. This requires huge investment and the reforms to optimize the uses of water. The need for certain reforms pertaining to water resources is augmented by the current challenging circumstances. It is also mandated in the Article 47 of Indian constitution, conferring the duty of providing clean drinking water and improving public health standards. Therefore the Government including State Governments need to reorient themselves for undertaking investments and reforms in the sector. As explained above water is not required only for drinking purposes, health and hygiene but also for farms to ensure food security and in industries for economic growth, for environment to maintain the ecosystem. Some of the reforms those may be needed to pace the development back on growth trajectory are:

21.2.1 Introduction of Water Markets and Basin Wise Water Management

Considering fresh water, a finite and vulnerable resource, Dublin Principles presented at Earth Summit in 1992 also recognized that Water is an economic good and failure to understand the economic value of water led to the exploitation of resource. In times of crisis where every drop of water is precious, there is a dire need to stimulate and trade the consumption of water among different stakeholders. There is need to introduce water markets (may firstly at the one basin level) to make more productive use of water and contributes to sustainable water management. The most successful water markets are found in Murray Darling Basin of Australia where Water trading has become a vital business tool and source of additional income for many irrigators (Murray-Darling Basin Authority 2020) (In India Maharashtra has also tried to implement this). This form of trading allows water users to buy and sell water in response to their individual needs. Under water markets, during the year, water is distributed by the basin authorities (or 'allocated') against entitlements in response to factors such as rainfall and storage levels. The entitlement holder can make the effective and efficient use of the water and sell rest of the quantity to the entities which is using more than their own allocation of water. Water markets create incentives for water to be moved to higher-value uses. The similar trading can be opted for treated waste water. But one of the pre-requisite for such trading is strong basin wise management of water which can be done by Water Regulatory Authority in each Basin (Murray-Darling Basin Authority 2020).

21.2.2 Making Water as Part of Economic Development (Priority Sector for Investment)

Improved water supply and sanitation and improved water resources management boost countries' economic growth and contribute greatly to poverty eradication. Investing in water is good business—improved water resources management and improved water supply and sanitation contributes significantly to increased production and productivity within economic sectors. Studies have revealed that the benefit-cost ratio (BCR) is significantly greater than 1, recording values in developing regions of between 4 and 32 for the water Millennium Development Goals (MDG), between 5 and 46 for the WS&S MDG and universal basic access, and between 5 and 41 for universal basic access with water disinfection at the point of use. The benefit-cost ratio for regulated piped water supply and sewer connection ranges between 2 and 12. Therefore it makes a strong case to increase the budget allocation to water and sanitation sector and along with agriculture, manufacturing and services sector, it should be focused as the priority sector for investment (Hutton 2007).

Also Government can promote Public Private Partnership in Water Sector by providing risk mitigation to long-term investment projects where it would result in more appropriate allocation of risks and their associated returns. India can take inspiration from USA which has initiated various steps in this direction. In United States, state revolving funds provide examples of a sustainable infrastructure financing model. Also the Water Infrastructure Finance and Innovation Act (WIFIA) in the United States established a new financing mechanism for water and wastewater infrastructure projects to be managed by the Environmental Protection Agency. The Act provides low interest rate financing for the construction of large dollar-value infrastructure (at least USD 20 million) of national or regional significance (OECD 2016). Credit assistance can be in the form of loans or guarantees. The programme attempts to fill a perceived gap left open by the State Revolving Funds by providing subsidized financing for large projects Similarly, public money can be used to cover parts of the risks that private financiers (debt or equity) are unable to take.

21.2.3 Institutional Mergers (CWC and CGWB)

Different aspects of water are dealt by different organizations with overlapping powers and functions. As a result of which when on ground implementation is needed, there are multiple players to take responsibility but none to implement the programme. Therefore it is highly needed to merge the institutions like Central water Commission and Central Groundwater Board so that the subject of water can be dealt at the common platform with full efficiency and without any delays.

21.2.4 Pollution Tax as the Remedy to Decrease Pollution in Water Bodies

The cost of water security has to be distributed to different stakeholders and the entities and communities which are harming the resources have to pay for their right to being polluter. We have to introduce the set water pollution charges for surface and groundwater use, and pollution or charges for wastewater discharge, at a sufficient level to have a significant incentive effect to prevent and control pollution and enhance water use efficiency. The pollution tax should be regarded as the part of Extended Producer Responsibility (EPR). Pollution taxes can lead to significant investment in pollution abatement and technological innovation, thereby lowering the overall cost to society of meeting environmental targets. Apart from this, these taxes or charges not only reduce the pollution and enhances the quality but also provide revenue to the government which can be used to further abatement of pollution.

21.2.5 *Expedite the Enforcement of Pending Legislations*

(a) **Inter-State River Water Disputes (Amendment) Bill, 2017** amends the Inter-State River Water Disputes Act, 1956. The Act provides for the adjudication of disputes relating to waters of inter-state rivers and river valleys. Under the Bill, when a state puts in a request regarding any water dispute, the central government will set up a Disputes Resolution Committee (DRC), to resolve the dispute amicably. The DRC will comprise of a Chairperson, and experts with at least 15 years of experience in relevant sectors, to be nominated by the central government. It will also comprise one member from each state (at Joint Secretary level), who are party to the dispute, to be nominated by the concerned state government. The DRC will seek to resolve the dispute through negotiations, within 1 year (extendable by 6 months), and submit its report to the central government. If a dispute cannot be settled by the DRC, the central government will refer it to the Inter-State River Water Disputes Tribunal. This Tribunal can have multiple benches. All existing Tribunals will be dissolved, and the water disputes pending adjudication before such existing Tribunals will be transferred to the new Tribunal.[1] The Bill also provides that the central government will appoint or authorize an agency to maintain such data bank (PRSIndia 2019).
***Current Status*:** Recommended by Loksabha for approval of the Rajya sabha.

(b) **River Basin Management Bill, 2018** proposes optimum development of inter-State rivers by facilitating inter-State coordination ensuring scientific planning of land and water resources taking basin/sub-basin as unit with unified perspectives of water in all its forms (including soil moisture, ground and surface water) and ensuring comprehensive and balanced development of both catchment and command areas. The draft Bill proposed to establish 13 River Basin Authorities for various river basins of the country. It is expected that enactment of the proposed legislation would result in optimum integrated development and management of inter-State River waters with basin approach and will result in change of environment from the one of conflicts to that of cooperation.
***Current Status*:** circulated for state consultation.

(c) **Dam Safety Bill, 2019** provides for the surveillance, inspection, operation, and maintenance of all specified dams across the country. These are dams with: (1) height more than 15 m, or (2) height between 10 m and 15 m and satisfying certain additional design conditions such as, reservoir capacity of at least one million cubic meter, and length of top of the dam at least 500 m. It constitutes

[1] https://www.prsindia.org/billtrack/inter-state-river-water-disputes-amendment-bill-2019

two national bodies and two state bodies whose functions include implementing policies of the National Committee, providing technical assistance to State Dam Safety Organisations (SDSOs), resolving matters between SDSOs of states or between a SDSO and any dam owner in that state, surveillance, inspection, and monitoring the operation and maintenance of dams within their jurisdiction. Dam owners as specified in the bill will be responsible for the safe construction, operation, maintenance and supervision of a dam. They must provide a dam safety unit in each dam. This unit will inspect the dams: (1) before and after monsoon season, and (2) during and after every earthquake, flood, calamity, or any sign of distress. Functions of dam owners include: (1) preparing an emergency action plan, (2) carrying out risk assessment studies at specified regular intervals, and (3) preparing a comprehensive dam safety evaluation through a panel of experts[2] (PRSIndia 2019).

***Current Status*:** Recommended by Lok Sabha for approval of the Rajya sabha.

(d) **Model Groundwater Bill,** 2018 provides a new template that states can use to adopt legislation capable of addressing the fast-increasing groundwater crisis faced by many states. It aims at addressing India's Groundwater problem: overdraft, delpetion, contamination to waterlogging. It states that the protection, conservation and regulation of groundwater shall be undertaken in such a way that it is integrated with the protection, conservation and regulation of surface water resources on a watershed basis, land and forest. The bill proposes a new regulatory framework based on the recognition of the unitary nature of groundwater pool, the need for decentralized control and the necessity to protect water at the aquifer level.

***Current Status*:** pending at the level of MoWR

21.2.6 Rainwater Harvesting

Capturing water available in the system and utilizing it effectively is the need of the hour. Assisted and protected water storage structures allow community to have at least basic access to water supply, reducing water related diseases such as Diarrhoea, Typhoid. Furthermore, apart from access to drinking water, storage also serves multiple purposes such as irrigation, water for livestock and many times also serve as an option for livelihood. With increasing incidences due to climate change, integrating water storage structures in landscapes in a planned and systematic manner also helps to create a "water buffer" reducing vulnerability to drought and seasonal variations in rainfall (FAO 2020). Watershed management is a holistic approach conserving varied natural resources from ridge to valley integrating different stakeholders in the process of capturing every falling raindrop for several

[2]https://www.prsindia.org/billtrack/dam-safety-bill-2019

tangible and intangible benefits including drinking water. Several government schemes such as MGNREGA, Har Khet Ko Pani (PMKSY), IWMP, Jal Jeevan Mission and other state schemes such as Kapil Dhara Yojna of Madhya Pradesh, Sujlam Suflam of Maharashtra, Neeru-Chettu of Andhra Pradesh and many more can be converged for ensuring sustainable access to water for domestic purposes.

Further, incessant urbanization has created an undue pressure on water resources with unsustainable use of water. Urban India consistent with rising population of India has nearly doubled since 1950, reaching up to 34% by 2018, where, as projected India will add another 416 million urban dwellers to cities by 2050 (UN 2019). Rapid population growth and the development of economic activity are accelerating the rate of resource consumption which has dissipated the basic necessities required for a living such as water for drinking and other domestic purposes. It has been estimated that municipal or domestic water demand in India will see higher growth trend in near future. For urban areas, Roof top Rainwater harvesting (RWH) can be seen as a simple and apt technique that can be used to collect rain water to supplement the municipal water available. It is a good alternative which not only conserve water and help in recharging ground water but also arouse sense of ownership on the water collected introducing water stewardship among residents. Legislatively, several state governments have mandated all government institutions, commercial complexes and residential high rise building to have rainwater harvesting.

21.2.7 Wastewater Treatment

Another major reform needed in water sector is to promote and build the system of wastewater. Degradation of natural resources, rising population and current pandemic makes it even more important to devise a circular path of water by using and reusing water in the system. Besides drinking water, there are innumerable non-potable uses of water in a domestic household which can be taken care of by treated wastewater. Wastewater sector; collecting, treating and reusing water prove to be an important sector in the present scenario. It requires an immediate and urgent attention from all the stakeholders including government, civil society organizations and communities. Currently, India generates approximately 61,948 MLD of sewage against the treatment capacity of 23,277 MLD i.e. 37% of wastewater generated only (CPCB 2015). Hence, there is an urgent need to promote and push the economy to inculcate the habit of reusing, recycling and treating wastewater in the system.

This will not only help in conserving and consuming water efficiently but can also create job opportunities in wastewater sector. Presently, the job creation is really important to push start the economy. Besides conventional jobs, wastewater treatment at decentralized levels can open new endeavors for entrepreneurs in the economy.

Besides this, wastewater also proved to be a good measure to understand and assess the extent of the outbreak of pandemic in the community. Several researchers around the world have proved that testing and monitoring of wastewater can help in identifying pandemic outbreak in early stages. This technique has been previously used to detect polioviruses and measles. Hence, there is a need to develop a proper system to collect and treat wastewater to identify the outbreak and treat sewage water to contain the spread of virus.

21.2.8 Water Purification

Water contamination is an important concern that poses huge health burden on community. As stated in Agenda 21, UNCED, 'An estimated 80% of all diseases and over one-third of deaths in developing countries are caused by the consumption of contaminated water and on average as much as one-tenth of each person's productive time is sacrificed to water related diseases' (UNCED 1992). Globally, 1.5 million children under five die and 200 million days of work are lost each year as a result of water-related diseases. It is estimated that around 37.7 million Indians (Khambete 2019) are affected by water borne diseases annually. While 'water borne diseases' such as Diarrhoea continue to take a heavy toll, 66 million Indians (CGWB 2014) are at risk due to excess fluoride and ten million (Ghosh and Singh 2009) due to excess arsenic in groundwater.

With limited availability of water for consumption, purifying and making water fit for drinking and other purposes is another way of reform to augment the supply of clean and potable water.

The prolonged, persistent and post-limit exposure to toxins may lay a heavy toll on health of human and livestock population. Identification and removal of contamination is highly important as water borne diseases are directly proportional to water quality. There is a wide range of microbial and chemical constituents of drinking water that can cause adverse human health effects. The idea is to identify and treat hotspots of contaminated water protecting population from the exposure. Along with Jal Jeevan Mission, ensuring piped water supply in rural households, Department of Drinking water and Sanitation, can also converge resources to purify and treat water from certain toxins. The idea is to leverage communities and private firms offering them a platform to invest in water purification plants at decentralized levels. Additionally, the advisories issued for ensuring safe drinking water during Coronavirus pandemic by Government of India also advised to issue sufficient field test kits to the villagers trained in their use to do periodic testing of water supplied and alert all concerned in the event of any contamination.

21.2.9 Water Metering

Water use largely depends on varied socio-economic and geographical factors such as level of affluence, caste, access to water source and more. It has been observed that only 10% of Bengaluru residents consume an average of 342 lpcd in comparison to 50% domestic consumers using less than 90 lpcd (Malghan 2017). Moreover, a study conducted to understand water consumption patterns in major Indian cities also revealed that socio economic class wise variations largely affects water consumption patterns where class defined as 'very poor' consumes only 78.9 lpcd in comparison to 'upper' class which consumes 102.1 lpcd (Shaban and Sharma 2007). To avoid and manage these stark variations in the consumption of water, the water utilities should promote water metering systems. Furthermore, as the pandemic has increased the demand of water for several other conventional uses of water in the households. Water metering can also be introduced at residential and commercial levels for relevant stakeholders. Generally, water utilities install bulk water meters for the residential or commercial buildings, charging uniform and nominal rate from the users. The system of bulk water meters disincentives the judicious water user over the carefree user. Here, an introduction of individual water meters can help water utilities to keep a check on water consumption of individual water users and charging them as per their water consumption. The system of water meters can also promote a sense of ownership among water users while promoting the idea of water conservation.

Besides this, the unrealistic tariff structures, poor collection efficiency and poorly directed subsidies (MoWR 2014) also pose a challenge for water utilities over revenue generation. Further, the lack of funds also restrains water utilities to invest in preventive maintenance of infrastructure. This leads to several anomalies in the system resulting into decrepit and ineffective water infrastructure. Therefore, water metering can also help water authorities to set a realistic and effective tariff structure and charge the premium fees from the imprudent users punishing them for water wastage.

In addition to this, while Jal Jeevan Mission is promoting piped water supply in rural households, early and apt choice of incorporating the installation and use of water meters for the use of piped water in rural households will be a futuristic decision keeping in mind the rise in water demand by 2030.

21.2.10 Leakage Management

Under water supply management, loss due to non-revenue water can be considered as a threat to water scarce economies of the world. Around the world, non-revenue water accounts for 25–50% of the total water supply (DWF 2016). More Specifically, Indian Utilities face huge distribution losses on account of the non-revenue water where it has estimated that about 40–70% of water distributed, lost on account of leakages, unauthorized connections; billing and collection inefficiencies (World Bank 2012). Moreover, the lack of relevant data at utility levels for non-revenue water also undermines the importance of issue. Reducing non-revenue water losses has considerable benefits from efficient management of water resources to revenue generating water utilities. However, with large water supply network, some level of technological intervention is required to support the leakage management. Real time monitoring of water supply infrastructure, by using GIS tools, installing smart devices and telemetry, offer utilities the scope to take timely action in repairing the leakages and finding out the illegal connections easily, thereby saving millions of litres of water along with time and energy. Leakage management needs expertise and funds to effectively take up suggested activities. However, utilities or water service providers lack relevant know-how and funds required for leakage management. In this scenario, seeking private sector participation can be seen as an option to distribute the financial burden and to integrate the expertise on the management.

21.2.11 Lake and Pond Rejuvenation

Urban and rural India has large number of lakes and ponds suffering from mis-management of water bodies. Incessant urbanization and accelerating population have also aggravated the problem of infringement and intrusion of water bodies. In 1960s, Bangalore had 262 lakes where the city now holds only 10 lakes. Similarly, New Delhi had 611 water bodies from which 274 have already dried up and as many as 190 have been lost forever and cannot be revived Similarly, Ahmedabad destroyed about 65 lakes out of 137 lakes listed in 2001. Hyderabad lost approximately 3245 hectares of its wetland. Lucknow with 964 tanks and ponds in 1952 has lost about 50% of these valuable water sources. These local ponds and lakes act as a sponge and thermo regulators, helping areas to accumulate rainwater, enhance groundwater and regulate micro-climate. However, the extinction and disappearance of water bodies will reduce economic and environmental services of lakes, making cities more vulnerable to flooding and more susceptible to climate change. Therefore, there is a dire need to rejuvenate, renovate and reclaim these valuable water bodies. There are few but relevant examples of lake and pond renovation which can be revisited and reconsidered while taking suitable steps for Lake Rejuvenation (NITI Aayog 2018; WDO 2018). Furthermore, Ministry of Urban Development has also issued an Advisory on Conservation and Restoration of Water bodies in Urban Areas (CPHEEO 2013) which should be followed to revive urban lakes and ponds.

21.2.12 Conclusion and Way Forward

COVID 19 could be an opportunity to address water related problems including optimum use of water. In India, dilapidated sewage system with open drains and untreated sewage is not ready to treat such viruses in water. Urgent and immediate response is required to build an infrastructure for wastewater treatment at local and centralized levels to be ready for such pandemics. With the Jal Jeevan Mission and Swachh Bharat Abhiyan, India has initiated some steps of creating an infrastructure for piped water supply including strengthening of solid and liquid waste management systems.

The Water Use Efficiency (WUE) in Indian agriculture ranges about 30–40%, which is very low in the world, against 55% in China. This requires paradigm shift in conservation and in agriculture policies, which should lead to saving of water, fertilizer and energy resulting in crop diversification and equitable distribution of resources. The objective of Pradhan Mantri Krishi Sinchai Yojana - Per Drop More Crop is the use of innovative water saving technologies in irrigation called Micro Irrigation. Micro irrigation saves irrigation water from 40 to 60% (Bahinipati and Vishwanathan 2019), conserves energy from 10–17%, reduces fertilizers consumption from 15 to 50%, and decreases labour cost of about 30–40% (FICCI 2016). However, the war has begun and the country needs to be more disaster and pandemic ready as early as possible. The Reforms and suggestion need to be implemented to reap the economic output of water and put the country on trajectory of Development.

Disclaimer The views and opinions expressed in this article are those of the authors and do not necessarily reflect the official policy or position of NITI Aayog, Government of India.

References

Bahinipati CS, Vishwanathan PK (2019) Micro-irrigation and groundwater use. India Water Portal. https://www.indiawaterportal.org/articles/micro-irrigation-and-groundwater-use

CGWB (2014) Concept note on geogenic contamination of ground water in India. Central ground Water Board, Ministry of Water Resources, Faridabad

CPCB (2015) Inventorization of sewage treatment plants. Central Pollution Control Board, New Delhi

CPHEEO (2013) Advisory on conservation and restoration of water bodies in urban areas. Ministry of Urban Development, New Delhi

CSE (1997) Northeastern Hill ranges. In: Agarwal A, Narain S (eds) Stae of India's Environment 4. Centre for Science and Environment, New Delhi, pp 64–65

DTE (2020) State of India's environment. Centre for Science and Environment, New Delhi

DWF (2016) Reducing urban water loss. Danish Water Forum

FAO (2020) Water harvesting and storage. Land & Water. http://www.fao.org/land-water/water/water-management/water-storage/en/. Accessed 11 Feb 2020

FICCI (2016) Strategy paper on future prospects of micro irrigation in India-accelerating growth of Indian agriculture: micro irrigation an efficient solution. FICCI, New Delhi

Ghosh N, Singh R (2009) Groundwater arsenic contamination in India: vulnerability and scope for remedy. e Special Session on Ground Water in the 5th Asian Regional Conference of INCID

Hutton GE (2007) Global cost-benefit analysis of water supply andsanitation interventions. J Water Health:481–502

Khambete AK (2019) When water kills. Retrieved March 5, 2020, from India Water Portal https://www.indiawaterportal.org/articles/when-water-kills

Malghan DS (2017) Domestic water consumption in Bengaluru and scope for demand-side management. In: Rethinking Bengaluru's water, wastewater and lakes. ATREE, Bengaluru

Mishra A (2020) Fighting coronavirus & saving water. The Financial Express

Mishra A, Sharma A (2020) Wastewater and the spread of novel coronavirus. The Times of India Blogs

MoWR (2014) Guidelines for improving water use efficiency in irrigation, domestic and industrial sectors. Ministry of Water Resources, New Delhi
Murray-Darling Basin Authority (2020) Water markets and trade. Retrieved from https://www.mdba.gov.au/managing-water/water-markets-and-trade
NITI Aayog (2018) Lake restoration: two successful models of lake restoration in Rajasthan and Karnataka. NITI Aayog, New Delhi
OECD (2016) Water, growth and finance. OECD, Paris
PRSIndia (2019) Inter-state-river-water-disputes-amendment-bill-2019. www.prsindia.org. https://www.prsindia.org/billtrack/inter-state-river-water-disputes-amendment-bill-2019
Shaban A, Sharma RN (2007) Water consumption patterns in domestic households in major cities. Econ Polit Wkly, 2190–2197. June 9, 2007
UN (2019) World urbanization prospects- the 2018 revision. Department of Economic and Social Affairs, New York
UNCED (1992) Agenda 21. United Nations Conference on Environment & Development, Rio de Janerio
WDO (2018) Kudikunta Lake Restoration project. https://wdo.org/kundikunta-lake-restoration/. Accessed 6 Mar 2020
WHO (2020) Infection prevention and control during health care when COVID-19 is suspected. WHO, Geneva
World Bank (2012) India: improving urban water supply and sanitation services. Ministry of Urban Development, New Delhi

Avinash Mishra is a General Civil Service officer, presently working as Adviser in Water and Land Resources, Tourism, Culture, Panchayati Raj, NITI Aayog (GOI). He is having an experience of 30 years of working in various positions in GOI. His educational qualifications are B.E., M.Tech. (Water Resources) from IIT Delhi. He has been working in the National Policy Planning formulation, Action Plans of the country. He is engaged in the Development Plans, Proposals of the State Governments relating to Water and Land Resources, Environment and Forests, Tourism, culture, Panchayati Raj sectors. He had been associated in various reports and publication and Presented many papers in National and International forums.

Part IV
Lessons and Recommendations

Chapter 22
Multi-Hazard Risk Management During Pandemic

Thinles Chondol, Shweta Bhardwaj, Ashish Kumar Panda, and Anil Kumar Gupta

Abstract Disasters not necessarily occur one at a time rather multiple disasters may occur amid another, in a form of secondary or cascading disaster or a new disaster may occur from a new origin due to the aggravated vulnerability factors. In past years, a number of cases of prevalence multiple disasters while dealing with another or a situation of crisis within a crisis emerged (e.g.: Haiti Earthquake, 2010; Cox Bazaar, Bangladesh; Kerala Flood, 2018; Typhoon Haiyan, 2013; Fukushima Nuclear Disaster, 2013). Considering these incidences as an opportunity and drawing lessons from them has led the global policy regime to emphasize largely on a holistic, proactive and an integrated multi-hazard disaster risk management approach. Despite the clear mention of multi-hazard in global policies, the practicality is often seen low and rather visualizes hazards in isolation. Despite the complexity in comprehending the multi-hazard risk assessments, the revamping of the approaches of risk management is the need of the hour when the frequency and severity of climate extreme events are increasing as an impact of climate change. The global pandemic of coronavirus has drawn the attention of the policymakers to the multi-hazard approach as a number of climate extreme events and other human-induced disasters occurred in chorus. This chapter underlines the need for a change in the disaster risk management approach to a multi-hazard approach with involvement of multi-stakeholders at regional level, highlighting some of the past case studies of concurrently occurring disasters.

Keywords Multi-hazard · Pandemic · Risk management · Crisis within crisis · COVID-19

T. Chondol (✉) · S. Bhardwaj · A. K. Panda · A. K. Gupta
National Institute of Disaster Management, Ministry of Home Affairs, Government of India, New Delhi, Delhi, India

M. K. Goyal, A. K. Gupta (eds.), *Integrated Risk of Pandemic: Covid-19 Impacts, Resilience and Recommendations*, Disaster Resilience and Green Growth,
https://doi.org/10.1007/978-981-15-7679-9_22

22.1 Introduction

The incidence of crisis unfolding each day when a global health crisis is already undergoing is an indication that the approach for disaster risk reduction needs to be reworked at practical end. The consequences of climatic change being already witnessed in the form of growing incidences and severity of climate extreme events, the occurrence of disasters in chorus is very much likely. It is evident from the recent several incidences of extreme events including tropical cyclones, floods, heat waves, etc. amid the pandemic and the locust attack, which studies have mentioned to have caused due to climate change El-Nino consequences. Similar incidences of multiple disasters occurring amid another or a disaster aggravating the vulnerability factors to cause a new disaster have occurred in past several times which is why policies have used the term 'multi-hazard' risk management. However, the complexity of looking at a multiple hazards at a time and comprehending the multi-hazard layers often gets challenging on practical ground. The hazards are habitually visualized and managed in isolation despite the clear emphasize in the policies about the multi-hazard risk reduction. Though, from the global policy regime, the gradual evolution of the disaster management approach is evident which gives a hopeful future of the multi-hazard approach. The India's National Disaster Management Policy, 2009 and National Disaster Management Plan, 2016 also mentions about a holistic, proactive, technology driven and multi-hazard oriented approach in all the phases of disaster risk reduction. Such approach is ideal for an efficient risk management since disasters many occur in a concurrent manner or even in parallel with other disaster as in the case on the current COVID-19 pandemic. This ongoing crisis of global pandemic must be visualized as an opportunity to identify the gaps in the disaster management approach and strategize a strong policy framework to efficiently tackle similar crisis in future. This chapter emphasizes on the revamping of the disaster management approach and its implementations that earlier was considering hazards in isolation, to a multi-hazard risk management approach which is the need of the hour since it is evident from the few case studies of cascading disasters and disasters occurring in parallel that have been discussed in brief in succeeding sections.

22.2 Global Policy Scenario: Promoting Multi-Hazard Approach

Over decades, multi-hazard approach for disaster risk management has been widely encouraged within international policy regime. The very first reference of it can be found in Agenda 21, the Rio declaration on environment and development (UNCED 1992). Agenda 21, endorsed the idea of sustainable development. The document called for "undertaking complete multi-hazard research" as an integral part of pre-disaster planning of human settlements, wherein assessment of multi-hazard risks and vulnerabilities associated with human settlements and infrastructures were encouraged to ensure that one type of risk reduction strategy might not increase vulnerabilities towards another type of risk. Further, the reference of this approach was made in Johannesburg Declaration on Sustainable development adapted by member nations of World Summit held in the year 2002. The document referred to an 'integrated multi-hazard and inclusive approach' for addressing disaster risk and vulnerabilities, which is an essential element for ensuring a safer world in future (United Nations 2002). The approach was further promoted through one of key global policy instruments on disaster risk reduction that is Hyogo Framework for Action, 2005–2015 (HFA), which was the outcome of Second World Conference on Disaster Risk Reduction, Japan (2005). The framework emphasized on building resilience of nations and communities towards disaster risks and for achieving so the framework provides sets out different priorities of actions. One the key consideration taken into the account while determining appropriate actions for implementing the framework included mainstreaming of integrated and multi-hazard approach into developmental planning and post-disaster or post-conflict phases across relief, rehabilitation and recovery activities (United Nations 2005). Recently, the approach has been highlighted through the successor instrument of HFA i.e. Sendai Framework for Disaster Risk Reduction, 2015–2030 (SFDRR), adopted at the third world conference on Disaster Risk Reduction, 2015, which builds on the insights of lessons, gaps and future challenges gained from the former instruments. Like HFA, SFDRR also advocates for multi-hazard approach for management of disaster risk through developmental planning and practices across all the sectors. Out of seven global targets proposed under SFDRR, one of the targets aims at *'substantially increasing availability and access to multi-hazard early warning systems as well as disaster risk assessments and information to people by the year 2030'* (SFDRR 2015). Also, one of the key guiding principles for the implementation of SFDRR emphases on promoting multi-hazard and inclusive risk-informed decision making for effective reduction and management of disaster risks. To achieve the set targets, SFDRR calls for promotion of comprehensive assessments of multi-hazard disaster risks at national and regional levels; enhancing support from and access to science and technology for multi-hazard and solution driven research and development; and investing in people-centric multi-hazard and multi-sectoral early warning and forecasting systems at local levels.

22.3 Multi-Hazard Disaster Risk Approaches

The concept of 'multi-hazard risk' has been widely referred and used across various international and regional policy frameworks, but it still lacks clarity in terms of its definition and scope in theory as well as practices. The rationale behind 'multi-hazard disaster risk' encompasses the need for considering multiple hazards and multiple vulnerabilities for effectively assessing the overall disaster risks. In this regard, different approaches have been adopted across various studies and literature for assessing the multi-hazard risk. The first approach is based on the spatial dimensions, which takes into the account all hazards within a defined area. This 'all-hazards' approach individually assesses various potential hazards and combines resultant single risk layers for defining the multi-hazard risks. There are number of studies that have used such independent analysis of multiple hazards for assessing the multi-hazard risk in a region, Grünthal et al. (2006) draws a comparative multi-risk assessment for city of Cologne by analyzing and combining different types of natural hazards (earthquakes, windstorms and floods), vulnerabilities and loss estimations associated with each types of hazards; and Van Westen et al. (2002) assessed the multi-hazard risk for a region by overlaying different hazards layers—eathquakes, landslies and floods and vulnerabilty layers (cost of infrastructure/buildings exposed to these hazards). In this particular approach where individual hazard or vulnerability layers are combined or overlaid to assess the overall multi-hazard risk, the interactions between hazards and vulnerabilities are often neglected and it also ignores different risks that arise from mutual interactions and interrelations between multiple hazard and multiple vulnerability sources occurring within the same space and time.

The second approach adopted for assessing multi-hazard risk takes into the account the conjoint as well as cascade effects that might result from multiple hazard and vulnerability interactions. In cascading scenario, the primary events subsequently trigger a secondary event or a sequence of events. Whereas, in conjoint scenarios, parallel disaster events originated from different sources occurs at the same space and time. In this type of approach, identifying different types of potential hazards and vulnerabilities is not the only defining characteristics rather it also emphasis on non- independent characteristics of hazards and vulnerabilities, in which it take note of interactions existing between individual hazards and vulnerabilities. There are varieties of frameworks, which can be found across literature proposed for categorizing hazards and vulnerabilities interactions and for integrating them into methodologies of multi-hazard risk. Han et al. (2007) classified potential hazards interactions into four hazard chains induced through: spatial and temporary conditions, exogenic geological processes, endogenic geological processes and anthropogenic activities. Gill and Malamud (2016) categorised possible hazard interactions relationship into three types viz. tigerring, increased-proababilty and catalysis/impedance. There are number of studies which have build on this particular approach of multi-hazard risk assessment identifying and assessing the potential interactions within natural hazards, De Pippo et al. (2008) carried out hazard risk asssesment and mapping for coastal region in Italy by investigating the primary hazards in the region and mapping the overall multi-hazard risks by ranking not just the hazads but also their interactions; Neri et al. (2013) estimated the multi-hazard risk assocaited with volcano Kanlaon, Philippines using an event tree method combining probabilistic frequencies of potential three categories of hazardous events and secondary hazards associated with them; and Kappes et al. (2010) analysed the multi-hazard risk for Barcelonnette Basin, Alps by analysing the relationship between different types of hazardstaking into the account disposition and tiggering concerns.

Single-hazard approach in which hazards are treated as independent phenomenon often leads to erroneous estimations of risk and hence the priorities of actions. In comparison to single-hazard approach, multi-hazard approaches provides an holistic idea of overall risk by comparing and ranking different elements of risk, which can help decision makers to prioritize their mitigation strategies while significantly reducing the cost associated with management of disaster risk and improving efficiency of risk management measures. However, unlike single-hazard risk approach, multi-hazard risk assessments lack standard or well-established approaches due to various reasons including: widely different characteristics of individual hazards (frequency, intensity, quantifying units etc.); associated vulnerabilities vary across hazards; and the number of different ways in two individual hazard and vulnerabilities relate and interact with each other. Though, these factors make implementation of multi-hazard risk assessments/approaches considerably difficult, but they do have the potential to influence the overall risk analyses significantly.

22.4 Case Studies of Some Concurrently Occurring Disasters

Disasters not necessarily occur one at a time rather may occur in a concurrent manner. Many a times a major disaster like earthquakes, floods, cyclones, etc. is followed by a secondary hazard also known as a collateral disaster or another disaster that occur from a new origin. Secondary hazards are more damaging and cause more problems than that of a primary disaster due to the exacerbated factors of vulnerability, and reduced adaptive capacity. This is majorly due to the displacement of a large number of population, unplanned and crowded shelters, poor personal hygiene, limited or lack of access to public healthcare facilities and change in physical and social environment (Kouadio et al. 2012). Though disasters are not directly linked with disease outbreaks but increased factors of vulnerabilities aggravates the spread of infectious diseases including diarrhea, acute respiratory infections, malaria, dengue, typhoid, cholera, meningitis, hepatitis, etc. These are some secondary disasters succeeding floods, cyclones, earthquakes, etc. (Watson et al. 2007; Kouadio et al. 2012).

Earthquakes are one of the most reported disasters followed by floods. Its affect varies depending on the intensity of the tremor, vulnerability of the area, population density of the area and the condition of the built-up locality. The direct casualties and deaths post-earthquakes is primarily due to the collapse of the buildings, mental trauma and in addition to that, disease outbreaks may occur due to the lack of hygiene in crowded shelters, poor nutritional care and disruption in WASH (Water, Sanitation and Hygiene) facility. Earthquakes are associated with multi-hazards including Tsunamis in coastal areas, volcanic eruptions, landslides, floods and cyclones. Some of the secondarily associated disasters with earthquakes in past are Diarrhea and influenza during Japan Earthquake (2011), Cholera during Haiti Earthquake (2010), Hepatitis E, Diarrhea, Acute Respiratory Infection (ARI), measles, meningitis and tetanus during Pakistan earthquake (2005), Diarrhea and Acute Respiratory Infection during Iran Earthquake (2003) and El Salvador (2001) (Kouadio et al. 2012). Floods are majorly associated with various water-borne and vector-borne diseases including diarrheal diseases like cholera, malaria, chikungunya, gastrointestinal infections, etc. These diseases are caused after floods as drinking water sources are often contaminated due to the damage to water treatment plants, failure of septic tanks and also due to poor WASH facilities in shelters, fecal-oral spread of pathogens. Outbreak of Cholera was reported during the Bangladesh flood (2004), Odisha flood (2007), Pakistan flood (2010) and Malaria after Surat flood (2006) (Bhuyan et al. 2016; Pawar et al. 2008; Schwartz et al. 2006). Similarly, disasters are occurring and are likely to occur despite the fact that worldwide nations are dwindling with the COVID-19 pandemic and due to which the response teams globally are already going through the strain. As the number of infected cases of coronavirus is rising, it is further complicating disaster management, overburdening the healthcare system, disrupting the supply chains and declining country's economy. To make the matter worse, when the countries were dealing with the pandemic, a series of emergencies occurred including the earthquake in Croatia, tornado in Southeastern US, plane crash in Pakistan, Cyclone Harold in Pacific countries, flood due to snow melting in Ottawa city, Canada, Ebola outbreak in Africa, Locust Attack in Africa and Middle East countries, Oil spill in Russia, Flooding of Mississippi, etc. Even in India amid the swelling cases of coronavirus pandemic, the NCT (National Capital Territory) Delhi experienced a couple of moderate earthquakes, Styrene Gas leakage from LG Polymer Industry, Vishakhapatnam, Cyclone *Amphan* in eastern states of India, Locust attack in Rajasthan, Punjab, Haryana and Madhya Pradesh and Cyclone *Nisarg* in western India led the panic amongst people to aggravate. It is forecasted to occur more climate extreme events in coming days including floods, heat waves, wildfires, cyclones and hurricanes in many parts of the world. It is challenging to manage disasters while struggling to contain the spread of corona virus pandemic, COVID-19. The incidences of disasters occurring in chorus posed a severe challenge to the containment of the pandemic since at this point of time the entire focus is on COVID-19 despite the fact that disaster management system in many countries are capable to tackle multi-hazard. But the odds here is when one disaster is highly transmissible from one person to another.

The occurrence of disasters while dealing with another disaster is not a new phenomenon. Such instances have been occurring in history as well. The following case studies highlight how network of hazard events or occurrence of multiple events in parallel can create complex emergencies. Such situations can cause serious disruption in normal functioning of society and can put huge strain the country's response capacities. In such scenarios, a multi-hazard risk approach provide an opportunity for developing better understanding and assessment of hazards and vulnerabilities, which can help in enabling inclusive and risk informed decision-making and holistic management of disaster risks.

22.4.1 Case Study 1: Haiti Earthquake 2010

On 12 January 2010, a magnitude of 7.0 occurred in world's one of the poorest country, Haiti, epicenter at 17 km from the capital city, Port-au-prince. It caused deaths of more than 316,00 people, 300,000 casualties, 403,176 buildings were damaged (including houses, hospitals and schools) and rendering more than 1.3 million homeless (Government of the Republic of Haiti 2010; Fritz et al. 2012; Bilham 2010; Lantagne et al. 2013). The earthquake tremor damaged and led to the collapse of most of the buildings since no building codes were followed and they were susceptible to damage (Lantagne et al. 2013). This disaster has heavily affected the economy of the country. The overall losses and damages incurred due to the earthquake was estimated to be approximately US$7.804 billion, surpassing country's GDP (Gross Domestic Production) for the year 2009 (PAHO 2010). To cause an additional loss to the people and infrastructure at risk, landslide and liquefaction posed a risk of secondary hazard. The earthquake generated two minor tsunamis along the Gulf of Gonave and south coast of Haiti, as warned by the Pacific Tsunami Warning Center (PTWC), which caused death of three individuals primarily due to the lack of awareness and information since the entire communication system were disrupted due to earthquake (Fritz et al. 2012). Had it been a major Tsunami, the losses to lives and property will be manifold. Whilst Haiti was dealing with aftermath of earthquake, disrupted transportation and communication services further complicating the humanitarian responses and another disaster of Cholera outbreak occurred in chorus after 9 months, post earthquake. It was first reported in Artibonite Province of Haiti on 19 October 2010 and spreading it to the rest of the country. It was very unlikely to witness an epidemic of Cholera in Artibonite province of Haiti where it was never reported before (Lantagne et al. 2013). More than 93,000 were sickened and it took more than 2100 lives (Chin et al. 2011). The infection caused due to *Vibrio cholerae* spreaded so rapidly, owing to the previously disrupted sanitation facilities, poor living conditions in shelters and high susceptibility of the people. Numerous studies was undertaken and concluded that the epidemic was caused because of the contamination of Artibonite River's tributary, Meye with a pathogenic strain (South Asian type *V. cholera*). It was found to be due to human activities coupled with poor sanitation facilities, huge dependency on Meye tributary, poor health facilities at that time when the country was already going through a major crisis (Chin et al. 2011; Lantagne et al. 2013). A series of complex emergencies were emerging for Haiti when Hurricane Tomas was hit at the island of Hispaniola. It was a very critical for Haiti when the country was already going through the aftermath of an earthquake and dealing with a severe epidemic of Cholera. The Hurricane Tomas hit the west coast of Haiti on 5–6 November 2010, caused heavy rains, strong winds, flooding, and took 35 Haitian lives (Shamir et al. 2013). The situation became more complicated with many taking shelter in relief camps when intense rain and flooding in Haiti turned the camps into swamps, thus further worsening the condition of living. With such situation of complex emergencies, the crisis response staffs and healthcare facilities were severely overwhelmed.

Despite of the limited knowledge about the cholera in Haiti, efforts were made to reduce its spread, treatment centers and stabilization centers were opened (Walton and Ivers 2011). Lessons of strengthening and reinforcement of healthcare facilities and capacity building of the medical staffs and emergency response teams to deal with multi-hazard can be drawn from the complex emergency that emerged in 2011 in Haiti.

22.4.2 Case Study 2: Cox Bazaar, Bangladesh

Cox Bazaar is known to be the world's largest refugee camp, located in the southeast of Bangladesh, which is housing nearly 855,000 Rohingya refugees (Vince 2020). These refugees fled from Myanmar due to the ongoing genocidal activity. They are living in a very deplorable condition in makeshift houses made of tins, woods, bamboos, mud, plastic sheets and straws (Mohiuddin and Latif 2013) that are multi-hazard prone including landslides, flash floods, cyclones, fire hazards (Ahmed 2020) and now SARS-CoV-2. These camps are densely packed with more than 10 families share a single toilet, limited drinking water supplying faucets, and open sewage drains all over, aggravating the factors of vulnerability. In such scenario, maintaining hygiene, frequently washing hands and keeping social distance is not only challenging but also nearly impossible. These refugees are so helpless since they are not allowed to move outside the camps and are still dependent on the humanitarian assistance and aid for food. More than 50% of the populations in these camps are children (Vince 2020). Despite the fact that mortality rate amongst children are lowest due to coronavirus, the children in the camps are vulnerable since they are malnourished and have existing medical conditions like respiratory infections, frequent diarrhea (Vince 2020). The pandemic have aggravated the risks to water-borne diseases and other infections due to the compromised immunization activities.

The entire country of Bangladesh is facing a severe challenge in adopting the covid preventive measures like social distancing due to its population density. The major challenges in Bangladesh against the containment of the infection are limited number of testing, lack of safety equipments (masks, gloves, PPEs), inadequate skilled personals, inadequate healthcare facilities and medical professionals. Another challenge in the country is the lack of public awareness and large number of vulnerable population (Islam et al. 2020). The n-CoV pandemic is a serious crisis to the vulnerable community of Cox bazaar that have already been living in a crisis like situation with poor safety, housing condition with limited basic facilities especially the healthcare facilities in the camps are inadequate and lack testing capacity. To make the matter worse, another catastrophe of super cyclone *Amphan* hit Bangladesh along with eastern states of India, generating heavy rain, severe winds which damaged the camps. There was shortage of space in cyclone shelters since these were being used as quarantine centers (Oxfam International 2020). However, the government and NGOs working in the area have taken several preparedness measures including broadcasting of cyclone messages, ambulance and medical team in place, stock piling the basic medicines, electrolytes and ready-to-use therapeutic food (RUTF) to deal with super cyclone amidst coronavirus pandemic (UNICEF 2020). In addition, every year the monsoon season causes an adverse condition in the camps causing destruction to their shelters, landslides and flooding. This year also the monsoon season will be an additional risk. The Cox bazaar is another case of crisis within crisis due to the prominent factors of vulnerability.

22.4.3 Case Study 3: Kerala Flood 2018

Flood in the Indian state of Kerala (2018) occurred just after the state faced the outbreak of Nipah virus (May–June 2018) and the district of Kozhikode was affected the most due to the infection, with clinical symptoms of acute respiratory distress, severe encephalitic symptoms, seizures, convulsions and coma (Spiropoulou 2019). Just after this virus outbreak, country's worst affected disaster, flood occurred that took the lives of more than 500 people and many went missing. The state lost 2.2% of its Gross Domestic Product due to the flood and related disasters (Vineesh 2019). This disaster is said to be caused primarily due to the excessive rainfall and reservoir storage exceeding its maximum capacity. According to the Indian Metrological Department, the state received 2346.6 mm rainfall during 1 June to 19 August while normally it receives 1649.5 mm of rainfall. Six out of 7 major dams were more than 90% full as of August 8, 2018 and when the dams reached its maximum level, the excess water was released from 35 dams out of 54 (Mishra et al. 2018; Nowfal and Sarath 2018). The district of Idukki, Palakkad, Wayanad, Kannur, Ernakulum, Kottayam and Pathanamthitta were severely affected as shown in Fig. 22.1. Other secondarily associated disasters, that worsened the situation and caused the death toll to increase was land sliding, food shortage and collapse of buildings (Nowfal and Sarath 2018). Just when the state was dealing with the flood aftermath with swelling river flows, the water levels in the rivers declined drastically that a drought like scenario occurred in the month of September. Some studies suggests that it is a usual phenomenon for the water levels in rivers and groundwater level to decline post fluvial flood as the topsoil have been flushed away along with the floodwater as also mentioned by The Kerala State Land Use Board (KSLUB). The major rivers such as Periyar, Bharathapuzha, Pampa and Kabani were depleted greatly. Post the heavy rainfall in August, the state have faced a long dry period of more than 20 days which contributed in further receding of the river flows andgroundwater table (Madhusoodhanan and Sreeja 2019). Post flood 2018, the state also witnessed numerous landslides. According to the Land Revenue Department, Kerala reported 331 landslides and 104 deaths (Fig. 22.2). The state has faced a number of concurrent disasters exacerbating the vulnerability factors that makes this flood (2018) as country's one of the worst hit disasters.

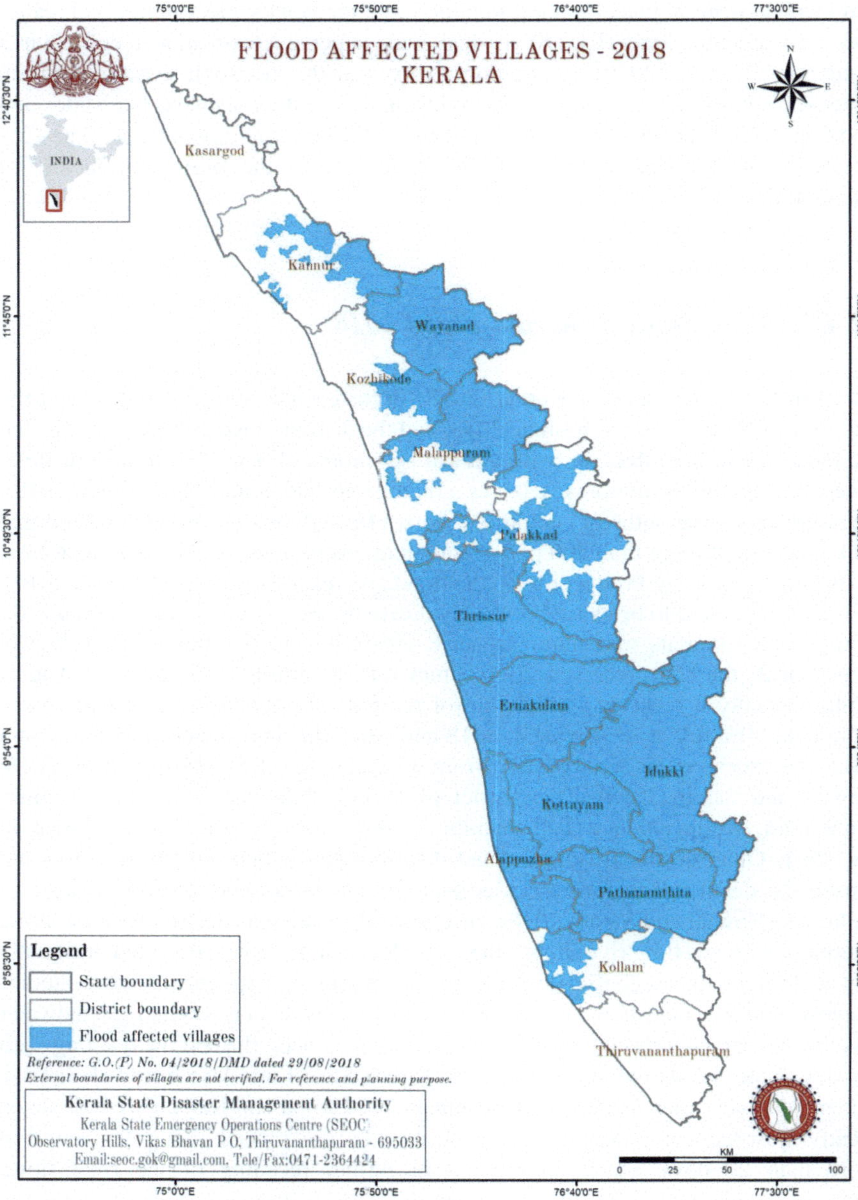

Fig. 22.1 Affected areas of Kerala flood 2018 (Source: Government of Kerala 2018)

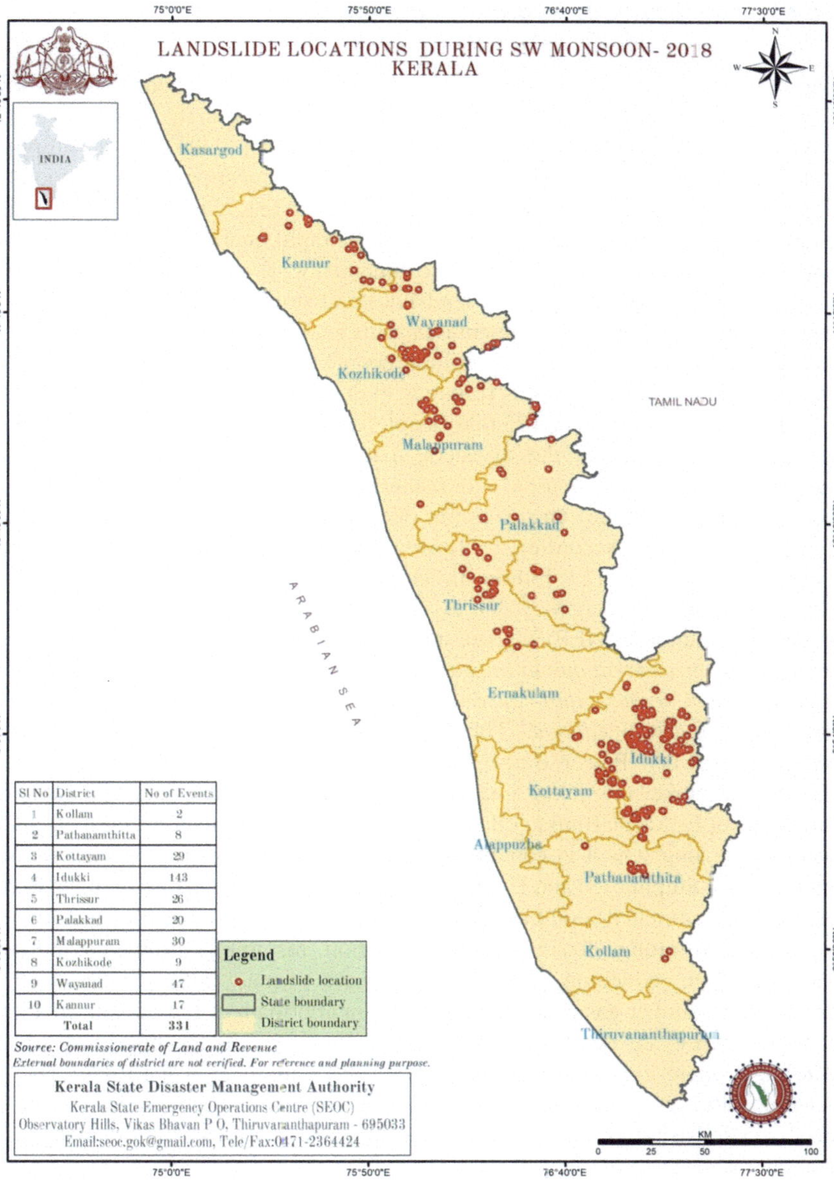

Fig. 22.2 Landslides due to Kerala flood 2018 (Source: Government of Kerala 2018)

22.4.4 Case Study 4: Typhoon Haiyan, 2013

Typhoon *Haiyan* or locally named as *Yolanda,* was a tropical cyclone that crossed a thickly populated and a disaster prone country, Philippines in November 2013 (Nguyen 2018). The typhoon *Haiyan* made its first landfall in the central Philippine Island on 8 November 2013. This typhoon is one of the most powerful typhoons in the history with maximum wind speed of 170 knots (312 kmph) and before the landfall having maximum gust of 205 knots (379 kmph) qualifying it to be category 5 typhoon (Lagmay et al. 2015). It was accompanied with very strong winds, intense rainfall, flooding in low-lying areas, landslides and storm surges causing an extensive economic, social and environmental impacts leading to 6300 live losses, 28,698 casualties, 1600 went missing, 3.43 million people displaced, 1.1 million damaged houses, and 628 damaged schools (Lagmay et al. 2015; Nguyen 2018; UN OCHA 2013) (Fig. 22.3). More than 125,000 people were evacuated (UN OCHA 2013). Even after the 3 years of the disasters, many continued to live in deplorable conditions in temporary houses and makeshift shelters. Livelihood sources of many were lost due to the storm surge in rice, sugarcane and corn producing areas and fishing grounds. According to Bowen (2016), the typhoon had an overall economic loss of US $ 12.9 billion. In addition to this, Typhoon *Haiyan* posed a major environmental emergency due to an oil spill incident. United Nations Environment Programme (UNEP) defines an environmental emergency as '*a sudden onset disaster or accident due to the natural, technological, human-induced factors or a combination of these, that cause severe environmental damage as well harm to human health and/or livelihood'*. The National Philippine's National Power Corporation (NPC) owned Power Barge No. 103 was hit by the typhoon and nearly 200,000 L of bunker oil spilled into the sea contaminating a several kilometers of the coastline of Estancia. Bunker oil is a low volatile distillery residual oil from an oil refinery processing which is contaminated with chemicals including Nickel (Ni), Vanadium (V), Sulphur (S) and Hydrogen Sulphide (H_2S). Exposure or ingestion of bunker oil contaminated food have health impacts including headache, irritation in eyes, nausea, vomiting, dizziness, throat irritation, respiratory problems and skin drying, irritation, reddening. Long time exposure to may cause cancer as some components of bunker gas is carcinogenic. As an aftermath of the oil spill incident, post typhoon Haiyan, nearly 200–500 houses that were in the proximity of the affected area were evacuated and many were left without power supply. The oil was washed and spread upto the shoreline of 10 km and the communities relying on these shores for fishing as their source of livelihood was affectedand the mangroves in that area was severely affected.

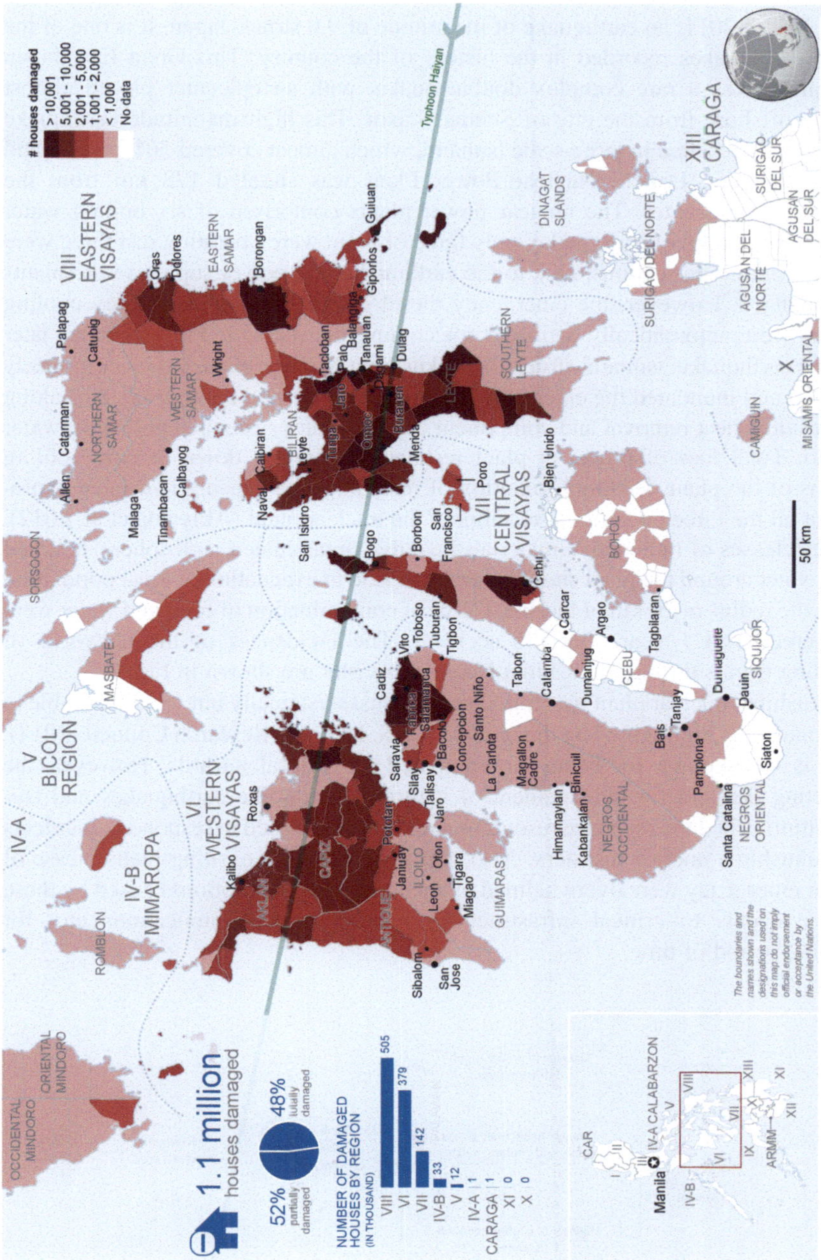

Fig. 22.3 Route of Typhoon Haiyan and damaged houses in the Philippines (Source: UN OCHA 2013)

22.4.5 Case Study 5: Fukushima Nuclear Disaster, 2011

On 11 March 2011, an earthquake of magnitude of 9.0 struck Japan. It is one of the largest earthquakes recorded in the history of the country. This Great East Japan Earthquake was a rare complex double quake, with an epicenter placed almost 130 km offshore from the city of Sendai, Japan. This high magnitude earthquake subsequently resulted in large-scale tsunami, which almost covered 561 km^2 of land area. Fukushima Daiichi Nuclear Power Plant was situated 178 km from the earthquake's epicenter. The nuclear power plants comprised of six boiling water reactors but at time of earthquake, only three of them were operating rest three were shut for routine inspections. Due to the earthquake, the power supply to the plants was damaged. However, the emergency diesel generators and emergency cooling systems were automatically activated for cooling the reactors. Forty minutes later after the earthquake, tsunami struck the Fukushima Nuclear plant, which completely damaged and inundated the emergency diesel generators and cooling systems along with residual heat removal and component cooling water systems (cooling seawater pumps). Total loss of power at plant resulted in shutting down of core cooling systems of the plant, leading to melting of reactors cores causing hydrogen explosions at all the three operational reactors of the nuclear plant (Akiyama et al. 2012). Offsite releases of radioactive materials heavily contaminated atmosphere, soil and ocean water around the plant site. The accident lead to evacuation of local population within the radius of 20 km of the site. Cases of contamination of food and water were discovered even 1 year after the accident. The incidences of the networks of cascading events during Fukushima Nuclear Disaster are shown in Fig. 22.4.

Fukushima nuclear plant was found to be robust seismically but was vulnerable to the tsunami. As highlighted by the report by U.S. National Research Council (2014), Japan is well-known for being well-prepared for natural hazards. However, the competing demand for management of natural hazards like earthquakes and tsunamis diminished the response resources and capacity needed to respond to incidents like Fukushima nuclear disasters. Also, implementation of exsiting plans incase of nuclear emergency were overwhelmed by the large-scale disruptions caused by these extreme events to critical infrastructures (electricity, communications etc) for extended period of time.

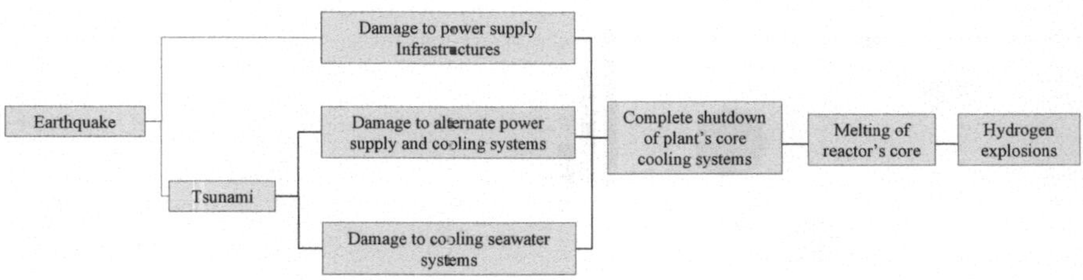

Fig. 22.4 Networks of events (cascades) during Fukushima Nuclear Disaster

22.5 Conclusion

Various policy frameworks have been emphasizing on the multi-hazard approach of disaster management that considers all the possible hazards of that particular area into consideration for an effective disaster management. However, the ground implementation and practicality of such approach was lacking largely. The current situation of coronavirus or SARS-CoV-2 pandemic has taught a big lesson and many have now started talking about dealing with crisis within crisis. Incidences of crisis during a crisis are not new, rather historical. However, what lessons have we drawn from these past crises and if those lessons are brought into policies and management strategies are very important.

Covid pandemic is a highly transmissible infection and dealing with another big disaster while the entire focus of every nation is on the pandemic is a huge challenge. We have to respond to a disaster in such a way that simultaneously abides by the covid protocols. Ideally, for an effective disaster response, responders, volunteers and community come together to provide immediate relief to the victims but the case during the outbreak of SARS-CoV-2, situation have become completely different as social distancing is to be maintained during the disasters and it is constantly posing an additional obstruction to the life-saving activities. Disaster responders are running short in many countries, as many have also been infected with the virus whilst taking care of the victims. At the same time, the training of new responders is also challenging despite efforts being made to train them online regardless of the inadequate essential resources, budget and technology.

The emergency responders worldwide are overwhelmed and coping capacities are reduced despite the fact that the likelihood of occurrence of additional disasters is amplifying the challenges. The humanitarian assistance for any disaster that occurs during this time needs to be modified and the emergency response teams must be trained adequately to tackle both simultaneously. The temporary relief shelters needs to be redesigned with proper water, sanitation and hygiene facilities, which earlier used to be a congested space. However, many nations have the capability to tackle multi-hazard simultaneously but the case of dealing a highly communicable pandemic along with another disaster is surely not easy. It is very necessary to be prepared to deal with multi-disasters simultaneously for disasters may not occur one at a time. The administration and the policy-makers must take the worst-case scenario in mind and plan accordingly for a multi-hazard risk reduction strategy.

Multi-hazard risk reduction is an approach to assess and evaluate a combined risk occurring from all types of potential threats in order to have an integrated planning strategy. Looking at the potential threats in silos may lead to a situation wherein the mitigation measures undertaken for one hazard may generate a new vulnerability for another hazard. A comprehensive and a holistic multi-hazard risk assessment must be carried out with the involvement of stakeholders at regional level and based on it propose a multi-hazard policies and strategize a multi-hazard community/property adaptive capacity. This multi-stakeholder involvement in multi-hazard risk management will improve disaster preparedness, mitigation measures, strengthen risk communication, and capacity to deal with multiple hazards at regional level.

References

Ahmed B (2020) Conflict, disaster, and disease: a colossal catastrophe looms in the Rohingya camps in Cox's bazar, Bangladesh. UCL Institute for risk and disaster reduction blog. https://blogs.ucl.ac.uk/irdr/2020/04/20/rohingya_cox_bd/. Accessed 29 May 2020

Akiyama N, Sato H, Naito K, Naoi Y, Katsuta T (2012) The Fukushima nuclear accident and crisis management: lessons for Japan-U.S. Alliance cooperation. The Sasakawa peace foundation

Bhuyan SK, Vairale MG, Arya N, Yadav P, Veer V, Singh L, Yadava PK, Kumar P (2016) Molecular epidemiology of Vibrio cholerae associated with flood in Brahamputra River valley, Assam, India. Infect Genet Evol 40:352–356

Bilham R (2010) Lessons from the Haiti earthquake. Nature 463:878–879

Bowen TV (2016) Typhoon Yolanda (Haiyan) and the case for building an emergency cash transfer program in the Philippines (no. 119393). The World Bank, Washington, DC

Chin CS, Sorenson J, Harris JB, Robins WP, Charles RC, Jean-Charles RR, Bullard J, Webster DR, Kasarskis A, Peluso P, Paxinos EE (2011) The origin of the Haitian cholera outbreak strain. N Engl J Med 364(1):33–42

De Pippo T, Donadio C, Pennetta M, Petrosino C, Terlizzi F, Valente A (2008) Coastal hazard assessment and mapping in northern Campania, Italy. Geomorphology 97(3–4):451–466

Fritz HM, Hillaire JV, Molière E, Wei Y, Mohammed F (2012) Twin tsunamis triggered by the 12 January 2010 Haiti Earthquake. Pure Appl Geophys 170(9–10):1463–1474

Gill JC, Malamud BD (2016) Hazard interactions and interaction networks (cascades) within multi-hazard methodologies. Earth Syst Dynam 7(3):659

Government of Kerala (2018) Kerala floods 2018: 1st August to 30th August 2018. Additional Memorandum

Government of the Republic of Haiti (2010) Action plan for National Recovery and development of Haiti. Port-au-Prince, Haiti

Grünthal G, Thieken AH, Schwarz J, Radtke KS, Smolka A, Merz B (2006) Comparative risk assessments for the city of Cologne–storms, floods, earthquakes. Nat Hazards 38(1–2):21–44

Han J, Wu S, Wang H (2007) Preliminary study on geological Hazard chains. Earth Sci Front 14:11–20

Islam MT, Talukdar AK, Siddiqui MN, Islam T (2020) Tackling the Pandemic COVID-19: the Bangladesh Perspective. Running Article: Tackling COVID-19 by Bangladesh. https://doi.org/10.20944/preprints202004.0384.v1

Kappes MS, Keiler M, Glade T (2010) From single-to multi-hazard risk analyses: a concept addressing emerging challenges. In: Mountain risks: bringing science to society: proceedings of the international conference, Florence, pp 351–356

Kouadio IK, Aljunid S, Kamigaki T, Hammad K, Oshitani H (2012) Infectious diseases following natural disasters: prevention and control measures. Expert Rev Anti Infect Ther 10(1):95–104

Lagmay AMF, Agaton RP, Bahala MAC, Briones JBLT, Cabacaba KMC, Caro CVC, Dasallas LL, Gonzalo LAL, Ladiero CN, Lapidez JP, Mungcal MTF (2015) Devastating storm surges of typhoon Haiyan. Int J Disaster Risk Reduct 11:1–12

Lantagne D, Nair GB, Lanata CF, Cravioto A (2013) The cholera outbreak in Haiti: where and how did it begin? In: Cholera outbreaks. Springer, Berlin, pp 145–164

Madhusoodhanan CG, Sreeja KG (2019) Can floods lead to drought? After the flood, severe drought looming over Kerala. Available at https://www.researchgate.net/profile/C_G_Madhusoodhanan/publication/331147298_Can_floods_lead_to_drought_After_the_flood_severe_drought_looming_over_Kerala/links/5c677775a6fdcc404eb458e0/Can-floods-lead-to-drought-After-the-flood-severe-drought-looming-over-Kerala.pdf. Accessed 21 August 2020

Mishra V, Aadhar S, Shah H, Kumar R, Pattanaik DR, Tiwari AD (2018) The Kerela flood of 2018: combined impact of extreme rainfall and reservoir storage. J Hydrol Earth Syst Sci Discuss:1–13. https://doi.org/10.5194/hess-2018-480

Mohiuddin M, Latif MB (2013) Housing condition of coastal area in Bangladesh: a case study of Kutubdia, Cox's Bazaar. J Environ Sci Nat Resour 6(1):15–20

National Research Council (2014) Lessons learned from the Fukushima nuclear accident for improving safety of U.S. nuclear plants. The National Academies Press, Washington, DC
Neri M, Le Cozannet G, Thierry P, Bignami C, Ruch J (2013) A method for multi-hazard mapping in poorly known volcanic areas: an example from Kanlaon (Philippines). Nat Hazard Earth Syst Sci 13:1929–1943
Nguyen HT (2018) Gendered vulnerabilities in times of natural disasters: male-to-female violence in the Philippines in the aftermath of Super Typhoon Haiyan. Violence Against Women 25 (4):421–440. https://doi.org/10.1177/1077801218790701
Nowfal N, Sarath SR (2018) Effective disaster management mechanism: experience from Kerala floods. Univ Rev 7(12):503–508
Oxfam International (2020) 'Crisis on top of crisis' as India and Bangladesh brace for super Cyclone Amphan. https://www.oxfam.org/en/press-releases/crisis-top-crisis-india-and-bangladesh-brace-super-cyclone-amphan-oxfam. Accessed 29 May 2020
PAHO, Goyet CV, Sarmiento JP, Grünewald F (2010) Health response to the earthquake in Haiti. Pan America Health Organisation, Washington, DC
Pawar AB, Bansal RK, Kumar M, Jain NC, Vaishnav KG (2008) A rapid assessment of mosquito breeding, vector control measures and treatment seeking behaviour in selected slums of Surat, Gujarat, India, during post-flood period. J Vector Borne Dis 45:325–327
Schwartz BS, Harris JB, Khan AI, Larocque RC, Sack DA, Malek MA, Faruque AS, Qadri F, Calderwood SB, Luby SP, Ryan ET (2006) Diarrheal epidemics in Dhaka, Bangladesh, during three consecutive floods: 1988, 1998, and 2004. Am J Trop Med Hyg 74(6):1067–1073
SFDRR (2015) Sendai framework for disaster risk reduction (2015–2030). United Nations Office for Disaster Risk Reduction Geneva
Shamir E, Georgakakos KP, Spencer C, Modrick TM, Murphy MJ, Jubach R (2013) Evaluation of real-time flash flood forecasts for Haiti during the passage of hurricane Tomas, November 4–6, 2010. Nat Hazards 67(2):459–482. https://doi.org/10.1007/s11069-013-0573-6
Spiropoulou CF (2019) Nipah virus outbreaks: still small but extremely lethal. J Infect Dis 219:1855–1857
UN OCHA (2013) Oil spill in estancia, Iloilo Province, Western Visayas, Philippines resulting from typhoon Haiyan (Yolanda) 8 November 2013. Joint Assessment Report of UNEP, UN OCHA & WHO. https://reliefweb.int/report/philippines/oil-spill-estancia-iloilo-province-western-visayas-philippines-resulting-typhoon. Accessed 2 Jun 2020
UNCED (1992) AGENDA 21: the Rio declaration on environment and development. United Nations, New York
UNICEF Bangladesh (2020) Cyclone Amphan situation report #1. UNICEF Bangladesh, Dhaka
United Nations (2002) Johannesburg declaration on sustainable development: from our origins to the future. United Nations, New York
United Nations (2005) Hyogo framework for action 2005–2015: building the resilience of nations and communities to disasters. United Nations, New York
Van Westen CJ, Montoya L, Boerboom L, Badilla Coto E (2002) Multi-hazard risk assessment using GIS in urban areas: a case study for the city of Turrialba, Costa Rica. In: Proceedings of the Regional workshop on Best Practise in Disaster Mitigation, Bali, pp 120–136
Vince G (2020) The world's largest refugee camp prepares for Covid-19. BMJ. https://www.ncbi.nlm.nih.gov/pubmed/32217520. Accessed 29 May 2020. https://doi.org/10.1136/bmj.m1205
Vineesh OK (2019) Impact assessment of Kerala flood 2018 & 2019. J Compos Theor 12 (11):168–174
Walton DA, Ivers LC (2011) Responding to cholera in post-earthquake Haiti. N Engl J Med 364 (1):3–5
Watson J et al (2007) Epidemics after natural disasters. Pediatr Infect Dis J 13:1–5

Chapter 23
Enhancing Epidemic Resilience: Planning and Institutional Resilience

Sonal Bindal, Pritha Acharya, Anil Kumar Gupta, and Jugal Kishore

Abstract The epidemics of COVID-19 is among the most expensive and devastating threats to millions of people worldwide. Policy makers generally have a linear world view, which can help the economic and social track of epidemic shocks and crisis by pulling the right levers Ebola, MERS, Zika, and SRAS, for example, have shown the limited will to respond quickly and appropriately in the last few years. This highlights the need to be more proactive, but existing epidemic response frameworks are either national disease-specific, category-specific or non-specific, but there is lack of integrated framework combining all components of institutional resilience. Moreover, COVID-19 pandemics require a holistic approach to address the challenges. In this study, the objective of the epidemic resilience for disease outbreaks is to develop a holistic framework through identification of institutional sectors in enhancing resilience planning. This framework can guide decision-making and priority making based on evidence. This chapter encapsulates a framework focuses on interaction between process, and their systemic properties, which leads to overemphasis on a limited number of features, particularly enhancing the resilience performances of institutions.

Keywords Institutional resilience · COVID-19 · Epidemic · Public health

---

S. Bindal (✉) · P. Acharya
National Institute of Disaster Management, Ministry of Home Affairs, Government of India, New Delhi, Delhi, India

A. K. Gupta
Department of Environment and Disaster Risk Management and Centre for Excellence on Climate Resilience, National Institute of Disaster Management, New Delhi, Delhi, India

J. Kishore
National Institute of Disaster Management, Ministry of Home Affairs, Government of India, New Delhi, Delhi, India

Department of Community Medicine, Vardhman Mahavir Medical College and Safdarjung Hospital, New Delhi, India

M. K. Goyal, A. K. Gupta (eds.), *Integrated Risk of Pandemic: Covid-19 Impacts, Resilience and Recommendations*, Disaster Resilience and Green Growth,
https://doi.org/10.1007/978-981-15-7679-9_23

23.1 Introduction

The epidemics of COVID-19 is among the most expensive and devastating threats to millions of people worldwide. As on 4th June 2020, the highest death rate of 315.54 per million is being faced by United States of America (USA) whereas 190.93 death rate per million people for Canada and a lower death rate of 3.91 per million for India (Roser et al. 2020). The first confirmed case of COVID-19 virus infection was in the state of Kerala, India on 30 January 2020. The carriers were from Wuhan, with a history of travel from China. Then onwards, India is facing an exponential increase in number of COVID-19 patients with symptomatic and asymptomatic symptoms (Pulla 2020).

Data driven frameworks of infectious diseases are increasingly used to provide actual or close to real time situational awareness during outbreaks of diseased conditions. In fact, despite the intrinsic limitations in the predictions of complicated processes, mathematical and computational models have been used to predict the size and magnitude of epidemics. This assists in estimation of the risk of case importation across the world and articulates the risk of uncurbed disease outbreak (Zhang et al. 2020). Despite conflicting opinions about the use of modelling techniques in epidemiological studies, a significant number of studies have been used in recent years to evaluate infection containment and mitigation strategies, as well as to warn emergency plans for pandemic preparedness. Model based epidemic scenarios focus in most cases on magnitude and durations of the epidemic or pandemic issues (Massaro et al. 2018). Furthermore, resilience, mitigation and containment policies are currently being evaluated in the modelling community by reducing the attack rate per hundreds, thousand or ten thousand people in the population. The assessment of vulnerabilities and implications of epidemics is a multi-dimensional problem that should address societal issues such as disruption of infrastructure and services, forgoing output, inflated prices, crisis-induced budget deficit and poverty (Massaro et al. 2018). It is therefore important to broaden the framework based approach to epidemic analysis. This framework will be including measures capable of assessing systemic and institutional resilience, such as the response of the overall system to disruptions, their consequences, the outcome of mitigation, as well as the recovery and retention of system functionalities (Massaro et al. 2018).

The COVID-19 outbreak, 2019–2020 is a reality check to the crucial importance of resilient systems by institutions. During public health emergencies, vulnerable institutions may become overwhelmed by adding to the human, economic and political burden. In this chapter, the general characteristics of the institutions that are resilient to these crises have been identified, and the next step is to define the particular capacities institutions have to build and sustain systems for enhancing institutional resilience. COVID-19 institutional resilience will involve multi-sector approach, such as health sector, industrial sector, educational sector, food and agriculture sector, hospitality and tourism sector, and transport sector. The impacts of COVID-19 outbreaks are multi-cascade in nature, impacting these sectors and may lead to global institutional crisis and negative economic growth. Understanding these sectors is importance for the sustenance of human life and their resilience will act as crucial indicators. Sustainable performance of institutions is partly dependent upon robust development but primarily on resilience: the ability to rebound, develop workflows that absorb losses and deficiencies, awareness and participation to evolving conditions and feedback, and methodically deal with the pattern of problems which will inevitably arise along the way (Ganin et al. 2016). The resilience will enhance the capacity to recover quickly from difficulties and toughness from pandemic situations of COVID-19.

Although the evaluation of adverse event resilience of complicated processes is the subject of extensive research, its integration into the analysis method of epidemic hazards remains largely unexplored. Most importantly, institutional resilience must include the time horizon; i.e. recovery and stabilization of functionality in the epidemic events (O'Brien 2008; Massaro et al. 2018; Wang and Wang 2020). Therefore, the identification and management of institutional resilience to epidemics must identify the critical functions of the system and assess the temporal profile of how they are retained or managed to recover in reaction to pandemic events. Development occurs primarily informally and unmanaged in developed and emerging economies.

23.2 Components of Institutional Resilience

The well-known phases of disasters such as mitigation, preparedness, response, and recovery do not work very well as the frameworks for the Institutional Resilience. Instead of communities, social media has an impact on the complexities of national and international institutions such as banking, trade and communication (Romaní et al. 2020). In order to determine the different activities of an epidemic response, the links between them and potential avenues for an impact have been developed for an epidemic response to this review. COVID-19 pandemics distinguished from epidemics by its more explicit policy objectives, presents similar challenges in terms of scope for developing frameworks for understanding. For example, many frameworks were aimed at designing suitable interventions or setting priorities during the implementation phase or some aspects of performance in many cases. A systematic, all-embracing approach has not been given to assess all stages of a life cycle intervention. Therefore, enhanced framework comprised of the preparedness, coping mechanism, adaptation, awareness and transformation have been found to enhance the institutional resilience (Fig. 23.1).

23.2.1 Preparedness

Countries such as UK, Greece, Spain, Poland, and Cyprus represent five EU member States have faced Middle East Respiratory Syndrome (MERS) and poliomyelitis.

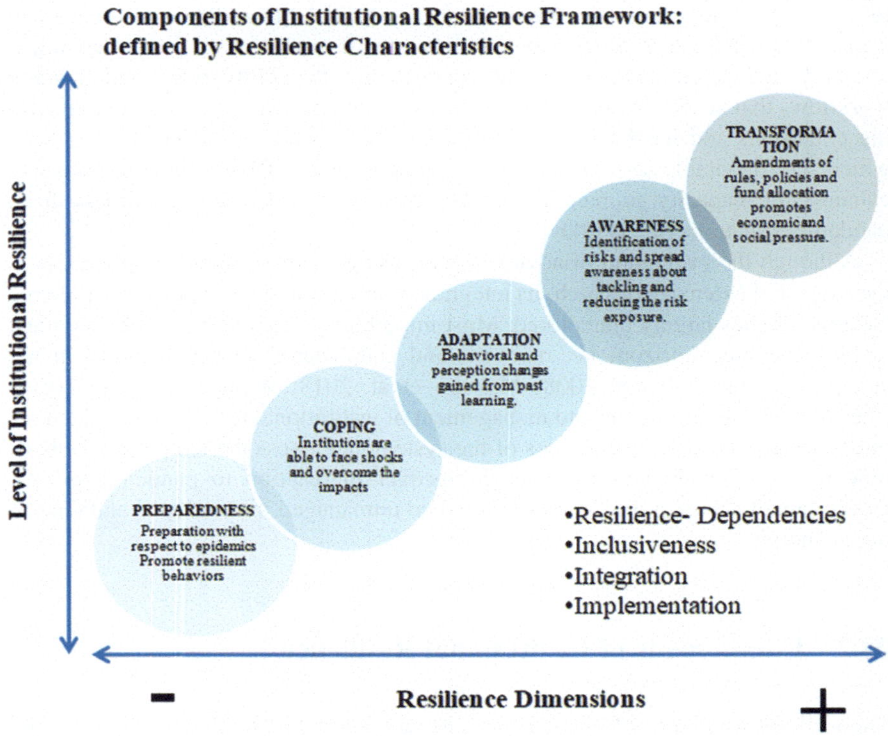

Fig. 23.1 The components of institutional resilience framework

Policies and strategies exist across all countries and valuable lessons have been learned and institutionalized on the basis of experience from recent public health emergencies, such as Ebola (2014–2016), the H1N1 pandemic (2009) and SARS (2003) (Bevc et al. 2014; Dausey and Moore 2014; Kinsman et al. 2018). There were also significant legacy benefits from mass gatherings, such as the Olympics, which took place in Greece in 2004 and the United Kingdom in 2012. However, new technologies, knowledge and experience from these events are vulnerable to political and policy prioritization, which is moving away from securing effective core public health capabilities in some of the participating countries (Kinsman et al. 2018). There was no preparedness regarding COVID-19, but the lessons learned from these previous epidemics have provided ways to institutions to increase their preparedness.

There are important practices to ensure the readiness of European public health preparedness programs to carry out assessments, record results and disseminate them widely through regular training of all relevant staff personals (Kinsman et al. 2018). Similar model is applicable in developing nations to enhance their institutional preparedness. Much more disaster preparedness planning is required, and institutional systems need to be included in it. The multi-scalar integration and governance structures, which allow coherent collaboration and cooperation within the system, are part of this preparedness for national and local institutional systems with shorter supply chains.

23.2.2 Coping

Resilience is more complicated and requires actions which may sometimes appear to be inconsistent to rationality. Literature review suggests the three key actions they can take to build their employees, teams and institution' resilience (Aguirre et al. 2005). The ability to cope up with the implications of economic losses, no work, job recession and instability in demand and supply chain makes the coping needs more important for enhancing the institutional resilience of individual organizations and governmental structures (Barton et al. 2020). Adopting a system approach that focuses on health policy for flexible, resilient, sustainable and equal resilient systems.

23.2.3 Adaptation

The occurrence of epidemics over the last 30 years, have resulted in increased adaptation strategies for tackling them. The outbreaks of such epidemics and pandemics leads to disruption of transportation, food, health, tourism, education and services, forgoing output, inflated prices, crisis-induced budget deficit and poverty (Warsame et al. 2020). The supply chain gets impacted the most which impacts all sectors of the society. Systemic shift is also observed across the globe both in developed and developing nations from expenditure from luxury to essentials such as life saving drugs, enhancing the resilience structures.

23.2.4 Awareness

Awareness is an integral component of enhanced to enhance the institutional resilience. During the Ebola outbreak spreading awareness worked strategically for moving beyond the crisis. It helped in Ebola epidemic resilience improvement for Liberia's health system (Ling et al. 2017). Additionally, enhancement of social media networks to guarantee viability would help in increasing the awareness. The centrality of the sharing of knowledge will increase the management of the consequences of epidemics. The sharing of value as a means of spreading good practices and learned lessons and building solidarity in order to amplify the voices of communities that need solutions. Science must be measured as these forms of learning. Small farmers, indigenous people, women and young people must also actively include our voices.

23.2.5 *Transformation*

The national-local relationship is vital to ensuring continuity between policy and implementation. The implementation of preparedness, scoping, adaptation and awareness causes the transformation of multiple concepts, design and analysis into the system (Ganin et al. 2016). Transformation in all forms of structural reformation and prioritization results into strengthening the implantation of resilient framework. These are specific to countries as each has different sets of resources and capacities of dealing with epidemics.

23.3 Sectoral Analysis

23.3.1 *Heath Sector*

After the 2014–15 West African Ebola epidemic, resilience has been widely identified as a key element in sounds of health systems. Lack of disease control and suspension in the operation of many health services in Liberia, Sierra Leone and Guinea has shown that health systems must plan for resilience long before the crisis occurs (Ling et al. 2017). However, the literature does not explain very well the operative elements of resilience and how a crisis experience will shape resilience. This is in agreement with the current pandemic of COVID-19 empathizing on the key lessons for disaster resilience. In addition community leaders and local health participants, numerous studies highlight that the prioritization roles of global and regional actors have changed in the course of the epidemic. Although the epidemic of COVID-19 has encouraged some positive adaptations to the health system in India, building a genuinely resilient health system will require long-term investments and long-term attention (Welby-Everard et al. 2020).

Resilience of the health system has been the ability of health workers, institutions and populations to prepare for and react effectively to crises; to retain key functions when crisis strikes; to reorganize, if conditions are required, in the knowledge of the lessons learned during crises. Healthcare staffs were 10–20 times more likely than members of the public to be diagnosed with the corona virus. Sickened healthcare workers were unable to treat COVID-19 patients further, and inadequate infection control in healthcare centres helped to spread COVID-19. In turn, risks associated with COVID-19 infection have increased public fear of hospitalization. Health care services have become a significant source of distress and worsen the human, economic and political cost of the epidemic, rather than helping to contain COVID-19.

23.3.2 Industrial Sector

Innovation is driven by the ever more global background of businesses but concern about the transformation of the supply chain is also growing. The COVID-19 pandemic clearly indicates that supply chain resilience and the effect of disruption on the global network scale struggle with individual supply chains and nodes. This cascading failure highlights the need in the existing literatures of the current supply chain for network analysis and advanced resilience analysis (Golan et al. 2020).

The acceleration of the local product economy through the communing institutional framework of self-reliant products has found roots in all regions throughout India, which ultimately, had evolved to becoming the swadeshi movement. This will bring employment opportunities for low and medium class workers, thereby reducing their migration in near future. While the COVID-19 crisis is just another crisis within the institutional system for producers and consumers alike, the COVID-19 crisis highlights several weaknesses within the industrial system that could, in the long term, have an effect in terms of ensuring our economic safety (Sharma et al. 2020). During the Post lockdown periods, the factories and industries are facing huge industrial accidents as a result of lack of proper implementation of guidelines. The current example of gas leakage in Visakhapatnam in India. This further worsens the situation of poverty and highlights the need for institutional resilience among industrial sector.

23.3.3 Educational Sector

The study examined how COVID-19 distressed the universe, its harmful factors in particular to the constituents and to teachers and students. However, it is also appropriate to extend the range of teaching methods, where the final scenarios that sometimes classes in classrooms are insignificant are more likely 21st pedagogues that can be conducted through the social media (online blogs, group chats, webinars or e-mail), except for the formative assessment. The study also demonstrated that methods of teaching based on research results are more desirable as if students are taught how to be a writer and a researcher (Masuhay 2020).

This COVID-19 is, however, reciprocally denoting conventional manoeuvres and thus practically does not need to participate; however on the other hand, the modular definition has resulted in the question being resolved. This COVID-19 is definitely an uprising to start with a new line of education, but it also endangers the educators, whose only path to the learning process is now in the online platform where they are exposed and forbidden to be safe. Sceptical excuses were given for those who were not concentrating, molested, or were studying or working, even challenging the self-government for security reasons, banning known infected citizens and economically suppressing the Local Government Unit (LGU) (Masuhay 2020).

23.3.4 Food and Agriculture Sector

Food and agriculture sector is one such sector on which more than half of the global population depends (FAO 2018). With COVID-19 crisis and imposition of lockdown measures globally, agriculture demand and supply has been badly affected. With no takers, the market and export demand are hampered. Farm produces are unable to reach market due transportation restrictions and even if they manage to reach the markets, there are no buyers. The Indian agriculture sector employs almost 50% of its labour task force (FAO 2018). With unavailability of labourers harvesting is delayed and crops are getting destroyed in certain areas. In addition the availability of a skeletal work force has made loading and unloading of market products a challenge with increased labour cost which is ultimately impacting the marginal farmers. Delayed transportation and crop resting in markets for long time is causing an increase in moisture content of the harvest and crop rots. As a result there is an increase in organic waste production and subsequent increase in emission of methane due to the waste generated (Henriques 2020). To sum up the agriculture sector globally is facing the following challenges:

- Market and farm price increase
- Supply chain shortage and slowdown
- Farmers health and well-being
- Reduced farmer workforce

However despite such Challenging situation the pandemic has also opened up new opportunities for the agriculture sector through:

- Improving/expanding the lesser developed sector
- Greening the supply chain
- Creation and revamping of rural infrastructure both farm and non-farm like godowns water management and further processing
- Restructuring reshaping of public distribution systems keeping in mind long-term strategies

23.3.5 Hospitality and Tourism Sector

- The famous tourist destinations which once used to be packed with people have become now shadows of themselves (Brouder 2020). The impact of the pandemic on tourism and the hospitality sector has been far reaching and countries like Maldives which primarily survived on tourism are facing the maximum blow. As countries try to contain the COVID-19 cases, travel restrictions have been imposed globally with airlines non operational and hotels, restaurants and home-stays closed. Almost by 96% (UNWTO 2020a) of the worldwide destinations have either partially or completely closed their borders for tourists, imposed travel restriction on specific destinations and suspended travel visas.

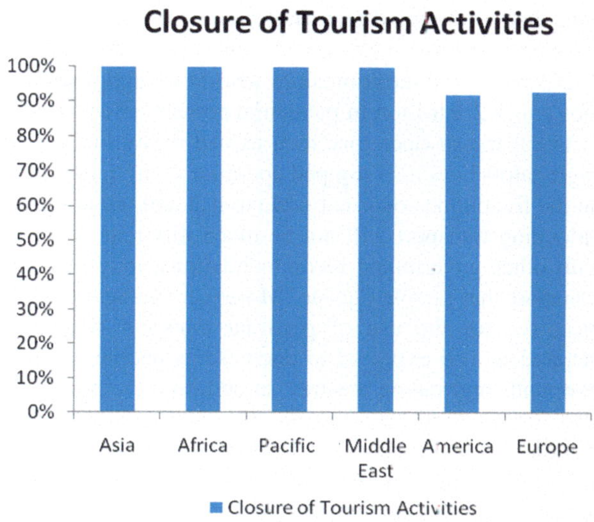

Fig. 23.2 Worldwide restrictions on travel activities

The current sensitive global situation may even lead to rolling back of the progress made towards sustainable development through equality and economic growth. Figure 23.2 below depicts the restrictions on travel and tourism imposed by countries worldwide.

The industry provides 10% to the global GDP, but with the pandemic situation in hand, a substantial dip in this figures are expected as 50 million jobs in the sector are at risk worldwide (Faus 2020). As the globe recovers from the impacts of COVID-19, the tourism and hospitality sector is slowly coming back on track. The recovery plans for the sector requires multi-stage action plans promoting 'hyper local tourism' first including intra district and inter district tourism, followed by interstate tourism.

However, the sector holds promising potential for new opportunities to unfold. There is a need to restructure the existing tourism scope to include new options like; (1) promotion of wellness centres' (2) conversion of small hotel facilities into quarantine centres, (3) promoting family travelling options and (4) promoting group tourism with small social bubbles maintaining proper hygiene standards (UNWTO 2020b).

23.3.6 Transport Sector

Public transport including airways, railways, roads and waterways, is seen as vectors for the transmission of Corona Virus. With the closure of many countries and the closing of national and international borders, the transport at sector has been severely affected by the impacts of COVID-19. In many countries, the national implementation of curfew regimes, the practice of social distancing and other policy regimes have also contributed to a sharp reduction in demand for transport facilities. The associated economic losses are enormous. According to the World Travel and Tourism Council the companies worldwide are projecting a global loss of 75 million jobs and $ 2.1 trillion in pandemic revenue (Becker 2020).

Local transit operators such as metro city busses auto rickshaws taxis and cab aggregators have also stopped operations within the city. This has also had an impact on the livelihoods of small-scale transit and travel operators. The future scenario of unlocking transport will not be much different. Social distancing standards, along with other government recommendations, may affect the frequency and mode of transport that people choose. Moreover, transport-related health risks increase the incentive for the use of personal transport (Papandreou 2020). Transportation demand is also expected to decrease as people avoid going to crowded locations like malls and movie theatres. In addition, the practice of working from home will also contribute to the situation. This will result in a behavioural shift and transformation of urban lifestyles (Chauhan 2020). Howsoever this could also be seen as a new opportunity for the transportation sector to shift from the business as usual scenario and pick up new business models which may include (Deloitte 2020):

- Long-term investment programmes may need to be re-planned and re-prioritised.
- Organisation planning needs to be restructured to imbibe flexibility in terms of business operations.
- New Community travel plans may be implemented which may include-
  - Greener modes of travel
  - Prioritising travel routes
  - Developing and Promoting regional rapid transit systems

23.4 Preventing Systems from Falling

With the growing concern of environmental challenges/climate change as well as emerging epidemic scenarios it is best to think about development of sustainable production systems. The planning necessity in an organisation for any business or activity is growing to be the incessant logical development facing the society today (UNEP 2020). The role of management tools is critical when business activity planning is considered. Effective utilisation of management tools are potential actors for sustainable planning and production practises (UNEP 2020). This section talks about the some key tools which are effective when organizational planning is concerned.

23.4.1 Live Cycle Assessment (LCA)

It is a methodology for assessing environmental impacts of all processes associated with all the stages of product, service and processes. LCA is an important tool for industries, businesses and policymakers to strategise their decisions and operations effectively. LCA uses cradle to grave approach which provides a comprehensive evaluation of the entire organisational structure including up and down chain elements as well as third-party elements involved in the system. By doing so it reduces the influence of external pressure on performance and action approaches of organisations thus increasing the organisational resilience. The Fig. 23.3 below depicts the benefits offered by LCA.

23.4.2 Cost and Benefit Analysis (CBA)

Cost benefit analysis can be seen as a systematic approach which compares the benefits versus the costs that is the weaknesses and strengths when considering any decision in hand when policy, organisational and institutional planning is concerned. It assists in choosing between the appropriate alternative by understanding whether to go forward with a decision or not. It does so by comparing the cost and benefits that will flow from an alternative as a guide to which choice will bring the greater margin of benefit over the cost. Comparing the total costs v/s total benefits, it also considers the extent to which the benefits will flow and also identifies the potential stakeholders who will get benefited. By doing so CBA provides a scope for taking informed decision early in the planning process.

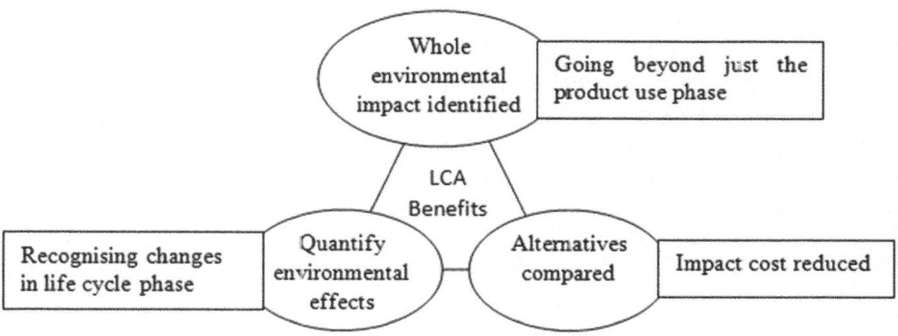

Fig. 23.3 Benefits of LCA

23.4.3 On-Site and Off-Site Planning

Emergencies call for immediate action and effective planning is essential to take up these actions. For any organisation to respond to emergencies on-site and off-site emergency planning becomes essential (Bachman 2010). On-site planning is specific to address impacts of any emergency situation that disrupt the internal operations, structures, or manpower. Of site planning addresses the emergency response need of the areas surrounding the organisation. Prior to developing operational planning, on-site and off-site studies serve as an important tool for prioritising the emergency action planning.

23.4.4 Impact Dependency Exercises

Businesses, organisations and institutions all depend on biodiversity (through the ecosystems and associated ecosystem services) for smooth functioning of their operations and activities. Any disruption in the essential ecosystems and their associated services on which the businesses depend will immensely impact their (structural and functional) operations. Conducting impact dependency exercises helps the business to future proof their operations through (LfN 2012):

- Safeguarding future supply of essential resources
- Enhance resource efficiency
- Secure future license to operate
- Better plan business decisions and actions
- It also helps in identifying the business risks and tap new opportunities for markets and products.

23.4.5 Vulnerability Assessment

Environmental challenges may it be disasters or epidemics hamper the organisational processes through interrupted services and damaged infrastructure. Vulnerability assessment is a tool which helps in identifying the hotspots i.e. zones/areas which are at maximum risk from hazards. Conducting vulnerability assessment helps to understand the required adaptive responses—whether the organisations structure and processes have the required coping and adaptive capacity to take up or go forward with specific activities.

23.4.6 Business Continuity Planning

Business continuity planning is a proactive approach to understand the weaknesses and threats of an organisation/institution when any form of crisis strikes. The main approach is to keep running their essential functions and ultimately recover with minimum downtime thus preventing systems from falling (Rouse 2020). An effective planning enables organisations to react quickly and efficiently in times of unpredictable events, establishing risk management processes to prevent disruptions in operations. A business continuity plan works around three principal components which are:

- Resilience: Considering multiple adversities during planning and designing of operations ensures organisational resilience for multiple hazards.
- Recovery: Focusing on only the operations including crucial/essential services during crisis should be the preference followed by setting up of recovery time objective for the rest of the operations.
- Contingency Plan: Ensures that appropriate procedures are in place for various internal work plans as well as value chain points.

23.5 Recovering from Disasters

23.5.1 Disaster Recovery Planning

It involves strategising operations recovery post-crisis events. It includes step-by-step procedures oriented to minimise the impact of disasters and regain operations with very little recovery time. Business Impact Analysis and Risk Analysis are the main components of the disaster recovery planning and must be performed before preparing and implementing disaster recovery plans (Rouse 2020). However testing of these plans is extremely important to identify and address the loopholes for successful implementation of the plan in need. Disaster recovery plan should be made with a integrative approach to include a broader spectrum of disasters. The Fig. 23.4 below provides the relation between business continuity and disaster recovery planning (Rouse 2020)

23.6 Maintaining Resource Flow

23.6.1 Value Chain Management

Maintaining resources during crisis/epidemics should involve implementing appropriate value change management strategies. It ensures the flow of materials and deliverables of end products by covering the entire range of supply chain. It will

Fig. 23.4 Business Continuity and Disaster recovery Planning

respond to multiple fronts including health safety of the workers and safeguarding operations viability with immediate as well as long-term responses (Company 2020).

23.6.2 Safety Audits

Conducting continuous evaluative safety audits are most effective ways for organisations to test their health and safety measures. This evaluates the organisation's working conditions and the organisation's attitude towards its health and safety measures. It further helps organisations to evaluate and optimise their occupational health and safety programmes and improve and improvise the internal safety management standards (Melo 2018).

23.6.3 Standard Operating Procedures

These are step-by-step instructions by organisations to carry out complex operations. SOPs provide quality outputs and performance uniformity which increases operation efficiency while reducing miscommunications and failures.

23.6.4 Circular Economy

The business as usual scenario works on the linear chain model. This has led to disruption of natural resources, increased social and is constantly pushing the extent of planetary boundaries. COVID-19 pandemic has exposed the fragility of the existing liner models and has provides an opportunity to establish the new normal that is a shift from linear to circular economy models (Wit and Haigh 2020).

Circular models redefine growth focusing on positive society wide benefits and shifts in approach from development based on consumption from finite resources and designing waste out of system to managing waste of one as input for the other. Circular economy models are guided by the notion of resilience and are build on circularity and can be instrumental in creating a sustainable model that prioritises reliability over growth (Wit and Haigh 2020). Circular economy has also been encouraged by the European councils who have put green transitions as part of their comprehensive recovery planning.

Transitioning towards circular economic models promotes inclusiveness and brings in a systematic shift in building long-term resilience by generating new business opportunities with social and environmental benefits improving the overall health of the systems (Macarthurfoundation 2017). The Fig. 23.5 below depicts the principles of circular economy.

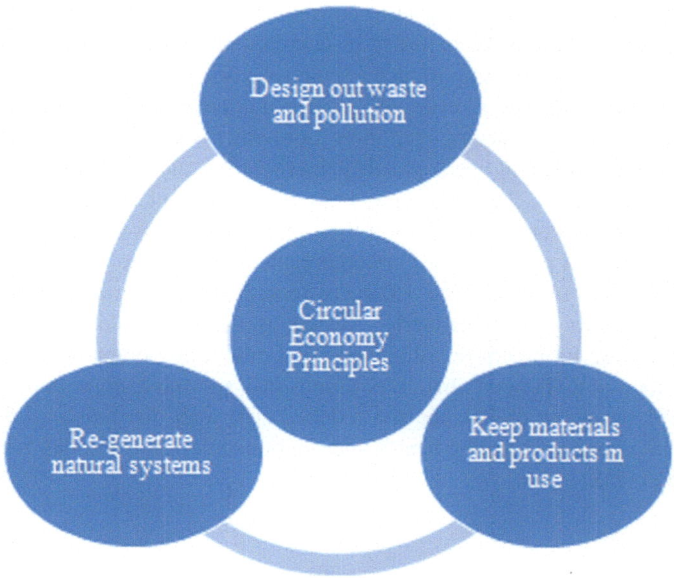

Fig. 23.5 Principles of Circular Economy

23.7 Policy Recommendations

Based on a review of our existing literature, we identified five major issues not addressed by the previous analysis including the need: the care standard to be decided when demand exceeds the resources; a remedial plan after the event; while the issues we search for are consistent with the five elements of a robust institutional framework. It is certainly a challenge to quantify practical threats and evaluate new disease epidemics' vulnerabilities and effects. We analyzed the institutional impact of the different epidemics of infectious diseases by considering a definition of system resilience that reflects not only the number of people who have been infected, but also the need to maintain certain key functions in society and time. We believe that the limitations and intervention in human mobility hinder the efficiency of the device in particular. It is observed that containment measures which restrict the mobility of individuals are beneficial in reducing risks but can deteriorate the functionality of the system for very long time. Though we have only considered two of the many dimensions of socio-technical systems functionality, we demonstrate that resilience studies enable stakeholders to measure and differentiate the impact of epidemic threat between the various managerial alternatives. It is simple for the vital features to be more practical. The components of critical functionality may be evaluated in terms of their importance to stakeholders through objective or subjective assessment. Organizations should offer operational flexibility in face adversities and emergencies, which can be achieved by effective driver models like business continuity planning. In the study of the social consequences of a disease that may lead to long-term consequences or even deaths of the people affected should eventually be included in cost-benefit analysis and moral implications. This study shows the importance for the selection of intervention strategies of resilience-oriented analysis. Conservative rational approaches to the management of risks associated with emerging and new epidemics may lead, for example, to unnecessary burden-saving and possibly inefficient actions such as quarantines and travel prohibitions. The emerging field of assessment and administration of resilience and its implementation could therefore assess cross-dominal alternatives, in order to identify a policy design which enhances the system's ability to (1) plan adverse events, (2) absorb stress, (3) recover and (4) prevent and prepare future stressors by adapting them properly. System planning should imbibe inclusive approaches, fitting in the entire value chain and integrating components of LCA in planning processes. To end this, the framework we have presented can be of potential use by trying to balance the reduced risk with disruption to the critical functions associated with public health interventions to optimize policy responses to a disease outbreak.

References

Aguirre B, Dynes RR, Kendra J, Connell R (2005) Institutional resilience and disaster planning for new hazards: insights from hospitals. J Homeland Secur Emerg Manage 2(2)

Bachman EG (2010) Off-site emergency response plans: a preparedness tool. FireEngineering. https://www.fireengineering.com/2011/11/01/222171/off-site-emergency-response-plans-a-preparedness-tool/. Accessed 3 Jun 2020

Barton MA, Christianson M, Myers CG, Sutcliffe K (2020) Resilience in action: leading for resilience in response to COVID-19. BMJ Leader, pp leader-2020

Becker E (2020) How hard will the coronavirus hit the travel industry? National Geographic. https://www.nationalgeographic.com/travel/2020/04/how-coronavirus-is-impacting-the-travel-industry/. Accessed 3 Jun 2020

Bevc CA, Simon MC, Montoya TA, Horney JA (2014) Institutional facilitators and barriers to local public health preparedness planning for vulnerable and at-risk populations. Public Health Rep 129(6_suppl 4):35–41

Brouder P (2020) Reset redux: possible evolutionary pathways towards the transformation of tourism in a COVID-19 world. Tourism Geogr 22:1–7

Chauhan M (2020) COVID19 and its effects on Rail and Transport Industry. Urban Transport News. https://www.urbantransportnews.com/covid-19-and-its-effects-on-rail-transport-industry/. Accessed 3 Jun 2020

Company M (2020) Value based management. mckinsey.com: https://www.mckinsey.com/business-functions/sustainability/how-we-help-clients/circular-economy. Accessed 3 Jun 2020

Dausey DJ, Moore M (2014) Using exercises to improve public health preparedness in Asia, the Middle East and Africa. BMC Res Notes 7(1):474

Deloitte (2020) Understanding COVID-19's impact on transport organizations. DELOITTE, London

FAO (2018) FAO in India. FAO of UN. http://www.fao.org/india/fao-in-india/india-at-a-glance/en/. Accessed 3 Jun 2020

Faus J (2020) This is how coronavirus will affect the travel and tourism industry. World Economic Forum. https://www.weforum.org/agenda/2020/03/world-travel-coronavirus-Covid19-jobs-pandemic-tourism-aviation/. Accessed 3 Jun 2020

Ganin AA, Massaro E, Gutfraind A, Steen N, Keisler JM, Kott A et al (2016) Operational resilience: concepts, design and analysis. Sci Rep 6(1):1–12

Golan MS, Jernegan LH, Linkov I (2020) Trends and applications of resilience analytics in supply chain modeling: systematic literature review in the context of the COVID-19 pandemic. Environ Syst Decis 1

Henriques M (2020) Will Covid-19 have a lasting impact on the environment? BBC Future. https://www.bbc.com/future/article/20200326-covid-19-the-impact-of-coronavirus-on-the-environment. Accessed 3 Jun 2020

Kinsman J, Angrén J, Elgh F, Furberg M, Mosquera PA, Otero-García L, Snacken R, Derrough T, Santisteve PC, Ciotti M, Tsolova S (2018) Preparedness and response against diseases with epidemic potential in the European Union: a qualitative case study of Middle East respiratory syndrome (MERS) and poliomyelitis in five member states. BMC Health Serv Res 18(1):528

LfN I (2012) Why follow the natural capital roadmap? Leaders for Nature. http://leadersfornature.in/get-started. Accessed 3 Jun 2020

Ling EJ, Larson E, Macauley RJ, Kodl Y, VanDeBogert B, Baawo S, Kruk ME (2017) Beyond the crisis: did the Ebola epidemic improve resilience of Liberia's health system? Health Policy Plan 32(suppl_3):iii40–iii47

Macarthurfoundation (2017) What is a circular economy? https://www.ellenmacarthurfoundation.org/. https://www.ellenmacarthurfoundation.org/circular-economy/concept. Accessed 4 Jun 2020

Massaro E, Ganin A, Perra N, Linkov I, Vespignani A (2018) Resilience management during large-scale epidemic outbreaks. Sci Rep 8(1):1–9

Masuhay EP (2020) Adversaries on Covid-19 set forth an argument onward to educational endeavor: resulting to develop a modular concept in the learning process. Am Int J Soc Sci Res 5(2):26–37
Melo S (2018) The importance of conducting safety audits. DataScope.com. https://www.mydatascope.com/blog/en/2018/11/16/the-importance-of-conducting-safety-audits/. Accessed 3 Jun 2020
O'Brien G (2008) UK emergency preparedness: a holistic response? Disaster Prev Manage 17 (2):232–243
Papandreou T (2020) Is the coronavirus the transportation industry's opportunity? FORBES. https://www.forbes.com/sites/timothypapandreou/2020/03/27/is-the-coronavirus-the-transportation-industrys-opportunity/#62f7002752b7. Accessed 3 Jun 2020
Pulla P (2020) Covid-19: India imposes lockdown for 21 days and cases rise. BMJ 368:m1251
Romaní J, Baselga E, Mitjà O, Riera-Martí N, Garbayo P, Vicente A et al (2020) Chilblain and acral purpuric lesions in Spain during COVID confinement: retrospective analysis of 12 cases. Actas Dermosifiliogr 111(5):426–429
Roser M, Ritchie H, Ortiz-Ospina E, Hasell J (2020) "Coronavirus Pandemic (COVID-19)". Published online at OurWorldInData.org. https://ourworldindata.org/coronavirus [Online Resource]
Rouse M (2020) What is Business Continuity and why is it important. Search Disaster Recovery. https://searchdisasterrecovery.techtarget.com/definition/business-continuity. Accessed 3 Jun 2020
Sharma A, Rangarajan D, Paesbrugghe B (2020) Increasing resilience by creating an adaptive salesforce. Ind Mark Manag 88:238–246
UNEP (2020) Benefits of life cycle approaches. Life Cycle Innitiative. https://www.lifecycleinitiative.org/starting-life-cycle-thinking/benefits/. Accessed 3 Jun 2020
UNWTO (2020a) COVID-19 response: 96% of global destinations impose travel restrictions, UNWTO reports. UNWTO.org. https://www.unwto.org/news/covid-19-response-travel-restrictions. Accessed 3 Jun 2020
UNWTO (2020b) COVID-19: putting people first. UNWTO.org. https://www.unwto.org/tourism-covid-19. Accessed 3 Jun 2020
Wang J, Wang Z (2020) Strengths, weaknesses, opportunities and threats (SWOT) analysis of China's prevention and control strategy for the COVID-19 epidemic. Int J Environ Res Public Health 17(7):2235
Warsame A, Blanchet K, Checchi F (2020) Towards systematic evaluation of epidemic responses during humanitarian crises: a scoping review of existing public health evaluation frameworks. BMJ Glob Health 5(1):e002109
Welby-Everard P, Quantick O, Green A (2020) Emergency preparedness, resilience and response to a biological outbreak. BMJ Military Health 166(1):37–41
Wit MD, Haigh L (2020) Covid-19: The tragedy to end all tragedies of the commons? Covid-19: the tragedy to end all tragedies of the commons?. https://medium.com/circleeconomy/covid-19-the-tragedy-to-end-all-tragedies-of-the-commons-1cb5d1be012e. Accessed 5 Jun 2020
Zhang Q, Deng H, Zhang C (2020) The value of early response by surrounding areas of epidemic center Hubei during COVID-2019 outbreak in China: a quasi-experiment analysis. Available at SSRN 3548372

Chapter 24
Lessons from Trans-Domain Assessment of COVID 19 Outbreak

Anjali Barwal, Atisha Sood, Anil Kumar Gupta, and Manish Kumar Goyal

Abstract The novel coronavirus pandemic has shown that humans are not impeccable and countries and communities are vulnerable and needed to be prepared. Coronavirus disease 2019 (COVID-19) is a global health emergency, and the ways countries across the globe are striving to hinder the disease's outbreak are wide-ranging. Worldwide nations are responding quite differently to the virus outbreak. The response to the coronavirus outbreak has been highly criticized by experts around the world. A delay to detect the onset of coronavirus symptoms and response by the government has been recorded, not only in China but in other major developed countries as well, which in turn threatened and overburdened the local healthcare systems. On the other hand, many other countries have put in place proactive efforts and effective strategies to track down and isolate the symptomatic and asymptomatic infections. Based on the experience of many countries, this chapter aims to draw out the key lessons learnt from successful and failed practices implemented against COVID-19 outbreaks across the globe.

Keywords Pandemic · Coronavirus · Covid-19 · Lockdown · Success stories · Failure stories

---

A. Barwal (✉) · A. Sood
National Institute of Disaster Management, Ministry of Home Affairs, Government of India, New Delhi, Delhi, India

A. K. Gupta
Department of Environment and Disaster Risk Management and Centre for Excellence on Climate Resilience, National Institute of Disaster Management, New Delhi, Delhi, India

M. K. Goyal
Department of Civil Engineering, Indian Institute of Technology, Indore, Madhya Pradesh, India
e-mail: mkgoyal@iiti.ac.in

M. K. Goyal, A. K. Gupta (eds.), *Integrated Risk of Pandemic: Covid-19 Impacts, Resilience and Recommendations*, Disaster Resilience and Green Growth,
https://doi.org/10.1007/978-981-15-7679-9_24

24.1 Introduction

Almost all the new contagious diseases that humans have visualized today emerged in animals, and so will the emerging diseases of tomorrow. Human immunodeficiency virus (HIV) jumped into humans from chimpanzees, severe acute respiratory syndrome (SARS) from Chinese horseshoe bats (*Rhinolophus sinicus*) and influenza from wild aquatic birds (Cohen 1999; Krauss et al. 2004; Lau et al. 2005; Rabaan et al. 2020). At some point the animal pathogens overcomes the natural species barrier to humans, an event public health experts call a ‘spillover event’. Spillover infections have always occurred, but the rapid and abrupt environmental change inflicted by human beings in recent past years has stimulated the disease outbreak (Carrington 2012).

Coronavirus disease 2019 (COVID-19) is more contagious than other coronaviruses such as SARS or Middle East Respiratory Syndrome—coronavirus (MERS-CoV). On January 30, 2020, the World Health Organization (WHO) declared the COVID-19 outbreak as the sixth public health emergency of international concern (PHEIC), following H1N1 (2009), polio (2014), Ebola in West Africa (2014), Zika (2016) and Ebola in the Democratic Republic of Congo (2019) (Lai et al. 2020a).

Disease knows no borders. In December 2019, the coronavirus was seemingly confined to China. But within few weeks, the virus became a global pandemic. There has been a continuous rise in the daily total number of COVID-19 cases worldwide, as confirmed cases has surpassed 7.2 million with the death toll exceeding 408,000 across over 213 countries and territories, on June 9, 2020 (Worldometers 2020). Figure 24.1 represents the world map showing coronavirus confirmed cases in various countries.

Microbes evolve about 40 million times as fast as humans do. The longer the virus spends in humans, the better the chance that it might mutate to become more contagious (Walsh 2017). This new coronavirus disease is less fatal—thousands of people showing mild or no symptoms (asymptomatic) have been spreading the virus unaware that they were even infected (Callaway et al. 2020). This means that before public health experts were aware of the disease and started to recommend control measures, the virus had already spread rapidly within and other countries.

China announced the first COVID-19 death of 61-year-old man who had purchased goods from the seafood market in Wuhan on January 11, 2020. On January 13, Thailand became the first country outside China to report patient with coronavirus having travel history to Wuhan city. Later, virus started to spread rampantly in several countries including the United States, Nepal, France, Australia, Malaysia, Singapore, South Korea, Vietnam and Taiwan reporting positive cases of coronavirus. Policymakers and medical communities around the world have been struggling to combat this rapidly escalating pandemic (Alandijany et al. 2020).

COVID-19 has changed the world forever. It has written new pages in world history. Worldwide countries are responding quite differently to the COVID-19 outbreak. A range of strategies have been adopted across the world based on the

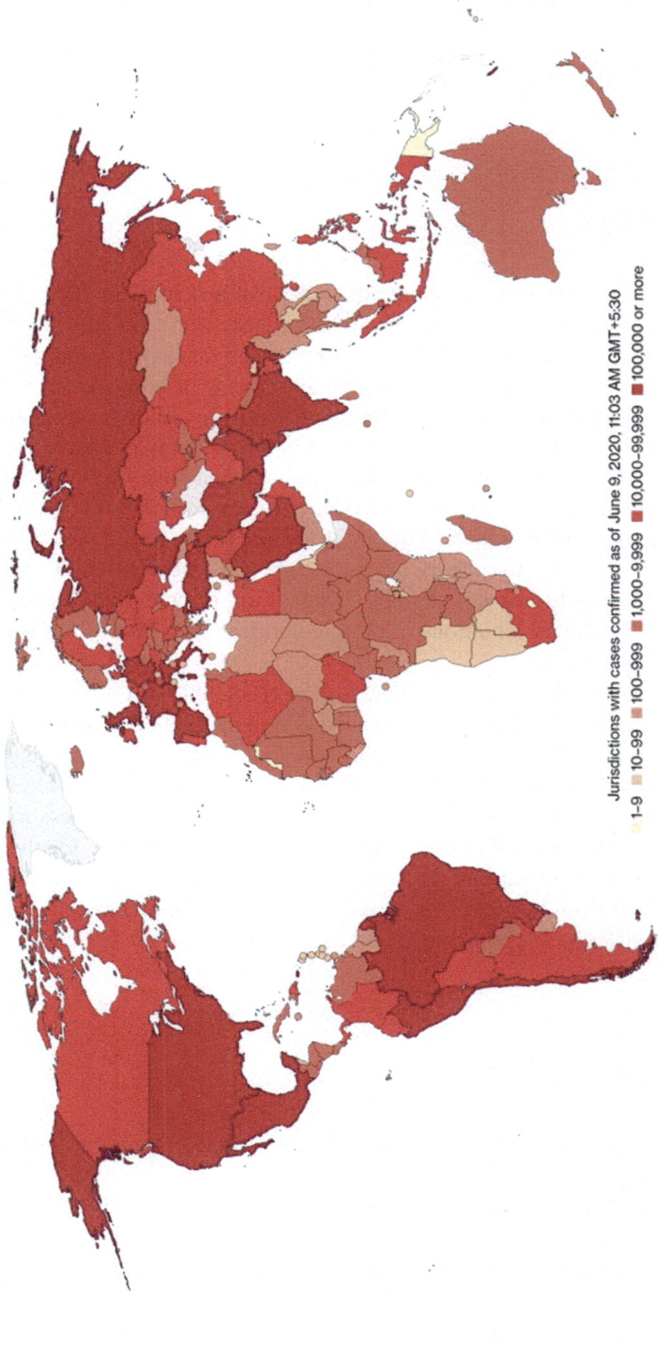

Fig. 24.1 World map showing countries with COVID-19 confirmed cases (as on June 9, 2020). (Sources: Johns Hopkins University 2020 and Bloomberg.com 2020)

population structure, political governance, socio-economic responses and the health care infrastructure of individual country. Efforts of many countries have been failing to tame this novel coronavirus. Whereas, many other countries have put in place much effective strategies and proactive efforts to track down and isolate the symptomatic and asymptomatic infections.

24.2 Failures

In many countries, prolonged nationwide lockdown was implemented to reduce and curb the spread of COVID-19, the respiratory illness caused by the novel coronavirus. But how exactly does a lockdown worked? The main purpose of the lockdown was to reduce virus infection by urging people to stay at home, close non essential activities or avoid public events and mass gatherings. Lockdown was also an attempt to flatten the curve. But it also proved to be a failure at many places where it was not imposed at the right time. One of the major failures that have been faced is the delay in the implementation of necessary and precautionary measures and poor healthcare infrastructure in many countries. The countries like China, Iran, Italy, Spain, the UK and the US were badly affected by coronavirus because they had delayed implementing full lockdowns, opting instead for only partial lockdowns. This enabled the COVID-19 situations in these countries to progress from the second to the dreaded third stage.

The lockdown also lead to many health risks that can render people more vulnerable to novel coronavirus and related complications. Many people are having a high burden of diabetes mellitus and hypertension, both conditions aggravated by cereals and wholegrain foods. The lockdown has certainly impacted the food security of large number of people and they are also unable to access medicines or screening tests. People with co-morbidity issues are at high risk for the complications of COVID-19 infections (Hussain et al. 2020). In effect, vulnerable communities experience higher risk for developing more complications and increased mortality because of the poorly-planned lockdown.

24.2.1 Mainland China

The Chinese government locked down Hubei Province on January 23, 2020, including Wuhan city, having population of 11 million people, where the outbreak started (Tanne et al. 2020). The unprecedented lockdown of 60 million of people was considered a "vast experiment". Government halted all public transport services and barred millions of people from working or going to school and closed all shops in markets and shopping complexes, except those selling medicines and grocery items.

The country has suffered socially and economically not just because of the virus but because of aggressive public health policies that government has enacted. Though China followed a high-tech mass surveillance system, but it was observed that it was not done with the concept of understanding people's participation. If China had executed control measures at the beginning of January, 67% of all cases could have been prevented; it would have reduced number of infection to 5% of the total (Khanna et al. 2020). The events in Wuhan show that for at least 3 weeks after the first case (originated in mid November) was reported, city authorities had been informed about the virus spreading but they issued orders to curb the spread of coronavirus. For instance, on January 18, roughly 6 weeks after coronavirus started to spread in Wuhan, they allowed the city's Baibuting district to organize its traditional annual mass banquet. Of the 40,000 families attending, more than 28,000 got infected and more than 560 succumbed to the virus in the later weeks. Just a few days later, more than five million people had traveled out of Wuhan for the upcoming Spring festival, just before Wuhan imposed lockdown, thus spreading COVID-19 across other provinces of China and abroad. The elevated connectivity of Wuhan with other international airports such as Singapore, Japan, and Thailand, facilitated the rapid spread to bordering countries (Khanna et al. 2020; Bogoch et al. 2020).

24.2.2 The United States of America and the United Kingdom

European countries like the United Kingdom, the United States, Italy and Spain delayed implementing containment measures and failed in anticipating the impact of the pandemic in their own boundaries. Unfortunately, throughout the United States and most of Europe, policymakers and medical communities have been struggling hard to keep up with the spreading pandemic.

The United States of America and the United Kingdom, both countries' national governments have made just about every possible mistake in addressing novel coronavirus. They ignored the warnings from health experts. When they were forced to act, they sent mixed messages to citizens on how to respond which encouraged many to act in ways which spread the virus. Both countries did not make any efforts to control and reduce the further spread of this contagious disease. They have failed to equip their hospitals and healthcare workers with the equipment they needed, triggering many preventable deaths.

The U.S. has maximum number of positive cases than any other country. As of June 9, there are more than two million confirmed cases with death toll exceeding 113,000 across its all 50 states (Worldometers 2020). For the first time in history every U.S. state is under a disaster declaration (Lampen et al. 2020). Figure 24.2 depicts the number of COVID 19 cases in the United States of America showing that New York has had the largest outbreak in the country.

The United States is the only industrialized country with no consistent national health policy (The U.S. Health Care System 2016). Policy on investigating contacts

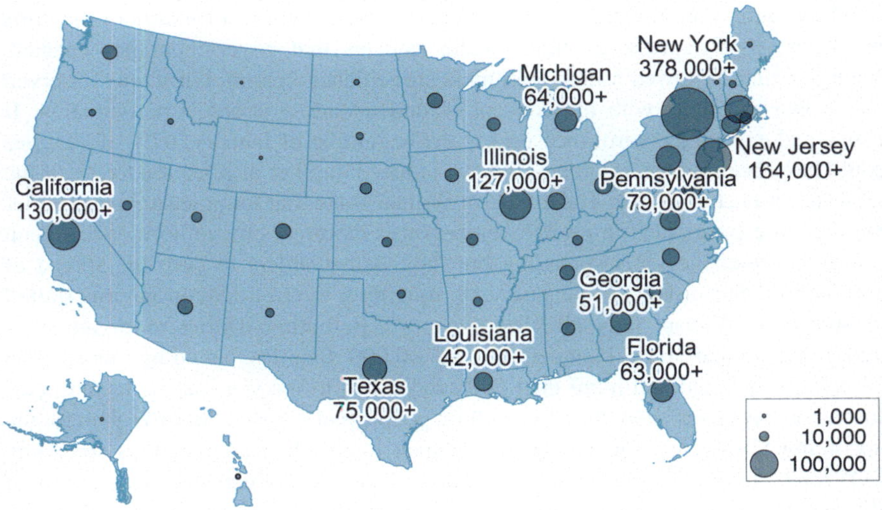

Fig. 24.2 Number of COVID 19 cases in the United States of America (as on June 8, 2020). (Sources: Johns Hopkins University 2020 and BBC News)

and quarantining varied from city to city. There was a lack of coordination in the national response, with an unclear message from the U.S. President with often a variance with information from the U.S. National Institutes of Health (NIH) and the Centers for Disease Control (CDC).

A scientific study from researchers at Columbia University found that if the U.S. had implemented social-distancing measures in early March, it could have prevented about 36,000 fewer Americans lives (Lampen et al. 2020; Glanz and Robertson 2020).

There was also an issue with testing due to a shortage of testing kits (Gostin and Hodge 2020). If the country had accurately tracked the spread of the virus, hospitals would have had the opportunity to be prepared and executed their pandemic plans effectively and on time. New York City hospitals were worsened by a shortage of PPE such as gloves, gowns, and masks and limited availability of ventilators and intensive care beds (Emanuel et al. 2020).

Similarly, the U.K. comes in the list of hardest hit countries, having more than 287,000 positive cases with death toll of 40,000. In U.K., authorities initially sought to trace and test every single person who had been in contact with virus-infected people. But later, they abandoned this strategy in mid-March as the number of infections overwhelmed the country's limited testing capacity (Deccan Herald 2020).

24.2.3 Other European Countries

Italy was the first western country to be slammed by the COVID-19 outbreak (Live Mint 2020). The infectious disease was first confirmed to have spread to Italy on January 31, 2020, when two Chinese tourists in Rome tested positive for the virus. A cluster of cases were later detected and by the beginning of March, the infectious disease had spread to the entire country. Italian cabinet took an early action and discontinued all flights to and from China and declared a 6-month state of emergency, prohibited people's movement within the whole territory, and the closure of all non-essential activities. In Italy, this pandemic turned into a disaster and has been the country's biggest crisis since World War II (Pisano et al. 2020).

The most effective time to take strong action is extremely early, when the threat appears to be small—or even before there are any cases. In its early stages, this infectious disease in Italy looked nothing like a crisis. Indeed, it was a systematic failure to absorb and act upon existing information rapidly and effectively rather than a complete lack of knowledge of what ought to be done. Moreover, in late February some notable Italian politicians were seen engaged in public handshaking in Milan to make the point that the economy should not panic and stop because of the virus (Pisano et al. 2020).

In Italy, doctors treating pneumonia in January and February were not aware about the novel coronavirus, since the infection symptoms were quite similar to those caused by flu and the virus was still believed to be largely confined to China. Even after the country recorded its initial COVID-19 case, medical experts didn't understand the unexpected behavior of coronavirus, with some patients experiencing shortness of breath.

The country was even forced to use 'home care' in part because of its low intensive care unit (ICU) capacity. Patients were treated and monitored at home by many primary care physicians but, this strategy proved fatal, and many died at home or soon after hospitalization, having waited too long to call an ambulance (Live Mint 2020). And this has probably be the determining factor of having a high mortality rate in Italy.

A steep increase in the number of cases has also been recorded in Spain, France, Russia, Germany just a few days later after the initial cases. Like Italy, the Spanish government also reacted late to the pandemic. The poor coordination and the delay in implementation of restrictive measures led an abrupt surge of cases in Spain before they enacted the lockdown on March 16, 2020 (Buescu et al. 2020). The shortage of hospital beds, ventilators, and health professionals became a concrete threat.

24.2.4 India

India reported its first COVID 19 positive case on January 30, 2020. Currently, India has maximum number of confirmed cases in Asia (Ray 2020). As of June 9, 2020, the Ministry of Health and Family Welfare (MoHFW) has confirmed a total of 267,643 cases and 7481 deaths in the country (MoHFW 2020). On March 22, India observed a 14-h voluntary public curfew. Further, from March 25, a nationwide lockdown was imposed for 21 days, affecting 1.3 billion population of the country. On April 14, the nationwide lockdown was extended till May 3, which was followed by further 2-week extension till May 17 with substantial relaxations. Later, from June 1, the Government started to unlock the nation (except containment zones) in three unlock phases (Banerjea 2020; Withnall 2020).

In India, lockdown has its effect in only to delay the entire curve of infections. It has shown that the lockdown, by itself, does not 'save lives' but only 'delays deaths' by a few weeks. Quickly escalated cases have caused significant problems for an already populous country, with a significant burden on the healthcare system. The 2011 Census reported that there are 1.77 million homeless people (staying in open or roadside, railway platforms, under flyovers, etc.) in the country, accounting for 0.15% of the nation's population (Goel and Chowdhary 2017; Banerjee and Bhattacharya 2020). Hence, the nationwide lockdown, the 'stay at home' order becomes a contradictory statement when it comes to the 1.77 million homeless individuals.

24.3 Success Stories of COVID-19

Some of the countries were successful in flattening the curve, including Singapore, Taiwan, Hong Kong, Vietnam, Bhutan and South Korea. Some of the important key factors in the success of these countries were continuous testing, isolation of infected, quarantining of contacts, closure of places of mass gathering, travel restrictions and case detections, were widely practiced. Apart from it, testing was scaled up aggressively and contact tracing was done through closed-circuit television (CCTV) and credit card transactions. The general community was also encouraged to inform about infected people.

24.3.1 China

Though China, where the virus emerged, had not implemented control measures at the very beginning. But the country has received praise for its handling of the crisis.

Chinese authorities built two dedicated hospitals—Huoshenshan and Leishenshan—in just over 1 week, adding 2300 beds to help ease an acute shortage in the central Chinese city of Wuhan (Zhu et al. 2020; Chen and Huang 2020). Healthcare workers from all over the country were sent to Wuhan city, the outbreak's epicenter. The government launched an unprecedented level of surveillance to track down the confirmed cases. In Wuhan alone, more than 1800 teams of epidemiologists, with a minimum of five or more people per team traced tens of thousands of contacts in each day (Kupferschmidt and Cohen 2020).

To tame coronavirus, China took very aggressive efforts and imposed an extremely harsh lockdown in the areas most affected. In Wuhan, armies of low-level enforcers were deployed to ensure citizens complied with the lockdown, and volunteers went door-to-door checking residents' temperatures. Those found to have fevers were sent to quarantine centers.

Amid the coronavirus outbreak, China has also banned the trade and consumption of wild animals, by saying that it would allow the country to win the battle against the coronavirus outbreak (World News 2020). Chinese authorities have announced new measures to prevent a second wave of the virus. Sick people are advised to wear a face mask or face covering in public, and anyone who does not cover their mouth and nose when coughing or sneezing could be fined. Social or physical distancing should be practiced by people to keep at least 1 m distance from each other and anyone who spits, defecate and litter in public would face a penalty. New set of rules have gone into effect from June 1.

China deterred low-income people pay for the COVID-19 tests and treatments. The risk of charging for testing, screening and treatment is that patients with early and mild-symptoms will not test, which could lead to further spread of the disease (Pinghui 2020). However, with all these measures, China prevented cases from increasing by 67-fold, i.e. without these measures there would have been eight million cases by the end of February (Lai et al. 2020b).

24.3.2 New Zealand

The south pacific nation of about five million populations has eliminated novel coronavirus after imposing 75 days of restrictions, including about 7 weeks of a stringent lockdown in which most businesses were shut, schools closed and everyone except essential workers had to stay at home. New Zealand managed the outbreak by imposing an 'Alert Level' system, shown in Fig. 24.3. Over 1500 people got infected and 22 died from the virus in New Zealand (Chaturvedi 2020). The country has lifted all the domestic restrictions on June 8, barring border closure restrictions, after its final COVID-19 patient was tested negative.

24.3.3 Singapore

When the most powerful countries in the world have failed to mitigate the spread of the virus, in spite of having a resilient health care systems, many Asian countries like Singapore, Taiwan, Vietnam, South Korea, Bhutan etc. have come up with very effective strategies to curb this contagious disease.

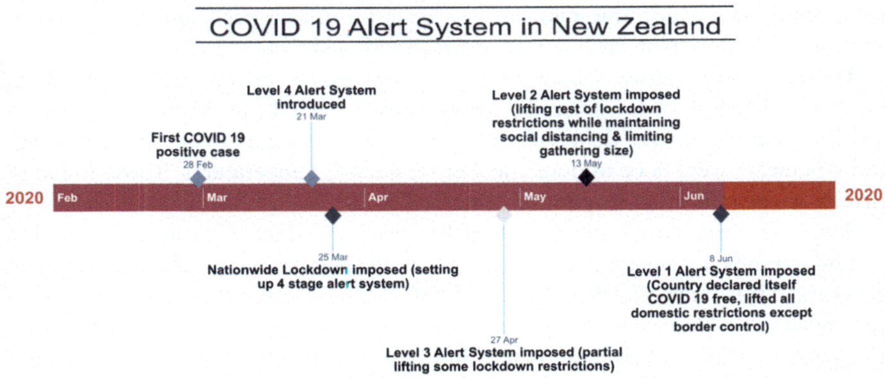

Fig. 24.3 COVID-19 alert system imposed in New Zealand

The unique and hard-to-replicate factors like, an outstanding health care system, stringent tracing and containment measures and a small population that's largely accepting of government's wide-ranging restrictive orders, are responsible for Singapore's containment success (Heijmans 2020). Country developed a robust infrastructure to deal with the virus and prepared itself well in advance to tackle the pandemic. As the news of coronavirus in China emerged in December 2019, Singapore banned all travelers from mainland China in late January. All incoming travelers and people entering schools and restaurant also had to check their temperatures. Schools remained open with staggered lunch time and break periods to avoid mass gatherings. It has also made efforts to develop a 'serological test' that is used to diagnose the antibodies produced in response to the coronavirus infection (Normile 2020). The government provided free testing facility to all its citizens. Mandatory quarantine was imposed to the people who have had close contact with confirmed cases (Carroll 2020).

The country has already experienced the past outbreaks, SARS 2003, in which 33 people died and swine flu known as H1N1in 2010, where more than 400,000 people were infected, meant that a very strong epidemiological surveillance system and a robust healthcare infrastructure was already in place to track and contain epidemics. These included ready-made government quarantine facilities and a state-of-the-art national center for managing contagious diseases that opened last year. Enhancement of border control measures were tightened with the first confirmed local case (on January 24, 2020), and later it was extended to land and sea checkpoints with the Immigration and Checkpoints Authority and Maritime and Port Authority, starting temperature screening for all travelers from noon of that day (Baharudin 2020).

Veredus laboratories, a Singapore biotech firm, launched a COVID-19 test kit for in-vitro diagnostic use. It could be directly used by laboratories or hospitals to test and confirm clinical diagnoses with accuracy more than 99% within 2 h (Poh 2020).

Singapore did very well with measures like social distancing. Many countries across the world congratulated Singapore for containing the spread. But unfortunately this novel virus made its comeback. The country saw a second wave and had to announce a lockdown.

24.3.4 Taiwan

Taiwan, an island nation, has seen relatively less than 450 confirmed COVID-19 cases (as on June 9, 2020), despite its proximity and numerous other links to the mainland China (Worldometers 2020). Taiwan, despite not being a member of World Health Organization (WHO), cannot stand alone and must be included in the battle against this epidemic. The country reacted quickly and effectively to deal with the novel coronavirus outbreak. The government's early intervention and prevention programmes have helped to prevent a public health crisis. The country has learned from its past experience with SARS epidemic in 2003, it helped the government to react immediately and effectively to deal with epidemics than many other countries. The government introduced a tighter travel restriction ban on visitors from countries like China, Hong Kong and Macau soon after the number of coronavirus cases began to increase in mainland China. The Taiwanese government has also used real-time alerts and data technology, including QR code scanning and online reporting of travel history and health symptoms to help medical personnel identify and trace suspected patients and high-risk individuals. A health declaration border pass was sent to the people with low risk via short message service (SMS) messaging to their phones for faster immigration clearance; those with higher risk were quarantined at home and tracked through their mobile phone to ensure that they remained at home during that period (Wang et al. 2020). Taiwan government has certainly set a good example of how a society can respond quickly to a public health crisis and protect the interests of its citizens.

The country has also fulfilled its national responsibilities as a global citizen and abided by the International Health Regulations 2005 (IHR 2005) in reporting WHO of confirmed COVID-19 cases. It has been sharing and communicated their epidemic prevention strategies with the United States, Canada and many other Asian and European countries, as well as the European Centre for Disease Prevention and Control (Shih-Chung 2020).

Taiwan's Academia Sinica (AS) successfully discovers first ever monoclonal antibodies for rapid virus screening and testing, aiming to test and quickly obtain results for the coronavirus infection within the short time frame of 15–20 min.

24.3.5 The Republic of Korea (South Korea)

Apart from China, South Korea was one of the worst-hit countries in the early stages of the coronavirus outbreak. South Korea has some of the world's most extensive and novel protective measures in place. The country has been able to dramatically lower the number of daily new cases and sustain a low mortality rate, making it a promising example of strong national response. The country has made great efforts to contain the virus spread by combining testing with contact tracing. Infected persons are required to go into isolation in government quarantine centers and their prior movements and contacts are traced by their phones, credit card transaction history and video footage from public surveillance cameras. Moreover, the Korean government also covered the testing cost for patients with suspected symptoms or in recent contact with confirmed cases (Lee and Lee 2020).

South Korea has made COVID-19 testing involving a method called the "phone booth". A hospital in Seoul has installed them around its building to offer easy and quick testing to people. In this glass-walled booth, one person can enter at a time at one of its side and grab a handset connected to a health worker standing on the other side of the glass-wall. After a consultation, the health worker can stick their arms into rubber gloves embedded into the booth to collect swab sample of the patient before the booth is quickly disinfected. And all it takes only 7 min to conduct the test in these special booths (Frank and Grady 2020).

24.3.6 Vietnam

Vietnam, a developing nation in south-east Asia, has also performed an outstanding work to prevent the spread of novel coronavirus infections rather than fighting against it. And this strategy has certainly proved to be very much effective to reduce the transmission and flatten the COVID-19 curve in the nation. Country is also cited by global media as having one of the best-organized epidemic control measures in the world, along with other Asian countries like South Korea, Taiwan and New Zealand (Walden 2020).

The country having 97 million population, 1444 km of land border with China, confirmed its first COVID-19 case on January 23, 2020, from a man having travel history to Wuhan (Nguyen 2009; World Population Review 2020). But, by then the country's emergency plan was in action. However, the country has reported less than 350 cases with no deaths and with no community transmission since mid-April (Alam 2020).

In the recent past, the country has also seen a lot of outbreaks, like SARS in 2003, avian influenza in 2010, and other large outbreaks of measles and dengue. Hence, the government and people know how to respond to these things and are used to deal with such infectious diseases, probably far better than the wealthier countries.

The main factors behind the success of Vietnam are its proactive and combined measures like early awareness, mass quarantine of tens of thousands of people, contact-tracing, appropriate people-centric approach, citizen support and testing to successfully contain relatively small and emerging clusters of COVID-19 outbreaks.

Country also restricted entry to Chinese tourists on January 28, when there was a rapid increase of COVID 19 cases in China. Schools were closed at the end of January and remained closed until mid-May (Jones 2020). The country also quickly developed its own testing kits for COVID-19 having a 90% rate of accuracy, according to the Ministry of Health.

24.3.7 *Bhutan*

In Bhutan, a south Asian country, health is a priority sector where all citizens receive free basic public health care services (Yangchen 2020). The country reported its first confirmed case on March 5, 2020. As of June 9, Bhutan has reported 59 confirmed cases of COVID-19 with no deaths. Most of the cases were either foreigners or recent returnees from affected nations. On March 22, Bhutan closed down their land border and 1 week later curbed public gatherings, transport and non essential services. On March 31, the government extended the quarantine period from 14 to 21 days, against the international WHO standard quarantine of 14 days (LeVine et al. 2020). Besides the free testing and other medical services, all meals and accommodation were also freely provided by the government at all the quarantine facilities (Yangchen 2020).

Using international references, Bhutan has prepared a four-stage COVID-19 alert system (ranging from Level 1 'Green' to Level 4 'Red') and is currently at level 3 'Orange' stage, which indicates confirmed cases without evidence of local transmission (Yangchen 2020). In comparison with other South Asian countries, Bhutan has certainly kept the public health crisis under control (Krishna 2020).

The Ministry of Health has also developed two mobile apps called 'Druk Trace' and 'Stay Home' for contact-tracing in public places, offices, and public transport and to monitor people placed in quarantine facilities.

24.4 Improving Prediction of the COVID-19 Outbreak

While COVID-19 may be devastating in every aspect of life, we should not lose the opportunity to learn and grow. This is the time to evolve strategies to think in terms of the whole healthcare system and population health.

It has been clear that lockdown has just pushed the number of cases into the future—without preventing them. The healthcare system of the countries needs to be strong and effective drugs that saves lives might soon be developed, tested and marketed quickly. Much hope is put in vaccines, but it would take time certainly (Giesecke 2020).

The need to plan for infectious disease outbreaks, whether naturally occurring or caused by bioterrorism, is now more than ever. Any such outbreak constitutes a threat to national and international security, with a potential to cause a health disaster. Legal frameworks, protocols and Health Adaptation Plans (HAP) need to be developed and adapted to handle such public health crisis. This can delineate the scope of the healthcare providers and government's responses to the public health emergencies (Sood et al. 2020).

The disease outbreak analysis looks at the location of an outbreak, the number of infections and the period of time—the where, what and when—to forecast the likelihood of the disease spreading in a short amount of time. Recent efforts have expanded to include many different data sources, which makes it possible to make disease outbreak predictions. With the advent of social media sites, more and more data can be linked with a location and mined for knowledge about an event like an outbreak.

References

Alam S (2020) Vietnam: a success story in fight against COVID-19, in interview with Anadolu agency, Vietnamese envoy to Turkey says his country managed to best virus with 'proactive prevention'. https://www.aa.com.tr/en/asia-pacific/vietnam-a-success-story-in-fight-against-covid-19/1866670. Accessed 6 May 2020

Alandijany TA, Faizo AA, Azhar EI (2020) Coronavirus disease of 2019 (COVID-19) in the Gulf cooperation council (GCC) countries: current status and management practices. J Infect Public Health 13(6):839–842

Baharudin H (2020) Wuhan virus: ICA implements temperature screening at Woodlands and Tuas checkpoints. https://www.straitstimes.com/singapore/health/temperature-screening-for-all-travellers-at-woodlands-and-tuas

Banerjea A (2020) Coronavirus lockdown extended till 31 May, says NDMA. https://www.livemint.com/news/india/covid-19-lockdown-4-0-coronavirus-lockdown-extended-till-31-may-says-ndma-11589715203633.html

Banerjee D, Bhattacharya P (2020) The hidden vulnerability of homelessness in the COVID-19 pandemic: perspectives from India. Int J Soc Psychiatry, pp 1–4

Bloomberg.com (2020). Available at https://www.bloomberg.com/graphics/2020-coronavirus-cases-world-map/

Bogoch II, Watts A, Thomas-Bachli A, Huber C, Kraemer MUG, Khan K (2020) Potential for global spread of a novel coronavirus from China. J Travel Med 27

Buescu J, Oliveira HM, Carlos A (2020) COVID-19 infection-to-death lag and lockdown-effect lag in Italy and Spain 1. Medicine 9(2):538

Callaway E, Cyranoski D, Mallapaty S, Stoye E, Tollefson J (2020) Coronavirus by the numbers. Nature 579:482–483

Carrington CV (2012) Viral genomics: implications for the understanding and control of emerging viral diseases. In: Genomics applications for the developing world. Springer, New York, pp 91–114

Carroll AE (2020) Lesson from Singapore: why we may need to think bigger. https://www.nytimes.com/2020/04/14/upshot/coronavirus-singapore-thinking-big.html

Chaturvedi A (2020) Zero active cases: how New Zealand eliminated Covid-19. https://www.hindustantimes.com/world-news/zero-active-cases-how-new-zealand-eliminated-covid-19/story-vTybkQoMZOp824nRAuYI7N.html

Chen W, Huang Y (2020) To protect health care workers better, to save more lives with COVID-19. Anesth Analg 131:97–101

Cohen J (1999) AIDS virus traced to chimp subspecies. Science 283:772–773

Deccan Herald (2020) United Kingdom government accused of coronavirus testing failures. https://www.deccanherald.com/international/united-kingdom-government-accused-of-coronavirus-testing-failures-839432.html. Accessed 29 May 2020

Emanuel EJ, Persad G, Upshur R, Thome B, Parker M, Glickman A, Zhang G, Boyle C, Smith M, Phillips JP (2020) Fair allocation of scarce medical resources in the time of Covid-19. N Engl J Med 382(21):2049–2055

Frank A, Grady C (2020) Phone booths, parades, and 10-minute test kits: how countries worldwide are fighting Covid-19. https://www.vox.com/science-and-health/2020/3/22/21189389/coronavirus-covid-19-pandemic-response-south-korea-phillipines-italy-nicaragua-senegal-hong-kong

Giesecke J (2020) The invisible pandemic. Lancet 395(10238):e98

Glanz J, Robertson C (2020) Lockdown delays cost at least 36,000 lives. Data Show. https://www.nytimes.com/2020/05/20/us/coronavirus-distancing-deaths.html

Goel K, Chowdhary R (2017) Living homeless in urban India: state and societal responses. In: Faces of homelessness in the Asia Pacific. Routledge, London, pp 47–63

Gostin LO, Hodge JG (2020) US emergency legal responses to novel coronavirus: balancing public health and civil liberties. JAMA 323(12):1131–1132

Heijmans PJ (2020) Singapore contained coronavirus. Could other countries learn from its approach? https://www.weforum.org/agenda/2020/03/singapore-response-contained-coronavirus-covid19-outbreak/

Hussain A, Bhowmik B, do Vale Moreira NC (2020) COVID-19 and diabetes: knowledge in progress. Diabetes Res Clin Pract 162:108142

Johns Hopkins University (2020). Available at https://hub.jhu.edu/novel-coronavirus-information/

Jones A (2020) Coronavirus: how 'overreaction' made Vietnam a virus success. BBC news. https://www.bbc.com/news/world-asia-52628283. Accessed 15 May 2020

Khanna RC, Cicinelli MV, Gilbert SS, Honavar SG, Murthy GS (2020) COVID-19 pandemic: lessons learned and future directions. Indian J Ophthalmol 68:703–710. http://www.ijo.in/text.asp?2020/68/5/703/282901

Krauss S, Walker D, Pryor SP, Niles L, Chenghong LI, Hinshaw VS, Webster RG (2004) Influenza a viruses of migrating wild aquatic birds in North America. Vector Borne Zoonot Dis 4 (3):177–189

Krishna SV (2020) Bhutan: a coronavirus success story, South Asia's Battle with the Coronavirus, Carnegie India. https://carnegieindia.org/2020/05/21/bhutan-coronavirus-success-story-pub-81856

Kupferschmidt K, Cohen J (2020) Can China's COVID-19 strategy work elsewhere? Science 367:1061–1062

Lai CC, Shih TP, Ko WC, Tang HJ, Hsueh PR (2020a) Severe acute respiratory syndrome coronavirus 2 (SARS-CoV-2) and corona virus disease-2019 (COVID-19): the epidemic and the challenges. Int J Antimicrob Agents 55:105924

Lai S, Ruktanonchai NW, Zhou L, Prosper O, Luo W, Floyd JR, Wesolowski A, Zhang C, Du X, Yu H, Tatem AJ (2020b) Effect of non-pharmaceutical interventions for containing the COVID-19 outbreak: an observational and modelling study. medRxiv

Lampen C, Gold H, Arnold A (2020) Everything to know about the coronavirus in the United States. https://www.thecut.com/2020/06/which-states-have-coronavirus-and-how-do-you-minimize-risk.html

Lau SK, Woo PC, Li KS, Huang Y, Tsoi HW, Wong BH, Leung SY, Chan KH, Yuen KY (2005) Severe acute respiratory syndrome coronavirus-like virus in Chinese horseshoe bats. Proc Natl Acad Sci 102(39):14040–14045

Lee D, Lee J (2020) Testing on the move South Korea's rapid response to the COVID-19 pandemic. Transport Res Interdiscip Perspect 5:100111

LeVine S, Dhakal GP, Penjor T, Chuki P, Namgyal K, Watts M (2020) Case report: the first case of novel coronavirus disease (COVID-19) in Bhutan. Am J Trop Med Hyg 102(6):1205–1207

Live Mint (2020) Covid-19: many failures combined to unleash death on Italy's Lombardy. https://www.livemint.com/news/world/covid-19-many-failures-combined-to-unleash-death-on-italy-s-lombardy-11587918526836.html

MoHFW (2020) Ministry of Health and Family Welfare, Government of India. mohfw.gov.in. Accessed 9 Jun 2020

Nguyen TH (2009) A line runs through it: Vietnam and China complete boundary marking process. Hanoi

Normile D (2020) Singapore claims first use of antibody test to track coronavirus infections. https://www.sciencemag.org/news/2020/02/singapore-claims-first-use-antibody-test-track-coronavirus-infections

Pinghui Z (2020) A coronavirus lesson from China: don't make patients pay for tests and treatment. https://www.scmp.com/news/china/society/article/3074506/coronavirus-lesson-china-dont-make-patients-pay-tests-and

Pisano GP, Sadun R, Zanini M (2020) Lessons from Italy's response to Coronavirus. https://hbr.org/2020/03/lessons-from-italys-response-to-coronavirus

Poh O (2020) Singapore biotech firm Veredus launches Covid-19 test kit for hospital, lab use. https://www.businesstimes.com.sg/garage/singapore-biotech-firm-veredus-launches-covid-19-test-kit-for-hospital-lab-use

Rabaan AA, Al-Ahmed SH, Sah R, Tiwari R, Yatoo MI, Patel SK, Pathak M, Dhama K, Singh KP, Bonilla-Aldana DK, Haque S, Rodriguez-Morales AJ (2020) SARS-CoV-2/COVID-19 and advances in developing potential therapeutics and vaccines to counter this emerging pandemic virus—a review. Preprints

Ray M (2020) India most infected by Covid-19 among Asian countries, leaves Turkey behind. Hindustan Times. Accessed 5 Jun 2020

Shih-Chung C (2020) Taiwan's coronavirus protocol shows how it is done. https://www.thehindu.com/opinion/op-ed/taiwans-coronavirus-protocol-shows-how-it-is-done/article31484681.ece

Sood A, Barwal A, Gupta AK, Kishore J (2020) Novel coronavirus epidemic: new dimension for disaster management and health resilience. Curr Sci 118(8):1149–1150

Tanne JH, Hayasaki E, Zastrow M, Pulla P, Smith P, Rada AG (2020) COVID-19: how doctors and healthcare systems are tackling coronavirus worldwide. BMJ 368:m1090

The U.S. Health Care System (2016) an international perspective—DPEAFLCIO. Factsheet. https://www.dpeaflcio.org/factsheets/the-us-health-care-system-an-international-perspective

Walden M (2020) How has Vietnam, a developing nation in South-East Asia, done so well to combat coronavirus? ABC News (Australia). Accessed 26 May 2020

Walsh B (2017) The world is not ready for the next pandemic. Time Magazine 189(18). https://time.com/magazine/us/4766607/may-15th-2017-vol-189-no-18-u-s/

Wang CJ, Ng CY, Brook RH (2020) Response to COVID-19 in Taiwan: big data analytics, new technology, and proactive testing. JAMA 323(14):1341–1342

Withnall A (2020) India coronavirus: Modi announces 21-day nationwide lockdown, limiting movement of 1.4bn people. https://www.independent.co.uk/news/world/asia/india-coronavirus-lockdown-modi-speech-cases-update-news-a9421491.html

World News (2020) China bans trade, consumption of wild animals due to coronavirus. https://www.reuters.com/article/us-china-health-wildlife/china-bans-trade-consumption-of-wild-animals-due-to-coronavirus-idUSKCN20J069

World Population Review (2020). Available at https://worldpopulationreview.com/countries/vietnam-population/

Worldometers (2020) COVID-19 Coronavirus Pandemic, (as updated on June 09, 2020, 07:50 GMT). https://www.worldometers.info/coronavirus/

Yangchen S (2020) Bhutan's preparedness and response to COVID-19. https://www.orfonline.org/expert-speak/bhutans-preparedness-and-response-to-covid19-64711/

Zhu H, Wei L, Niu P (2020) The novel coronavirus outbreak in Wuhan, China. Glob Health Res Policy 5(1):1–3

Correction to: A Review of Pandemics

Lorenz Hilfiker and Shashwat Ganguly

Correction to:
Chapter 2 in: M. K. Goyal, A. K. Gupta (eds.), *Integrated Risk of Pandemic: Covid-19 Impacts, Resilience and Recommendations*, Disaster Resilience and Green Growth, https://doi.org/10.1007/978-981-15-7679-9_2

Chapter 2 of the book was inadvertently published with incorrect affiliation of the author Lorenz Hilfiker. This has now been corrected.

The updated online version of this chapter can be found at
https://doi.org/10.1007/978-981-15-7679-9_2

M. K. Goyal, A. K. Gupta (eds.), *Integrated Risk of Pandemic: Covid-19 Impacts, Resilience and Recommendations*, Disaster Resilience and Green Growth,
https://doi.org/10.1007/978-981-15-7679-9_25

The manufacturer's authorised representative in the EU is Springer Nature Customer Service Centre GmbH, Europaplatz 3, 69115 Heidelberg, Germany. If you have any concerns regarding our products, please contact ProductSafety@springernature.com

Printed and bound by CPI Group (UK) Ltd, Croydon, CR0 4YY
30/09/2026
02225506-0014